Space: Technologies, Materials, Structures

Welding and Allied Processes
A series of books and monographs on welding and other processes of metal treatment
Edited by B.E. Paton, E.O. Paton Electric Welding Institute, Kiev, Ukraine
and P. Seyffarth, Welding Training and Research Institute, Rostock, Germany

Volume 1

Laser-Arc Processes and their Applications in Welding and Material Treatment
P. Seyffarth and I.V. Krivtsun

Volume 2

Space: Technologies, Materials, Structures
Edited by B.E. Paton

This book is part of a series. The publisher will accept continuation orders which may be cancelled at any time and which provide for automatic billing and shipping of each title in the series upon publication. Please write for details.

Space: Technologies, Materials, Structures

Edited by B.E. Paton

E.O. Paton Electric Welding Institute,
National Academy of Sciences of Ukraine,
Kiev, Ukraine

CRC Press
Taylor & Francis Group
Boca Raton London New York

CRC Press is an imprint of the
Taylor & Francis Group, an **informa** business
A TAYLOR & FRANCIS BOOK

First published 2003 by Taylor & Francis

Published 2019 by CRC Press
Taylor & Francis Group
6000 Broken Sound Parkway NW, Suite 300
Boca Raton, FL 33487-2742

First issued in paperback 2019

ISBN 13: 978-0-367-45452-4 (pbk)
ISBN 13: 978-0-415-29985-5 (hbk)

Visit the Taylor & Francis Web site at
http://www.taylorandfrancis.com

and the CRC Press Web site at
http://www.crcpress.com

British Library Cataloguing in Publication Data
A catalogue record for this book is available from the British Library

Library of Congress Cataloging in Publication Data
A catalog record for this book has been requested

First published as:
*Space: Technologies, Materials Science, Structures. Collection of
scientific papers.* Edited by B.E. Paton.

© 2000 E.O. Paton Electric Welding Institute of the NAS of Ukraine

Compiled by E.S. Mikhajlovskaya, S.V. Pavlova, V.F. Shulym
and A.A. Zagrebelny
Design I.V. Petushkov
CRC preparation T.Yu. Snegiryova

Translated by S.A. Fomina, I.N. Kutianova, T.K. Vasilenko (English)
and M.I. Zykova (German)

Contents

Foreword to the English Edition of Book

The process of international integration and formation of collaborative space projects has initiated a new turn of space exploration. The beginning of the third millennium has been marked by a strong surge in the progress of the space technology. For man, space is not just a fantasy or dream, but reality and one of the spheres of science intensive production. The unique properties of space offer the possibility of space production-technological complexes. Space technology has marched with confidence into the XXI century.

In 2000 the E.O. Paton Electric Welding Institute of the National Academy of Sciences of Ukraine (Kiev) published the book «Space: Technologies, Materials, Structures». The book was in Russian. It covered the period from 1964 to 2000 and included scientific publications taken from highly specialised journals, which were almost unavailable for a wide circle of specialists involved in space technology studies.

The book contains approximately 70 papers and articles written by scientists and experts from the PWI and colleagues from different institutions of the former Soviet Union, who in the 1960s of the last century had laid the basis for a new scientific area — space technologies. The book covers issues associated with technology and materials science, research into instrumentation, ergonomic problems of training cosmonauts/astronauts, i.e. operators, to conduct both IVA and EVA technological experiments, near-earth space monitoring, repair and maintenance operations on space vehicles and their systems, construction of various-application large-sized structures in orbit and their maintenance during long-time operation under extreme conditions. Naturally a book covering such a wide spectrum of scientific information was prepared by a large team of authors, experts in their professional areas, i.e. technologists, material scientists and engineers, designers, mechanical and instrument engineers, investigators, etc. A tremendous role was played by cosmonauts who conducted investigations under space conditions.

The book generated high interest among Russian-speaking scientists and engineers specialising in different areas, and all its copies printed in a limited number were quickly sold out. In 2001 the Publishing House «Taylor & Francis, Ltd.» (Great Britain) suggested publishing the English version of the book, which we are now doing with great satisfaction. I will add that the English version of the book has been substantially supplemented. Now it includes new scientific information published during the period from 2000 to 2002, as well as a number of articles and papers not included in the previous edition. In the new edition we corrected mistakes we noticed and expanded the reference data.

Hopefully, this book will attract the attention of English-speaking specialists and allow them to appreciate the efforts of the contributing authors. We also express our deep appreciation to the publishers of the book.

B. Paton
June, 2002

Editorial

In the entire spectrum of intensively and successfully developing theoretical and applied space research, advanced material processing technologies are steadily gaining in importance. Their emergence at the dawn of the space era (1969) was marked by the experiment on welding and cutting of metals in «Vulkan» unit. The issues of fast industrialisation of space which was S.P. Korolyov's dream, should certainly be addressed allowing for the features of this working environment. This is primarily necessary for the reason that many physical phenomena proceeding on the molecular level (surface tension, adhesion, wetting, capillary pressure) manifest themselves more actively in space environment that on Earth.

Successful performance in orbit of such complex technological processes as welding, brazing, coating deposition, etc., where liquid metal is present, required a broad range of experts in theory and practical application to become involved in the investigations. They laid down the foundations of the new field of science — space technology.

One of the important issues in the field of thermal treatment of materials is the correct selection of the method for their heating. In the 1960s, when the first space technology experiment was prepared, the method of direct and indirect electron beam heating was recognised to be the most advantageous. The process flexibility and power efficiency allows one and the same all-purpose equipment to be used to perform in space a sequence of various technological operations in different modes. The «Universal» hardware developed in 1991, which consists of four manual electron beam tools has successfully passed ground-based testing and is ready for use as part of the long-term space systems of the future. The versatility and advantages of electron beam utilisation in space are further illustrated also by the ability to apply an injected flow of accelerated electrons to study the magnetosphere of Earth, which was theoretically predicted and experimentally confirmed. Use of electron beam heating for a number of processing operations in orbit, does not, however, eliminate the application of other heat sources, for instance, the radiant energy of Sun, which can still be justified in some cases. Creation of long-term orbital systems gave an impetus to performance of systematic research in the priority areas of space technology.

Experiment optimisation on the ground is preceded by verification of various hypotheses and theoretical development, in particular mathematical simulation of various processes. Solving applied problems of space technology will require knowledge of the laws of behaviour of multiphase systems (solid body–liquid, liquid with solid and gas inclusions). The processes running in such multiphase systems can be intensified and controlled by external impact (vibration and ultrasound). So, for instance, the vibration effect of resonance stirring of liquid immiscible media and formation of stable periodical structures can be used in fabrication of unique materials (foam materials, composite materials, etc.), and the effect of localising and controlled displacement of gas bubbles in oscillating media can be applied in degassing of liquids and liquid metals. It is intended to conduct theoretical development in the field of forecasting the service life of structural materials and their joints under the integrated impact of space factors by physical simulation on scaled thin-walled models, allowing both the surface and three-dimensional processes of material degradation to be studied with relatively short exposure times.

Prolongation of operational life of long-term space vehicles is closely related to application of welding processes in orbit for repair of large-sized structures, as well as non-destructive testing of metal structures, including the acoustic emission method, in difficult of access places.

Accelerating exploration of space at the start of the 1970s with a simultaneous increase in the scope of work and experiments performed in raw space, led to a radical change in operator activity, due to the fact that performance of a number of technological processes in orbit raises problems related both to the specific conditions of space and substantial professional training of operators. The ability to perform precisely co-ordinated movements under microgravity and conduct extra-vehicular activity in a space-suit should form the basis for qualification training of operators.

Performance of space experiments is always preceded by their optimisation on the ground under conditions simulating those of space (microgravity, vacuum, variable lighting, temperature gradient, ultraviolet radiation, etc.). Used on the ground with this purpose are various simulation facilities: thermal, altitude simulation, echo and other chambers, flying laboratories or «weightlessness towers», providing short-time microgravity, and various test rigs reproducing certain factors of the space environment. Such a retrofitting, while being costly, is still worthwhile as it confirms the ability to perform the defined task, allows correction of the procedure of conducting the planned experiment and the consideration of the possibility of an emergency situation arising during its performance in space.

For ground-based retrofitting of the manual processing operations under conditions simulating the space environment, the E.O. Paton Electric Welding Institute (PWI) developed special testing facilities which have been used for more than 25 years now and became an indispensable tool in training operators to use sophisticated engineering hardware functioning outside the space vehicle.

Thorough retrofitting on the ground and testing in the flying laboratory which allowed determination of the optimal process parameters of the «Vulkan» unit, were followed by numerous process experiments in orbit, which were the basis for development of all-purpose equipment for performance of repair and erection work directly in orbit, which has been successfully tested in raw space.

The many years of the PWI experience on development and improvement of the hardware for conducting processing operations in space, which meets the requirements of safe operation in a space vehicle, high operational reliability, compatibility with the space vehicle systems and capabilities of its crew, are the foundation of successful development of control and information systems of a new generation designed for operation on board the International Space Station which is under construction now.

Manned missions in orbital stations demonstrated that during long-term operation in the raw space environment, the surface of the space vehicles is degraded to a depth of up to 10 μm, thus impairing the functional (optical, electrophysical, physicomechanical, etc.) properties of structural materials. Therefore, deposition of coatings for various purposes in space is one of the most important practical problems in terms of the ability to perform repair-restoration operations directly in orbit.

Starting from 1979, the PWI has conducted development of special hardware for coating deposition in space. Over this period not only the stationary «Isparitel», «Isparitel-M», «Yantar» laboratory units, but also all-purpose manual hardware have been successfully tested in «Salyut» and «Mir» orbital stations. Investigation of the properties of the produced coatings from pure metals (silver, gold, copper) demonstrated that their functional properties (adhesion, residual internal stresses, morphology, microstructure, optical and radiation characteristics) are not inferior to those of their analogs produced on the ground, and meet the requirements of industrial standards. The ability to produce various coatings, in particular, of binary alloys in space opens up wide possibilities for their application in industry.

Over the years of manned cosmonautical development, the greatest number of process experiments in various space vehicles have been conducted in the field of space materials science. And it is no coincidence. Absence of gravity really enables production of materials with a high «structural purity», a specified distribution of impurities, as well as implementation of container-less technology, etc.

The multi-faceted utilisation of space for fulfilment of research programs necessitated a considerable increase in the overall dimensions of the space vehicles and development of their infrastructure in service. Thus, the construction of the currently orbiting «Mir» station features a multitude of all kinds of various-purpose superstructures and truss structures located on its outer surface. In addition, over the 15 years of the «Mir» orbital complex functioning, a great number of large-sized structures or their elements and components have been delivered on board the station for further fitting of its service systems, as well as experimental retrofitting of the promising structures or procedures of their construction directly in orbit.

Brazing is an attractive technology for mounting of the truss structure elements. Therefore, the issues related to this technology, are being continuously followed by specialists, starting with theoretical calculations of the process parameters and ground retrofitting in the space simulation test chamber up to testing of brazed joints in raw space. The issues of construction of extended large-sized structures should be addressed allowing for the specifics of their operation in space, namely costly delivery to orbit, limitations on the weight of the delivered structure, service life, simple assembly and repairability during service. Several basic principles have been proposed for construction of large-sized structures in space.

As far back as the 1970s, development of various kinds of transformable structures, also pressurised, was looked upon as one of the most promising areas, allowing working, storage, and docking modules, aerials, etc. to be erected in orbit. A number of interesting approaches were developed to the fabrication of convenient easy-to-transport structures (metallic transformable shells). The transformable structure of the docking module of the orbital station serves as an illustration of such a development.

Another area of research which is still continuing, is the development of deployable frame structures allowing deployment of large aerial systems in orbit, mounting «power fields», etc. Creation of this kind of structure and the related issues of rigidity, and dynamic stability, required not only theoretical calculations, but also extensive ground-based and flight testing.

This collection, including about 70 of the earlier published papers with indication of their source, certainly cannot cover all the aspects of space technology. The editors tried to sum up the many years of cooperation of the specialists of the PWI and other organisations in this, still emerging field of science, in order to trace the evolution of some of what we consider its priority areas, and share the gained (although not always positive) experience of staging various experiments in orbit, and analysis of their results. Moreover, the collection provides abundant photos that not only recreate the episodes of the major events, but also pay tribute to those who are no longer with us. The specialists who performed this work, are real enthusiasts of their cause. The importance of their efforts aimed at the practical implementation of the space programmes, can scarcely be exaggerated.

The XX century, the era of great discoveries and deeds, has slipped away. One of its greatest achievements is man's breaking beyond the bounds of Earth into infinite space. Analysing the traversed path, it can be stated that significant progress has been made. The space missions, no longer causing a sensation, have moved into the sphere of routine scheduled activities. Nonetheless, a lot is still to be accomplished.

Looking into the future, one can anticipate that the progress of science and technology, just as the on-going global information revolution, will be closely connected with further space exploration. In this context the advanced high technologies of processing and production of materials in space, biological and agricultural technologies, etc. have a special, probably, even primary role. Their priority is confirmed by the continuously growing pace of their development. Close is the day when large-scale production vitally important for mankind, will be deployed in the near-earth orbits.

Prof. Boris E. Paton

June, 2000

Authors' Affiliations

	[1] **PWI** E.O. Paton Electric Welding Institute of the National Academy of Sciences of Ukraine, Kiev, Ukraine
	[2] **S.P. Korolyov RSC «Energiya»** S.P. Korolyov Rocket Space Corporation «Energiya» (former OKB-1, TsKBEM, NPO «Energiya»), Korolyov, Moscow district, Russia
	[3] **Yu.A. Gagarin CTC** Yu.A. Gagarin Cosmonauts Training Centre of the Russian State Research Institute, Shchelkovo, Moscow district, Russia
	[4] **KhIRE** Kharkov Institute of Radio Electronics of the National Academy of Sciences of Ukraine, Kharkov, Ukraine
	[5] **SRI** Space Research Institute of the Russian Academy of Sciences, Moscow, Russia
	[6] **KhPI** Kharkov Polytechnic Institute, Kharkov, Ukraine
	[7] **S.P. Timoshenko IM** S.P. Timoshenko Institute of Mechanics of the National Academy of Sciences of Ukraine, Kiev, Ukraine
	[8] **IPSC SD RAS** Institute for Physics of Semiconductors of Siberian Division of the Russian Academy of Sciences, Novosibirsk, Russia
	[9] **«Fonon»** Research Centre «Fonon», Kiev, Ukraine
	[10] **B.I. Verkin ILTP** B.I. Verkin Institute for Low Temperature Physics and Engineering of the National Academy of Sciences of Ukraine, Kharkov, Ukraine
	[11] **Company «Boston-Animation»** Kiev, Ukraine
	[12] **IED** Institute of Electrodynamics of the National Academy of Sciences of Ukraine, Kiev, Ukraine
	[13] **I.M. Frantsevich IMSP** I.M. Frantsevich Institute for Materials Science Problems of the National Academy of Sciences of Ukraine, Kiev, Ukraine

Erno Raumfahrttechnik GmbH **Erno Raumfahrttechnik GmbH**	[14] **ERNO** Raumfahrttechnik GmbH, Bremen, Germany
	[15] **SLV Halle** Halle, Germany
	[16] **DB «Yuzhnoye»** Design Bureau «Yuzhnoye», Dnepropetrovsk, Ukraine
	[17] **A.A. Bajkov IMet** A.A. Bajkov Institute of Metallurgy of the Russian Academy of Sciences, Moscow, Russia
	[18] **IPSC** Institute for Physics of Semiconductors of the National Academy of Sciences of Ukraine, Kiev, Ukraine
Paul Sabatier	[19] **CESR** Universite Paul Sabatier, Toulouse, France
	[20] **L.G.E.,** Universite Paris VI, France
	[21] **ITMIRAS** Institute of Terrestrial Magnetism, Ionosphere and Radio Wave Propagation of the Russian Academy of Sciences, Moscow, Russia
	[22] **I.V. Kurchatov AEI** I.V. Kurchatov Atomic Energy Institute of the Russian Academy of Sciences, Moscow, Russia
	[23] **Ecole Polytechnique** Paris, France
Polar Geophysical Institute	[24] **Polar Geophysical Institute of the Russian** **Academy of Sciences** Murmansk, Russia
	[25] **University of Houston** Texas, USA
	[26] **A.I. Mikoyan KIFPI** A.I. Mikoyan Kiev Institute for Food Processing Industry, Kiev, Ukraine
	[27] **KICE** Kiev Institute of Construction Engineers, Kiev, Ukraine
	[28] **SDB GPI** Special Design Bureau of the Georgian Polytechnic Institute, Tbilisi, Georgia

List of Abbreviations

ARAKS	Artificial Radiation and Aurora between Kerguelen and Sogra
ASEF	«Apollo–Soyuz» Experimental Flight
CNES	National Centre of Space Exploration, France
EBW	Electron Beam Welding
EVA	Extra-Vehicular Activity
FCC	Flight Control Centre
IAF	International Aerospace Federation
ISS	International Space Station
ISWE	International Space Welding Experiment
IVA	Intra-Vehicular Activity
NASA	National Aeronautics and Space Administration
NSAU	National Space Agency of Ukraine
OC	Orbital Complex
OL	Orbital Laboratory
OS	Orbital Station
PWI[1]	Paton Welding Institute
RM ISS	Russian Module of International Space Station
VHT	Versatile Hand Tool

Chapter 1

SPACE TECHNOLOGIES

ARE A REALITY

Chapter 1. Space Technologies are a Reality

The history of science and technology is rich with examples of how the unbelievable may with time become reality. Fantasy has always been and will be an important drive of the creative scientific work. It took space technology a bit more than 30 years not just to get on its feet, but make the first resolute steps on the way of advancement of space fabrication. Results of experiments conducted in the field of the space technology became an obvious and convincing proof of inception of the space industry.

Welding in Space[*]

B.E. Paton[1]

Trying to imagine welding in an interstellar space, I would like, first of all, to look into the future of our planet. Welding is known to be originated in our earth conditions and, here, it will have to reach the unknown heights.

However, why am I saying «welding»? Even now, welding is closer and closer to adhesion, when the bond is made between the «foreign» atoms in metals without their preliminary melting. In these cases the term «welding» will sound as anachronism.

The challenging trend in welding progress is the refusal from metal melting and the wider application of ultrasonic oscillations, friction forces and explosion energy. It does not mean that metal melting will be excluded entirely. In parallel with arc, electron and light beams ion beam and hot plasma will also find application.

Today, the new type of welding with a concentrated flux of electrons in vacuum, so-called electron beam welding, is widely used. The very high concentration, not used earlier, makes it possible to melt the massive metal with a «spike» weld.

The scientists have in their mind the superpowerful beams of light energy, namely lasers. The first results of experimental welding of refractory metals have been already obtained in the laboratories. Using lasers it would be possible to perform welding at large distances and even through transparent surfaces.

Now, let us return to our problem of space welding. And, here, surely, its importance cannot be overemphasised for the creation of orbital stations, towns on the Moon and interplanetary vehicles. The absolute equal strength of welded joints allows designers to join new materials for space vehicles. The strong joints of metals with ceramics, metals with films (for self-sealing) will be subjects for the future works of the scientists. And these are not distant projects. At present, there is a theoretical feasibility under laboratory conditions to weld all the metals between themselves and in different combinations, and also to weld metals with non-metals.

Further, scientists have investigated for several years the different methods of welding plastics with metals. The reliable method turned out to be «nuclear welding». Its principle is as follows. A thin layer of lithium or barium is deposited on the two surfaces being welded. Then, the butt edge is radiated with slow neutrons. Nuclear reactions, occurring during radiation, are accompanied by high temperatures and their duration is only a billionth fraction of a second. However, even this time is sufficient to join these surfaces.

Undoubtedly then, after space exploration, mankind will learn to use the cost-free energy of solar beams for welding of corpuscular particles of space.

I would like to dwell also on one more possible application of space welding.

Science-fiction writers like to describe in their works the critical moments when even small fragments of meteorites are piercing the skin of spacecrafts. And, here, the dramatic situations have been always appeared in which cosmonauts escape either by superstrong spacesuits or different screens of cosmic protection. As you see, safety is also envisaged in space.

Safety rules cannot be disregarded. But it seems to me that when the interplanetary vehicles set out on the unknown routes, mankind will have developed safer means of protection. But, if a meteorite does enter one of the compartments, then a mechanical robot-welder will help cosmonauts to eliminate the accident. Electronic monitors will determine in the hundredths of a second the level of the accident, will define the size of

[*] (1964) *Nauka i Zhizn*, No. 11, p. 13.

damage from the rate of the pressure drop, and will give instantaneously the task for the robot-welder.

The major role to the works of this kind will be given in the construction of all space objects without an exception — space rockets, stations, towns. To make their skins protect man reliably from all harmful effects, the development and joining of superstrength materials under space conditions will be required.

Thus, the welding is required to play a large role in space exploration. In the world of the future, it will occupy a worthy place among other creations of the human mind.

Space Technology Issues[*]

B.E. Paton[1]

Space exploration by man provided a powerful impetus to progress in various fields of science and technology, including the technology of processing and production of various materials. The time has come when one can anticipate technological-production facilities being installed in space. In the near future not only performance of various technological processing operations related, in particular, to heating and melting, but also application of a number of technologies for production of materials and items with fundamentally new physical-chemical properties, will be required in space. Work in this area has been successfully going on both in the Soviet Union and the United States of America.

In October, 1969 the first experiment on welding and cutting of metals in space was conducted in the USSR on the «Soyuz-6» space vehicle, which practically marked the emergence of space technology; in 1973–1974 an experiment on melting and welding of metals was also performed in the USA in the OL «Skylab». In 1975 during the joint flight of «Soyuz–Apollo» space vehicles a number of process investigations of metallic and non-metallic materials were also performed as part of «Versatile Electric Furnace» experiment. Even this brief chronological sequence is a vivid proof of the acute interest of scientists and engineers in conducting experiments with molten materials in space.

This is attributable, primarily, to the fact that a number of conditions present in space, such as zero gravity, deep vacuum, special temperature mode, cannot be reproduced on Earth. These conditions can have both a positive and a negative impact on technological processes. Therefore, the scientists' goal is to thoroughly study the features of conducting various processing operations in space, so as to learn to make use of the positive influence of space environment and to determine the methods of controlling its adverse impact. Numerous fundamental investigations will probably be required, aimed at studying the behaviour of molten materials under these conditions, and determination of the main methods of their processing in space. It is possible to outline at least three promising fields where the space technology will probably develop.

1. Behaviour of liquid and gaseous materials under space conditions is to be thoroughly studied, so as to be able to actively use the accumulated information in the future. Both single-phase systems and multi-phase systems, i.e. containing the solid, liquid and gaseous phases, should be investigated. The basic principles of melt cooling and solidification in space in the free state or with forced heat removal along one or several axes; principles of separation of phases and methods for their control; action of surface tension forces; adhesion and wettability in different combinations of phases and materials should be studied, as well as other issues. This will require development of special research equipment.

[*] (1977) *Space materials science and technology*, Moscow, Nauka, pp. 6–12.

2. Technological studies will, it is predicted, allow selection of a number of materials and items which it will be rational to produce in orbital stations or laboratories.

Firstly, various composite materials can be selected, namely joints of metals with each other and metal oxides, ceramic and organic substances immiscible on the ground, light alloys reinforced with high strength fibres or whiskers, etc. It is anticipated that such materials can have unique structural, mechanical and electro-physical properties.

The possibility of uniform mixing of gases with metals will enable creation of high strength foam metals with a large number of gas bubbles. Thus, it will probably be also possible to produce high-temperature resistant materials with a low specific weight (2–3 g/cm^3).

Remelting of optically transparent materials in space will allow production of glasses and transparent single crystals with minimal possible content of impurities or with specified distribution. The optical properties of such materials produced in space can be significantly better, than the properties of materials currently used by industry.

Finally, production under the conditions of space of high-strength high-purity single crystals with semiconductor properties is important, this allowing an essential improvement in the effectiveness of numerous electronic devices.

So, for instance, $0.1 \times 0.4 \times 2.5$ cm single crystals of germanium silicide have already been produced in space, their dimensions being several times larger than the dimensions of the same single crystals grown on the ground. A single crystal of indium antimonide produced in space has a highly perfect structure that is even theoretically unattainable under the impact of the gravity of Earth, whereas the strength of a sapphire single crystal grown in space is $\approx 2 \cdot 10^5$ kgf/cm^2, this being by an order of magnitude higher than the regular value.

Of considerable importance is conducting in space a number of chemical and technological processes such as electrophoresis or growing bacteria cultures with a high purity of the product.

Moreover, it will apparently turn out to be rational to manufacture in space various special products which are so far impossible to produce under any other conditions (for instance, solid and hollow spheres of an ideal shape, unique castings from reactive metals, or parts in the form of single crystals).

Thus, even at this point one can say that the space environment opens up broad possibilities for creation of fundamentally new types of instruments or apparatus based on unique materials and parts, primarily for radio engineering, instrument-making, aircraft, aerospace and medical industries.

3. Development of methods and means for performance of repair-restoration and construction work in space is of great importance. The experience of the most recent space flights showed that space objects as, for instance, the OL «Skylab», can be repaired, if need be. In this case severing of structures and their joining by welding, brazing or adhesion bonding may be required. This work acquires a special importance for operation of heavy long-term space laboratories and stations erected from individual models that will be taken to the assembly zone separately.

One of the most important issues in this respect is study of the ergonomic and physiological capabilities of the cosmonaut-operator. Development of various auxiliary devices facilitating the performance of work and providing for their safety, is of interest.

To ensure a successful progress of these most promising fields of space technology, development and testing of the equipment and units designed specially for performance of research technological operations in space are the primary problems which are urgent even now.

Our more than ten years of experience of working in this field allowed the determination and in the majority of cases successful verification of the main requirements for the hardware for performance of processing operations in space. This is primarily a high reliability of hardware operation. The need to provide complete safety of work performance is closely related to this. This is especially important in the development of hardware controlled directly by the cosmonauts. Some of the main requirements are small weight, small overall dimensions and insignificant power consumption of the hardware. This is achieved through the use of highly effective heat sources, such as, for instance, the electron beam or artificially constricted high-temperature low-pressure arc. Resistance electric furnaces and exothermic heat sources are quite effective in the space environment due to their simplicity and ability for flexible control of temperature. Radiant energy of the Sun is promising for the space environment. It can be used either as a source of direct heating with application of special concentrators (for instance, for melting), or in combination with converters of radiant energy into electric energy for powering the above highly concentrated heat sources.

Rational design and use of special materials allow the weight and overall dimensions of the units to be minimised. The units designed for operation in space must be as versatile as possible. It is not rational to use in space separate units for performance of each operation or a series of operations of one type. Therefore it is necessary to try to ensure that one unit provides the performance of an intensive programme of work on board of a certain space vehicle or outside it, through re-equipment or replacement of individual components.

The following categories of units for performance of processing operations in space can be singled out:

1. Fully automatic units with program control. Such units can be used in cases where the nature of work performance is known in advance; they are mounted and set up already on Earth and are switched on through radio and TV systems.

2. Automated units with remote and program control are used in cases where the nature of work is not known in advance. In this case the cosmonaut determines the programme of work performance prior to switching on and is able to correct it during the process.

3. Units directly controlled by the operator. When working with these units the cosmonaut takes an active part in the performance of a particular processing operation in all its stages.

4. Manual not automatic processing tools.

Various combinations of the above equipment categories are also possible.

It is generally known that performance of each experiment in space involves significant expense. Therefore it is rational to conduct all the initial retrofitting on Earth using space environment simulators. This statement is completely true also for the process experiments.

An entire arsenal has been created of the methods and means allowing simulation of the space environment to varying degrees, such as, for instance, large-volume high-vacuum chambers, solar radiation simulators, cooling screens, zero gravity simulators based on powerful magnetic fields, etc.

However, from the view point of the technologist, space conditions can be most fully reproduced in a flying laboratory which carries special vacuum or thermal vacuum chambers. In spite of the short duration of the zero gravity state (25–30 s), the experiments in the flying laboratory provide the most extensive and rich research material. However, these conditions do not allow the study of such long-term technological processes as single crystals growing or zone cleaning. On the other hand, the features of the majority of the processes (melting of small amounts of metal, welding, brazing, coating and cutting) and of physical phenomena (adhesion, heat and mass transfer, surface tension, distribution and separation of phases) can be studied fundamentally. Moreover, the experiments in the flying laboratory permit outlining some ways of active control of the processes of materials processing at zero gravity, development of the procedures of processing units operation in space and more precise determination of their ergonomic characteristics.

The effectiveness of the flying laboratory as a means for studying space technology processes is well illustrated by the fact that the majority of the observations made in it were later on fully confirmed in performance of experiments directly in space in «Soyuz-6», «Skylab» and «Soyuz–Apollo» vehicles.

Experience of work performance gained in a flying laboratory is especially valuable, if an experimenter wearing a space suit directly participates in the investigations. In this connection a special testing facility was also developed which allows the work of a man in a space suit to be analyzed during his performance of various processing operations.

Despite the value of the work carried out in the flying laboratory, direct experiments in space certainly are the most important stage of the investigations. In this case at different stages of investigations the work can be performed using ballistic missiles or unmanned orbital objects with re-entry vehicles and eventually in manned vehicles, for instance, in specialized orbital stations. Naturally in the unmanned objects the experiments can be performed only in automatic units with program control with or without feedbacks. Information about the progress of the experiments in this case is received through the channels of telemetry, TV or filming.

The largest scope of investigations can be performed in manned objects. The direct participation of the operators in this case ensures the acquisition of the maximum amount of information.

The first process experiment in a manned space vehicle «Soyuz-6», as was already mentioned, was conducted by pilot-cosmonauts in 1969. During the experiment welding of samples was performed by various processes using a versatile automatic «Vulkan» unit. The experimental results were described in detail in Soviet and foreign publications.

The small-sized welding units operating as part of the «Vulkan» system demonstrated sufficient reliability and serviceability under the conditions of space. The fundamental decisions taken during their development and the data derived in the experiment formed the base for designing special units for performance of specific processing operations in space.

In order to investigate the processes of melting and production of pure metals in the US OL «Skylab», special processing hardware was developed and manufactured, incorporating, in particular, a 1.6 kW electron beam unit, an all-purpose electric furnace and an exothermic reactor. This hardware was used to conduct extensive investigations related to production of single crystals, super pure metals, investigation of a number of physical-chemical properties of materials. Investigations in the field of welding, cutting, brazing, etc. were also conducted with extremely encouraging results. A similar all-purpose electric furnace was also used in a Joint Soviet-American ASEF Programme.

Thus space exploration by man gave a new impetus to technological and materials science investigations in the field of metallurgy, welding and production of new materials. This made the designers and researchers start development of new highly efficient processes and reliable small-sized equipment, the first trials of which have been successfully completed both here and in the USA. Apparently, new experiments in space technology can be anticipated in the near future, already having a specific applied importance for production of various new materials and supporting space object flights. So, it will be rational to envisage the performance of scheduled repair-restoration operations for the structures of space vehicles and long-term orbital stations, which will significantly improve their operational reliability and viability.

In the future one can anticipate development of special process modules incorporated in the multipurpose space objects and eventually creation of specialized processing orbital stations fitted with production and research installations. In other words, progress of space technology will result in an even closer connection between space research and the needs of the national economy in the near future, which will indubitably promote the further advance of science and technology.

Ten Years of Space Technology*

B.E. Paton[1] and V.N. Kubasov[2]

October 16, 1969 may be considered as the date of birth of space technology. This day the technological process of welding and cutting of metals in space was performed for the first time in the world on board the spaceship «Soyuz-6». During this experiment the peculiarities of automatic EBW and arc welding of aluminium and titanium alloys, and stainless steel were investigated, the feasibility of cutting these materials was tested and the behaviour of molten metal in solidification of welds and cuts was studied. The experiments were conducted using the research equipment «Vulkan», which was designed and manufactured at the PWI.

Space technology has evolved dramatically in the years since that time. At present there are two most important trends of investigations: firstly, the repair and construction of space vehicles, instruments and equipment directly in space and, secondly, producing materials in space with quite new or unexpected improved properties.

The origin of the first of these trends was caused by the intensive progress in space technology. The modern space orbital stations are, in principle, well-equipped research laboratories, functioning reliably in orbit for a long time. Therefore, the problems have to be considered of construction of different experimental space vehicles, such as orbital stations, radio telescopes, antennas, reflecting screens and systems of solar power engineering, not already on Earth, but in space. In addition, with the increase in functionality and life of these vehicles, interest in the problems of their repair and restoration is growing. Also the increase in the mass and dimensions of such vehicles raises the problem of assembly and installation in open space.

In this connection, it is interesting to recall that as early as the 1960s Sergey P. Korolyov predicted this trend of space technology. He supported and promoted the investigations on welding and cutting in space. It was his initiative to start the works which were completed by the experiment with the unit «Vulkan» in October, 1969.

Later the scope of these works was widened. In 1973 in the USA the experiments on automatic electron beam welding and cutting of metals were repeated in the OL «Skylab». Simultaneously, in the USSR systematic investigations of peculiarities of different methods of permanent joining of materials under space conditions were started. The main volume of investigations was fulfilled at the PWI in collaboration with the Yu.A. Gagarin CTC[3] on board the flying laboratory TU-104, allowing short-term creation of relative weightlessness. To reproduce the other factors of space (vacuum and temperature) special test benches were mounted in the laboratory. It was established that under these conditions such versatile methods of fusion welding as arc and electron beam welding, widely-used on Earth and promising for space, have a number of unfavourable technological peculiarities (unstable low-constricted arc discharge, unstable coarse-drop transfer of the electrode metal, high porosity of welds, etc.). During experiments the difficulties, encountered with these peculiarities, were overcome. However, the development of special welding equipment and procedures was required for this purpose.

Assembly and permanent joining of structures manually by the operator in a spacesuit were of special interest for the development of advanced space vehicles. To study this problem, a special test bench containing a part of a spacesuit was designed, manufactured and installed. A large volume of experiments carried out with the help of this equipment during 1973–1978 showed the feasibility of most assembly-welding jobs in space. Technological, ergonomic and medical-biological peculiarities of manual welding, brazing and

---
* (1979) *Avtomaticheskaya Svarka*, No. 12, pp. 1–3.

cutting in space were studied; the effect of environment, protective clothing and equipment, systems of fastening and training of operators for the quality work were analysed. The fundamentals of procedures for education and training of operators have been developed.

It is interesting that, in the course of these investigations, not only the negative, but also the positive aspects of the space environment were revealed. Due to the absence of gravity forces, the danger of burn-outs in welding thin-sheet metal decreases drastically and can be reduced almost to zero. The manipulation of the welding tool becomes much easier so that even low-skilled welders at their good fastening can produce quality welds, even with much larger gaps and edge displacements than those on Earth. The feasibility of increasing gaps and edge displacements was also observed in brazing, explained by abruptly increasing capillary effect under conditions of weightlessness.

In some cases (on stainless steels and titanium alloys) a significant increase in strength (up to 30–40 %), obtained in weightlessness, was observed. It is, probably, associated with peculiarities of the metal solidification, but this effect has to be further studied under the conditions of long-term weightlessness.

In 1974 the PWI in collaboration with other organizations performed the first experiments on manual EBW in a large-volume space stimulation test chamber, showing the feasibility of performance of these works in space.

Based on the results of all the investigations, a kit of manual tools was designed for welding, brazing and cutting in space, which could be used in space stations for erecting and repairing future space vehicles.

At the same time some organizations of the country were performing experiments on permanent joining of materials without melting (explosion welding, magnetic-impulse welding and diffusion bonding). These methods of joining are also very promising for technological works in space, in spite of some limitations (low versatility and the need for a careful preparation of the surfaces being joined). Most of these investigations, as well as experiments on flash-butt welding, were conducted on Earth, because weightlessness should not greatly influence these technological processes.

In the USSR studies were also carried out on exothermal brazing in space. Equipment created for this challenging process was tested many times in ballistic rockets, and in 1976–1977 — in the OS «Salyut-5». In the course of experiments the technology of brazing tubular joints using refractory brazing alloys was tested.

Peculiar to all these investigations is the fact that the results obtained have always been implemented in the national economy, in different fields of science. Thus, small-sized highly reliable equipment for EBW and electric arc welding has found an application in some branches of industry. Special accelerators for conducting space geophysical experiments «Zarnitsa» (Soviet) and ARAKS (Soviet-French) were based on electron beam guns and special power sources developed for space welding. These experiments were aimed at the simulation of ionizing phenomena in terrestrial space using an injection of electrons from ballistic rockets. In experiment «Zarnitsa» injected flows of electrons of about 12 keV energy and up to 3.3 kW capacity caused in the Earth's magnetosphere the phenomena of magnetic and atmospheric backward scattering, polar aurora, interaction of waves with particles and other similar effects observed in the zone of the injection.

The experiment ARAKS was more complicated. In the course of this experiment the injection of electrons of energy up to 27 keV and capacity up to 14 kW was made from a ballistic rocket launched from Kerguelen island, located in the south part of the Indian ocean, while the phenomena caused by the injection were observed in a conjugate point of the magnetosphere of Earth (village Sogra of Arkhangelsk region, Russia).

At the beginning of the 1970s interest was first taken in the deposition of different coatings in space. In 1975 in the OS «Salyut-4» the first experiment was performed for the restoration of the reflecting metallic coating of a telescope mirror. Subsequently, the task was set to develop more sophisticated technology and equipment for the deposition and restoration of different coatings (thermoregulating, protective or optical).

The «Isparitel»unit operating on the principle of electron-beam thermal evaporation of the molten metal in vacuum was designed and manufactured at the PWI. The metal to be evaporated is heated in a refractory crucible by its bombarding by a beam of electrons forming by a special electron gun. The directed flow of vapours is condensed in vacuum on samples of materials, thus forming a coating. The samples are arranged in a manipulator which makes it possible to change them by a remote control. Silver was used as a model evaporating material.

The «Isparitel» installation was delivered to orbit by a transport spaceship «Progress-7» in June, 1979 and mounted on board the OS «Salyut-6» by cosmonauts V.A. Lyakhov and V.V. Ryumin for conducting the experiments. The operating unit of the installation was located in a lock chamber, and a remote control panel in the working compartment of the station. To conduct experiments, the lock chamber was depressurized and the installation was switched on after reaching the required vacuum. Several samples were exposed successively to the vapour stream using a manipulator. During experiments the exposure duration was varied, the temperature of the material evaporated and the evaporation rate were adjusted. The main parameters of the condition of the installation operation were recorded in telemetry channels.

As a result of several series of experiments, 24 samples were produced, which were delivered to Earth and examined comprehensively. Thus, the feasibility of achievement of repair-restoration works involving deposition of metallic coatings under space conditions was shown and the initial data for the development of specialized onboard equipment were obtained.

In the course of preparation of this experiment, during tests in the flying laboratory in particular, the problems of safe maintenance of the molten metal and stable division of phases in open crucibles in weightlessness were solved successfully, which is also very important for the development of another trend of space technology, namely the production of unique materials. A boost to the development of this trend was also given by experiments in the «Vulkan» unit in 1969, where the feasibility of successful realization of technological operations with molten metal in space was shown for the first time.

It should be explained that the main premises for the development of this trend are the suppression of the processes of convective stirring in liquids and gases, special conditions for heat-mass exchange, capillary phenomena and adhesion, and also the feasibility of use of microacceleration to control the technological process under conditions of the weightlessness.

The first experiments on producing materials by melting in space were conducted in 1973–1974 in the OL «Skylab» using a special versatile furnace. Then, in 1975 a major series of experiments in a furnace of similar design was performed within the scope of the Joint Soviet-American ASEF Programme during the flight of the orbital complex «Soyuz–Apollo».

In the Soviet Union during 1975–1978 experiments were performed on melting using exothermic heat sources which were arranged on the ballistic rockets. At the existing functioning OS «Salyut-6» the electric furnaces «Splav» and «Kristall» were mounted. During 1978–1979 numerous technological experiments were performed which would make a large contribution to the development of space technology. Finally, at the same station the above-described «Isparitel» installation was mounted and operated. Experiments made using this installation in July–August, 1979 showed the radical feasibility of producing unique materials by their evaporation and condensation in weightlessness.

Current investigations will, probably make it possible to define clearly the range of materials and products which it will be rational to produce in space in the near future.

However, even today it is possible to say that space technology which was born ten years ago, has taken its place and is progressing successfully as one of the important trends in space exploration. Soon the creation of special technological units will be incorporated into the composition of multipurposeful space stations. Specialized technological orbital vehicles equipped with research and production units and designed for study of the behaviour of substances in space and producing unique materials, can be expected.

It is impossible now to imagine the future progress of cosmonautics without creation of large-modular orbital stations having numerous changeable crews and functioning for many years in orbit. Undoubtedly the construction, periodic repair and assurance of long life of these stations, will require, as the technical-economical calculations showed, the use of most different types of permanent joints of the materials. Even the crews of the OS «Salyut-6» restored many times the performance of its separate units using brazing. With improvement of long-term functioning orbital stations, these repair-technological operations will become, without any doubts, a routine, and with the appearance of transport shuttles, it would be impossible to imagine their normal operation without a wide use of welding, brazing, cutting and coating.

There are projects for the interplanetary manned spaceships which will be in autonomous flight for several years. Their separate modules will land on different planets and take-off from them. The furnishing of these spaceships with highly-reliable versatile repair equipment is one of the main conditions of their failure-free operation during all the flight.

Finally, over recent years solar power engineering is progressing more and more actively, being very attractive from the point of view of environmental protection. Solar power engineering can be justified economically only in the creation of receivers of beam energy in space. They would be tens and hundreds of square kilometers in area. Naturally, the construction of these systems will require the development of new technology and equipment for coating and permanent joining of elements. But this is already the future of space technology.

It is certainly difficult now to define all possible problems which will face space technology by scientific-technical progress. One thing is certain: space technology will still encounter great success, unforeseen difficulties, and unrealisible hopes. But progress and improvement are irreversible, as well as the process of space exploration by mankind.

Tested in Orbit[*]

B.E. Paton[1], V.A. Dzhanibekov[3] and S.E. Savitskaya[2]

TASS INFORMS

On July 25, 1984, in accordance with the program of flight of the orbital research complex «Salyut-7»–«Soyuz T-11»–«Soyuz T-12» cosmonauts Svetlana Savitskaya and Vladimir Dzhanibekov performed extra-vehicular activity. The task of EVA was to test a new versatile hand tool designed for complex technological operations...

For the first time in the world, EVA was performed by a woman-cosmonaut Svetlana Savitskaya. Her successful performance of unique experiments under space conditions demonstrated the feasibility of effective activity of women in fulfilment of complex works not only on board the manned orbital complex, but also in open space.

* (1986) *Nauka i Zhizn*, No. 2, pp. 2–7.

These short phrases, taken from this report of TASS and spread all over the world, were preceded by many years of intensive, and at times dramatic, work by a large team of scientists, engineers and cosmonauts. It concerns the creation of the versatile hand tool (VHT) and its testing in orbit. Six cosmonauts working on board OS «Salyut-7» participated in this test, namely V. Dzhanibekov, S. Savitskaya, I. Volk, L. Kizim, V. Soloviov and O. Atkov.

Let us recollect the event which took place fifteen years before. On October 16, 1969 during the flight of spaceship «Soyuz-6» the pilots-cosmonauts G. Shonin and V. Kubasov performed, another first in the world, the welding and cutting of metal in space. This unique experiment can be considered as the beginning of the era of space technology. This experiment was realized in the unit «Vulkan», created at the PWI and made it possible to perform welding using different methods: electron beam, plasma and arc consumable electrode methods. All these experiments were automatic.

Soon a new step in the development of space technology was made. The PWI developed a new unit «Isparitel». With its use thin-film metallic coatings were deposited on samples from structural materials under conditions of weightlessness and outboard vacuum using the method of thermal evaporation and condensation of materials.

Cosmonauts V. Ryumin and V. Lyakhov in 1979, and then L. Popov and V. Ryumin in 1980 and, finally, the following year, V. Kovalyonok and V. Savinykh carried out experiments onboard the OS «Salyut-6», which confirmed the high efficiency of the unit «Isparitel».

This diversion into history is highly relevant because it is the experience gained in the process of design of «Vulkan» and «Isparitel», and the experiments made with their use on Earth and in space, which made it possible to create the VHT.

From the automatic unit to hand tool. The tool, designed and manufactured at the PWI, can be used to perform welding, cutting, brazing and metal deposition in open space. The versatility of this tool is its most important merit. Using this one tool the cosmonaut will be able to perform the necessary operations during the regular repair of the space vehicle, to restore and improve its service life, to improve reliability of different sub-assemblies, to deposit protective coatings, to assemble orbital scientific platforms from modular structures, to construct large radio telescopes, reflecting screens, etc.

And if the versatility of the tool is undoubtedly seen as its most impressive feature, then it would seem strange that it is realized in a hand-operated version. It is known that «Vulkan» and «Isparitel» were automatically-operated units. Why within 15 years did people arrive not at a robotic manipulator, but a hand tool?

Nothing strange has happened. Automatic units or robotic manipulators are appropriate, as a rule, to those cases where the system of technological operations is clearly defined, some of them repeated many times and where it is possible to provide rigid space orientation of the parts and products being treated. The most unexpected situations may appear during work in open space, that will require cutting, welding or deposition of different coatings. And often, the type and volume of operations will be defined by a cosmonaut directly « in site». Thus, V. Ryumin had to cut a steel rope with cutting pliers before he disengaged an antenna from the docking unit. L. Kizim and V. Soloviov had to cut out a piece of the spaceship skin on the accessory bay and a piece of the solar battery. In short, the cosmonauts will have to work on different areas of the external surface of the space vehicle and to deal with structural materials having different physical and chemical properties.

Even in the period of testing the unit «Vulkan», the scientists of the PWI pondered over the creation of a compact, haversack versatile tool with a self-contained power source which could allow cosmonauts to perform works, such as repair or assembly, on any areas of the space object surface. The most frequently required operations for these purposes were specified as cutting, welding, brazing and coating deposition.

OS «Salyut-7». Working station in open space. Experiment is made by S. Savitskaya.

Technological unit «Isparitel» for deposition of thin-film metallic coatings. *Below*, two monoblocks are seen, each having own electron beam gun. When VHT was designed it was managed to arrange two guns in one monoblock: one — for welding, cutting and brazing, another — for coating deposition.

When selecting the means of action for the fulfilment of all these operations the electron beam was chosen. Why?

Using an electron beam, it is possible to concentrate energy easily on a small «spot», which is necessary in welding and cutting, and it is also not difficult to distribute uniformly its energy over the entire relatively large surface without the beam focusing that is required in brazing and evaporation of metals from a crucible for the deposition of coatings. The selection of electron beam technology led to its wide spread use under Earth conditions where in many cases it is fully automatic.

Perfection can be the enemy to progress. As sometimes happens, when complex problems are being solved, there appeared not only enthusiasts for the idea of creation of the versatile tool, but sceptics as well. The enthusiasts began to embody the idea into drawings, schemes and hardware, to test separate sub-assemblies, while the sceptics began to reveal new scientific and technical problems, as if without their solution it would be impossible to create the tool. And, it should be noted, that scepticism was not a «bare negation», but it had serious grounds connected with complexity of the arising problems, which have not been yet solved by anybody before.

The most difficult task was the removal of heat, which was evolving inevitably during equipment operation, in particular from that part of the tool which the cosmonaut should hold in his hand. Here, the maximum admissible temperature is defined by the materials of gloves and the spacesuit. This is not the problem for Earth conditions, as it is solved very simply by using a forced cooling system. However, this is not acceptable in space conditions, because it would lead to making the tool bigger in mass and dimensions and, as a consequence, to decrease in mobility and increase in energy consumption.

No less complicated problems occurred due to the fact that the formation of the flow of electrons in electron beam guns, their acceleration and beam control are performed using high voltage (more than 10 kV). Due to this, secondary (braking) X-ray radiation occurs in treatment of materials, against which reliable protection for the operator is required.

The drops of molten metal which are formed in operation may cause serious injuries if they enter the spacesuit.

The complexity of the above-mentioned technical problems was due, first of all, to the fact that the traditional solutions used by the Earth equipment for close or similar conditions, were not suitable here. Space and specifics of activity in it require the provision of the highest safety of the equipment, complete safety of people who use it, and elimination of any possible damage of the station. In addition, the equipment being designed must be characterized by compactness, and low energy consumption, and be easy and convenient in operation.

At the first stage an Earth model, similar to the future space tool, was designed and manufactured. This made it possible to test (1974) the principal reproducibility of the idea of the design of an electron beam tool with manual control. However many problems connected with assurance of reliable and safe operation with this tool, including protection from secondary X-ray radiation, remained unsolved.

By that time the designs of crucibles, capable to maintain the molten metal in weight-lessness began to be tested in the flying laboratory on planes TU-104, IL-76. The miniature high-voltage converters were tested in thermal vacuum chambers.

A special attention was paid to the design of a short-focus electron gun with a low (to 10 kV) accelerating voltage. It is these guns that could allow solution of two problems at once. The short focusing made the beam safe for the remote objects, in particular hulls of spaceships, while at an accelerating voltage to 10 kV the level of secondary X-ray radiation becomes below the allowable level, thus the challenging problem of protection from it is eliminated.

At the very same time the unit «Isparitel» was being designed in parallel. It was managed to design an electron gun for this unit with characteristics which corresponded completely to the requirements specified to a gun for a hand tool. It is a monoblock unit, consisting of a short-focus gun and a miniature high-voltage converter, which became a unified unit of space technological equipment.

This monoblock was tested many times in a lock chamber (under open space conditions) of stations «Salyut-6» and «Salyut-7» in the operation of «Isparitel». Data were obtained about the serviceability of the monoblock unit in real service conditions. These data having not only scientific, but also great practical value, were used in designing and manufacture of the VHT space variant.

Finally, the tool was manufactured and passed successfully trials according to many private and integrated programs and was submitted to the strict judgement of scientists and leading specialists. The tool design caused genuine interest, and the test results aroused some amazement and even some confusion in sceptics. Nevertheless, almost all of them continued to insist in their doubts concerning the tool safety and, therefore, considered naturally the work in space using this tool untimely. And again, auxiliary devices were suggested, «to avoid burning through the station board and damage to the spacesuit». These would be intricate electric blockings, mechanical limiters of the tool movement, and entire sensor systems reacting to the type of material in front of the tool and recording the distance to the surface treated and its temperature and so on. All these devices should have the feedback in supply circuits.

It turned out that to create a versatile tool required for routine cosmonautics, is many times easier than to give it a start in life. It was necessary to take a decision: either to continue modification of the VHT taking into account the «new» ideas, not seeing the end, or to transfer it for finishing tests and training pilots-cosmonauts to operate with it. Taking into account all arguments «for» and «against», the decision was taken to transfer it to specialists in space engineering after certification of this tool as a quite safe device in operation.

VHT as such. What is the versatile hand tool itself? All the tool sub-assemblies are arranged in a $400 \times 450 \times 500$ mm size container, welded of tubular elements, that made it rigid at a small mass. In this container, called also « knapsack» (can be carried on shoulders) and convenient for fastening on the external surface of the orbital station, are mounted: secondary power source with a panel; cables connecting the power source to a board socket and hand tool, and the working tool itself in a special recess. A board with samples of materials for welding, cutting, brazing and coating deposition is fixed to the container.

The working tool is a monoblock consisting of a high-voltage electric supply source and two electron guns. One of them is designed for the fulfillment of the technological operations of cutting, welding and brazing. Another gun, where the focusing system is replaced by a crucible with an evaporating metal, is designed for coating deposition. The monoblock has a handle of a shape which is convenient for operation of a person in a spacesuit.

The mass of the VHT is somewhat more than 30 kg, and the monoblock which is used by cosmonaut is slightly more than 2.5 kg. The consumed power is 750 W and it can be adjusted depending on the operating conditions and the material treated.

At the Yu.A. Gagarin CTC tests started from a space stimulation test chamber. A work station for the operator, similar to that of the cosmonaut in the OS «Salyut-7», was mounted in it. Several test operators, a woman among them, were equipped into cosmonaut's suits, entered the vacuum chamber and fulfilled operations of welding, cutting, brazing of metals and coating deposition under the supervision of specialists. The operators noted at once the convenience of operation with VHT. This was very important. However, of more importance was that even test operators who did not deal with welding earlier, could, from the second or third trial, fulfil any of the above-listed technological operations with «good» and «excellent» marks. In the vacuum chamber not only the tool and technology were tested, but also the convenience of operation in the spacesuit and safety measures were evaluated and the conditions of labour and rest were selected. From the test results the program of

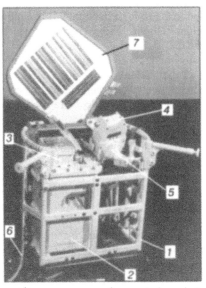

VHT: *1* — container; *2* — secondary power source; *3* — control panel; *4* — working tool; *5* — handle
with a trigger; *6* — cable; *7* — 6-sample board.

experiments in space was worked out and the procedure of training cosmonauts was
developed.

Eventually, the versatile hand tool was included in the research equipment of the
OS «Salyut-7».

Cosmonauts V. Dzhanibekov and S. Savitskaya mastered very quickly the tool and
began training with it in the space simulation test chamber, neutral buoyancy tank and at
short-time zero-gravity (25–30 s) in the plane-laboratory. During training they tested all the
movements associated with a VHT being used in space, its fastening to the handrail, and
preparation for operation.

Work station in space. On July 25, 1984. V. Dzhanibekov and S. Savitskaya carried
out EVA. Dzhanibekov fastened the VHT carefully to a flap handrail and it turned out to be
a very convenient work station for the operator-welder, providing all requirements to safety.
The cosmonaut-welder is standing on a special foot rest-«anchor», his feet are fixed. The

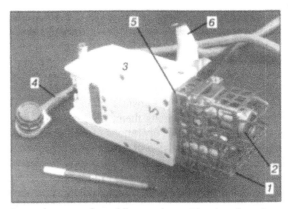

Working tool: *1* — electron beam gun for welding, cutting and brazing; *2* — electron beam gun (with a
crucible) for coating deposition; *3* — high-voltage converter; *4* — supply cable; *5* — screen, protecting
hand from heat emission of metal being treated; *6* — handle.

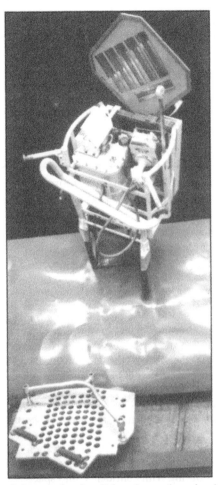

Working station in space simulation test chamber. *Below*, the «anchor», in which the cosmonaut's feet are fixsed; VHT with an open sample board is fixed on a flap handrail of a fragment of OS «Salyut-7».

VHT is mounted in front of him on the handrail. The control panel, working tool and sample board are easily accessible. The lower part of the cosmonaut's body is protected by the container itself. The left hand of the cosmonaut is located on the control panel and the panel cover protects it from occasional injury by the electron beam. The cosmonaut takes a working tool in the right hand, directed from the cosmonaut to the side of the sample board.

To ensure safety of the station the work station is oriented relative to it in such a way that if the beam passes beyond the sample board it will not be able to touch the station body. If in case of any unforeseen circumstances the tool «falls out» from the hand, then the trigger at the handle will switch off the electrical supply automatically. In case of the trigger jamming the cosmonaut will switch off the tool supply by the button «stop» at its control panel. To avoid the uncontrolled operation of the tool, if the cosmonaut is not able to do this by any reasons at the work station, the facility to cut off the supply by the second or third crew members is provided. They can do this by breaking the split connection of the cable, disconnection of the board socket and, finally, from the central control panel.

Having equipped the weld station, V. Dzhanibekov prepared the tool for operation and let S. Savitskaya have his place. First she performed the operation of cutting, then welding, brazing and coating deposition. She commented on all actions in detail.

Here the fragments of radio communication of cosmonauts with the Flight Control Centre (FCC), which describe the performance of this unique experiment.

— I start the work, — reported Savitskaya. — I switch on tool. There is supply. There is a trace. Weld is not so uniform, but beautiful. I close the sample board. I pull out the second sample board. I switch on conditions, take the tool. The welding of metal is going on.

— Weld is forming. It is uniform and beautiful. I see it well... Now I will try the third condition... It is convenient for me to make the fifth sample... I take it... There is a red spot. I am pressing upward and downward...

— We recommend to start coating, — operator prompted.

— Well, — answered Savitskaya calmly.

— In a minute You will come to the shade, — operator reminded.

In the next communication session Savitskaya informed:

— The first sample board was coated intensively — this was seen well... During brazing a bright drop was formed...

She did not want to interrupt the works, but the program is the program, and Savitskaya let Dzhanibekov have her place.

— I have completed brazing, — reported Dzhanibekov, who continued work with VHT. — I am coating. I am working as if with a brush. Generally, I became a painter... Excellent tool. I think it will have a great future in cosmonautics.

V. Dzhanibekov dismantled VHT and carried it to the transition bay of the station. The hatch was closed.

The operation with VHT in the open space was performed during 3 hours. Specialists at the FCC watched carefully the work of the cosmonaut. Everything was done accurately and at one breath. After return of the crew of cosmonauts to Earth the samples were given to the specialists for examination.

Their results prove convincingly that the traditional Earth operations of metal joining, cutting and coating can be done successfully in space for the fulfilment of any repair and site jobs.

The versatile hand tool is available now at the OS «Salyut-7». Specialists are preparing interesting programs for welding and brazing of connections of the truss structures.

The new technology is pacing with confidence into space.

Milestones in Space Exploration (1969–2000)

Technologies in space are a reality that is confirmed by the results of works of the E.O. Paton Electric Welding Institute of the NAS of Ukraine. The main stages of the works are presented below.

October 16, 1969
«VULKAN»
Customer OKB-1[2]
Main performer PWI[1]
Co-performer IED[12]
Space object «Soyuz-6»
Crew V. Kubasov and G. Shonin

Experiments on welding and cutting of metals using three methods:
- low-pressure arc of consumable electrode
- low-pressure constricted arc
- electron beam

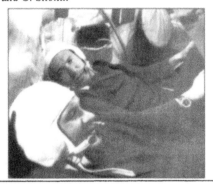

1973–1975
«ZARNITSA»
Customer ITMIRAS[21]
Main performer PWI[1]
Co-performer IED[12]
Space object Ballistic rocket MR-12

Study of the near-earth space using a modulated electron beam, injected from onboard a ballistic rocket

1975
ARAKS

Customers CNES (France); ITMIRAS[21]; SRI[5]
Main performer PWI[1]
Co-performer IED[12]
Space object Ballistic rocket «Eridan» (France)

Soviet-French experiment in magnetoconjugating regions (Kerguelen–Sogra) on injection of electrons and plasma jet from onboard rockets and examination of accompanying effects in magneto- and ionosphere of Earth

1979, 1980, 1981
«ISPARITEL»

Customer TsKBEM[2]
Main performer PWI[1]
Co-performer IED[12]
Space object OS «Salyut-6»
Crews V. Ryumin and V. Lyakhov; V. Ryumin and L. Popov; V. Savinykh and V. Kovalyonok

Series of experiments on deposition of thin-film metallic coatings on samples of structural materials using the methods of electron beam evaporation and condensation

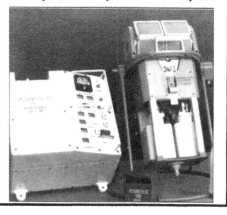

1980, 1982, 1983, 1987
«MODEL»

Customer NPO «Energiya»[2]
Main performers PWI[1] (1980–1983);
Ukr. Research Institute of Polymeric Fibre (1987)
Co-performers Ukr. Research Institute of Polymeric Fibre (1982, 1983);
PWI[1] (1987)
Space object «Progress-11»–OS «Salyut-6»;
«Progress-14»–OS «Salyut-7»;
«Progress-18»–OS «Salyut-7»;
«Progress-28»–OC «Mir»
Crews G. Strekalov, O. Makarov and L. Kizim;
A. Berezovoj and V. Lebedev;
V. Lyakhov and A. Aleksandrov;
Yu. Romanenko and A. Lavejkin

Experiment on deployment of large-sized flexible framed antennas to confirm the feasibility of creation of the space system of communication in the ultralow-frequency range of radio waves

1984
«ISPARITEL-M»
Customer NPO «Energiya»[2]
Main performer PWI[1]
Co-performers IED[12]; KhPI[6]; Leningrad State Optical Institute
Space object OS «Salyut-7»
Crew V. Soloviov, L. Kizim and O. Atkov

Experiments on evaporation and condensation of binary alloys to evaluate the peculiarities of the process in the conditions of weightlessness

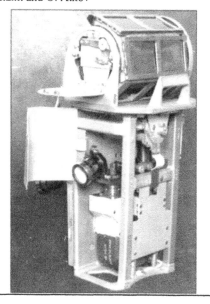

1984
«ISPARITEL-M»—«OVERCOOLING»
Customer SRI[5]
Main performer PWI[1]
Co-performer IED[12]
Space object OS «Salyut-7»
Crew R. Sharma (India), G. Strekalov and Yu. Malyshev

Soviet-Indian experiment on electron beam melting and solidification of silver-germanium alloy

1984
VHT

Customer NPO «Energiya»[2]
Main performer PWI[1]
Co-performer IED[12]
Space object OS «Salyut-7»
Crew S. Savitskaya and V. Dzhanibekov

Experiment on manual electron beam welding, cutting, brazing and coating deposition in conditions of open space using VHT

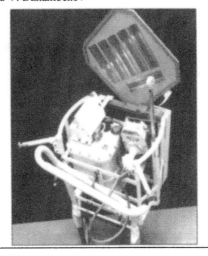

1986
VHT, «MAYAK»

Customer NPO «Energiya»[2]
Main performer PWI[1]
Co-performers IED[12]; Moscow Aviation Institute
Space object OS «Salyut-7»
Crew V. Soloviov and L. Kizim

Manual electron beam welding and brazing of members and fragments of truss structures at the external surface of the station using VHT. Experiment «Mayak» for deployment and folding of hinge-framed truss structures of 13 m length

1987, 1989
«YANTAR»
Customer NPO «Energiya»[2]
Main performer PWI[1]
Co-performers IED[12]; KhPI[6]
Space object OC «Mir»
Crews A. Lavejkin and Yu. Romanenko; S. Krikalyov and A. Volkov

Evaporation and condensation of metals and alloys on the moving polymeric film to produce samples of foils and to study the dynamics of evaporation of binary alloys

1989
«KRAB»
Customer NPO «Energiya»[2]
Main performer Ukr. Research Institute of Polymeric Fibre
Co-performers PWI[1]; A.I. Mikoyan KIFPI[26]; SDB GPI[28]
Space object «Progress-40»–OC «Mir»
Crew S. Krikalyov, A. Volkov and V. Polyakov

Deployment of circular large-sized framed structures using metals possessing effect of a shape memory

1990
RSB
Customer NPO «Energiya»[2]
Main performer NPO «Energiya»[2]
Co-performers PWI[1]; UkrProektstalkonstruktsiya; All-Union RI of current power sources;
Central RI of machine-building
Space object TM «Kristall»–OC «Mir»

During the autonomous flight of the technological module «Kristall» the deployment of two reusable solar batteries (RSB) for 10 m length each was made in automatic conditions. After module docking to OC «Mir» a remote final deployment of batteries for a full 15 m length was realized

The work was realized by FCC in a command radio line

1991
«SOFORA»
Customer NPO «Energiya»[2]
Main performer NPO «Energiya»[2]
Co-performers PWI[1]; A.I. Mikoyan KIFPI[26]
Space object OC «Mir»
Crew S. Krikalyov and A. Artsebarsky

Erection of truss structure using an effect of shape memory of metals to provide a clearance-free joining of the truss connections

1995, 1997
RSB
Customer S.P. Korolyov RSC «Energiya»[2]
Main performer S.P. Korolyov RSC «Energiya»[2]
Co-performers Yu.A. Gagarin CTC[3]; PWI[1]; FCC
Space object OC «Mir»
Crews G. Strekalov, V. Dezhurov and N. Thagard (USA);
A. Soloviov, P. Vinogradov and D. Wolf (USA)

Using a step-by-step condition, the folding of two reusable solar batteries (RSB) in technological module «Kristall», their transfer, erection and unfolding of one of them in module «Kvant» were realized.

Next crew conducted folding and dismantling of used battery in module «Kvant». On its place a new, delivered to Shuttle «Atlantis» battery, was mounted and unfolded

1992–1997
ISWE
(International Space Welding Experiment)
Customer NASA (USA)
Main performer PWI[1]
Co-performers IED[12]; Teledyne Brown (USA)
Space object ITSS «Columbia» (USA)

US-Ukrainian experiment on welding technologies in space using hardware «Universal» for manual EBW was prepared. Experiment was not realized, but was planned to be carried out in OC «Mir» in a new version according to project «Flagman»

1997–2000
«FLAGMAN»
Customer S.P. Korolyov RSC «Energiya»[2]
Main performer PWI[1]
Co-performers IED[12];
Regional R&D Centre of Voronezh State University of Technology; «Zvezda» Company;
Yu.A. Gagarin CTC[3]
Space object OC «Mir»
Crews G. Padalko and S. Avdeev; A. Kaleri and S. Zalyotin

Equipment and crews are prepared for conducting experiments on manual electron beam welding, cutting and brazing using samples of different materials from George C. Marshall Centre (NASA, USA). Welding equipment was delivered onboard the station in May 1998. The platform with a work station for the operator was not delivered to the station because of the decision to interrupt scientific experiments and subsequent sinking OC «Mir»

State-of-the-Art and Prospects
for Development of Aerospace Engineering in the USSR[*]

B.E. Paton[1]

Speaking about the prospects of space welding development one should dwell on one more problem. The progress of applied cosmonautics urgently needs to cover the use of welding processes directly in space. The first idea about welding application was expressed by Prof. S. Korolyov, the outstanding Soviet scientist and designer. On his initiative, as far back as 1964 a special scientific program was developed, and the research works were started. Since that time the volume of studies on space welding has continuously increased (Figure 1).

At first the studies were only of a searching nature. During 1965–1969 the peculiarities of behaviour of different welding methods were studied and the melting and solidification of metallic materials in welding, brazing and evaporation in conditions simulating space conditions were investigated in the flying laboratory, in conditions of short-time weightlessness and in on-land thermal vacuum chambers. Simultaneously, the behaviour of arc discharges and electron beams were analyzed under these conditions.

The end of this period was marked by the implementation of the first in the world welding experiment in space in October 16, 1969 using the automatic equipment «Vulkan»

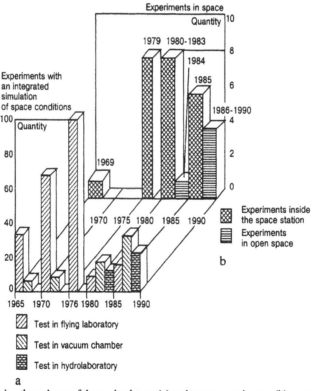

Figure 1 Changing the volume of the on-land tests (a) and space experiments (b) on welding in years.

* (1991) *Proceedings of Conf. on Welding in Space and the Construction of Space Vehicles by Welding*, USA, Miami, Sept. 24–26, 1991. Miami, pp. 1–12. (Published in a shortened version.)

Figure 2 Microplasma arc welding is performed by an operator in a spacesuit.

and the analysis of the results obtained. This experiment showed that the process of melting, welding and cutting by electron beam is stable and the necessary conditions for a good weld formation or cutting are provided. The small-sized welding units incorporated in «Vulkan» displayed the required reliability and serviceability in the space conditions.

In 1970–1974 studies of welding methods (Figure 2), materials and metallurgical processes were continued in the flying laboratory and thermal vacuum chamber. By the end of this period the electron beam was selected as the most promising welding heat source for the conditions of space.

Special jigs were designed, manufactured and tested (Figure 3), which made it possible to perform safe welding processes manually in the conditions close to those of space.

During the period from 1975 till 1979 research-applied studies were conducted, directed to testing the processes of bonding of specific structural materials and deposition of thin-film coatings on them in space. By this time the study of general problems of space welding was mainly completed. The applied technological problems were tested in the flying laboratory.

By the end of this period the studies were transferred to the OS «Salyut-6», where the processes of coating deposition on different substrates were investigated in the unit «Isparitel».

By 1980 the study of general regularities typical of the different welding processes in space was mainly completed. This made it possible to reduce significantly the volume of experimental testing in the flying laboratory in 1980–1984, but great attention had to be paid to works in the thermal vacuum chambers and to the start of testing the equipment in buoyancy conditions. The space experiments on coatings, including those of multicomponent alloys, were intensively continuing.

In 1984 a very important experiment was performed outboard the OS «Salyut-7». In this experiment cosmonauts S. Savitskaya and V. Dzhanibekov first used in space the manual electron beam VHT, performing the processes of welding, brazing, cutting and coating. In the course of preparing and conducting these experiments all scientific information accumulated during the recent years was fully used.

Figure 3 Operator in spacesuit performs the consumable electrode arc welding.

The period from 1985 till 1990 was characterized by a rapid growth of applied works made in space. The works were continued in unit «Yantar» on coatings, brazing alloys and welding of metals. In open space were conducted integrated experiments (Figure 4) on deployment of a 12 m truss structure using welding and brazing of its separate members using VHT. And finally, in 1990 two 15 m truss structures, being the load-carrying base of reusable solar batteries of technological module «Kristall», which was docked to the

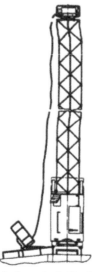

Figure 4 Scheme of deploying extended load-carrying truss structure.

Figure 5 Testing of the kit of manual electron beam tools for OC «Mir» in vacuum chamber.

OC «Mir», were deployed. Naturally, the conductance of these works required a large number of on-land tests. Figure 4 shows only the volumes of space experiments and on-land tests. But due to them, special test jigs, models of equipment for on-land space simulation test chambers and basins of buoyancy tests, and systems of weight loss providing complete simulation of the weightlessness in erection and deployment of extended truss structures, were manufactured and modified.

The large volume of preparatory works was justified by the fact that in the flying laboratory from 5 to 12 conditions of weightlessness of 25–30 s duration were checked for each flight. So, the test pilots and researchers from the PWI had in total 12 flying hours in weightlessness. The hours spent by the team of the Institute in space simulation test chambers and basins of buoyancy are no less impressive.

All these contributions make it possible to state today that we are completely ready to perform assembly-erection, welding and repair works in space. We have at our disposal the proven technology of making permanent joints, cutting and coating deposition for many structural materials used in the aerospace industry. A standard series of electron beam multi-purpose technological tools of up to 3 kW capacity have been developed, which can be used both in automatic units and for manual operations.

The kit of these tools in the manual version for OC «Mir», passed the on-land tests at the end of 1990 and were then prepared for the transportation to the station (Figure 5).

Since the hardware and technological problems have already been solved, we are now paying the main attention to the fabrication of large-sized structures in space and selection of optimum structural materials. In this direction many problems are still to be solved.

Chapter 2

SPACE AS
A PROCESSING ENVIRONMENT

Chapter 2. Space as a Processing Environment

The environment is very important for processes of joining materials. There-fore, the main factors of the space environment (vacuum, microgravity, high and low temperatures, etc.) provide a unique possibility for using space as a processing environment. The impact of these factors on the processes of materials joining has a positive effect on performance of structural materials and their permanent joints intended for long-time operation in space. This Chapter includes studies dedicated to the following issues:

- *technological aspects of the welding process in space;*

- *prediction of service life of structural materials and their joints;*

- *mathematical modelling and experimental investigation of technological processes under microgravity conditions;*

- *scientific-and-technical substantiation of space experiments for materials science and technology research;*

- *peculiarities of design, manufacture and operation of process equipment.*

The result of many years of the theoretical and experimental studies in space provides convincing proof of the possibility of extensively employing the space environment for performance of various technological processes.

Technological Aspects of Space Welding*

B.E. Paton[1]

The exploration of near-earth space envisages the fabrication and assembly of powerful orbital stations and platforms directly in space. These projects are being developed both in the USSR and in the USA. For example, the platforms in the form of truss structures of more than 100 m length and several ton mass cannot be transported into the orbit by one rocket. This delivery is possible only in the form of separate sub-assemblies which will be further erected in site, i.e. in space. Therefore, welding, which is necessary the same as brazing and thermal cutting in repair of long-service structure, is used in parallel with other methods. In this connection, many of new scientific and technical problems arise. For example, the following questions should be clarified: external shape of structures; methods, mechanisms, equipment, etc., which are necessary for their fabrication and erection; safety rules; peculiarities which should taken into account in the assembly and service of the structures.

In the USSR these questions were coming under consideration more than 25 years ago, and a number of problems has already been solved. The investigations resulted in the conclusion that excellent joints may be produced in space [1]. Below, some technological, physical and materials science aspects will be described.

Technological, physical and materials science problems. Four features characterize space as a working environment.

Zero-gravity (weightlessness). It suppresses buoyancy, convection and a number of other physical effects. It affects also, for example, densities of materials and their phases and has a great influence on the surface tension of liquids. During welding under space conditions the cosmonaut cannot rest on something as he does on Earth, thus making assembly and welding more difficult.

Vacuum. It should be assumed that large orbital stations are constructed with allowance for 10^{-2}–10^{-4} Pa pressure. This range of pressures has been mastered in electron beam and diffusion welding. However, space vacuum is characterized by extremely fast, almost «infinite» rate of evacuation and very high (up to 80 %) content of atomic oxygen. It is also necessary to work in spacesuits which creates additional difficulties.

Temperature. The difference in temperature range in sunlight and in shade is high. It means that the structure is subjected to the effect of temperatures approximately between −120 and +220 °C. The reduced heat- and mass transfer in space leads to the fact that closely adjacent zones of the same component may have a large drop in temperatures.

Rigid ionizing ultraviolet radiation. During long-term action this radiation degrades properties of materials and welded joints. It turned out that different methods of welding, brazing, thermal cutting and coating can be used in space, having their own advantages and drawbacks. At first glance, in parallel with coating, the methods of solid-state welding and cutting (explosion welding, diffusion and cold pressure welding and explosion cutting) would seem challenging. However, they require careful edge preparation and fit-up of the butt surfaces. Therefore they have not been yet used in space. Under some circumstances, it is possible to use explosion cutting, but, because of safety, only aboard unmanned vehicles. The electron beam method turned out to be the most beneficial [1], as it can be used not only for welding, but also for brazing, cutting and coating. Due to its versatility, high reliability and high efficiency factor, it has no competitors now in space. However it is not improbable that this may change in the future.

The processes of welding and thermal cutting in space differ from those used on Earth, firstly, when the liquid phases appear. The liquid molecules possess, besides kinetic (chemical and heat

* (1990) *Schweissen und Schneiden*, **42**, No. 3, pp. 117–120.

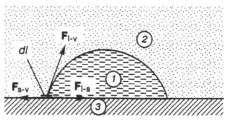

Figure 1 Scheme of interaction of molten metal (*1*) with vapour (*2*) and solid phase (*3*).

energy) energy E_k, potential energy E_g, depending on the gravity force, which is much lower than the kinetic force. For many physical phenomena at the «molecular level» (surface tension, cohesion, wetting, capillary pressure) the energy characteristics are comparable with potential energy. Therefore, these phenomena in the fields of low gravitation are proceeding more intensively than those on Earth. In zero-gravity, surface tension and wetting prevail in welding. Figure 1 shows the general case of interaction between the molten metal and a solid phase, and also forming vapours of metal. The following forces are acting on the element of interface of length *dl*:

$$\mathbf{F}_{l-v} = \sigma_{l-v}dl; \qquad \mathbf{F}_{l-s} = \sigma_{l-s}dl; \qquad \mathbf{F}_{s-v} = \sigma_{s-v}dl,$$

where σ_{l-v}, σ_{l-s}, σ_{s-v} are the coefficients of surface tension at the liquid–vapour (l–v), liquid–solid body (l–s) and solid body–vapour (s–v) interfaces. The relationship of these forces depends on the distribution of temperatures at the interphase interface. If at the solid body–liquid and solid body–vapour interfaces, the value σ does not depend on temperature, then at the liquid–vapour interface the value of the coefficients of the surface tension is determined mainly by the temperature. The dependence of σ_{l-v} on temperature is described approximately by the equation

$$\frac{d\sigma_{l-v}}{dT} = -k\left(\frac{\rho}{M}\right)^{\frac{2}{3}}, \tag{1}$$

where ρ is the liquid density; M is its molecular mass; k is the constant. It is seen that $(d\sigma_{l-v}/dT) < 0$. If the gradient of temperatures is dT/dl along the liquid–vapour interface, then the gradient of the surface tension will be

$$\Delta\sigma = \int\left(\frac{\partial\sigma_{l-v}}{\partial T}\frac{dT}{dl}\right)dl. \tag{2}$$

It leads to the appearance of the gradient of the surface tension force $d\mathbf{F}_{l-v}/d\mathbf{T}$, whose vector is directed towards the temperature reduction and initiates the mass exchange v (Figure 2).

Another distribution of forces takes place at the liquid–solid body interface. In this case the cohesive force is proportional to a free energy E_l. The change of this parameter in relation to the surface unity is described by the equation

$$\Delta E_l = \sigma_{l-v} - \sigma_{l-s} - \sigma_{s-v}. \tag{3}$$

As σ_{l-s} and σ_{s-v} do not depend on the temperature, and σ_{l-v} decreases with temperature increase, then a gradient of free energy dE_l/dT occurs at the liquid–solid body interface in the presence of the temperature gradient. This leads to the occurrence of the gradients of

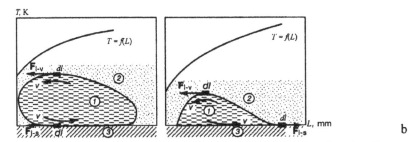

Figure 2 Scheme of processes at liquid–solid body interface at non-wetting (a) and wetting (b).

cohesion dF_a/dl. Value and direction of the adhesion force depend on the relation of σ_{l-v}, σ_{l-s} and σ_{s-v} which is determined by the angle of contact υ:

$$\cos \upsilon = \frac{\sigma_{s-v} - \sigma_{l-s}}{\sigma_{l-v}}. \tag{4}$$

In the region of $\pi \geq \upsilon \geq \pi/2$ (non-wetting) the vector of this force (the same as at the liquid–vapour interface) is directed towards the lower temperature and changes from the maximum value (at π) to zero (at $\pi/2$) (Figure 2a). Between $0 \leq \upsilon \leq \pi/2$ (wetting) it is, on the contrary, directed to the higher temperature and its maximum value corresponds to $\upsilon = 0$ (complete wetting) (Figure 2b).

The case when the liquid has more than one interface with a solid phase is of interest (Figure 3). Besides the distribution of temperature, it is necessary to consider here the angle between the solid surfaces. It follows from the equations (3) and (4) that the sum of adhesion forces, acting in the direction of the angle divergence, occurs on the non-wetting surface. It is different on the wetting surface. The situation with the direction of the mass transfer in liquid is similar. It is possible to create similarly the conditions for the directed shear and localizing the liquid phase in a definite region. Thus, it is clear that both one and two heat flows may occur in the molten metal pool: one flow is directed to the cold zone, while another — to the hot zone.

The described effects are observed and investigated in zero gravity on transparent salt models [2].

Peculiarities of welding in space. The above-described effects are taken into account in the development of the technology of welding in space. These features are manifested also, to a certain degree, on Earth. However their role is to an extent suppressed by the force of gravity. Only stirring of the molten metal under the action of the electron beam and effects of dissipation, connected with viscosity and heat conductance, can prevent or weaken the great influence of the surface forces. Most metallic materials in a molten state have a low viscosity, moderate heat conductivity and very high surface tension. At a small volume of

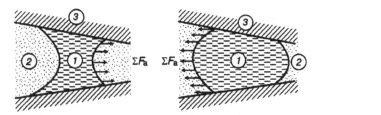

Figure 3 Scheme of interaction between the liquid and several surfaces of a solid body at wetting (a) and non-wetting (b).

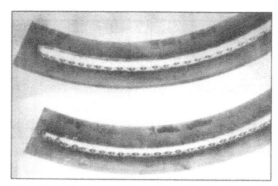

Figure 4 Concentration of metallic drops on the cut lower edge in electron beam cutting in space.

the melt, several seconds or even fractions of a second are enough under the zero-gravity conditions to form stable flows of mass exchange, initiated by the surface forces.

It is difficult to melt a thin sheet in space due to low concentration of the heat energy. The diameter of the molten pool can be many times larger than the workpiece thickness. During the electron beam cutting it can lead to both parts joining again behind the cut. This can be prevented, however, to the detriment of formation of an optimum temperature field. During space welding the molten metal is not forced out from a gap (that could be very hazardous), but it is accumulated at the lower side of the workpiece in the area of the cut edges. Owing to the proper formation of the temperature field it was possible to concentrate the solidifying metal on one of the cut edges (Figure 4).

The behaviour of the molten metal in brazing and evaporation can be controlled. Thus, it is possible to guarantee that the brazing alloy will flow in the proper direction or the vapour will be separated reliably from the liquid phase.

If all the peculiar features of molten metal behaviour in space are taken into account, then the welded or brazed joints performed there will not be inferior to those made on Earth as regards to tightness, toughness, strength and hardness. The same applies to the quality of the coatings produced. Only the difference in microstructure and gas content in welds will remain. The structure of stainless steel welded joints made in space differs from the on-land analogues by the finer grain boundaries and somewhat lower content of α-phase. It is much reduced in hydrogen, but it is somewhat enriched with oxygen. For example, in the titanium alloy the hydrogen content in the parent metal (initial state) was 0.0056–0.0072, while in welds — only 0.0010–0.0015 %. The oxygen content was 0.058–0.061 and 0.071–0.084 %, respectively.

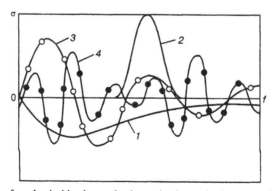

Figure 5 Scheme of mechanical loads on a load-carrying large-sized structure in space: *1* — static; *2* — shock; *3, 4* — loads, occurring under the action of control or stabilizing systems.

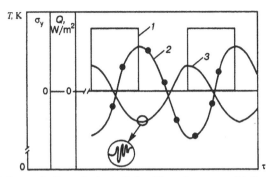

Figure 6 Scheme of effect of solar radiation (*1*) on the nature of deformation (*2, 3*) of elements of the space structure.

Peculiarities of service of the structures in space. The service life of welded structures in space should amount to several decades. During service the structure elements are subjected to different external effects. Figure 5 shows a typical scheme of loads of a large-sized load-carrying structure in space. Static loads (curve *1*) which are determined by the gradients of the field of terrestrial gravity and aerodynamic resistance of the residual atmosphere, are rather low. Their values depend on the shape of the structure and its orientation. However, it is impossible to determine the static loads when designing the space

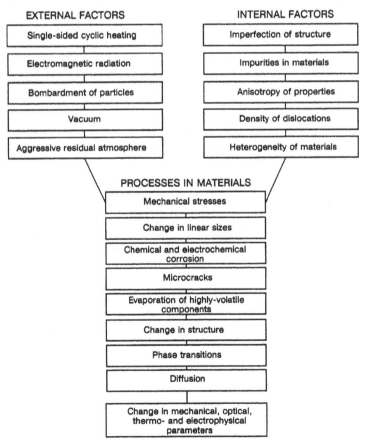

Figure 7 Factors and processes causing the degradation of materials in space.

structures. In parallel with this, it is impossible to avoid rare shock loads (curve *2*), occurring in erection or docking two or several structures. This should be taken into account during calculations, because these loads can lead to complete or local buckling of the structures or their separate elements. The significant constant load effect originates from the very function of space stations, for example, under the action of control or stabilizing systems or a crew (curve *3* and *4*). These (alternating) loads refer to a low-frequency range (10^{-3} up to 10 Hz) and could lead to a brittle fracture in decades.

A welded structure, moving in a near-earth orbit, crosses twice the boundary between the sunny and shade sides. Solar radiation is changed from maximum (≈ 1400 W/m^2 [3]) to almost to zero (Figure 6, curve *1*). All the structure, as well as its separate elements, are subjected to these variations in temperature (curve *2*) that lead to high deformations. However, as the structure is very rigid, then thermal stresses occur (curve *3*). These long-term cyclic thermal stresses together with other unfavourable factors of space effects lead to the constant deterioration of characteristics of materials, and, consequently, also of welded joints, thus resulting in a gradual fracturing of the entire structure.

Figure 7 classifies the most important factors, influencing the progressive deterioration of properties of the materials and joints of space structures. Under the action of these factors a number of physical-chemical processes proceed in structural materials, which, finally, influence negatively the operating parameters of the structure as a whole. The integrated nature of this effect has not yet been studied sufficiently. Firstly it would be necessary to evaluate the effect of external factors on the degradation of the materials. The available scarce data indicate that the fracture is started from the surface and consists, at least, of two stages. At the beginning, these processes are proceeding slowly and comparatively uniformly. Gradually, they penetrate inside. At the second stage clearly expressed «zones of decay» are formed, thus resulting in deterioration of the operating properties of structural materials by 10–100 %.

Conclusions

The lack of knowledge requires a purposeful, integrated program for investigation of the behaviour of materials under space conditions, which will predict the mathematical and physical simulation of space conditions for the elements of large-sized structures, made from different materials using welding, brazing, adhesion and thermal spraying. Physical simulation should be performed on thin-walled scale models. The processes of decay should be studied both in space (to evaluate the simultaneous action of all the factors), and also in laboratory conditions on Earth (to investigate the effect of separate factors). The first steps have already been made. The available results make it possible to define the criteria of selection of optimum materials, shapes of structures, service life of space vehicles and to develop methods of control. All this will make knowledge in the field of materials science and welding engineering more profound.

1. Paton, B.E. (1989) *Schweissen im Weltraum*, Düsseldorf, DVS, pp. 112–115.

2. Paton, B.E., Lapchinsky, V.F., Bulatsev, A.R. et al. (1989) *See pp. 468–473 in this Book.*

3. Belyakov, I.T. and Borisov, Yu.D. (1974) *Technology in space*, Moscow, Mashinos-troyeniye, 290 pp.

Special Factors and Future Development of Space Welding[*]

V.F. Shulym[1], V.F. Lapchinsky[1], D.L. Demidov[1],
V.P. Nikitsky[2] and L.O. Neznamova[2]

The concept of «welding operations» includes realization of the following technological processes (Figure 1): heating, brazing, welding, coating deposition and cutting of materials. On Earth each of these processes can be easily performed by using various methods. In space their performance is more complicated, that is due to the extraordinary characteristics of the environment [1]. Three main characteristics of space can be distinguished which highly affect the welding process: weightlessness, ultrahigh space vacuum and the presence of abrupt boundaries of transition between light and shade (Figure 2). Each of these factors changes the nature of many physical phenomena occurring in the process of welding. The effect of space conditions depends greatly on the method which is used for joining and cutting of materials.

Space conditions (Figure 2) influence to the least degree those processes which are not accompanied by evolution or use of the liquid phase, gases or vapours. Consequently, explosion welding and cutting, diffusion bonding, cold welding and flash-butt welding to a certain extent, are quite suitable for their application under conditions of weightlessness (zero gravity conditions) and space vacuum. Difficulties in use of these methods can arise only in the case where the object to be treated is located for a long time in the shade and became cold. In this case the additional preheating of the elements welded can be required.

Explosion welding and cutting will not be considered in detail in this paper. It should only be noted that the common drawback of the above-mentioned methods is the necessity of careful preparation and fit-up of the objects treated and low flexibility of the process. This limits the capabilities and fields of their application in space. From the point of view of flexibility and simplicity, the methods of electron beam, plasma and consumable electrode arc welding, used widely in industry, are more challenging. They can be used both in the assembly and also in the repair of the space vehicles.

Therefore, since 1965 these methods have been investigated comprehensively at the PWI in the flying laboratory and directly in space. Under the conditions of the flying laboratory it is possible to reproduce continuously the state of weightlessness during a short time (25–30 s). To perform investigations, a test complex was developed and arranged in the cabin of the flying laboratory TU-104 (Figure 3), including special thermal vacuum chambers with a system of measuring and recording units.

The comparative tests carried out in the flying laboratory in vacuum and at a short-time weightlessness showed that all the above-mentioned methods are suitable for these conditions, though they are subjected to the effect of space factors. The rate of evacuation of the surrounding atmosphere is important in case of use of constricted (plasma) low-pressure arc. Its increase reduces drastically the stability of exciting the arc discharge and arc constriction at low currents.

In consumable electrode arc welding the decrease in atmospheric pressure and increase in the rate of evacuation also degrade the arc characteristics, which requires the use of methods of forced compression of the arc discharge.

No significant effect of the rate of evacuation and vacuum was observed on the processes of EBW.

Weightlessness affects greatly all the above-mentioned methods of welding. In consumable electrode arc welding at low currents (to 100 A) the molten metal drops can reach large

[*] (1991) *Proceedings of Conf. on Welding in Space and the Construction of Space Vehicles by Welding*, USA, Miami, Sept. 24–26, 1991. Miami, pp. 12–24.

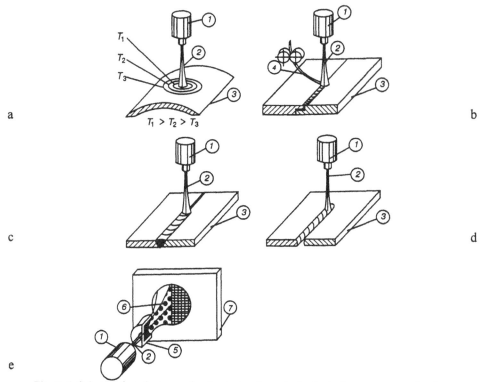

Figure 1 Scheme of conductance of main electron beam technological processes: a — heating; b — brazing; c — welding; d — cutting; e — coating deposition; *1* — electron gun; *2* — electron beam; *3* — object to be treated; *4* — filler wire; *5* — crucible with evaporating material; *6* — vapour flow; *7* — substrate.

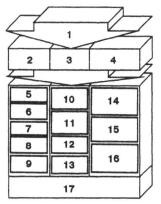

Figure 2 Typical phenomena caused by main space factors: *1* — factors of space environment; *2* — weightlessness; *3* — high space vacuum; *4* — abrupt light–shade boundaries; *5* — absence of gravitation; *6* — absence of buoyancy forces; *7* — weakened convection; *8* — increase in activity of surface phenomena; *9* — absence of support reaction; *10* — high rate of evacuation of gases and vapours; *11* — absence of convection heat exchange with environment; *12* — good protection of molten metal; *13* — presence of operator's spacesuit; *14* — wide temperature range of workpieces treated (150–500 K); *15* — forced thermal control; *16* — changing solar illuminination; *17* — physical-chemical processes.

Figure 3 Arrangement of test vacuum benches in the cabin of the flying laboratory TU-104.

sizes (Figure 4a). To avoid this, the process of melting and transfer of the electrode metal should be controllable (Figure 4b). At the same time weightlessness, as a medium, improves the quality of welds and prevents the appearance of undercuts. Some increase in weld porosity, in particular in welding of aluminium alloys, was observed. Difficulties were also encountered in electron beam and plasma cutting.

However, as was mentioned earlier, all the technological difficulties observed during investigations in the flying laboratories are not of a fundamental nature and, consequently, they can be overcome. Moreover, the mentioned methods of welding were tested directly in space using the unit «Vulkan» [2]. The aim of the on-Earth flying-lab and full-scale tests is the selection of one or two potentially optimum methods of welding for open space [3].

Figure 4 Electrode metal transfer in consumable electrode arc welding in weightlessness (shielding gas — argon, welding current — 50 A, arc voltage —15 V, speed of wire feeding — 180 m/h, electrode diameter — 0.8 mm): a — uncontrollable; b — controllable.

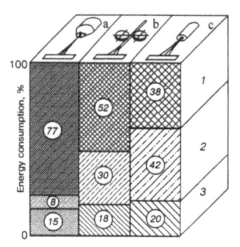

Figure 5 Distribution of consumed electric energy between the circuits of power source, tool and workpiece treated using sources of heat by electron beam (a), arc (b) and plasma with a hollow cathode (c) (vacuum, weightlessness): *1* — workpiece treated; *2* — tool; *3* — power source and cables.

However, the two following characteristics were decisive: flexibility and energy effectiveness.

It was demonstrated that the electron beam is capable of realizing all five technological processes successively in space (see Figure 1). Versatile equipment was designed on this basis. Thus, the selection was made to the benefit of the electron beam processes.

The energy effectiveness of the method is also an important characteristic. Figure 5 shows the share of energy consumed by different methods of welding in vacuum. It is seen that in EBW the percentage of the energy consumed is higher as compared with that of the processes of plasma or arc welding. It is characteristic that in this case only 8 % of heat energy enters the tool, while in arc welding it is up to 30 % and in plasma welding — up to 42 %. This makes it possible to save the electric energy, to avoid forced heat dissipation and to perform the technological processes manually with direct participation of the operator, that is especially important for the repair works.

Finally, all these factors were decisive in selection of the process of space welding. Further investigations were based on use of the processes of EBW [4].

Let us consider briefly the peculiarities of performance of each of the above-mentioned technological processes in space using the electron beam.

Heating. The local surface heating (see Figure 1a) of the objects is performed by a defocusing beam. It is natural, that the temperature field in the object in this case is non-stationary and non-isometric. The isometric temperature field can be obtained on relatively small areas of the surface. For this purpose the beam should be scanned according to a definite program. The absence of convective heat exchange with the environment facilitates the process of surface heating in space, and also its calculation and programming.

All other processes are characterized by the presence of the liquid phase in the working zone and are subjected to a certain extent to the effect of weightlessness. This is firstly due to the fact that surface phenomena play an important role under these conditions. The principles of these phenomena are described in detail in [1]. Let us consider their concrete embodiment.

Brazing. The quality of brazing (see Figure 1b) depends greatly on the wetting angle. When a well-wetted brazing alloy is used, the force $F_{1\to}$, whose vector is directed to the side of the higher temperature, is acting on the element of the boundary layer dl between the liquid and solid phases (see Figure 1 on p. 36 and Figures 2 and 3 on p. 37). In parallel,

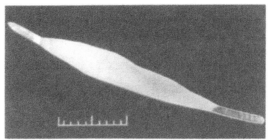

Figure 6 Macrosection of very large weld pool obtained in filler welding in weightlessness.

thermal capillary convection causes stirring in the volume of the molten alloy. In weightlessness these forces lead to intensive displacement of the alloy to the high-temperature zone. Consequently, at a good wetting in space, the lack of adhesion is not practically observed.

In case of multicomponent alloys, when the concentration capillary convection is proceeding, a high degree of microheterogeneity can be observed in the structure of the brazed joints. Some heterogeneity of the structure can be caused by the heterogeneity of the temperature field, as the heating is performed by a local heat source.

The small amount of brazed joints produced to date in space, does not make it possible to define their peculiar differences from on-land joints. It is important that strength and density of these joints are not lower than those of the on-land joints.

The brazing technique may be different. Usually, the brazing alloy is deposited preliminary on the parts being joined. However, it can also be fed in the form of the filler wire or rods.

Welding. The process of EBW in space (see Figure 1c) of thin-sheet products of thickness of 3 mm and more differs negligibly from the on-land process. On Earth pencil-type beams are used rarely for these purposes from an economic point of view and because of the burn-outs.

In space the risk of burn-outs is significantly less [5] as the flowing out of the weld pool under the action of the gravity forces is not observed. Nevertheless, at some types of the works it is necessary to perform some beam defocusing purposefully. When the gaps are large, it is necessary to use the filler wire. Moreover, it is possible in space to produce a weld pool of large size even on very thin metal (Figure 6). This method is indispensable in welding-in of holes and other repair works.

The technology of welding and quality of welded joints in space for most materials (heat-resistant and stainless steels, titanium alloys) do not significantly differ from on-land technologies. This concerns also the inner structure of welded joints [6].

The largest difficulties are encountered in welding aluminium and other alloys with an increased content of dissolved gases or easily-volatile components. In these cases the welds made in weightlessness are characterized, as a rule, by a high porosity (Figure 7) [7]. To prevent it, there are a number of technological measures which are based mainly on electron beam modulation.

Cutting. Under on-land conditions, the process of electron beam cutting (see Figure 1d) is used not frequently. In space this process is indispensable, as it does not lead to the formation of chips, drops or metal spattering, which are inadmissible in these conditions. Well-focused electron beams are used usually for cutting. However, even in this case serious difficulties arise often in space. The effective action of surface tension forces in welding which prevents burn-outs, sometimes, in cutting leads to the cut filling with molten metal. In these cases the forced shifting of the molten pool to the cut edges is required by using a special configuration of the heat dissipation. Here, one more positive effect is reached. The drops of the molten metal are not solidified randomly on both edges (Figure 8a), but only

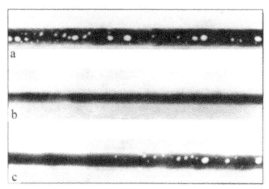

Figure 7 Formation of pores in welds in EBW of Al–Mg–Mn alloy in weightlessness (a),
on Earth (b) and in transition from overload >1.5 *g* to weightlessness (c).

on one colder edge (Figure 8b). This is important when the cut sheets are to be welded after cutting.

Coating deposition. The greatest differences between the on-land and space technologies are observed in deposition of coatings by using the method of evaporation and condensation (see Figure 1e). This is explained by the fact that the coating material at the beginning of the evaporation is in the liquid phase for a long time, and then it is transformed into the vapour phase and, finally, condensed on the substrate in the form of a hard film. At all the stages of this process, two-phase systems are used where the action of surface forces are mostly expressed. In the crucible with evaporating materials active interaction is observed in two regions of the interface between the solid body–liquid and solid body–vapours and near substrate — the region of interface between the vapours and solid body.

At the first stage it is necessary, on the one hand, to provide safe maintenance of the molten material in the crucible, and, on the other hand, to provide safe separation of the vapour stream without admixtures of the drop phase from the molten surface. This is attained with the help of crucibles of a special design in which a temperature field of a special shape is formed. Here, the liquid maintenance and separation of phases are initiated by the thermal capillary and concentration-capillary convection, i.e. the surface forces are active [6].

The coatings, produced according to this technology, are not inferior to those of on-land technologies. In some cases, for example, when multicomponent systems are evaporated, it is possible to obtain very interesting results which are impossible on Earth.

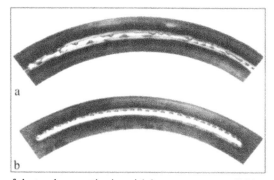

Figure 8 Specimens of electron beam cutting in weightlessness at uncontrollable (a) and controllable (b)
localizing of the melted metal.

It should be noted in conclusion that all the above-mentioned technological processes have already passed experimental-industrial trials under space conditions and can be used in construction, repair and periodical maintenance of existing and new space vehicles.

1. Paton, B.E., Lapchinsky, V.F., Bulatsev, A.R. et al. (1989) *See pp. 468–473 in this Book.*

2. Paton, B.E. and Kubasov, V.N. (1970) *See pp. 154–160 in this Book.*

3. Dzhanibekov, V.A., Zagrebelny, A.A., Gavrish, S.S. et al. (1991) *See pp. 184–190 in this Book.*

4. Paton, B.E., Dudko, D.A. and Lapchinsky, V.F. (1984) Welding processes in space, In: *Welding and special electrometallurgy,* Kiev, Naukova Dumka, pp. 121–129.

5. Lapchinsky, V.F. (1984) *See pp. 278–282 in this Book.*

6. Paton, B.E. and Lapchinsky, V.F. (1990) Materials science investigations of samples of permanent joints and coatings made in space, In: *Transact. of IIW Congr. on Studies in the Field of Producing Joints,* Montreal, Champan&Hall, pp. 236–240.

7. Ganiev, R.F. and Lapchinsky, V.F. (1978) *Problems of mechanics and space technologies,* Moscow, Mashinostroyeniye, pp. 34–36.

Formation of Electron Beams for Process and Research Work in Space*

B.E. Paton[1], O.K. Nazarenko[1], S.K. Patsiora[1], Yu.V. Neporozhny[1],
V.N. Bernadsky[1], V.I. Chalov[1], V.D. Shelyagin[1], V.K. Mokhnach[1],
O.A. Metallov[1], P.V. Bliokh[4] and A.A. Puzenko[4]

High energy concentration (up to 10^8 W/cm^2), high efficiency of electric energy conversion into thermal energy (up to 80 %), ability of precision control of position and parameters, practically complete absence of reactive forces make a beam of accelerated electrons one of the most promising process and research tools in space [1].

High energy beams of charged particles can be used for pressurizing the space vehicle structures, reduction of air leakage, elimination of skin damage due to the impact of small particles and meteorites, as well as dents and other defects arising in a space vehicle landing on other planets. The electron beam can be used for welding space stations and vehicles.

Injecting flows of accelerated electrons allows the initiation in the Earth magnetosphere of the phenomena of magnetic and atmospheric reverse scattering, northern lights, and interaction of waves with particles similar to those occurring there naturally. By changing the causal impacts and measuring the appropriate effects, scientists will be able to comprehend the physical processes of magnetospheric dynamics. It will become possible to control for the first time wave generation in the magnetosphere (magnetospheric plasma cannot be created in the laboratory, where a low density of electrons with an almost complete absence of neutral particles corresponds to a small magnetic field). The first active experiments with electron beams in space were conducted in 1969–1970 [2].

The main features of space as a natural environment for formation of charged particle beams and performance of process and research operations are weightlessness, high rate of removal of gases formed in the molten metal zone, broad range of possible temperatures of the metal being welded and the hardware, the presence of a magnetic field and plasma in

* (1975) *Space research in Ukraine,* Issue 6, pp. 3–7.

near-earth space, and directional movement of the electron source in relation to the environment.

The development of electron beam guns, power units and control systems for welding, heat treatment and coating, as well as for investigation of the physical phenomena in the upper layers of the atmosphere of Earth is related to satisfying the requirements of orbital space vehicles. Such hardware was developed at the PWI and was tested on the ground, in flying laboratories and near-earth space [1].

Formation of beams of charged particles in near-earth space taking into account the influence of the Earth's magnetic field on the beam trajectory and focusing, is of special interest.

Let us consider a converging electron beam focused by a short electromagnetic lens. Let us assume that the beam axis (in the absence of a magnetic field) is inclined at angle α to the magnetic field **H** (Figure 1). Field **H** coincides with axis z, beam axis L belongs to plane y, z. Initial coordinates of particles are characterized by distance **R** from beam axis and angle β, and coordinates of the point of crossover image in the target plane in the absence of the magnetic field, are **r** and β', respectively. The initial values of beam velocity components v_{0x}, v_{0y}, v_{0z} are determined taking into account the scatter of velocities by the angles at the expense of the final radius of the crossover. The electron projector and the lens are screened from the external magnetic field **H**. The following equations will not allow for the influence of the beam spatial charge.

The magnetic field introduces two instabilities, namely beam axis is curved and it does not get into the calculated point; beam section is deformed and focusing is violated.

Curving of the beam axis is described by values of x, y, z averaged by angles β and β':

$$\langle x \rangle = \frac{v_0}{\omega} \sin \alpha \, (1 - \cos \omega t),$$

$$\langle y \rangle = \frac{v_0}{\omega} \sin \alpha \sin \omega t,$$

$$\langle z \rangle = v_0 \, t \cos \alpha.$$

The average value of the electron trajectories does not depend on **R**, i.e. it coincides with the central electron trajectory. The characteristic scale of beam axis deviation is the Larmor

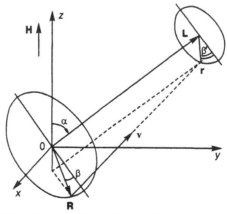

Figure 1 For calculation of the optimal position of the focusing lens.

radius of the central particle $\rho_0 \sin \alpha = \dfrac{v_0}{\omega} \sin \alpha$ (for non-relativistic electrons $\rho_0 [\text{cm}] =$

$= \dfrac{10^2}{H\,[\text{gf}]} \sqrt{U\,[\text{kV}]}$, $\omega t = L/\rho_0$).

With fixed α, **H**, γ (half-angle of beam divergence), r_{cr} and **L**, there exists the optimal value of the position of the electromagnetic lens a_{opt} in which the mean root square deviation of particles from the axis in the target plane $\langle \Delta^2 \rangle$ is a minimum:

$$a_{opt}^2 = \frac{r_{cr} L}{\gamma} \sqrt{\frac{B}{A}}, \tag{1}$$

$$\langle \Delta^2 \rangle_{min} = 2\gamma r_{cr} L \sqrt{AB}, \tag{2}$$

where the target parameters

$$A = 1 + \frac{\rho_0^2}{L^2}\left[\frac{1}{2}\sin^2\omega t\,(1 + \cos^4\alpha) + (1 - \cos\omega t)^2\cos^2\alpha + \frac{1}{2}(\omega t)^2\sin^4\alpha + \omega t \sin\omega t \cos^2\alpha \sin^2\alpha \right]$$

$$- \frac{\rho_0}{L}[\sin\omega t\,(1 + \cos^2\alpha) + \omega t \sin^2\alpha], \tag{3}$$

and

$$B = \frac{\rho_0^2}{L^2}\left[\frac{1}{2}\sin^2\omega t\,(1 + \cos^4\alpha) + (1 - \cos\omega t)^2\cos^2\alpha + \frac{1}{2}(\omega t)^2\sin^4\alpha + \omega t \sin\omega t \cos^2\alpha \sin^2\alpha \right]. \tag{4}$$

In a particular case at $a = 0$ expressions (1) and (2) take the form of

$$a_{opt}^2 = \frac{r_{cr}\,\rho_0}{\gamma} \sqrt{\frac{2(1 - \cos\omega t)}{1 - 2\dfrac{\rho_0}{L}\left[\sin\omega t - \dfrac{\rho_0}{L}(1 - \cos\omega t) \right]}}, \tag{5}$$

$$\langle \Delta^2 \rangle_{min} = 2\gamma r_{cr}\,\rho_0 \sqrt{2(1 - \cos\omega t)\left\{ 1 - 2\frac{\rho_0}{L}\left[\sin\omega t - \frac{\rho_0}{L}(1 - \cos\omega t) \right] \right\}}. \tag{6}$$

At $\alpha = \pi/2$

$$a_{opt}^2 \cong \frac{r_{cr} L}{\gamma} \sqrt{\frac{1 + (\omega t/\sin\omega t)^2}{(1 - \omega t/\sin\omega t)^2}}, \tag{7}$$

$$\langle \Delta^2 \rangle_{min} = \gamma r_{cr} L \sqrt{\left(1 + \frac{\omega t}{\sin\omega t}\right)^2\left[1 + \left(1 + \frac{\omega t}{\sin\omega t}\right)^2\right]}. \tag{8}$$

Calculation using equations (5)–(8) of the dependence of the minimal beam diameter on the crossover–lens distance shows that this dependence is sloping and that it is not rational to increase a by more than 50 cm.

The specific beam power also depends on the extent of the defocusing impact of the spatial charge. As the spatial charge cannot be compensated by ions under the conditions of near-earth space with the transverse displacement of the electron beam relative to the environment [3], guns with a small perveance should be used for welding performance at long distance from the gun. However, the possibility of increasing the accelerating voltage

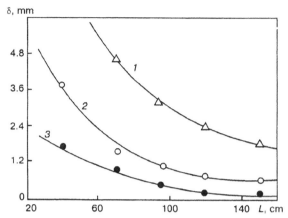

Figure 2 Remote penetration of metals ($W = 3$ kW, $U = 15$ kV, $v_w = 17$ m/h): *1* — commercial titanium; *2* — Al–Mg–Mn alloy; *3* — stainless steel.

with the aim of reduction of the gun perveance is limited by the increase of the X-ray radiation intensity. Therefore, in each individual case it is necessary to select the optimal parameters of the electron beam and limit the distance to the point of welding.

Experimental investigations were conducted into the possibility of remote welding of different metals. Dependence of the penetration depth on the distance to the target sample is shown in Figure 2. Reduction of the penetration depth with increase of the gun–target distance is related to lowering the specific power of the welding beam electrons. In this case the total power of the electrons remained constant, and the position of the focusing electromagnetic lens was the same. Change of the spatial position of the item and the beam does not influence the geometry or stability of the process of penetration. The molten metal is contained by the forces of adhesion and surface tension due to the small volume of the pool.

The PWI developed several designs of small-sized electron beam hardware for performing processing operations and studying the physical phenomena in space (Table 1).

A combined block of power source and gun for experiments on electron beam welding of metals in space [1, 4] is shown in Figure 3a. The twin system of guns is applied with the

Table 1

Hardware type	Brief description of hardware	Purpose
«Vulkan»	Electrons energy of 10 keV. Beam power of 0.5 kW, specific beam power of 1 kW/mm^2 at 40 mm distance from gun edge. Cathode is directly heated tantalum. Diode projector. Gun weight of 450 g. Weight of gun with power unit is 6.5 kg	Experiments on EBW of metals in space [1]
«Zarnitsa»	Electron energy of 10 keV. Beam power of 5 kW. Cathode is directly heated tantalum. Diode projector. Weight of power unit with guns is 9 kg	Electron injection into space
ARAKS	Electrons energy of 15 and 30 keV. Beam power of 13.5 kW. Cathode is lanthanum boride with electronic heating. Triode projector. Pulse modulation of beam current. Beam deflection in one plane in the range of 120°. Weight of gun with deflecting system is 2.6 kg. Weight of power unit with guns is 40 kg	Electron injection from board of «Eridan» rocket launched from Kerguelen island [2]

a

b

Figure 3 Combined blocks of power unit–gun for joining 1–3 mm thick metals (a) and for injection of 5 kW electron beam (b).

aim of duplication of the directly heated cathode. Figure 3b shows the injector of «Zarnitsa» hardware with three guns.

Use of a lanthanum boride cathode with electronic heating in ARAKS hardware improved the gun reliability and decreased the power consumed in cathode heating. The general view of the gun of ARAKS hardware with the deflecting system is given in Figure 4a and the photo of the luminescence of 15 kW beam generated by this gun under vacuum of $5 \cdot 10^{-4}$ mm Hg is shown in Figure 4b. The deflecting system allows control of beam position up to an angle of not more than 150°. At injection from a rocket, three angles of injection (0, 60, 120°) evoke different physical phenomena. At 60° angle the phenomenon of magnetic backward scattering can be implemented, namely the particles will oscillate between mated points following a line of force. In the case of downward injection (120° angle) the electrons can be reflected from the low layers of the ionosphere (atmospheric backward scattering).

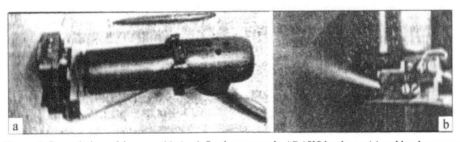

Figure 4 General view of the gun with the deflecting system in ARAKS hardware (a) and luminescence of 15 kW beam (b).

Figure 5 Electron beam gun of 3 kW power for remote welding of sheet metals.

If the injection angle is 0°, fast electrons can penetrate into the upper layers of the atmosphere and excite the atoms making it up. These atoms will return to the main state having emitted the light radiation which is observed in the northern lights.

Application of triode projector allows pulse modulation of beam current and stabilization of its value with ≈ 1 % accuracy. The electron beam gun of 3 kW power for remote welding of sheet metals which is a modification of the ARAKS gun, is shown in Figure 5.

1. Paton, B.E., Nazarenko, O.K., Chalov, V.I. et al. (1970) *See pp. 160–168 in this Book.*

2. Goldmen, P. (1972) ARAKS: accelerators in magnetosphere, *Recherche*, October 27.

3. Bliokh, P.V. (1971) Ion focusing of the electron beam in a transverse gas flow, *Radiotekhnika i Elektronika*, No. 11, pp. 7–10.

4. Shelyagin, V.D., Lebedev, V.K., Zaruba, I.I. et al. *Device for electron beam welding.* USSR author's cert. 206991, Int. Cl. B 23 K, priority 22.02.66, publ. 08.12.67.

Some Results of Works Made by the Soviet Scientists in the Field of a Space Technology*

A.S. Okhotin[5], V.F. Lapchinsky[1] and G.S. Shonin[3]

Space exploration by mankind required progress in different branches of science and engineering, including space materials science. It has become evident already, that it would be necessary soon to perform not only separate operations of treatment of materials under space conditions, for example, heating and melting, but also all the technological processes to produce materials with quite new physical and chemical properties. In this connection scientists face the problems of investigating peculiarities of conductance of different technological operations in space, associated with such specific conditions as microgravity, deep vacuum, variety of heat processes [1].

Fundamental studies are required for the development of principles of control of metal properties under space conditions, and also development of technological processes directed to the creation of new materials and support for flights in space. These studies will probably be developed by the following main trends.

* (1980) *Space technology.* Ed. by L. Steg, Moscow, Mir, pp. 244–250.

1. Comprehensive study of single- and multiphase liquid and gaseous materials under microgravity conditions. For this purpose it is necessary to investigate the following processes of cooling and solidification of melts in space, both in a free state and also with forced heat dissipation, bases of phase separation, methods of control of transfer processes [2], surface tension, adhesion and wettability of different phases and materials, and processes of diffusion and heat transfer, in particular the natural convection at a free surface and in a closed volume.

2. Selection of materials, whose manufacture in space is rational and economically justified. Within the scope of this trend, special attention should be paid to different composite materials, such as materials of metal–metal and metal–metal oxide types, ceramic and organic substances which are not mixed in on-land conditions, and light alloys which are reinforced with fibers of high-strength materials. The feasibility of uniform stirring of gases by metals supports the idea of creating high-strength heat-resistant foam-metals with a low specific weight in space conditions. Probably, it would be possible in these conditions to produce optically-transparent materials (glasses or single crystals) with a minimum content of impurities or preset their distribution, as well as structurally-perfect semiconductor single crystals of a high purity.

3. Development of methods for conducting repair-restoration and erection works in space. In conductance these works, cutting of structures and joining of separate elements by welding, brazing or adhesion will probably be necessary. The most important tasks, connecting with this trend of investigation, will be the study of ergonomic and physiological capabilities of a cosmonaut-operator and development of different auxiliary devices which will facilitate the successful completion of works and ensure their safety.

4. Development and testing of special equipment and units for the study of the behaviour of substances and technological processes in space. The results which can be obtained in other investigations, greatly depend on the extent of progress of these studies. Systematic works in these areas have been developed widely in the Soviet Union. The present article deals only with those investigations which are of the greatest interest.

The experiment «Versatile Electric Furnace» was conducted according to the Joint Soviet-American ASEF Programme. The aim of this experiment was to produce composite and semiconductor materials in microgravity conditions. The preliminary results of processing of the data obtained revealed a number of interesting anomalies, which are now being studied comprehensively.

One of the main peculiarities of technological processes in space is their realization under conditions of the microgravity. On Earth the energy of gravitation of one mole of liquid has approximately the same order as the energy of one mole in reactions of the growth of crystals and formation of bubbles and drops in wetting and catalysis, and also in biological reactions. The presence of microgravity should influence the above-mentioned reactions and phenomena, and the technological processes in space associated with treatment of materials in the liquid state should have definite peculiarities compared with the same processes on Earth.

When external forces are absent, the mechanical properties of the liquid are defined only by the interaction of forces of surface tension and intermolecular bonding (viscosity). Therefore, any liquid tries to take a spherical shape in the absence of any disturbance. However, the surface tension forces are very sensitive to small amounts of impurities which can enter the free surface of the liquid from the natural atmosphere of the space vehicle. This atmosphere is rather non-homogeneous in space and time, its composition depending on degassing and sublimation of structural materials, gas leakages from sealed compartments and exhaust products of jet engines. Moreover, a portion of the molecules and atoms of the atmosphere of the space vehicles is in an ionized state, and some particles have also an electrical charge.

Consequently the concentration of impurities in the surface layer of the melt, and also the local value of the surface tension can change, thus influencing the melt properties. Therefore, due to the high force of the surface tension (1500 dyn/cm) and relatively low viscosity (0.04 P), the drops of liquid iron acquire the shape of a perfect sphere at a high surface rate (370 m/s). It is believed that it is possible to produce spherical bodies with an accuracy of surface geometry of 1 Å order in microgravity conditions at the expense of the molecular forces. This is probably possible only in the case when the surface tension in the metal drop is similar over the entire sphere. However, if, for example, the a larger amount of by-pass substance from the natural atmosphere of the space vehicle (for instance, glycerine or paraffin, which have a low surface tension, ~ 20 dyn/cm), enters the right semi-sphere of the molten metal, than enters the left semi-sphere, then the right semi-sphere will buckle, and the metal drop, similar to a mercury ball, will jump away to the right. Thus when stable-nonhomogeneous flow of free molecules of carbon-containing impurities is present, the molten metal drop will not have the shape of a perfect sphere. The resultant force of the surface tension local forces, occurring in this case, is directed opposite to the main flow of the carbon-containing impurities. If the direction of this flow is changed with time, then the drop of metal in free space, in the absence of external forces, will move similar to a Brownian particle. By changing the local value of the surface tension the non-homogeneity of concentration of impurities in the surface layer will lead to displacement of separate elements of the heated liquid volume relative to each other and, consequently, to the convective flows which hinder producing of quality metal free of flaws (dislocations, twins) and inner stresses. This will mean a partial or complete liquidation of advantages in the use of the microgravity for the elimination of natural convection.

The specific substances, contained in the natural atmosphere of the space vehicle (especially oxides of such metals as iron, molybdenum, platinum, etc.), can be catalysts of some endothermal reactions and violate the rate which is selected for them. When the extra high-purity materials are produced in space, the impurities of the natural atmosphere of the space vehicle can also reduce abruptly the purity of vacuum and, consequently, the prospects of space technology.

Thus the presence of the natural atmosphere of the space vehicles can prevent the realization of many advantages of fulfilment of technological processes in space.

In the Soviet Union many studies have been undertaken into processes of welding under conditions of the microgravity and creation of different equipment for this purpose. When designing this equipment it is necessary to take into account a number of requirements for its design and service, specified by the peculiarities of operating in space. Safe service of the equipment in the space vehicle depends on the correct allowance for such factors as the destructive action of the heat sources, the presence of a pool with molten metal and spatters of molten metal, high voltage of the power sources and by-pass phenomena of heat or X-ray radiation. For example, in the unit of the «Vulkan» type, designed for the EBW, the accelerating voltage was selected to be less than 15 kV, as the possibility of appearance of a braking X-ray radiation in this case is eliminated. The successful selection of the arc welding conditions made it possible to prevent the metal spattering. In the same unit, the high-voltage elements and circuits, as potential hazardous sources, were enclosed into one block and potted with an epoxy resin. A special protective jacket was used in the «Vulkan» unit for localizing the metallic dust, heat and light radiations. The control of process parameters and their maintenance at the required level were provided by a system of electrical and mechanical protection.

Analysis of different methods of welding showed that the relative simplicity of accomplishment of EBW, high efficiency of the process, and feasibility of its use for all metals makes this process one of the most challenging in space technology. It can be taken into account, however, that almost all the operations connected with maintenance and repair of space systems, when it is not known beforehand what kind of work has to be done and what are the parameters of parts and sub-assemblies being treated, will be performed either completely manually or manually with a partial mechanization. In the installation for EBW, manufactured for this purpose, a serial diode electron beam gun with indirect heating of 0.5 kW capacity and beam operating current up to 100 mA was used. The optical system provides stable focusing of the beam at distances of several dozens and even hundreds of millimeters from the gun. The high-voltage power source represents a monolithic block, potted with an epoxy resin and similar to the block operating in the welding unit «Vulkan». The block with a high-voltage cable is enclosed into a metal housing connected to a common circuit of earthing of the electron beam equipment. To ensure the safety of works the accelerating voltage of the equipment does not exceed 15 kV. During testing this equipment the welding of strips of stainless steel, aluminium and titanium alloys of thickness from 1.5 to 3.0 mm, located at 200–220 mm distance from gun, was performed. The welding speed was kept within the 7–10 m/h ranges at beam current 55–75 mA. The change in distance from the gun to the sample in the range of ± 20 mm did not cause any noticeable change in the depth of penetration and weld width. The experiments showed that the quality welding of butt and overlap joints with and without edge flanging was achieved using equipment.

The investigations on the study of weldability of aluminium alloys using an electron beam in microgravity conditions were performed in the welding unit «Vulkan» of 0.5 kW capacity. Welding was performed at 28–36 m/h speed. It was stated from the results of the data analysis that the microgravity does not influence greatly the formation of the weld and the depth of the metal penetration. Welds made by welding of samples in the space vehicle, are characterized by a higher porosity (approximately by 40 times) as compared with welds of samples which were welded in on-land conditions. This increase in porosity can be explained by the fact that the rate of separation of the gaseous and liquid phases in the conditions of the microgravity is lower than that on Earth (the pores are formed due to the presence of vapours of easily-volatile elements of metals). It was also stated that a common pool of the molten metal is formed in conditions of the microgravity if the gaps in the welded joints are large. The comparative analysis of content of alloying elements in the parent metal and weld showed that their evaporation does not, in principle, depend on the gravity value.

To decrease the pore content in the weld it is possible to perform the process using different vibrating actions on the zone of welding, The use of vibrations, which are changed according to a non-linear law, is one of the most promising methods of such actions having positive effect not only on welding processes, but also on other technological processes. Using vibrations, it is possible to perform degassing of materials, to form the surfaces of ingots and crystals, especially those solidifying in conditions of crucible-free melting.

The investigation of processes of consumable electrode welding was aimed at the study of melting and transfer of the electrode metal in microgravity conditions. During these investigations the welding was performed in argon using 1 mm diameter electrodes of stainless steel, titanium and aluminium alloys. The same material as the electrode material was used for the parent metal being welded. It was established from the results of the experiments that the relationship between the forces acting on the drop of the molten electrode changes in microgravity conditions compared with on-land conditions. At low values of welding current (up to 40–50 A) the drops of the electrode metal in conditions of the microgravity are enlarged, their transition to the weld zone becomes irregular and the process of the weld formation is disturbed. This undesirable effect can be avoided by the reduction of the arc gap length or by the imposition of current pulses on it.

Microgravity can influence greatly the crystallizing of metal alloys, and, consequently, their structure, in particular the structure of alloys of eutectic composition [3]. The investigation of these alloys made it possible to state that the effect of microgravity on the formation of their structure consists mainly of changing the dispersity of phases in the process of alloy crystallization. Here, the eutectic alloys, consisting of two components being different in composition and properties with an unlimited reciprocal solubility in a liquid state and negligible solubility in a solid state, are the most sensitive to the effect of the microgravity.

In our country the investigation of the effect of controllable external actions, in particular ultrasonic vibrations, on the crystallizing of multiphase media has commenced. Theoretical and experimental works are being developed mainly in two directions: creation of such forms of motion of liquid media and inclusions which provide dynamic steadiness and stability of the mixture in order to prevent the effect of external disturbances on the process of crystallizing, and the degassing and purification of liquids or melts in conditions of microgravity. The first results of the works showed that it is possible to use a number of effects in space technology, such as resonance stirring of liquid media, which are not mixed in on-land conditions, formation of steady periodic structures, steady unidirectional movement of drops of molten metals and bubbles of vapour, and localizing and controllable mixing of gas bubbles.

Thus, it can be concluded that space materials science is developing in different directions. The works conducted in this field inspired researchers and designers to deal with the development of new highly-effective processes and reliable small-sized equipment, whose first tests were completed successfully both in the Soviet Union and in the USA. Probably in the near future new technological experiments in space are to be expected which will have definite applied importance both for producing of different materials with new properties, and for the support of flights of space vehicles.

1. Belyakov, I.T. and Borisov, Yu.D. (1974) *Technology in space*, Moscow, Mashinos-troyeniye, 290 pp.

2. Ganiev, R.F., Lapchinsky, V.F. and Okhotin, A.S. (1980) Using monitored external effect to control of technological processes under microgravity conditions, In: *Space technology*. Ed. by L. Steg, Moscow, Mir, pp. 64–72.

3. Okhotin, A.S., Livanov, L.K., Tsviling, M.Ya. et al. (1980) Effect of the power field on the structure of eutectic alloys, *Ibid.*, pp. 314–318.

Mathematical Modelling and Experimental Study of Processes of Evaporation of Alloys in Microgravity Conditions[*]

A.L. Toptygin[6], L.O. Neznamova[2] and I.S. Malashenko[1]

The development of space technology and materials science has led to the definition and realization of the problem of producing different-purpose film coatings used in space vehicles, providing functionality and increasing service life. The search for radically new possibilities for the development of promising industrial technologies of new systems of film coatings in orbit, using deposition in vacuum, is a priority for microelectronics, cryoelectronics and other fields of engineering. The solution of these problems requires the study of the impact of space factors, in particular microgravity, on the mechanisms of formation of the structure and properties of coatings in the process of their manufacture. This is necessary both for producing materials with a preset composition and properties, and also

---

[*] (1990) Paper was presented at the 5th Korolyov Readings of the 2nd Republ. Conf. on Fundamental and Applied Problems of Cosmonautics, Kiev.

for the understanding of fundamental laws of physics of condensed thin films and physics of surface.

In 1981 various peculiarities [1, 2] were revealed in the course of experiments on deposition of two-component film coatings on the base of Ag–Cu alloys in space using the «Isparitel» unit, changing the processes of evaporation and condensation of binary alloys in conditions of flight of orbital stations «Salyut-6» and «Salyut-7». They were stipulated by the effect of microgravity on the nature of mass transfer in the melt, determining the kinetics of evaporation of the alloys.

The present work deals with the kinetics of evaporation of binary alloys Ag–Cu in microgravity conditions using mathematical modelling of the effect of processes of mass transfer in a crucible on the process of evaporation and comparative analysis of thickness and composition of coatings produced in space and on Earth.

The study of these problems is also important for the problems of on-land application, as the analysis of the literature shows that, in spite of comprehensive understanding of the process of evaporation of alloys from a common source in vacuum, the literature is confined, as a rule, to isothermal conditions. The change in the kinetics of evaporation in special non-isothermal conditions was discussed in [3] in connection with the problem of producing coatings of a constant composition. However, to describe the behaviour of systems in producing coatings of complex composition in non-isothermal conditions of evaporation of a general nature, it is necessary to create a mathematical model to describe this type of evaporation process.

The diffusion mechanism of mass transfer in the melt being evaporated has also been discussed in a limited number of works, mainly for the perfect case of an infinite column of the melt. The numerical solution of diffusion problems for real processes is especially important in statement of experiments in microgravity conditions and processing of their results for producing coatings of a preset composition in the condition of orbital station flight and data about diffusion characteristics of the systems studied.

Producing of coatings and methods of their examination. Coatings of alloys Ag–Cu were produced onboard the OS «Saluyt-7» and OC «Mir» using the equipment «Isparitel-M» and «Yantar», designed and manufactured at the PWI. The procedure of deposition, design of the evaporating unit and the crucible of the unit «Isparitel-M» are described in [1–3], and unit «Yantar» — in [4]. It should be noted that the condensation of coatings in unit «Isparitel-M» was performed in a discrete way on separate glass substrates at different temperature of evaporation T_{ev}. In the unit «Yantar» the condensation of the vapour flow was made on a polyimide strip, moving continuously during a specific time (67 s). The source of evaporation (crucible), unlike the previous case, had a cylindrical shape, flat bottom and was open (without a cover). The melt in the crucible was maintained at the expense of good wettability of walls of the evaporator with molten metal. The alloy weighed quantity was 2 g.

The full-scale experiments in the OC «Mir» were fulfilled by cosmonauts Yu. Romanenko and A. Lavejkin in May–June 1987 using the «Yantar» unit. The condensed coatings from copper and alloys Ag–Cu were obtained on the moving strip.

Similar experiments on Earth were performed with allowance for the real parameters of space experiments during the period from October 5 to 14, 1987 at the PWI. The «Yantar» unit was arranged in an operating vacuum chamber, evacuated by pumps. The pressure of residual gases was here $2 \cdot 10^{-4}$ mm Hg.

The thickness of condensation of the coatings was determined using methods of interferometry and X-ray fluorescent analysis. The element composition was determined using X-ray fluorescent analysis in units VRA-30 and SPARK-1 and Auger-spectroscopy. In the latter case, for the three-dimensional analysis of the surface the equipment LAS-2000 of Riber Company was used with a microprobe of the Auger-type cylindrical mirror.

Mathematical model and calculation of kinetics of evaporation of alloys in on-land conditions. The kinetics of the process of evaporation of alloys Ag–Cu in on-land conditions was calculated by the approximation of convective mass transfer in the crucible using the known concepts of the evaporation of alloys [5–8]. Here, the following assumptions were made: the behaviour of the alloy is governed by Raoult's law for perfect solutions; the composition of the surface and volume of the alloy evaporated is similar in any moment of time; the volume of melt is equal to the sum of volumes of the components; the temperature in all the points of the melt is similar; the rate of evaporation of a pure component is proportional to its partial pressure and described by the Langmuir equation:

$$W_1 = P_i \, (M_i / 2\pi RT)^{1/2}. \tag{1}$$

In the binary alloy with a mass content of components N_i the loss in mass of each component for a time $\Delta\tau$ is

$$dN_i = -KP_i f_i S\sqrt{1/M_i T}\, n_i\,(\tau)d\tau, \tag{2}$$

where M_i are the molar masses of pure components; P_i are the pressures of vapours of pure components at temperature T; $n_i\,(\tau)$ are the molar fractions of components; f_i is the coefficient of activity of components in the alloy; S is the area of surface of evaporation; $K = 0.0585$.

The solution of this system has the form

$$\sum_{i=1}^{2}\left(\frac{N_i - N_{i0}}{K_i}\right) = \tau; \quad N_i = CN_{i+1}^\alpha, \tag{3}$$

where

$$\alpha = P_1/P_2(M_2/M_1)^{1/2}; \ C = N_{10}/N_{20}^\alpha; \ K_1 = KP_i S f_1(1/M_i T). \tag{4}$$

The percentage of alloy in crucible

$$C_i = N_i/N_1 + N_2; \ C_1 + C_2 = 1. \tag{5}$$

The percentage of vapour phase

$$C_{vi} = \left(1 + \frac{1 - C_i}{C_i}\alpha\right)^{-1}. \tag{6}$$

In any moment of time τ the thickness of coating h is determined by the mass of the condensed substance. At the coefficient of condensation, equal to unity,

$$h = \left(\frac{N_{10} - N_1}{R_1} + \frac{N_{20} - N_2}{R_2}\right)\lambda, \tag{7}$$

where λ is the coefficient, taking into account the geometry of the experiment. These relations are valid for the isothermal process of the evaporation.

In modelling of the real process of evaporation with a change in melt temperature, this model was modified as follows. The relation $T(\tau)$, known from the experiment, was divided into several regions, i.e. cycles of evaporation, so that each cycle could be approximated by a linear dependence. Further, each cycle was divided into several temperature intervals inside of which an isothermal approximation was used. Thus the linear approximation was

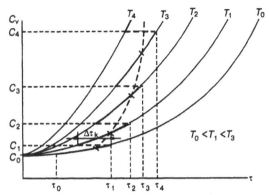

Figure 1 Diagram of modelling of temperature change in a convective mass transfer.

replaced by a step approximation. The selection of the step length, corresponding to one value T_{ev} was realized so that the error of approximation was not more than 10 %.

The following data were used for calculations: T_{in} is the initial temperature of the experiment; ΔT_k is the interval of temperature change at k-th region ($k = 1, 2, \ldots, N$); N is the number of regions; τ_i is the continuity of i-th region; $\Delta\tau_k$ is the duration of isothermal intervals at the k-th region.

In the transition between the regions, N_{k0} at the end of the previous region was taken as the initial content of the component in the melt N_{k-1}.

In transition between the isotherms the real dependence $C_v(\tau)$ will have a saw-shaped form. This is explained by the fact that at different T, a different time is required for reaching similar concentrations $N_k = N_{k-1}$. Actually, there is a rough linking of solutions of the system of equations (2) by the initial condition: $N_{i0} = $ const.

To smooth the solution, the i-th isotherm is shifted for $\Delta\tau = (\tau_i - \tau_{i-1})$ so long as and until $\Delta\tau \geq 0$ at a constant content of N_i (or C_{vi}). Also the movement is going on by a current isotherm. The resulting curve acquires a continuous nature (Figure 1).

The calculation of the kinetics of evaporation in the isothermal condition was performed for preliminary estimation of parameters and selection of conditions for the evaporation of the alloys. The calculation of the non-isothermal kinetics of evaporation was performed for a calculation of experimental relations, obtained in on-land conditions, and their comparison with experimental curves obtained in full-scale conditions.

Mathematical model and calculation of the kinetics of evaporation under full-scale conditions. The equations and conditions describing the process of vacuum evaporation of the flat surface of the binary solution, both of whose components are volatile, are given in [9–13]. The transfer of the substance to the evaporation surface from the melt depth is realized at the expense of diffusion, and the movement of the evaporation boundary is taken into account. The main equation, describing the process, is the equation of diffusion

$$\frac{\partial C}{\partial \tau} = \frac{\partial}{\partial x}\left(D\frac{\partial C}{\partial x}\right), \tag{8}$$

where $C(x, \tau)$ is the volume fraction of one of the melt components. Usually, the calculation is made using a fusible element, which is marked by an index 1 ($W_1 > W_2$). Everywhere $C_1 + C_2 = 1$, where C_1, C_2 is the concentration of the first and second component of the alloy, respectively.

Calculations are made for the cylindrical column of the melt with a single finite cross-section [9–11] and main length [12, 13].

The initial distribution of the first component is taken:
for the limited melt column of height h and

$$C(x, \tau) = C_0 = \text{const}, \quad \tau = 0, \quad 0 \le X \le h; \qquad (9)$$

for finite

$$C(x, 0) = C(\infty, \tau) = C_0 = \text{const}, \quad 0 \le X_\infty.$$

In both cases the movement of the boundary of the liquid–vapour interface is defined by the behaviour of the substance in the crucible at the expense of evaporation of the components. The approaches of different authors differ by the arrangement of the beginning of coordinates and coordinate axis and some peculiarities of writing the equation of balance of masses. The models for the limited melt column without a source with boundary conditions at the surface bottom are given in [9, 10]. The boundary condition at the surface is linearized for the case of small surface concentrations. In [9] the equation of mass balance is written by changing the concentrations in the melt volume, unlike other works, where only the change in concentration in a subsurface layer is considered.

Analysis of models, described in [9–13], showed that none of them in the initial form is suitable for the calculation of real processes of evaporation. We believe that the approaches of authors [9–13] are closer. In our case the beginning of coordinates was fixed at the mobile melt–vapour boundary. The axis of the coordinates is directed to the bottom. Coming from equation (8), where D is taken independently of the concentration, we shall write the initial condition at the region

$$0 \le \tau \le \tau_{\text{full}}, \quad \text{as } C(x, 0) = C_0 = \text{const} \qquad (10)$$

and boundary (for $0 \le X \le h$)
    on the surface

$$\frac{\partial C}{\partial x}\Big|_{x=x^*(t)} = \frac{W_1 - W_2\psi}{\psi D}; \qquad (11)$$

on the bottom

$$\frac{\partial C}{\partial x_1}\Big|_{x=0} = 0, \qquad (12)$$

where $\psi = M_1\rho/M_2\rho$.
    The equation of balance has the form

$$\frac{\partial x^*}{\partial t} = w_1 + (w_1 - w_2) C(x^*, t). \qquad (13)$$

The full time of evaporation is determined similarly to the convective case:

$$\tau_{\text{full}} = \left(\frac{C_{10}R_1}{w_1} + \frac{C_{20}R_2}{w_2} \right). \qquad (14)$$

The calculation is performed in volume fractions of an evaporated component. The concentration of the substance in a vapour phase is also determined similar to the convective case.

For the numerical realization of the model, it was reduced to a dimensionless form by variables x, τ:

$$\tilde{x} = x/h, \qquad \tilde{\tau} = \tau'/\tau_{full}. \tag{15}$$

In accordance with (12) the reduction of C to dimensionless coordinates, i.e. to the interval (0, 1), does not influence the adequacy and convergence of the diagram. However, in the case where C is measured not in percents, but in fractions, then this condition is kept automatically.

The model was solved in finite differences using the method of running using a Krank–Nicholson scheme with a weighed mean ($\lambda = 0.5$). This difference scheme is implicit and stable absolutely by $\Delta X/\Delta \tau$. The system of equations

$$C_{i+1,j} = \Delta X/\Delta \tau D\, (C_{i,j+1} + C_{i,j-1}) + [(1 - 2D)\Delta X/\Delta \tau]C_{i,j} \tag{16}$$

(indices i refer to x, j — to t) are reduced by simple transformations to a three-diagonal matrix which is solved by the Thomas's algorithm [14]. The network, in which the solution of the equation is realized, is double-layer, with a constant pitch in x and variable in t. The system of equations is solved by the method of running using the boundary conditions and equations of balance.

To set the stationary condition it is necessary to follow some geometric relations in the crucible design. To calculate the real systems of evaporation it is necessary to take into account the area of the evaporation surface. To apply the models of unidimensional diffusion, it is necessary to have the relationship between the column sizes in the form of $h > d$. However, this cannot be realized in practice. The larger area of the evaporation surface, the lower probability of existence of the stationary condition and the nature of the curves will more resemble the convective evaporation.

It should be also noted that the given model is not stable in the boundary condition (12). The criterion of stability will be written as

$$\frac{w_1 - w_2\psi}{\psi D}\Delta X > 1. \tag{17}$$

It should be taken into account in selection of ΔX.

To provide an effective start of calculation it is necessary to preset $\Delta \tau$, being in the vicinity of a true value. Otherwise, the system will choose a pitch τ for a long time. For this purpose, it is necessary to provide the link between ΔX and $\Delta \tau$ by the rate of the melt boundary movement at the initial moment of time:

$$\Delta X = [w_1 + (w_1 - w_2)\, C_0]\Delta \tau. \tag{18}$$

Comparison of kinetics of evaporation of alloys under on-land and real conditions. The experimental dependence of the composition of coatings produced in on-land conditions, on the time of condensation, is given in Figure 2. The same Figure shows the calculated curves of dependence of the composition of coatings on the time of melt evaporation, plotted on the basis of temperature conditions taken for the evaporator in the unit «Yantar» (Figure 3).

The calculation of kinetics of evaporation was made based on the approximation of convective mass transfer in the crucible. It is seen from Figure 3, that the calculated and experimental curves are similar with close enough values of concentration in the second half of the evaporation period. Both the experimental curve and the theoretical curves have a

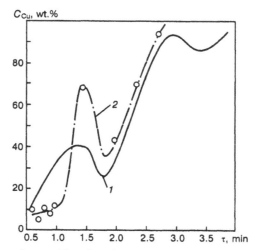

Figure 2 Dependence of composition of coatings, cut for samples in on-land conditions, on time of evaporation: *1* — calculation; *2* — experiment.

non-monotonous nature, defined by a change (decrease) in the melt temperature due to stop of the strip and disconnection of the gun in transition from deposition of one layer to the condensation of the next layer. Here, the temperature is reduced almost by 300 °C (see Figure 3). When the gun is switched on the strip rewinding is started, the curtain is open, the melt temperature and the evaporation intensity are both increased, which is reflected on the nature of a concentration curve.

Figure 4 gives the experimental and theoretical dependencies of the composition of «space» coatings Ag–Cu on the time of evaporation of the melt from the crucible. The theoretical curves were calculated on the model of the diffusion mass transfer in the melt, but without allowance for a change in the evaporation temperature (isothermal approximation). It was assumed for the calculation that the coefficient of diffusion of silver is described by the expression $D = D_0\exp(-E/RT)$, where $D_0 = 2\cdot10^{-3}$, $E = 1.5\cdot10^4$ cal/mol.

A comparison of experimental and calculated curves shows that the change in composition of the vapour phase in space conditions during the process of evaporation can be described within the scope of the diffusion model, as the nature of the experimental curves and appropriate values of concentrations for definite time intervals have a good correlation with the calculation.

Thus, it can be concluded that the process of evaporation of alloy Ag–Cu in space conditions is determined to a great extent by a diffusion mechanism of mass transfer in the melt. Comparing the data of dependence of composition of films on the evaporation time,

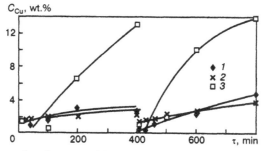

Figure 3 Dependence of coating composition on time of evaporation: *1* — Earth; *2* — calculation; *3* — space.

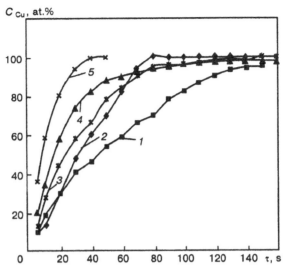

Figure 4 Experimental (*1*) and theoretical (*2–5*) dependencies of composition of space coatings Ag–Cu on time of evaporation.

which were obtained during an experiment in the unit «Yantar», with similar data, obtained earlier in unit «Isparitel-M», it can be noted that the diffusion mechanism of mass transfer was manifested more clearly in the unit «Yantar» due to the change in crucible design. Actually, a crucible with a cover and concave bottom was used in unit «Isparitel-M», while a crucible without a cover and with a flat bottom was used in unit «Yantar». In the first case the condensation of alloy vapours occurred on the cover and crucible walls, which was accompanied by drop formation and repeated evaporation from the drops. This led to the redistribution of vapour flows and levelled the effect of evaporation directly from the melt. In addition, high gradients of the temperature were observed in the crucible of an intricate shape, that contributed, in turn, to the melt stirring due to convection of Marangoni.

These effects in the crucible with a flat bottom are weakened significantly and the manifestation of the diffusion mechanism of mass transfer can be decreased due to purely geometric reasons, connected with typical sizes of the evaporating system (*h* and *d*) and also with the non-isothermal nature of evaporation. The latter circumstance should be taken into account on the basis of the development of the mathematical model of evaporation in the conditions, when the diffusion of components from the melt depth is decisive.

1. Lukash, E.S., Shulym, V.F., Grigorenko, G.M. et al. (1986) *See pp. 340–344 in this Book.*

2. Toptygin, A.L., Arinkin, A.V., Savitsky, B.A. et al. (1986) Composition and structure of films of Ag–Cu alloys, condensed under conditions of orbital station flight, In: *Problems of space technology of metals,* Kiev, PWI, pp. 43–51.

3. Toptygin, A.L. and Sikar, D.B. (1989) About feasibility of producing condensates of a constant composition in vacuum evaporation of alloys, *Fizika i Khimiya Obrab. Materialov,* No. 2, pp. 55–58.

4. Zagrebelny, A.A., Neznamova, L.O., Nikitsky, V.P. et al. (1988) *See pp. 319–321 in this Book.*

5. Pazukhin, V.A. and Fisher, A.Ya. (1969) *Separation and refining of metals in vacuum,* Moscow, Metallurgiya, 204 pp.

6. (1977) T*echnology of thin films.* A handbook. Ed. by L. Mayssel and R.Glang, Moscow, Sov. Radio, 465 pp.

7. Zinsmeister, G. (1964) Die direkte Verdampfung von Legierungen, *Vacuum Tech.*, **13**, No. 8, pp. 223–240.

8. Kostrzhitsky, A.I., Pribbe, S.A., Fedosov, S.N. et al. (1976) Fractioning of binary alloys in evaporation from a common crucible, *Fizika i Khimiya Obrab. Materialov*, No. 3, pp. 50–54.

9. Ziling, K.K. and Pchyolkin, V.Yu. (1976) Evaporation of binary alloys, associated with diffusion of components in the alloy volume, *Izv. AN SSSR, Metally*, No. 3, pp. 51–57.

10. Ziling, K.K. and Pchyolkin, V.Yu. (1975) Evaporation of binary alloys, controlled by diffusion in liquid phase, *Fizika i Khimiya Obrab. Materialov*, No. 1, pp. 15–19.

11. Dolidze, G.F. and Yakashvili, D.V. (1981) Theoretical investigation of conditions of producing films of a preset composition, *Ibid.*, No. 5, pp. 66–77.

12. Pchyolkin, V.Yu. (1979) Analysis of processes proceeding in evaporation of components of solid melts, *Ibid.*, No. 1, pp. 95–100.

13. Zolotaryov, P.P., Zhukhovitsky, A.A. and Pokhvisneev, Yu.V. (1983) Towards the theory of process of vacuum evaporation from the surface of the binary solution, *Ibid.*, No. 8, pp. 68–72.

14. Klein, S.J. (1988) *Similarity and approximated methods*, Moscow, Mir, 431 pp.

On New Problems of Technological Processes Control in Space[*]

R.F. Ganiev[7], D.A. Dudko[1], A.A. Zagrebelny[1], V.D. Lakiza[7], V.F. Lapchinsky[1] and A.S. Tsapenko[7]

This paper is devoted to discussion of some issues related to the development and implementation of technological processes under conditions of space, based on analysis of new phenomena and mechanical effects discovered in experimental studies of dynamic behaviour of multiphase media (mixtures of liquids with gaseous and solid inclusions) under conditions close to those of zero gravity under the impact of controllable periodic disturbances.

In our opinion addressing the scientific and practical problems of controllable processes in space technology is highly urgent now. This implies a purpose-oriented use of various controlling impacts, such as vibration, ultrasound, electric and magnetic fields, etc., both for intensification of the existing processes and development of fundamentally new technological operations which cannot be implemented without the controlling impacts. The latter should be based on certain physical and mechanical effects due to the specific features of space environment.

It is known [1–3] that the conditions of absolute zero gravity are difficult to implement in practice, as space objects almost always are exposed to the effect of different kinds of external disturbances. In this connection a number of unforeseen difficulties arise in implementation of some technologies involving materials processing in the liquid state. The technological processes may proceed not at all the way it was anticipated under conditions of ideal zero gravity in the absence of external disturbances. For instance, in fabrication of materials with uniformly distributed inclusions of other phases, external disturbances may lead to development of inhomogeneities inside the material (i.e. the zero gravity effect is,

---

[*] (1977) Paper was presented at the 6th Gagarin Scientific Readings on Cosmonautics and Aviation, Moscow.

essentially, nullified). In this case the problem of dynamic stabilization of the mixture arises, which is solved by controllable impacts. Thus, the technologies to create composite and foam materials under actual conditions of space are hardly practicable without the use of controlling external impacts, as in the presence of external disturbances static stability of mixtures is impossible to attain.

This is just one of the aspects which make it necessary to address the problem of controllable processes in space technology. Furthermore, in addition to the task of mixture stabilization, in many cases it is necessary to perform an opposite operation in a certain sense, namely formation of periodical structures in a multiphase media, localizing and separation of different phases, for instance, gas separation from liquid, etc. Such a task arises, in particular, when solving the problems of degassing and cleaning of liquid media under zero gravity.

One of the illustrative examples is a task also important in practical terms, of containment and directional displacement of liquids or molten metals in space. Preliminary theoretical analysis shows that contactless containment and displacement of liquid media at zero gravity can be implemented by means of periodical impacts on the medium.

Theoretical and experimental studies performed in the laboratory on Earth lead to the conclusion that, in our opinion, use of external controlling impacts and, in particular, of periodic disturbances, such as vibration, ultrasound, variable electric and magnetic fields, is rational, in order to solve the above-mentioned and many other similar problems. In the first stages of investigations it appears logical to use model liquids and inclusions in order to determine the most general laws. This essentially facilitates the staging and performance of investigations, particularly under special conditions. Mathematical models of multiphase media at periodic impacts can be used in a number of cases to provide a theoretical solution of space technology problems. Solving many problems important for practice of controllable technological processes, in particular vibration, is reduced to study of the stability of equilibrium states or periodic movements of the multiphase medium [4–6]. Such a definition of the problems allows visual representation of the conditions under which technological operations can be implemented as, for instance, displacement of liquids and various inclusions; uniform distribution of gas bubbles or solid particles in liquid metals (production of foam materials and composites); and translational movement of gas relative to the liquid (degassing of fuel or liquid metals), etc. Thus, various mechanical effects can be established which arise from the specifics of zero gravity and periodic impacts on multiphase media, which can be used as a basis for proposing new technological processes for space applications.

Experimental investigations of dynamic behaviour of multiphase media at zero gravity can be conducted by different methods [3, 7]. This paper gives some experimental results, demonstrating the essence of problem definitions being considered and clarifying the dynamic behaviour of multiphase media in relation to the technological processes in space. Experiments were conducted in a small-sized vibration unit developed for the purpose. It allowed inducing, measuring and recording periodic movements of the studied objects in a broad range of oscillation amplitudes and frequencies (Figure 1). The unit was mounted on board a flying laboratory, which made manoeuvres along a Keppler trajectory [1–3]. In this case conditions close to zero gravity (residual overloads were equal to ~ $5 \cdot 10^{-2}$ g) were created. The used objects of study were transparent cylindrical vessels filled with model liquids (water, capacitor oil, water with alcohol, hydrochloric acid solution, etc.) and containing solid and gaseous inclusions. In order to reveal the specific features of multiphase media behaviour at zero gravity, experiments were conducted both in the flying laboratory and on Earth. The aim of investigations was study of stable equilibrium states and movements of multiphase media under conditions close to zero gravity with periodic external impacts; the main focus being the study of the resonance movement modes.

Figure 1 Appearance of small-sized vibration unit.

Given below are the results for several vibration effects, which can turn out to be useful for solving the appropriate applied problems of space technology.

When dynamic behaviour of two immiscible liquids (water and capacitor oil) with solid particles (poppy seeds) was studied at zero gravity (Figure 2a), it was found, that the effect of fast mixing of the media and formation of a uniform foamed suspension is manifested in a rather narrow range of oscillation frequencies (Figure 2b), the process of chaotic movement of the mixture being of a steady nature. Note that with the same parameters of external excitation (amplitude and frequency) in the laboratory on Earth, only intensive movement and breaking up of the free surface of the liquid, as well as subharmonic low frequency oscillations of the media interphase are found, but mixing does not occur as such (Figure 3).

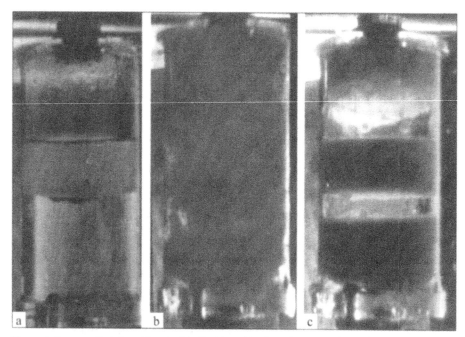

Figure 2 Process of mixing of immiscible liquids and laminated structure formation at zero gravity: a — absence of external excitation; b — vibration impact; c — resonance vibration impact.

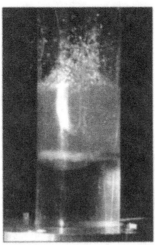

Figure 3 Periodic laminated structure mixing and forming at external impact on Earth.

The process of vibration mixing of liquid media is of a resonance nature. Even with insignificant variations of the frequency of external excitation in the vicinity of the resonance, the mixing intensity and quality markedly deteriorate. Moreover, if after complete mixing of the media the excitation frequency is to be increased, discontinuities and compacting develop in individual locations of the suspension. This eventually leads to formation at a certain frequency of a stable periodical laminated structure, which consists of alternating layers of the suspension and air (Figure 2c), the solid particles in the suspension layers being «almost uniformly» dispersed.

When external excitation parameters are changed, the formed laminated structure becomes unstable. With a change in frequency, oscillations of the suspension layers arise, which lead to its disintegration, and lowering of the amplitude leads to «blurring» of suspension–air interphase. This points to the resonance nature of the phenomenon of laminated structures formation and to their vibrational stability (in the absence of vibrations, breaking up of the suspension layer boundaries occurs).

This series of experiments had the aim of studying the behaviour and some possible forms of dynamic equilibrium of the free surface of the liquid with different types of excitation (longitudinal and transverse) of carrier bodies in a broad range of oscillation frequencies.

It was found that in the low frequency range (~10 Hz) oscillations of the free surface occur which mostly are of a chaotic nature (Figure 4). Unlike the ground-based experiments ($g = 1$), where one clearly defined form is usually observed, simultaneous excitation of several forms of oscillations occurs at low gravity.

Of special interest are the resonance effects, arising on the free surface of the liquid at high frequency ($\approx$ 1000 Hz) transverse oscillations of the carrier bodies. Their essence consists in formation of deep cavities on the surface of a low-viscous liquid (Figure 5a and b) and appearance of peculiar geometrical patterns on the free surface of a viscous liquid (Figure 5c). These phenomena are of a pronounced resonance nature and are only manifested in a narrow range of oscillation frequencies. This kind of effect, which cannot in principle occur on Earth because of the considerable magnitude of the forces of gravity, can be interpreted as the possible forms of dynamic equilibrium on the interphases of the liquid and gaseous media under conditions close to zero gravity.

We need to pose the problems of study of the dynamic behaviour of gas bubbles at controllable periodic external impacts primarily because of the needs of current engineering

Figure 4 Shape of a free surface of liquid at low frequency impact at zero gravity.

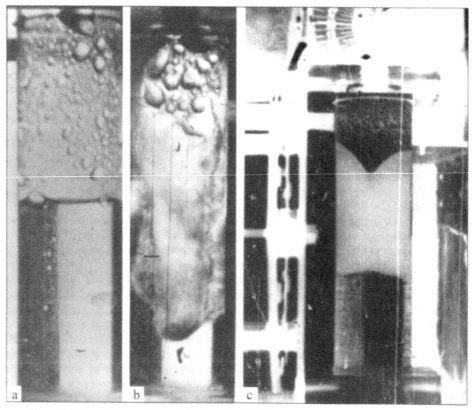

a   b   c

Figure 5 Shape of free surface of the liquid at high-frequency transverse resonance impacts:
a, b — water; c — capacitor oil.

and space technology. The tasks of localizing the gas inclusions arise, when the issues of degassing of the liquids and metal melts, both on Earth and in space, are considered. The problem of degassing of the fuel and molten metals at zero gravity is especially urgent. The vapour or gas bubbles, formed in liquids during their boiling or at the expense of evolution of the dissolved gas, because of the absence of the Archimedean force, remain suspended in the medium, and this leads to undesirable effects in many cases. So, for instance, in welding in space, the welds in a number of cases contain a large number of pores, which essentially lower their strength and reliability [8].

The problem of experimental investigation of gas bubble stable equilibrium states and controllable motions in oscillating media at lower and zero gravity, was posed to clarify their dynamic behaviour. Studies were based on the known phenomenon of gas bubbles localizing under the conditions of terrestrial gravity at oscillations of a vessel with liquid [9]. As shown by the experimental results, similar processes also proceed at zero and low gravity. With the impact of high frequency (~ 1000 Hz) periodic disturbances on the system, an intensive evolution of dissolved gas is found in the form of individual quickly growing bubbles, which then coalesce into groups near the shell bottom (Figure 6). If a small positive overload is applied to the system, after localizing of a sufficiently large amount of gas, when the Archimedean force exceeds the vibration forces, keeping the group in an equilibrium position, it floats to the surface and a new group starts forming at the bottom. With unchanged parameters of external excitation, such a process goes on periodically, until the zero gravity mode is over. Now, if the zero gravity is «pure», the groups remain suspended in the liquid at certain levels which can be adjusted by smoothly changing the external excitation frequency, i.e. controllability of gas bubbles movement in liquid is in place. So, with increase of the excitation frequency, the group of bubbles moves upwards to a free surface, and towards the vessel bottom with its decrease.

In addition to investigations of the dynamic behaviour of gas bubbles in vibrating liquid media, the problem was also considered of movement of a drop of molten metal under conditions close to zero gravity with the impact of vibration. In this case a cylindrical transparent shell, filled with 20 % solution of hydrochloric acid and containing a drop of mercury (drops of various dimensions were studied) was used as the model. In these

Figure 6 Bubble cluster formed at zero gravity at high-frequency vibration impact.

a   b

Figure 7 Resonance vibration impact on a mercury drop (a) and large gas bubble (b).

experiments, where zero gravity was sufficiently «pure», capillary «leaps» of the drops were observed, which are known from earlier research [7]. In the majority of cases, however, the drop acquired a spherical shape, while remaining at the vessel bottom. When low frequency vibrations were applied, it made small «leaps» hitting the vessel bottom and pulsated while being in a suspended state. When the frequency of external excitation became close to the frequency of the drop pulsations, the «leaps» amplitude increased, and when these frequencies coincided, the effect of monotonic unidirectional motion of the drop was observed, due to its resonance interaction with the environment (Figure 7a). Similar phenomena also occurred in investigation of the motion of large gas bubbles formed as a result of closing of the free surface of the liquid in the transition of the liquid–gas system into zero gravity (Figure 7b).

Thus, experimental studies led to the discovery of a number of new vibrational effects, which in our opinion can be used in development and implementation of technologies in space. For instance, the phenomena of mixing and formation of periodic laminated structures can be recommended for use in fabrication of foam and composite materials and materials having a laminated structure, whereas it is rational to use the effects of localizing and controllable motion of gas bubbles when addressing the problems of liquids degassing and molten metal solidification under conditions of low and zero gravity. Furthermore, posing and solving such problems can form the ideological basis of one of the scientific fields, the so-called mechanics of controllable technological processes under conditions of space. The results derived here lead to the conclusion of the rationality of raising the issue of controllable, in particular, vibrational technological processes in space and of the need to expand both the theoretical and experimental investigations in this area.

1. Ganiev, R.F., Lakiza, V.D. and Tsapenko, A.S. (1976) Behaviour of a liquid metal drop under the conditions close to zero gravity at vibration effects, *Doklady AN SSSR,* Series A, Issue 4, pp. 329–332.

2. Paton, B.E., Paton, V.E., Dudko, D.A. et al. (1973) *See pp. 118–122 in this Book.*

3. Belyakov, I.T. and Borisov, Yu.D. (1974) *Technology in space*, Moscow, Mashinostroyeniye, 290 pp.

4. Ganiev, R.F. and Ukrainsky, L.E. (1975) *Dynamics of particles under the effect of vibration*, Kiev, Naukova Dumka, 167 pp.

5. Ganiev, R.F., Puchka, G.N., Ukrainsky, L.E. et al. (1975) On non-linear vibration effects in multiphase media, In: *Proc. of 6th Int. Symp. on Non-Linear Acoustics*. Moscow, pp. 5–11.

6. Ganiev, R.F. and Tsapenko, A.S. (1975) Dynamics of gas bubbles in liquid subjected to vibration effects, In: *Problems of math. physics and theory of oscillations*, Issue 3, pp. 18–21.

7. Kirko, I.M., Dobychin, E.I. and Popov, V.I. (1970) Phenomenon of capillary «play ball» under the conditions of zero gravity, *Doklady AN SSSR*, No. 2, p. 31.

8. Ternovoj, E.G., Bondarev, A.A., Lapchinsky, V.F. et al. (1976) Some peculiarities of electron beam welding of aluminium alloys in zero gravity, In: *Space research in Ukraine*, Issue 9, pp. 5–11.

9. Apshtejn, E.Z., Grigoryan, S.S. and Yakimov, Yu.L. (1969) Stability of air bubble cluster in oscillating liquid, *MZhG*, No. 3, pp. 10–12.

Prediction of Life of Structural Materials and Their Joints Exposed to Long-Term Service in Space*

S.V. Bakushin[1], D.L. Demidov[1], Yu.D. Morozov[1], V.P. Nikitsky[2] and I.V. Churilo[2]

The conditions of service of large-sized structures in space differ radically from the usual Earth conditions. The main differences concern the forces acting on the structure and service environment [1].

A typical scheme of loading of the load-carrying large-sized structures is given in see Figure 5 on p. 38. Static loads (curve *1*), acting on the structure, are quite low and reduced to the effect of the gradient of the Earth gravitation field and aerodynamic resistance of the rarefied ambient atmosphere. The value of these loads depends on the area of the midsection and the structure orientation. In the majority of cases these loads are not taken into account during the structure design. Rare single pulse actions are possible (curve *2*), for instance, in docking, separations or other similar operations. These actions should be taken into consideration, as they can lead to the local or general buckling of the structure or its separate elements. The constantly acting loads are low-frequency (0.01–10 Hz) cyclic alternating vibrations of the structure which are caused by the operation of systems of orientation, control and stabilizing, different service mechanisms and crew (curves *3* and *4*). The effect of these loads on the structure during the many years of service leads to the occurrence of fatigue fractures [2].

Many differences are observed in the environment. There is a number of space factors that have a negative influence on the structural materials [2]. Moreover, it should be noted, that the degree and nature of the space environment effect cannot be considered separately from the properties and quality of the materials themselves.

Factors of the combined influence of space and the natural properties of the structural materials on physical-chemical processes proceeding in them and, thus, causing degradation, are given in Figure 7, p. 39. Let us describe briefly some of the factors shown in Figure. Any space vehicle located in a near-earth orbit, twice intersects a terminator per one revolution.

---

* (1991) *Proceedings of Conf. on Welding in Space and the Construction of Space Vehicles by Welding*, USA, Miami, Sept. 24–26, 1991. Miami, pp. 299–306.

Here, the light-and-shade situation is changed (see Figure 6 on p. 39). The flux of full solar radiation (curve *1*) is changed from maximum ($\approx$ 1400 W/m^2) almost down to zero. The structure as a whole or its separate elements (curve *2*) are subjected to corresponding cyclic changes, leading to cyclic deformations of non-rigid structures or occurrence of temperature stresses (curve *3*) in the connections of rigid structures. Moreover the occurrence of secondary harmonics of stress variations at relatively high amplitude is possible in the surface layers of low-heat-conducting materials.

The long-term action of these thermal cycles in combination with other unfavourable factors of the space medium lead to the monotonous systematic deterioration of service characteristics of structural materials and their joints. Underestimation of this phenomenon in long-term service of the structure in space may lead to its unpredicted failure.

The effect of electromagnetic radiations, firstly, ultraviolet and radio radiation of the space, and also the significant effect of high-energy particles (cosmic rays), are already apparent in the working near-earth orbits of the modern space vehicles. Moreover at altitudes of 300–500 km there are both primary cosmic rays and also secondary particles, occurring during interaction of the primary rays with the Earth's atmosphere. Hard particles of a meteoritic dust type, destroy the surface of materials. Finally, the action of electrons and protons, available in background (below 500 km) regions of the inner radiation belt of Earth, should not be neglected [3].

The space vacuum characterized by a high rate of evacuation of gas evolutions leads to the intensive degassing of materials, their depletion with easily-volatile components, and increase in specific density of slip planes in local volumes of materials at cyclic loads. With long-term stay in space vacuum, these phenomena are observed not only in surface layers, but they also penetrate into the inner volumes due to diffusion.

The atmosphere surrounding the space vehicles affects significantly the materials [1]. At altitude of 300–500 km the content of atomic oxygen in the atmosphere amounts to 70 to 90 %. Therefore, in spite of low density ($1.6 \cdot 10^9$–$1.6 \cdot 10^8$ atm/cm^3), it interacts intensively with the structural materials. The intensity of interaction is initiated by high rate of incoming flow of atmosphere and the ultraviolet radiation of the Sun.

Atomic oxygen interacts dually with the surface of materials. With chemical interaction (oxidation) the mechanical and optical characteristics of materials are changed. Also the incoming stream of oxygen atoms can cause mechanical damage of surface layers at the expense of the mass loss.

It should be noted that the atmosphere, directly surrounding the space vehicle differs remarkably from the estimated value for the above-mentioned altitudes. It is explained by the exchange interaction of the atmosphere with the space vehicle. The atmosphere is usually rich in nitrogen, carbon monoxide, CO_2, fuel remnants, and products of the crews' bodily activities. Besides, the plasma density around the vehicle is significantly higher than the free-stream space density. The volume density of plasma around the vehicle is not uniform and depends on the vector of its velocity.

The long-term effect of the mentioned factors due to physical-chemical processes proceeding in the structural materials, leads to change in their service characteristics. The rate of change depends mainly on the structural integrity and the quality of materials and their joints.

These are the present general approaches to this problem which are based on theoretical considerations and rather scarce experimental data. Nevertheless, even these scarce data prove that the risk of degradation of the properties of materials, operating for a long time in space is quite real. In addition, the degradation starts usually from the surface and undergoes at least two phases. At the beginning the processes are delayed and running uniformly, penetrating gradually into the deeper layers. At the second phase, the local sources of a more intensive degradation are usually initiated at which up to 100 % service quality of the material can be lost with time [2].

These peculiarities of change of the material properties, after a long time in space, become especially important in connection with the problem of construction and service of large-sized structures in space. The impact of the problem is intensified by the fact that in the design of space structures, mass reduction is a traditionally important requirement. Therefore, only small allowances on strength and rigidity of the structure are usually specified. This approach is quite admissible for small-sized low-loaded structures launched into space for a short period (up to several years). Increases in structure dimensions, payloads and duration of stay in space up to several dozens of years, require a proved safety factor (taking into account the degradation of structural materials) and rigidity.

With this in view, a purposeful integrated program of materials research has been developed at the PWI and NPO «Energiya»[2].

The program envisages the parallel mathematical and physical simulation of space service conditions of fragments of large-sized structures, manufactured from different structural materials using welding, brazing, adhesion and coating.

The physical simulation is to be realized on the thin-walled scale-models that make it possible to study both surface and internal processes of degradation at comparatively short exposures. The exposure of specimens is to be conducted both directly in space for evaluating the integrated effect of the above factors, and in terrestrial conditions for their separate evaluation.

The space experiments will be conducted in a specially designed set of equipment (Figure 1). This set, with pre-arranged experimental specimens, will be launched to a near-earth orbit, fastened at the external surface of the space vehicle and deployed. It is planned to expose specimens under different conditions of illumination, different conditions of loading during different periods of time. Specimens similar to those launched into space, will remain on Earth as reference specimens. The as-exposed specimens will be delivered to Earth to continue the studies. The terrestrial studies will consist of mechanical tests of the as-exposed specimens, structural and spectral analyses, fractography, evaluation of local

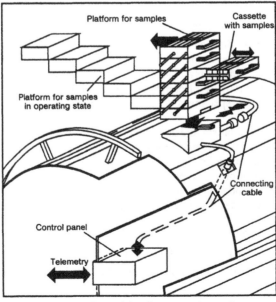

Figure 1 Scheme of conductance of materials science experiments in a manned space vehicle.

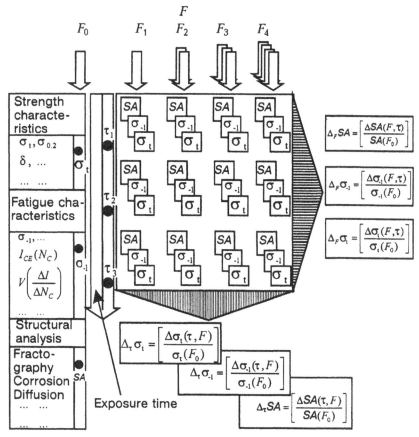

Figure 2 Scheme of adding the database on degradation of structural materials and their permanent joints in space (F_0 — on-land examination; F_1, F_3 — non-loaded material, subjected to the shade and in the sun, respectively; F_2, F_4 — loaded material, subjected to the shade and in the sun, respectively).

deformations of specimens during tests, etc. The tests will result in quantitative relationships by using various units of measurement in the majority of cases. In plotting of kinetic curves of degradation of studied materials and joints, a dimensionless database will be used, whose elements are formed as a ratio of difference of single-type experimental results of testing of as-exposed and reference specimens to the experimental result of the same reference specimens (Figure 2).

In the course of development and continuation of space and terrestrial experiments, the database will be added and widened, thus improving the accuracy of prediction of the degradation of materials and joints.

The complete realization of the program will make it possible to work out a system of criteria for the well-grounded selection of optimum structural materials, technology and methods of joining and to develop a procedure to monitor and control their condition [2].

1. (1969) *A handbook on space engineering*, Moscow, Voenizdat, 693 pp.

2. Paton, B.E. (1990) *See pp. 35–40 in this Book.*

3. (1986) *Physics of space. Small encyclopedia*, Moscow, Sov. Entsiklopediya, pp. 12–16.

Application of Welding for Repair of Space Objects*

B.E. Paton[1], D.A. Dudko[1], V.N. Bernadsky[1], V.E. Paton[1], V.V. Stesin[1], V.F. Lapchinsky[1] and A.A. Zagrebelny[1]

The problem of repair of the space objects arose with the development of projects of orbital stations, designed for many-year functioning. Authors of existing publications on this subject consider it necessary and rational to perform preventive and repair works with a crew, not specifying the nature and content of these works. The present article describes the considerations about expediency of creation of repair welding stations in the orbital stations, and also about capabilities of these stations.

The objective of repair welding is reliable and long-lived metal structures. From US data, 40 % of failures in space systems occur in electronic systems, and 60 % on electromechanical systems. Damages to metal structures were observed only in particular cases (accident in «Apollo-13», cracks in stabilizers and tanks which were detected before launch). Damage to metal can be caused by accidental factors (strikes of meteorites, collisions in docking, explosions on board) and by the long-term action of service in the environment (corrosion, erosion and ageing of metals).

In all the above-mentioned cases the probability of accidents is a function of the objects service time. This makes it possible to state that damage to metal structures will take place many times in long-time orbital stations.

Let us follow the evolution of measures which support the serviceability of space vehicles. In the spaceships of «Vostok» and «Mercury» type with 1–5-day mission regular works were not envisaged, but already US astronaut W. Schirra had had to repair the valve of his spacesuit.

On board the spaceships «Soyuz», «Gemini» and «Apollo» with up to 20-day missions, the materials and accessories for elimination of the simplest troubles (adhesive tape, pliers, etc.) were available, which were used for different purposes. In «Gemini» the tools were also tested for the fulfilment of more complicated works outside the spaceship.

OS «Salyut» and OL «Skylab», which were on mission for about 200 days, were equipped with sets of spare parts and a tool for preventive repairs, including the replacement of faulty units. The actual volume of repair works in «Skylab» exceeded that which was envisaged. The crew of «Salyut-4» performed the first technological operation for the recovery of optical properties of coatings.

Thus, a tenfold increase in life of the space object was accompanied by the transition to the new stage of repair-preventive works and each time the repair capabilities were behind the needs.

The next logical stage in the development was the creation of a workshop on board the station for the repair of damaged units. This workshop had to be equipped with versatile equipment for electric wiring and metal works. It can be assumed that this workshop should appear in stations which will undertake 2000-day missions. The presence of a kit of work tools is, at the same time, a obligatory condition of performance of welding jobs, which usually involve by the removal of damaged metal, cleaning of non-metallic coatings and other similar operations.

In the authors' opinion, there is a true analogy between orbital space stations and submarines. At the beginning of the XX century the problem of repair of submarines at sea by the crew had not arisen. Modern submarines designed for long periods at sea are equipped with equipment for offboard welding jobs and welding inside the compartments, including

* (1976) *Space research in Ukraine*, Issue 9, pp. 3–5.

in the submersed state [1]. The problems of fire safety and contamination of the gas medium were solved successfully.

In repair of space objects, welding works will probably be reduced to three types: restoration of the leak-tightness of pipelines; restoration of leak-tightness of compartments and vessels; and restoration of configuration and strength of non-sealed three-dimensional structures. The use of adhesives and sealants does not provide long life of the repaired element.

In existing structures the body elements are manufactured mainly of aluminium and magnesium alloys of up to 5 mm thickness, and pipelines and vessels are manufactured from steels or titanium alloys of up to 2 mm thickness. Consequently, the technology of welding and welding equipment should be sufficiently versatile with respect to these structural material.

It can be assumed that works on restoration of leak-tightness will be made mainly in vacuum, i.e. outboard the station or in an unsealed compartment. And here, it is necessary to take into account those limitations which will be induced by the protective means on the operator's actions.

In restoration of the leak-tightness of the damaged compartments it is supposed that the station will still be also function effectively as a whole. This requirement will be, probably, fulfilled on the next stations, whose life-spans are envisaged to be about 10 years.

In the authors' opinion, when selecting the method of welding, preference should be given to consumable electrode arc welding and EBW. Both these methods are sufficiently versatile and, as the experiments showed, are suitable for an operator in a protective spacesuit. The advantage of arc welding is the presence of an additional metal for filling the gaps between the parts, while the electron bean can also be used for cutting and coating of metals. Power of not less than 2 kW is required for the above-mentioned thicknesses, and the welding speed which can be provided by the operator is 100–150 mm/min. For the accomplishment of welding jobs, convertors of energy and power units can be used designed for carrying out metallurgical experiments.

The high stability of a weld pool in zero gravity, decreasing the risk of burn-outs, is one of the factors which will help to achive good welds.

Special attention should be paid to the protection of surrounding instruments and electrical circuits from the heat action of the welding arc and products of its combustion. This effect can be reduced using technological procedures, for example, welding with a pulse property of the arc burning or using a protective means. German engineers suggested protection using a foam layer, deposited on the surrounding objects before welding [2]. Engineers from Chelyabinsk city performed the process of arc surfacing under the layer of a foam, but they did not manage to provide complete protection from the arc combustion products.

The most convenient procedure of welding is, in the authors' opinion, the use of specialized devices, for example, rigid captive chambers with a controllable atmosphere, developed at the PWI. In industry, including the aircraft industry, argon arc welding of 8–100 mm diameter pipelines, made from stainless steels, aluminium and titanium alloys, is realized using these chambers. The captive chamber, illustrated in Figure, provides orientation and fixation of pipes being welded, electrode displacement, and protection of welding zone from oxidation and of the environment from the combustion products of the arc and its heat action.

Obviously the above-described information does not cover all aspects of the problem. The above statements are based on the results of experiments performed in short-time zero gravity in airoplanes, as well as on the analysis of the data in the literature.

The next stage should be experimental testing of the technology of welding and welding equipment under the the conditions of space flight. For this testing a working

Captive chamber for pipeline welding.

chamber with a built-in fragment of a spacesuit, developed at the PWI, can be initially used as a test bench [3].

So the need for welding to repair of damage in metal structures, caused by accidental factors or gradual degradation of materials, is expected in orbital stations with a term of functioning of more than 2000 days. Consumable arc welding and EBW are suitable for the repair. The products of welding arc and its heat action can be localized. The technology and equipment for repair welding require testing under conditions of the space flight.

1. Rimkovich, V. (1972) Electric welding and cutting in submarine at sea, *Morskoj Sbornik*, No. 3, pp. 20–23.

2. (1971) *Metallhandwerk + Metalltechnik*, **73**, No. 3, p. 162.

3. Paton, B.E., Dudko, D.A., Bernadsky, V.N. et al. (1975) *See pp. 137–140 in this Book.*

Scientific-Technical Substantiation of «Tyulpan» Space Experiment*

V.I. Berzhaty[2], L.L. Zvorykin[2], A.V. Markov[2], I.V. Churilo[2], A.A. Zagrebelny[1], A.O. Pchelyakov[8] and L.V. Sokolov[8]

Considerable experience of conducting applied research and technological studies at zero gravity on low near-earth orbits has been gained over the last 20 years. In «Salyut» and «Mir» stations more than 450 experiments were performed altogether for implementation of basic space technologies, and unique technologies of assembly and deployment of truss and film large-sized structures in space by orbital station crews were tried out. In addition methods and means used to perform scheduled maintenance and repair-restoration work in open space were retrofitted for servicing the hardware and extending the life of orbital complex and its systems.

When astrophysical, geophysical, materials science and engineering experiments were conducted on board «Salyut» and «Mir» it was established, that an ambient outer atmosphere (AOA) forms around these objects, having a negative influence on the measurement results and condition of measuring instruments. AOA is a complex dynamic formation, including

* (1999) Paper was presented at the 1st Russian Conf. on Space Materials Science, Russia, Kaluga.

Table 1 Composition of AOA gaseous phase.

M	Neutral particles	Ions	M	Neutral particles	Ions
2	H_2 (P)		28	N_2 (P, IF, L)	N_2^+ (P, IF, L)
4	He (P)		28	CO (P)	
14	N (P, IF)	N^+ (P, IF)	30	NO (P)	NO^+ (P)
16	CH_4 (P)		32	O_2 (IF)	O_2^+ (IF)
16	O (IF)	O^+ (IF)	40	Ar (IF)	
18	H_2O (P, G)	H_2O^+ (P, G)	44	CO_2 (P, L)	CO_2^+ (P, L)
19		H_3O^+ (P)	103	Freons (21 and 12)	
			121	(L)	

the gaseous, aerosol and finely dispersed phases. The main sources of AOA formation specifically are:

• desorption and diffusion of gases and vapours, adsorbed and absorbed by the outer coating materials;

• destruction and evaporation of outer coating materials;

• gas jets of operating engines, gas and vapour evolutions during functioning of the life support system;

• gas leakage from pressurised compartments;

• evolutions of aerosol and disperse particles in engine starting and cut off and through drainage systems;

• evolutions of disperse particles in structural element vibrations and mechanical impacts.

On OS «Salyut-7» during «Astra-1» experiment [1], mass-spectrometric measurements revealed the presence of gases and vapours in AOA, which are listed in Table 1. Here M are the molecular weights of gases and vapours, while abbreviations G, P, IF and L correspond to AOA sources (gas evolution, propulsion systems, incident flow and leakage from orbital complex compartments and systems). The Table does not have data on «heavy» components, because of the limited resolution on mass numbers of the mass spectrometric systems, used in «Astra-1» experiment.

As observed by the crews of «Soyuz», «Salyut» and «Mir» the AOA disperse phase initially forms after separation of the object taken to orbit from the last stage of the launcher, when a large number of disperse particles are seen in the field of vision (about several thousand particles of 1–2 mm size by visual estimates). Various mechanical impacts on the structural elements of the space vehicles (vibration, docking and undocking operations) result in appearance of a multitude of disperse particles in AOA, whose dimensions, judging by their brightness, are from 0.1 mm up to several millimeters.

These disperse particles are registered, as a rule, at up to 15 m distance from the OC surface [2]. Moreover, aerosol and disperse particles can consist of condensates and products of incomplete burning of fuel components during engine starting and cut off. In pulse engines the transient processes take up to 50 % of their operation time, and therefore, exhaust products can contain up to half of the used fuel mass.

In AOA stationary («background») state, when no dynamic operations have been conducted for a long time, the AOA gaseous phase pressure at OC surface is estimated to be $\sim 10^{-6}$–10^{-5} mm Hg [1], which is confirmed by the data of flight measurements using a sensor. The data of these measurements also indicated that the pressure in AOA at 3 m distance from OC surface is 10^{-7}–10^{-6} mm Hg, and at about 10 m distance it does not exceed values of the order of 10^{-7} mm Hg [3].

During dynamic operations, when propulsion systems are running, the pressure in AOA rises abruptly by 2 to 4 orders of magnitude, compared to the background conditions, and then is relaxed to the initial condition. The time of pressure disturbance relaxation is determined by the engine operation cyclogram, as well as the dimensions of the area of the structural element surface, exposed to these engine jets. So, for instance, propulsion units of OC correction and orientation systems mostly operate in the pulsed mode, while the directions of the jets outflow are selected, based on the admissibility of the force and thermal impact of just the peripheral zones of these jets on the outer surfaces of the structural elements. Not more that 2–3 % of the total mass of consumed gases flow through these peripheral zones, and, therefore, in this case the time of AOA relaxation to its background condition, as shown by flight measurements, is from several up to tens of minutes. Now during operation of the docking and separation system engines, a considerable part of the space vehicle outer surface is exposed to the propulsion system jets, and the characteristic time of AOA relaxation back to its background condition rises up to several tens of hours. In the dynamic condition AOA composition is determined mainly by the products of propulsion system exhaust.

These circumstances point to the need to apply special extending units that allow the experimental and measuring instruments to function beyond the AOA during performance of astrophysical, geophysical, materials science and process studies in the Russian Module of the International Space Station (RM ISS).

A prototype of such a extending unit can be based on the experience and practical results, obtained during development and retrofitting of «Opora» type deployable trusses. A deployable extending unit can be fitted with a rotary support platform for mounting experimental and measuring instruments on it. To provide additional protection from AOA, and also protection from the incident flow, if required, rotary support platform can further accommodate a deployable protective shield (DPS). A draft design of extending unit with a deployable truss and rotary support platform is shown in Figure 1.

In this connection, conducting «Tyulpan» space experiment is of significant scientific and practical interest. The intent is to perform this experiment in the RM ISS in a special experimental set up «Tyulpan-T» (T-1) between the years 2000 and 2006. The main components of T-1 unit include:
- deployable truss extending unit (EU);
- rotary support platform, mounted on extending unit;

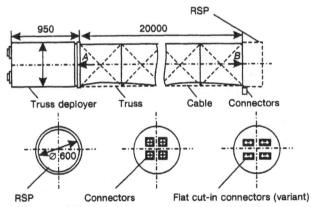

Figure 1 Schematic of extending unit with a deployable truss.

- DPS, installed on rotary support platform;
- instrumentation and research instruments also mounted on rotary support platform.

T-1 unit is installed on the outer surface of one of the RM ISS, or in one of the docking compartments, provided in RM ISS. In addition another possible variant would be to mount the unit in on-board vacuum chamber of VC-2 docking compartment. The seat for T-1 unit on RM ISS outer surface is selected from the condition of guaranteed delivery of DPS, instrumentation and research instruments into the undisturbed incident flow (or the area of AOA minimal influence near ISS) at complete deployment of extending unit. The main goals of «Tyulpan» space experiment are as follows:

- trying out in RM ISS the deployment of truss extending unit with rotary support platform for T-1 unit, as well as the modes of controlling rotary support platform orientation relative to the incident flow in the orbital flight conditions;
- measurement of the parameters and determination of the flow pattern (composition, concentration, rarefaction levels, presence of finely dispersed particles, etc.) in the near DPS aerodynamic wake at cross-flow;
- experimental confirmation of the appearance in orbital flight conditions of a high-vacuum zone in DPS aerodynamic wake at cross-flow (with vacuum levels of $\sim 10^{-14}$–10^{-12} mm Hg and practically infinite pumping speed, unattainable in ground-based commercial high-vacuum units);
- experimental confirmation of the principal possibility of using this zone to implement high technologies (for instance, growing semiconductor structures of 3-5, 4-4 and 2-6 type by the method of molecular-beam epitaxy that require rarefaction levels of not less than 10^{-12} mm Hg for their synthesis);
- preparation for conducting scientific and process studies in the high vacuum zone in DPS wake, both under government programmes and under commercial contracts;
- conducting scientific and engineering experiments (astrophysics, geophysics, electron beam, etc.) using the instrumentation and scientific instruments, mounted on rotary support platform.

Thus, DPS deployed on extending unit is used in this experiment to create in DPS aerodynamic wake a high-vacuum area (with rarefaction levels of not less than 10^{-12} mm Hg) at incident cross-flow at orbital altitudes of RM ISS flight.

Change of flow parameters in DPS aerodynamic wake is shown in Figures 2 and 3. Figure 2 gives the calculated position of the isobars, mm Hg, in the flow field, and in Figure 3 the pressure variation along the flow axis is plotted allowing for gas evolution from the DPS inner surface.

When the achieved rarefaction levels were estimated, it was assumed that sorbed impurities had been first removed from DPS working («shadow») surface, while the rates of gas evolution into the wake zone are not more than 10^{-12} cm^{3}cm/cm^{2}s atm. Partial pressure of $\sim 10^{-14}$ mm Hg corresponds to this level of gas evolution, and it is characteristic of degassed metals applied in vacuum engineering.

Estimation results demonstrated that mostly «fast» He and H molecules penetrate into the rarefaction zone behind DPS from the environment. The velocities of their thermal motion are essentially higher than the orbital flight velocity, and their partial pressures at altitudes $H \approx 250$–500 km are 5 to 6 orders of magnitude lower, compared to the above partial pressure of gas evolution molecules.

Therefore, the pressure behind DPS in its independent flight in the upper atmosphere depends only on the degree of degassing of the «shadow» surface, and the rarefaction level behind the DPS can be 2 to 3 orders of magnitude higher than the limit vacuum achievable in the high-vacuum ground-based process units. As shown by calculations, dimensions and position of the high vacuum zone in the aerodynamic wake at $H = 250$–400 km only slightly

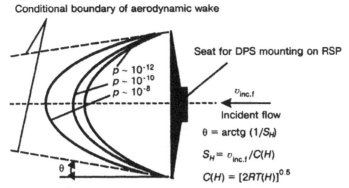

Conditional boundary of aerodynamic wake

Figure 2 Levels of rarefaction behind the DPS at cross-flow in an orbital flight (P = const, $v_{inc.f}$— incident flow velocity, $C(H)$ — average speed of thermal motion of atmospheric gases, R — disc radius).

depend on the flight altitude (Table 2). At the same altitudes values of parameters S_H and θ mainly depend on the level of solar activity and usually vary in the range of $S_H \approx 4$–8, and $\theta \approx 10$–15°.

With DPS transverse dimensions of ~ 3 m, the technological zone for implementation of molecular-beam epitaxy process can be conditionally represented as a cylindrical area of ~ 0.7 m diameter and length of ~1.5 m from the shield bottom, i.e. it readily allows placing an molecular-beam epitaxy process unit behind the DPS.

After a series of ground-based materials science studies and preliminary modelling, the intent is to use the following orbital flight factors in the presented programmes of flight experiments:

1. *Deep vacuum (~ 10^{-12}–10^{-14} mm Hg) and practically complete absence of oxygen and carbon-containing components* in the technological zone behind the DPS. (Epitaxial layers, produced under such conditions, can have record levels of purity and concentration of

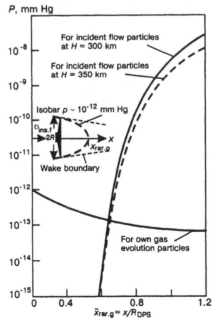

Figure 3 Pressure distribution along DPS aerodynamic wake axis at cross-flow in orbital flight.

Table 2 Position of $\rho \approx 10^{-12}$ mm Hg isobar on the aerodynamic wake axis.

H, km	250	300	350	400
$\tilde{x}_{\text{rar.g}} = x/R_{\text{DPS}}$	1.058	1.078	1.115	1.135

non-radiating recombination sites. This was experimentally demonstrated in a US program in 1996.)

2. *Rates and efficiency of pumping* of the components of the working molecular beam close to the ultimate possible ones, creating a unique possibility for a super fast change of the gaseous phase composition in the growth zone on the substrate surface. (Above-mentioned factors enable producing heterojunctions with ideally sharp profiles.)

3. Practically complete *absence of the working vessel walls and ability to essentially reduce the total surface of elements of process fixtures* in the epitaxial growth zone. (This allows eliminating the accumulated earlier coated substances («memory« effect) and their uncontrollable transfer to the film with subsequent growth of other materials and simultaneously producing dissimilar materials without the traditional moving of the substrate from one growing vessel into another one. Furthermore, it also becomes possible to increase the number of independent individual sources of molecular beams and produce supersharp heterojunctions and interfaces between the epitaxial layers, as well as form multilayer structures, containing a large number of layers of different composition.)

4. *Ability to greatly increase the distance from the substrate to the molecular beam source.* (This factor is the main parameter that determines the homogeneity of the layers across the substrate area, and has a special role in increasing their diameter.)

5. Practically *unlimited heat removal at the expense of radiation* into space allows reducing the thermal inertia of the source and substrate heaters.

6. *Ability to use toxic volatile liquids and gases* (hydrides, metal-organic compounds) as initial materials for film synthesis without environmental contamination. (These compounds quickly disperse to safe concentrations and easily dissociate into safe components under the impact of ionising solar radiation.)

7. *Microgravity* is a factor that is not used directly, but *can provide a stable operation* of crucibleless sources of molecular beams.

1. Zvorykin, L.L., Kotov, V.M. and Krylov, A.N. (1991) Modelling the interaction of a highly rarefied gas with the surface, around which it flows, In: *Proc. of 10th All-Union Conf. on Rarefied Gas Dynamics*. Moscow, MEI, pp. 31–39.

2. Klimuk, P.I., Zabelina, I.A. and Gogolev, V.P. (1983) *Visual observations and optics contamination in space*, Leningrad, Mashinostroyeniye.

3. Guzhva, E.G., Neznamova, L.O. and Rogachyov, A.S. (1992) *Determination of the parameters of ambient outer atmosphere of 27KS vehicle. «Diagram» experiment*, STA RSC «Energiya», P-30996.

Diagnostics of Space Objects
Using the Method of Acoustic Emission*

A.Ya. Nedoseka[1]

Operating space objects require periodic inspection for evaluating their condition and feasibility of further service. The presence of insulation, different parts and equipment, connected to such objects as extended complex pipework, elements of the body load-carrying structures, truss girders, etc., make the places of inspection inaccessible and partial or complete dismantling is required for the fulfilment of control operations. Under space conditions it is difficult to perform these operations. Therefore, methods and appropriate equipment are necessary to evaluate physical condition without interruption of service using a centralized system from the main panel.

One of these methods is based on revealing and analyzing the places where flaws in the material of structures are initiated and propagated gradually during service under load, both under the conditions of space and also in different aggressive media (in active liquids and gases, at increased and high temperatures, hard radiation, accompanying vibration, etc.).

As the investigations and practices of conducting the inspection works showed, defects of the material begin to radiate elastic waves after reaching a critical value and further propagating. It is possible to detect and record these waves by special transducers, to decode the obtained information and to determine the coordinates of places of activity of the defects. There are no secret zones or inaccessible places for this method. Using this method it is possible to obtain the necessary information from the material, located within 5 m of an installed transducer.

This is, undoubtedly, one of main advantages of the method and made it possible to solve, for the first time in practice, the problem of short-time and continuous diagnostics at 100 % of the material which is subjected to deformation during service.

Special monitoring equipment, which is oriented to this procedure, receives and decodes the signals about the condition of the material. Moreover, long before fracture, when the defects have not yet reached a critical level, the system informs the operator about the coming hazard and allows the operator to take the necessary measures for accident prevention.

The method offered by us is illustrated in Figure 1 and is called acoustic emission (AE). An elastic wave, generated by a propagating defect of a crack type, moves along the plate surface and reaches the place where the AE transducer is set. Special equipment records and processes the signal received. The equipment of this class of EMA2 type is shown in Figure 2.

More than 10 years of tests at the enterprises of Ukraine, Russia and Poland showed that the method and equipment developed make it possible to realize both continuous and also periodic diagnostic control with a high validity of results.

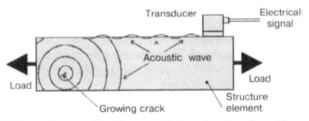

Figure 1 Initiation and propagation of waves of deformations in a plate with a growing crack.

* (1999) *Avtomaticheskaya Svarka*, No. 10, pp. 122–123.

Figure 2 Diagnostic system of the EMA2 type.

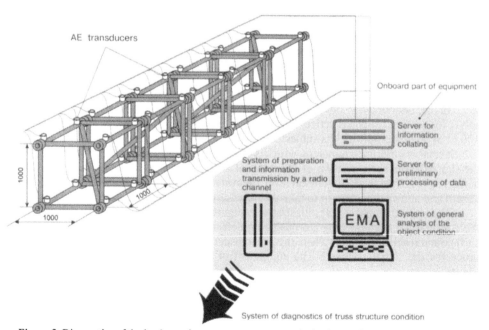

Figure 3 Diagnostics of the load-carrying space truss structure during its service using the AE method.

One of the possible variants of use of diagnostic equipment using AE for the analysis of the truss structure condition in space is shown in Figure 3. The AE transducers were placed in the connections of the truss elements. The operating structure is under continuous control by the system. The information, which is accumulated by the transducers, is processed and sent via the communication systems to Earth.

Peculiarities of Equipment for Technological Works in Space*

D.A. Dudko[1], A.A. Zagrebelny[1], V.E. Paton[1] and V.V. Stesin[1]

The generating new trend of science and engineering, space technology, requires the creation of different equipment for work with molten metal under conditions of space flight. Based on the experience of development of the «Vulkan» unit [1], this article describes briefly the specific problems occurring in designing similar equipment and the feasibility of their solution.

Technological equipment to be used on a space object, requires minimum dimensions and weight, dynamic strength, reliability, good serviceability in vacuum and with temperature variations. Therefore, this equipment must be considered, firstly, as a space instrument. To fulfil the above-mentioned general requirements it is necessary to use all the store of schematic and design solutions developed in a aircraft and space engineering using advanced materials and standardized units, meeting the requirements of service in space objects. Manufacture, inspection and testing of this equipment should be regulated by the technical specifications used at the enterprises which are manufacturing the space equipment.

At the same time, this technological equipment differs greatly from other space instruments by the higher consumed power and the presence of new factors (concentrated heat sources, molten metal, by-products of processes in the form of radiations, gases, metal spatters, etc.) which can present a potential hazard for the systems of the space object and its crew in case of a faulty technological process control.

From the above it may be possible to distinguish the following requirements which are specified for the technological equipment: safety of service in the space object; high reliability of functioning; feasibility of control of process parameters and equipment condition; compatibility with the spaceship systems and capabilities of its crew. The requirements are, undoubtedly, reflect the design of the technological equipment, the object systems and procedure of works. In turn, the systems of the space object are required to keep strictly the physical conditions, i.e. weightlessness and vacuum, providing the desirable result of the technological process.

Weightlessness (zero-gravity) is the main factor, predetermining the arrangement of the technological equipment at space objects. Therefore, when working with melts, they will require protection from disturbing forces. Coming from these considerations, preference in the technological equipment should be given to heat sources which cause the minimum mechanical effect on the object being heated (for example, heating with beam energy). It should be taken into account that the presence of a gradient of temperatures at the melt surface causes stirring under the action of a thermal capillary convection [2, 3].

To reduce to a minimum the disturbances caused by rotation and manoeuvring of the space object it is desirable to arrange the working zone of the technological equipment near the centre of mass of the object and to not switch on the engines during performance of the technological process.

Weightlessness is known to be manifested in a free drift of subjects in space. Probably it would be necessary to limit their drift during crucible-free crystallization of melts, for example, by using low-intensity electric and magnetic fields.

It is clear, that it is necessary to provide protection of mechanisms from the entry of foreign subjects, means of installation of interchangeable elements of the structure, as well

---

* (1977) *Space materials science and technology*, Moscow, Nauka, pp. 12–16.

as accessories such as handrails, handles, etc., which make the maintenance of the equipment in weightlessness easier.

Vacuum as a factor of the technological process can be characterized by three parameters, namely pressure, composition of residual atmosphere and evacuation rate.

Unfortunately, the limiting values of pressure in the working zone will always be higher than the «pure» space vacuum at the flight altitude due to the presence of a gas cloud around the object.

A proper selection of materials for components of the working zone on the basis of known requirements of vacuum hygiene and with allowance for their heating during operation will contribute to a decrease in contamination of the residual atmosphere. Hatches, through which the evacuation is realized, should not be located in the region of the coating jets of the engine units or other sources of contamination.

The rate of evacuation of gases from the working zone is determined by the capacity of the vacuum line. To maintain the limiting low pressure in the working zone it is necessary to design vacuum lines of maximum section and minimum length such that their capacity will exceed the total entry of gases into the working zone as a result of gas evolution from the melts and from heated elements of the structure, inleakage through the packing and other sources (for example, feeding of a plasma gas).

Safety. Potential sources of hazard in the technological processes are as follows: destroying capability of the heat source (electron beam, arc discharge, exothermic mixtures, etc.); pool and spatters of molten metal; increased voltage of power sources; secondary phenomena (heat and X-ray radiation, power-line interferences, radio noises, etc.).

Safety assurance starts from the selection of such parameters of the technological processes at which a minimum level of hazardous factors during the normal functioning of the equipment can be reached. Thus an the accelerating voltage of not less than 15 kV was selected in «Vulkan» unit for EBW at which there is no hard braking radiation. The proper selection of the arc welding conditions made it possible to eliminate the formation of metal spatters.

The principle of localizing the source of the potential hazard is very effective. The design of the EBW unit can serve as an example of realization of this technological solution [4], where all the high-voltage elements and circuits are joined and isolated by potting with an epoxy compound. A housing enclosing the welding devices and samples served as a means of localizing the metal dust, thermal and light radiation in «Vulkan».

The branching system of electrical and mechanical protection should prevent the exceeding of safe conditions of the equipment operation.

Reliability. The peculiarity of technological processes consists in the reliability of their fulfilment due in general by to two components: the reliability of realization of the physical processes on which the technology is based, and the reliability of the equipment functioning.

The reliability of realization of the physical process depends on its sensitivity to the variations in environmental conditions and the stability of the equipment operation. The role of the developer here is to evaluate properly the possible range of these variations, to select optimum parameters for a stable fulfilment of the process and to predict the measures for compensation of undesirable variations of the conditions.

The reliability and stability of the equipment functioning are provided by conventional procedures known from the theory of reliability and used widely in radio electronics, aircraft and rocket construction.

The independent supply to the working parts and control circuits from different sources promotes, in particular, reliable and failure-free operation of the technological equipment.

Control of the process parameters and condition of equipment helps with two important functions: receiving by the operator of express information about the normal proceeding of the technological process in the form of light or sound signals; recording and transfer to Earth of the quantitative values of process parameters through the installed telemetry system.

In «Vulkan» the light panels on the control panel, which are supplied with a voltage proportional to the welding current value, served as a means of express-information.

The concept of **compatibility** of the technological equipment with a space object means: arrangement of the technological equipment in the zone providing the best physical conditions for the technological processes; absence of mutual interference in functioning of the technological equipment and other systems of the object; use, where possible, of stationary equipment of the object; convenience in maintenance and conformity of the appearance to the object interior.

The absence of mutual interference is attained by a proper compiling of the cyclogram of the space object operation, autonomous energy supply, and the presence of devices for suppression of interference in the set of technological equipment.

The supply voltage of the technological equipment should correspond to the voltage of the object onboard electrical mains. Signals from the system of measuring technological parameters should enter in a form suitable for transfer via the space objects system of telemetry.

The available onboard TV, cinema and photographic equipment should be used to the maximum. It is desirable to avoid the use of recording systems requiring special training of the crew.

The design of the control units and panels should correspond to the anthropometric characteristics of the crew and peculiarities of the protective cloth, if any. The working places should be equipped with devices to attach and the fix the operator and accessories facilitating as manipulations in weightlessness.

Check-outs and tests. Success in fulfilment of works in space is guaranteed in many aspects by comprehensive on-land tests of the equipment. The tests are made usually in several stages: factory, development and prelaunch.

At the stage of the development tests the functioning of the equipment is checked under conditions which are close to real conditions existing in flight, and conformity of its design to the capabilities of a man in weightlessness is also evaluated. These works are complicated because it is necessary to provide integrated or a partial simulation of the service conditions. For this purpose, expensive test equipment is used, for example, the flying laboratory, large-sized space simulation chamber, etc. Sometimes, it is necessary to create also special self-contained stands [5].

The programs of development tests of the technological equipment should include the study of problems of the safety of the crew and of the object under unforeseen situations. During safety tests it is necessary to record the behaviour of equipment and to define the effectiveness of protection against at artificially-induced interferences, caused by failures of separate units, errors of operators and so on.

A mandarory feature of prelaunch tests is an integrated check-out of operation of the technological equipment together with other systems of the object, functioning simultaneously in a real flight to study their mutual effect.

1. Paton, B.E. and Kubasov, V.N. (1970) *See pp. 154–160 in this Book.*

2. Povitsky, A.S. and Lyubin, D.Ya. (1972) *Fundamentals of dynamics and heat- and mass exchange of liquids and gases in weightlessness,* Moscow, Mashinostroyeniye.

3. Babsky, V.G., Sklovskaya, I.M. and Sklovsky, Yu.B. (1973) Thermal capillary convection under weightlessness conditions, In: *Space research in Ukraine,* Issue 1, pp. 63–68.

4. Paton, B.E., Paton, V.E., Dudko, D.A. et al. (1973) *See pp. 118–122 in this Book.*

5. Paton, B.E., Dudko, D.A., Bernadsky, V.N. et al. (1975) *See pp. 137–140 in this Book.*

Experience of Design of Process Equipment for Performance of Work in Space Vehicles[*]

B.E. Paton[1], V.V. Stesin[1] and A.A. Zagrebelny[1]

Space technology as an independent branch of science is almost three decades old. Many prototypes of equipment have been built in different industrialised countries during this period for realisation of technological processes on board space vehicles, intended mostly for in-depth materials research, investigation of the occurrence of physical processes in zero gravity, as well as optimisation and application of tools and techniques for manual operations on assembly and repair of space vehicles and their systems.

The unprecedented capabilities for systematic research and practical work in this area on board the ISS, which is being constructed now, has generated much interest from scientists and engineers of different countries, who had never worked in this field before. Experience accumulated during the previous years, both successes and failures greatly facilitate the achievemnt of the final transformation of space into a permanent research and development base for obtaining a unique technological product.

The equipment for space technology built and tested so far, can be conditionally subdivided into two categories: for materials science and metallurgy research and for repair-assembly operations. These two types of equipment are intended for handling different problems and for operation mostly under different conditions. Accordingly, this requires highlighting different key points in their design and ground experimental optimisation.

Equipment of both categories share the same requirements for space instruments and devices, such as minimum dimensions and weight, dynamic strength in transportation, reliability and safety, high performance in vacuum and at temperature fluctuations, compatibility with systems of a vehicle, and high informatiom potential [1–4].

Service and research equipment, as well as devices installed on manned and unmanned vehicles certainly have differences in their requirements. These differences concern mostly particular matters, whereas basic requirements for the entire space equipment are uniform. The CIS enterprises involved in space engineering developed regulatory documents with validated specifications for the equipment, procedure for ground optimisation of space equipment, and ranges and methods for pre-flight tests. Unfortunately, the experience accumulated is mainly confined to in-house documents, and has just scanty or fragmentary coverage in technical literature.

Collaboration with US space centres on building the equipment for ISS presents a new stage in its design and ground optimisation. Sharing the experience, which is natural in collaboration, inevitably ends with working out of uniform approaches and standards, which was the case, in particular, in the creation of a unified docking module under the Joint Soviet-American ASEF Programme by NASA and RSC «Energiya»[2] [5]. The «Universal» hardware for manual EBW built at the PWI passed the round-robin tests at different US space centres following the procedures accepted by NASA.

Environment. Developers of space equipment are very interested in purely space environments, where technological processes are to be performed, as they are the key points for technology developers and designers.

Vacuum has a different effect on occurrence of the technological processes and performance of the equipment. There may be cases when vacuum exerts no effect, as the process investigated occurs inside a vehicle in a controlled atmosphere, e.g. sealed ampoule. Sometimes vacuum serves as an auxiliary environment (e.g. in heating of the ampoule with

* (1999) *Avtomaticheskaya Svarka*, No. 10, pp. 59–65. (Published in shortened version.)

a material inside it in a vacuum resistance furnace) or a parameter of the process (e.g. in deposition of a coating by thermal evaporation).

Vacuum as a factor of the technological process is characterised by three parameters: pressure, composition of a residual atmosphere and evacuation rate.

Normally, pressure on the surface of a space vehicle at an altitude of flight is higher than the residual atmospheric pressure because of the vehicle's own atmosphere formed due to leakage from pressurised modules, exhaust products of orientation engines and sublimation of structural materials [6].

However, after initial degassing of the vehicle and with idle orientation of the engines, the pressure is close to atmospheric ($1 \cdot 10^{-4}$–$1 \cdot 10^{-6}$ Pa) even at a distance of about 1 m from the surface of the vehicle. If special measures are taken, for example, if a screen oriented across the velocity vector of the vehicle is formed, pressure behind such a screen may amount to $1 \cdot 10^{-9}$–$1 \cdot 10^{-12}$ Pa. Therefore, conducting experiments in open space involves no fundamental problems associated with achievement of vacuum.

If a process unit is located inside the evacuated module of the spacecraft, pressure in the work zone is determined by the relationship between ingress of gases (leak-in plus gas release) and capacity of vacuum ducts. Under such conditions a high vacuum can hardly be produced.

For example, analysis of samples coated by thermal evaporation using the «Isparitel» hardware showed that residual pressure in the zone of location of the samples was in a range of $1 \cdot 10^{-1}$–$0.5 \cdot 10^{-2}$ Pa. Experiments were conducted in a docking module of the OS «Salyut», which had a diameter of about 800 mm and an outside hatch in the form of a diaphragm with a diameter of about 300 mm. The samples were located at a distance of about 250 mm from the exit section of the hatch. In the first American experiment on the automatic EBW the pressure in a chamber with a volume of $3 \cdot 10^7$ mm^3, equipped with a vacuum duct of about 100 mm in diameter and about 300 mm long, was also not higher than $1 \cdot 10^{-1}$ Pa, as the electron beam exhibited good functioning [7].

In experiments with vacuum resistance furnaces, where vacuum is a heat insulator and a shielding atmosphere for heaters, it is sufficient that pressure in the chamber is below 1 Pa.

Design calculations confirm that universal chambers for experiments, scheduled to be part of composition of specialised technological modules, should be equipped with vacuum ducts not less than 120–160 mm in diameter and not more than 500–1000 mm long.

With high-voltage equipment operating in such chambers it is necessary to carry out the break-down tests, allowing for the following dependence: $U_b = f(Pd)$ (Paschen's law), which has a minimum at Pd = 133.3÷1333 Pa·cm. In the general case this dependence has the following form: $U = f(\rho d)$ [8] (U_b is the break-down voltage, ρ is the gas density and d is the distance between electrodes), which is indicative that tests should be conducted at several temperature values.

Experiments on manual EBW involved no problems with the quality of vacuum as a process environment. Focusing of the electron beam and the gas content of the weld were indicative of a high vacuum in the welding zone, despite the «exhaust» and leak-in from the spacesuit.

Some technological processes are very critical from the standpoint of residual atmosphere. To a larger degree this is applicable to production of semiconductors and new materials with unique physical and mechanical properties. These problems are described in detail in [6, 19].

The rate of evacuation of evolved gases as one of the factors that characterise the vacuum situation is a technological process parameter. In particular, this makes successful realisation of plasma-arc processes in EVA doubtful. For example, in preparation of an experiment using the «Vulkan» unit the test welds were made in vacuum chambers with a volume of 0.1, 1 and 30 m^3. Each time a feed rate of the plasma gas had to be increased 2–3 times. Prior to the full-scale experiment the gas feed rate was additionally increased 2.5–3 times, but no ignition of the arc was achieved in space.

Also, it is well known that vacuum is a factor that decreases reliability of mechanical systems, which is caused by violation of the operation of parts in friction. As a rule, this problem is not very pressing for process equipment because of low forces and low travel speed, and because of the small number of work cycles. Nevertheless, the design approaches used in all cases are based on the maximum utilisation of experience accumulated by space engineering enterprises in the field of selection of materials for friction pairs, designs of bearing assemblies, special grease and lubricants [9].

Zero gravity (microgravity) is the main technological parameter on which the whole of space materials science is based.

Background values of acceleration of gravity in a non-manoeuvring space vehicle are $1 \cdot 10^{-4}$–$1 \cdot 10^{-6}$ g, depending upon the distance of a measurement location to the centre of mass. This level increases to $1 \cdot 10^{-2}$–$1 \cdot 10^{-3}$ g at actuation of the orientation engines. Another disturbing factor is vibration caused by operation of mechanisms, displacements and actions of the crew.

By the way, parameters of the external atmosphere of the OC «Mir», as well as vibration disturbances and micro accelerations inside its modules, have been regularly measured since 1989. According to the data of [10], residual pressure of the external atmosphere of the complex in the direct proximity of it near modules «Kvant» and «Priroda» is within the range of $5 \cdot 10^{-9}$–$1 \cdot 10^{-8}$ mm Hg. Depending upon the type of sophisticated dynamic operations (adjustment of orbit, turn, docking of transport vehicle, etc.), the background level of density of the residual atmosphere of the complex can be increased for a short time (for the period of performance of an operation) to $2 \cdot 10^{-5}$–$2 \cdot 10^{-7}$ mm Hg. It should be noted that owing to the infinite evacuation rate, which is natural for space vacuum, the external pressure at a distance of 0.2–0.4 m from the complex surface returns to values characterising open space ($\sim 10^{-7}$–10^{-11} mm Hg at an altitude of 250–500 km).

Vibration disturbances in the «Mir» are determined [11] by service conditions and specifications of flight airborne systems and facilities. The mean capacity of the vibration processes for typical service conditions is 3–15 *mg*. However, at certain moments it may be in excess of $\sim$100 *mg* (in the main frequency range of 20–75 Hz). Micro accelerations inside the «Mir» with respect to g_0 (gravity on Earth) are $\sim 10^{-5}$–10^{-7} g. However, this amount may be increased to $1 \cdot 10^{-3}$–$4 \cdot 10^{-4}$ g by controlling its orientation, maintaining constant position in the inertia coordinate system, turning, etc. [12].

At a macro level the effects caused by zero gravity show up at as low as 10^{-2} g and manifest themselves in a free drift of items in space, «no support» for a human, displacement of liquids under the effect of surface tension forces, etc.

General design rules include protection of equipment from ingress of foreign items, means of fixing replaceable members, presence of handles and handrails for an operator, etc.

It should be borne in mind in designing of repair-assembly equipment controlled and moved manually in EVA, that maintaining a melt in the crucible is achieved by capillary forces by forming a temperature gradient in the crucible or using capillary structures, e.g. grids [13]. The safety of an operator in a spacesuit and accuracy of his operations with work tools are ensured by a stationary or mobile work station fitted with a foot anchor, handrails, devices for location of process equipment, etc. [14]. Special consideration should be given to the attachment of temporary cable networks and flexible main ducts installed on the vehicle surface.

In materials science experiments, a low level of gravity opens up new possibilities, which have not yet been realised, for monitoring of the melt drift outside the crucible using low electric and magnetic fields (electromagnetic and electrostatic levitation devices) and controlling the shape of a single crystal being grown.

Experiments under short-time zero gravity conditions prove [15, 16] that microgravity also widens the possibilities of controlling the distribution of impurities, fractions and gas inclusions in a liquid using such known approaches as the vibration or electromagnetic effect.

It has been proved experimentally that the distribution of micro impurities in single crystals is greatly affected by force fields of super low intensity (below 10^{-2} g), which are generated as a result of operation of mechanisms of the space vehicle, activity of the crew and free rotation of the space vehicle about the centre of mass. To neutralise such effects, it is advisable to locate the process equipment near the centre of mass of the vehicle on vibration-protective platforms with a low (less than 1 Hz) frequency of natural oscillations.

It is likely that the minimum possible level of microgravity can be achieved by locating the process equipment on a free drifting platform connected to the space vehicle only through a flexible connection (or even without it).

The space technology experiments conducted in the 1970–1990s allowed an insight into the physics of metallurgical processes under microgravity conditions, development of general approaches to designing of process equipment and optimisation of designs of its main parts.

Effect of heat conditions and thermal regulation of process equipment are considered in several ways, e.g. from the standpoint of heating by solar radiation and cooling in the shadow of the equipment operating on the external surface of the space vehicle (from –70 to +120 °C under a pressure below $1.33 \cdot 10^{-2}$ Pa), and heating of the equipment by heat used for realisation of the technological process. High-temperature heaters used for space technology consume much more power (0.5–2.0 kW) than any other facility or device on board the space vehicle.

Equipment located on the external surface of the vehicle uses devices which are traditional for space engineering, such as heat-stabilising coatings and multilayer screen-vacuum insulation.

Cooling of airborne equipment subject to prolonged operation is provided by connecting it to the fluid thermal-regulation system of the spacecraft. A particularly difficult problem is the thermal regulation of small hand tools for repair-assembly operations, which are independently used in EVA and not connected to the thermal-regulation system of the spacecraft. For example, the following package of design and methodical approaches for limitation of heating was used for electron beam hand tools developed by the PWI (VHT, «Universal»):

• heat insulation of hot parts of the tool (crucible, cathode unit), which provides the maximum possible removal of heat by radiation at high temperatures (300–1500 °C);

• using the external surface of the tool body covered by a heat-stabilising enamel for removal of part of the heat by radiation at a moderate temperature (below 100 °C);

• limitation of duty cycle of the tool (DC ≥ 20 %), according to real physical capabilities of an operator working in a spacesuit;

• limitation of maximum power of the tool (to 1 kW) to a value actually required for repair of typical structural members of the space vehicle.

Functions and structure of process equipment. Composition and design of process equipment are determined primarily by a set of functions it has to perform and by service conditions.

Equipment which is intended for the realisation of metallurgical processes should be designed to realise technologies involving melting of a certain part of an ingot, maintaining it for a long time in the isothermal or gradient state and automated displacement of the molten zone at a speed of a few millimetres per hour. Based on considerations of safety and cleanness of an experiment, the process is normally performed in a closed volume (chamber, ampoule) inside the space vehicle. The process occurs without direct participation of an operator.

The technology to be performed using devices or tools for repair-assembly operations provides for local heating and melting of metal, as well as uniform displacement of the heat spot manually or using mechanised devices at a speed of several millimetres per second. The process may be accompanied by mechanised feeding of a consumable in the form of wire or strip. The work is performed on the external surface of the space vehicle with participation of an operator dressed in a spacesuit. Composition of the process equipment depends upon its functions.

Equipment intended for realisation of metallurgical processes usually comprises a heater (resistance, electron beam, light, etc.), fixtures and auxiliary devices (ampoule with a sample, sample holder, etc.), mechanical equipment (work chamber, device for movement of the heater or sample, device for replacement of samples, etc.).

However, process equipment cannot function without the flight systems of the spacecraft which supply power to it, remove heat generated, evacuate the work chamber, stabilise the spacecraft during an experiment and transmit telemetry information.

In some cases extra devices may be required, such as those which adapt the process equipment to the space vehicle. Such devices include, for example, a vibration-protection platform, as well as a secondary power supply, which is part of the process equipment control system and converts power fed from the spacecraft power system into a form required to feed the heater and control its capacity.

The control system also includes units which provide programming of the work cycle, realisation of the programme, monitoring of the technological process and collection of information to transmit it via telemetry channels of the vehicle.

There is no doubt that the space vehicle and the process equipment located in it should be considered as one system. Such systems called technological modules (e.g. module «Kristall» on board the OC «Mir») have been applied in modern cosmonautics.

Therefore, during a period of 30 years we have passed a long route from off-line equipment for pioneering experiments that proved the feasibility of the technological processes in zero gravity, to specialised modules used as part of the composition of long-time orbital stations, allowing not only comprehensive research to be conducted, but also semi-commercial production of materials with special characteristics to be arranged.

The generalised structure of the technological system is given in Table 1.

In our opinion and according to our experience, ground-based hardware consisting of the process equipment and a rig used to promptly perform a control experiment, which makes analysis of the results obtained in space much easier, should also be considered part of such a technological system.

Equipment for repair-assembly operations in definition should not be «fixed» to a specific point on the surface of the space vehicle, while in emergency cases it may be necessary that it operates in the off-line mode. Therefore, such equipment is not included into the technological modules, although in fact it is also part of the entire system of the space vehicle.

The peculiarity of the technological complex comprising equipment for repair-assembly operations is the active participation of an operator working in a spacesuit. The operator is an integral part of the complex. This gives grounds to consider such complexes as a variety of the system «man–machine». The problem of adaptation of a man in a spacesuit to the spacecraft is an important area of development.

Adaptation devices, facilitating the work performed by the operator in a spacesuit, are permanent and mobile work stations equipped with anchoring platforms. Mobile work stations can be moved using flight manipulators, which are part of equipment of the spacecraft, as it is the case of the space shuttle.

Table 1 Structure of process hardware.

Hardware composition, application of equipment	Process equipment					Spacecraft flight systems	Earth-based equipment
	Work unit			Control system	Adaptation devices		
	Heater	Auxiliaries and fixture	Mechanical equipment				
Metallurgical processes of producing alloys and growing single crystals	Electric resistance furnace. Electron beam gun. Artificial light source. Laser light source	Ampoule for sample. Sample or ampoule holder	Pressurised work chamber. Heater or sample displacement device. Sample replacement device	Secondary power supply (power converter). Programming control unit. Process parameter sensors. Telemetry unit. Programming control panel	Vibration-protection platform. Vacuum screen. Free-floating platform	Power supply system. Thermal-regulation system. Telemetry system. Stabilisation system. Work chamber evacuation system	Process equipment backup. Control experiment rig. Training mock-up
Repair and assembly of metal structures using welding, cutting, brazing and coating by evaporation in vacuum	Electron beam gun	Focusing device. Crucible for evaporated material. Deflection device	Device for fixing equipment and cables. Filler metal feeding device. Device for mechanised displacement of heater	Secondary power supply (power converter). Control unit. Telemetry unit. Operator's panel. Control panel	Anchoring platform. Mobile work place	Electric power system. Telemetry system. Cosmonaut suit. Airborne manipulator. Extra electric connections at different points on the vehicle surface	Process equipment backup. Process equipment zero buoyancy mock-up. Training mock-up

An example of an integrated approach to optimisation of the process equipment and its adaptation to capabilities of a man in a spacesuit is the layout of equipment for the planned space experiment «Flagman», which comprises a hand tool for electron beam welding and cutting of metals, mobile work station of an operator and a bench in the form of a drum containing replaceable cassettes with samples to be treated.

In commercial application of EBW the high power density achieved in a small (diameter 0.5–1.0 mm) heat spot at an accelerating voltage of about 30 kV and more requires precise (0.1–0.3 mm) guiding of the electron beam to the joint to be welded and maintaining constant welding speed, as any interruption or slowing down leads to burn through. Experiments conducted using a photo-optical electron beam simulator showed that a man in a spacesuit cannot perform manual operations at the above precision level and cannot ensure the required smoothness of the tool movement. At the same time, welding experiments conducted on a rig with a fragment of the spacesuit proved that with a decrease of about an order of

Table 2 Arrangement and performance of a technological experiment in space.

No.	Develop-ment stage	Performer	Content of work	Materials (hardware) support	Final document	Basis for work continuation
1	Research work	Research laboratory – developer	Development of base line procedure, iden-tification of experi-mental conditions and process equipment parameters	Laboratory mock-up of equipment	Scientific substantiation and work statement for the expe-riment. Technical proposal for process equipment	Results of competition of scientific de-velopments
2	Optimisation of experi-mental design	Designers of spacecraft with partici-pation of equipment developers	Evaluation of capabi-lities of vehicles in ensuring experimental conditions. Identification of limi-tations on design of equipment		Draft design of space vehicle. Source data for develop-ment of work statement	Approval of draft design of space vehicle.
3	Development work. Draft designing of process equipment	Design Bureau of equipment developer	Development of work statement. Base line designs of equipment and control system. Selection of compo-nents. Tentative prog-ramme and procedure of the experiment. Identification of information flows. Identification of cargo flows. Programme for ensuring reliability. Integrated program of experimental optimi-sation. Range of mock-ups and auxiliary equip-ment. Requirements to the crew. Work sche-dule. Estimation of funding	Laboratory mock-up of equipment (with modi-fications)	Draft design of process equipment. Draft work statement. Work schedule	Coordination of draft design of equipment and work schedule with vehicle de-veloper. Opening of funding
4	Development work. Detail designing of process equipment	Same	Study of regulatory documents of the space vehicle developer. Designing of mecha-nical units of equip-ment. Development and design of control system. Making mock-ups of mechanical units and control system elements	Laboratory mock-ups of mechanisms and control system elements	Working do cuments for process equipment and control sys-tems (letter O)	Coordination of mechanical and electric fixation to space vehicle, coordination of design of critical assemblies with space vehicle developer

Table 2 (cont.)

No.	Development stage	Performer	Content of work	Materials (hardware) support	Final document	Basis for work continuation
5	Development work. Detail designing of mock-ups and auxiliary equipment	»	Designing of mock-ups required for tests, special devices and tools required for ground service. Designing of tare and packing. Designing of special rigs		Working documents for auxiliary equipment	Coordination of documents with test services of developer of space vehicle
6	Manufacturing of pilot set of process equipment	Manufacturer and Design Bureau of equipment developer	Manufacture of parts, assembly, factory quality inspection. Author's supervision of manufacture, current adjustment of drawings. Preliminary (factory) tests	Commercial equipment of manufacturer	Protocols of preliminary (factory) tests	Bilateral certificate of acceptance of pilot set of equipment by developer
7	Manufacture of mock-ups of auxiliary equipment	Same	Manufacture, assembly, factory quality inspection. Author's supervision of manufacture		Protocol of factory tests	Bilateral certificate of acceptance of mock-up of equipment by developer
8	Development of test programmes and procedures	Design Bureau of equipment developer	Development of procedure of each type of tests provided for by integrated programme of experimental optimisation. Refining of programmes of laboratory-optimisation (LOT), design-optimisation (DOT) and special tests		LOT program and procedure. DOT program and procedure. Special tests program and procedure	Coordination of DOT and special tests programs and procedures with space vehicle developers
9	LOT	Developer's laboratory, standard tests laboratory	Testing of pilot set of process equipment functioning under flight conditions, compliance with requirements of work statement for resistance to flight factors effects	Pilot set of equipment, versatile rigs for standard tests, testing devices	Test protocols and conclusions of testers. LOT report	Approval of LOT protocol by customer's representative
10	DOT	Developer's laboratory, test service of vehicle developer	Testing of pilot set of process equipment functioning under extreme values of flight factors, testing to reliability in emergency situations, service life tests	Pilot set of equipment. Versatile and special rigs	Test protocols and conclusions of testers. DOT report. List of remarks	Approval of DOT protocol by customer's representative. Approval of work statement

Table 2 (cont.)

No.	Development stage	Performer	Content of work	Materials (hardware) support	Final document	Basis for work continuation
11	Updating of design documents	Design Bureau of equipment developer	Remedial work on concerns and drawbacks revealed during LOT and DOT. Optimisation of design in compliance with amendments made in work statement		Updated set of documents with letter 01	
12	Manufacture of flight sets of process equipment	Manufacturer and Design Bureau of equipment developer	Manufacture of parts, assembly, factory quality inspection and setting up with participation of developer. Acceptance by the customer's representative according to the preliminary (factory) tests. Optimisation of pilot and mock-ups by the LOT and DOT results	Commercial equipment. Manufacture	Protocols of preliminary (factory) tests signed by customer's representative	Certificate of acceptance by customer's representative
13	Special tests according to integrated programme of experimental optimisation	Test service of developer together with equipment developer	Trial of layout of equipment on board the vehicle, tests to functioning with power supply, thermal-regulation and control systems of vehicle, checking of electromagnetic compatibility of the vehicle, checking performance of flight operations on equipment maintenance, checking func-tioning of equipment in short-time zero gravity, special safety tests (if necessary)	Pilot set of equipment. Set of mock-ups: dimension-weight (DW), training (TM), electric, thermal mock-up for neutral buoyancy tests (NB), set of service documents	Protocols of tests with testers' conclusions approved by customer's representative	Positive conclusion of customer's representative

Table 2 (cont.)

No.	Develop-ment stage	Performer	Content of work	Materials (hardware) support	Final document	Basis for work continuation
14	Teaching and training of the crew	Cosmonauts Training Centre together with test service of developer of the vehic-le and equip-ment deve-loper	Acquainting the crews with scientific content of purpose of experi-ment, design of equipment, experimental procedure and work program. Training in practical habits for preparation and conducting of experiments and maintenance of equipment. Optimisation of actions of the crew in emergency situations. Safety requirements. Work timing	Pilot set of equipment, set of mock-ups. Log docu-ments	Training reports. Remarks of the crew	Positive results of tests (exams) of the crew
15	Acceptance tests	Input inspection service of vehicle developer	Verification of comp-leteness of process equipment, service and accompanying docu-ments; checking the presence of positive conclusions of custo-mer's representative. Checking the presence of seals and absence of external damage in equipment. Vibration tests to check fixation elements and absence of resonance at low frequencies. Checking serviceability of electric circuits and resistance of insulation. Checking the fact of functioning in the rated mode. Checking tare and packing	Flight sets of process equip-ment. Rigs or auxiliary devices (if necessary)	Input inspection protocol	Certificate of acceptance of equipment
16[*]	Comprehensi ve tests of equipment as part of the vehicle of its elec-tric mock-up	Test service of vehicle developer (prior to flight) or crew (during flight)	Checking mating of electric circuits, che-cking the fact of func-tioning of equipment, checking functioning of information channels	Flight set of equipment installed on board the vehicle	Logging, telemetry signals	Positive results of checking

Table 2 (cont.)

No.	Development stage	Performer	Content of work	Materials (hardware) support	Final document	Basis for work continuation
17	Ground-based supervision	Laboratory – equipment developer	Direct communication from Flight Control Centre with the crew performing the experiment	Flight set of equipment installed on board the vehicle. Synchronous performance of experiment in laboratory. Synchronous control of experiment via telemetry channels	Reports of the crew and logging on board the vehicle, pilot set of equipment in laboratory	Failure-free performance of equipment

_* After assembly of equipment on the vehicle, prior to or during the flight.

magnitude in the power density (heat spot diameter 2–3 mm) a man not only can make a continuous weld, but can also control the beam focusing and changing the distance from the tool to the sample. Such «poor» focusing, according to the earth standards, can be provided by the simplest electron beam gun of the diode type with a straight-channel cathode and electrostatic focusing, operating in this case at low accelerating voltage (up to 10 kV). This made it possible to build a reliable and compact tool of the simplest design, avoiding a hard X-ray radiation and requiring no sophisticated control system.

Long-time service and functioning of the process equipment on board a space vehicle lead to the necessity to make preliminary provisions for its upgrading. The problem of upgrading and ensuring repairability can be solved through using a modular design. For example, the materials research unit «Karat» was initially completed with heating elements (furnaces) of six types, and an allowance is made for using other heating devices and expanding the control system on the base of the native micro computer. The repair-assembly hardware «Universal» [17, 18] is completed with four types of work tools — for heating, welding and cutting using a filler, and for coating by evaporation of metal from crucibles and evaporation of a filler metal. Each tool comprises an electron beam heater and required fixture. Available also are the developments of specialised tools for position butt welding of pipes, deposition of patches about a contour by welding, etc.

Therefore, it is a long and difficult route from scientific substantiation to realisation of a technological process on board a spacecraft. The procedure for arrangement and performance of a space experiment is based on regulatory requirements in force in space engineering (Table 2), although some stages in their content are not characteristic of the system of space vehicles and result from specific features of unique process equipment, which makes it more expensive. Nevertheless, as shown by experience, «jumping» over the stages leads to a dramatic increase in the probability of disruption of the research programme and may involve dangerous consequences for the spacecraft and its crew.

Conclusions

1. True values of factors of space flight and specifications for process equipment, allowing for their possible effect, are defined.

2. The need has been realised for and experience has been accumulated in building on board a spacecraft of complexes comprising special process equipment, spacecraft flight systems, a cosmonaut–operator as part of the system «man–machine», as well as auxiliary devices to adapt the equipment to the spacecraft and capabilities of cosmonauts.

3. The demand has been realised for and experience has been gained in using replaceable final control elements and special fixtures for widening of potentialities of the process equipment.

4. A procedure for testing the process equipment, which guarantees reliability and safety of the equipment, allowing for the specific character of its application, has been developed and checked.

1. Andreyanov, V.V., Artamonov, V.V., Atamanov, I.G. et al. (1973) *Automatic planetary stations*, Moscow, Nauka, 280 pp.

2. Paton, B.E. (1977) *See pp. 4–7 in this Book*.

3. Dudko, D.A., Zagrebelny, A.A., Paton, V.E. et al. (1977) *See pp. 85–87 in this Book*.

4. Paton, B.E., Dudko, D.A., Bernadsky, V.N. et al. (1976) *See pp. 75–77 in this Book*.

5. Syromyatnikov, V.S. (1984) *Docking devices of spacecraft*, Moscow, Mashinos-troyeniye, 216 pp.

6. Akishin, A.I., Dunaev, N.M. and Konstantinova, V.V. (1977) Natural atmosphere of spacecraft and its effect on airborne instruments and space technology, In: *Space materials science and technology*, Moscow, Nauka, pp. 65–78.

7. Feret, J.M. and Mazelsky, R. (1973) Skylab furnace system provides precise thermal environment for materials experiments, *Westinghouse Engineer*, No. 11, pp. 174–179.

8. Mick, J. and Crags, J. (1960) *Electric tests in gases*, Moscow, Inostr. Literatura, 215 pp.

9. (1982) *Planet rovers*. Ed. by A.L. Kemurdzhian, Moscow, Mashinostroyeniye, 319 pp.

10. Svetchkin, V.N., Nikitsky, V.P. and Neznamova, L.O. (1997) Spacecraft atmospheric environment as one of the space factors affecting operation, life and reliability of systems, In: *Proc. of 7th Int. Symp. on Space Environment*, Toulouse, 1997. Noordwijk, pp. 75–77.

11. Ryabukha, S.B. and Kiselyov, S.V. (2000) Investigation of vibration disturbances on board the orbital complex «Mir», In: *Abstr. of pap. of 70th Russian Symp. on Mechanics of Zero Gravity, Results and Prospects of Gravity-Sensitive Systems*, Moscow, 2000. Moscow, pp. 7–8.

12. Sazonov, V.V., Ermakov, M.K. and Ivanov, A.I. (1998) Measurement of micro accelera-tions on the orbital station «Mir» during the ALICE hardware experiments, *Kosmich. Issledovan.*, **36**, No. 2, pp. 156–166.

13. Zagrebelny, A.A., Lapchinsky, V.F., Stesin, V.V. et al. *Method and device for melting materials under zero gravity conditions*. USSR author's cert. 816070, Int. Cl. B 64 G 1/00, priority 10.01.80, publ. 21.11.80.

14. Paton, B.E., Kryukov, V.A., Gavrish, S.S. et al. *Astronauts work station device*. Pat. 5.779.002, USA, priority 14.02.97, publ. 14.07.98.

15. Ganiev, R.F., Lapchinsky, V.F. and Okhotin, A.S. (1980) Using monitored external effect to control of technological processes under microgravity conditions, In: *Space technol-ogy*. Ed. by L. Steg, Moscow, Mir, pp. 64–72.

16. Popov, V.I. (1986) Electrodynamic control of mass transfer in heterogeneous systems under microgravity conditions, In: *Problems of space technology of metals*, Kiev, PWI, pp. 107–116.

17. Paton, B.E. and Lapchinsky, V.F. (1998) *Welding and related technologies in space*, Kiev, Naukova Dumka, 184 pp.

18. Paton, B.E., Lapchinsky, V.F., Stesin, V.V. et al. *Device for manual electron beam processing of materials in space.* Pat. 5.869.801, USA, priority 14.02.96, publ. 09.02.99.

19. Berzhaty, V.I., Zvorykin, L.L., Ivanov, A.I. et al. (1999) Prospects for realisation of vacuum technologies under orbital flight conditions, *Avtomatich. Svarka*, No. 10, pp. 108–117.

Hardware for Control
of Onboard Technological Equipment*

P.P. Rusinov[1], V.V. Demianenko[1], T.E. Palamarchuk[1], I.O. Shimanovsky[1], V.V. Tochin[9] and B.I. Yushchenko[9]

The continuous progress in science and engineering, including cosmonautics, has led inevitably to the problem of use of the unique peculiarities of space, i.e. combination of weightlessness and high vacuum.

Research and development problems which have been solved by the specialists of the PWI, are described below [1].

The initial search for technological solutions with experimental testing in on-land conditions has been finalised by a series of experiments on board the space objects using units «Vulkan» and «Isparitel». To evaluate the feasibility in principle of operation with an electron beam in space, it was sufficient to use equipment with maximum simplified remote control and control on relay elements. However for the new technologies, it was necessary to create more sophisticated equipment with sufficient information capabilities which were convenient, reliable and safe for the operator-cosmonaut. The developers of research-technological equipment faced the following task: on the one hand, it was necessary to increase the specific science intensity of the equipment being developed taking into account the cost of delivery of an efficient cargo to orbit; and on the other hand, the operator should not be overloaded with unknown operations. This could be realized by using new design solutions in combination with maximum automation of technological processes on the more sophisticated element base for the information-control part of the equipment. These were the peculiarities which were taken consideration in the design of the technological unit «Isparitel». A group of specialists, having experience of creation of onboard military-purpose hardware for operation in severe conditions, with high requirements for reliability, energy consumption and mass characteristics, greatly helped the designers of the PWI.

The multiple functions of the unit «Isparitel» were realized by an information-control system made on integral microcircuits of CMOS technology of series 564 (CMOS are the complementary elements, i.e. additional transistors with a metal–oxide–semiconductor structure). By this time, the CMOS technology was recognized around the world [2, 3]. It possesses a high noise-resistance, microconsumption of energy (i.e. soft heat conditions), a wide range of operating temperatures ($-60 - +125$ °C), supply voltage from 3 to 15 V, sufficiently high reliability, in particular when the elements with a special acceptance for the onboard hardware are used.

The control system of the equipment provided up to eight variants of programs of the technological experiments with the feasibility to preset different operation conditions, and the information system provided control and imaging of up to fifteen parameters.

* (1999) *Avtomaticheskaya Svarka*, No. 10, pp. 66–73.

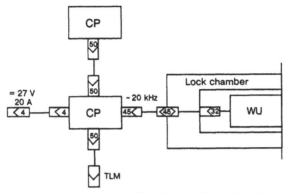

Figure 1 General scheme of unit «Isparitel-M» with indication of a number of contacts in the connector.

As to the design, the equipment consisted of a control panel (CP), a working unit (WU), a programming panel (PP), changeable constant memory devices (CMD) (Figure 1).

CP included a power transistor inverter (PTI), converting the direct current voltage of 27 V of the board mains to the voltage of alternating current of square shape of 20 kHz frequency, and a unit of control and telemetry (TLM), whose functional circuit is given in Figure 2. The output voltage of PTI was supplied to the working unit on filament transform-ers.and to the high-voltage unit for the supply of the electron beam guns of the equipment.

PP was used to preset the parameters of the experiment program in accordance with the technological task. The program contained binary codes of sequence of parameters, which are preset by a set of key switches. Several standard programs representing in-cut connectors of RS50 type, were sealed-off to CMD. One such connector could contain two programs of different kinds of works. Separate CMD were designed for test check-outs. The check-out of performance of executive mechanisms was envisaged in the conditions of atmosphere of TEST-1, i.e. without switching on of PTI and vacuum control. The check-out of TEST-2 was performed using a shorter program in conditions of controllable vacuum with a

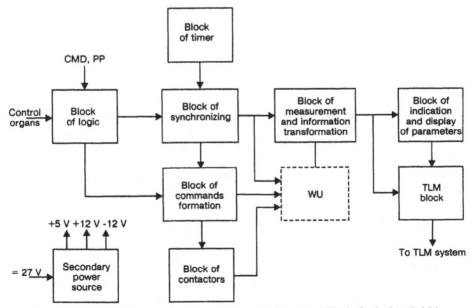

Figure 2 Functional diagram of a block of control, monitoring and TLM of unit «Isparitel-M».

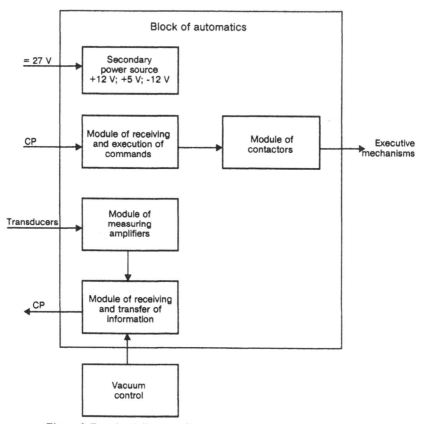

Figure 3 Functional diagram of block of automatics of unit «Isparitel-M».

switching on of PTI at low conditions and this provided control of the performance of the whole equipment.

The working unit consisted of a two-gun high-voltage block, three changeable working heads, a block of vacuum and a block of automatics.

The block of automatics (Figure 3) provided receiving and execution of commands, preliminary processing and transfer of information. The equipment operated automatically after the presetting of the parameters of the program, control of vacuum in the zone of the working unit and giving the command «Start» by a key at the front panel of the CP.

Figure 4 shows CP and PP of the unit. On the front panel of the CP are located: control organs, light-diode and digital indicators, which make it possible to control the PTI parameters, temperature at five points of the unit, vacuum (with a blocking of command «Start» at insufficient level of vacuum), operation of a tape-drawing mechanism (or the position of the positioner with samples treated) and time parameters of the program. Moreover, three light-diode indicators showed the operation of shields from overheating and overloading in power and control circuits with an automatic interruption of the process. Here, the operator could reset the «emergency» indication by a key «Stop» and then switch off the panel and primary supply. The digital indicator showed preferably the effective values of the anode current.

The primary supply voltage at the PTI input, anode current (electron beam current). output voltage of the anode inverter, filament current, filament voltage (output voltage of the filament inverter) and the voltage of the anode current control were displayed according

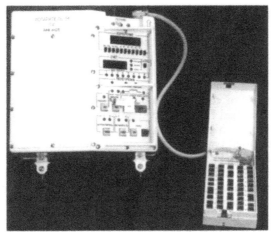

Figure 4 Control panel with a programming panel of unit «Isparitel-M».

to the operator's request by pressing the key switches located on the panel below the indicator. The information was renewed at a periodicity of 4 s.

The functional diagram of the block of transformation and imaging of the information is given in Figure 5.

A serious problem was the noise-resistance of the control system due to the close vicinity of power unit to the powerful source of noises, PTI, operating at 20 kHz frequency of the square-shaped alternating current, forming a wide spectrum of radiation. Moreover, the communication of CP with a technological working unit located outboard, was realized by a cable from the space station, but not of the unit. This cable transported both the power circuits of 20 kHz frequency and also all the circuits of the information control.

The limited amount of lines of communication stipulated coding of fifteen commands from CP into a working unit and alternate synchronizing of information being transferred (six analogue and five discrete parameters) in opposite direction.

The principal drawback in the development of the equipment was the separation of tasks for the PTI development and high-voltage block between different groups of specialists. PTI was designed and manufactured by the IED[12], while a high-voltage block was designed by one of the departments of the PWI. Each designer was qualified to fulfil their own part, but, as the single power system was created, then even negligible mismatching in all the working range of parameters led to significant noise formation. Practice showed that the single power system for the electron beam technologies should be designed by one group of specialists, which is also confirmed by the rule of prevention of noise in the zone of their formation.

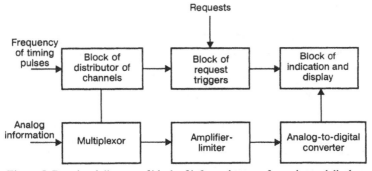

Figure 5 Functional diagram of block of information transformation and display.

Apart from the problem of noise prevention, the «control operators» had to solve two main tasks. The first task consisted in minimizing of schematic solutions without loss of reliability at the conditions of providing all preset functions. The second task concerned the arrangement of all the electron equipment in the dimensions of the earlier unit «Isparitel», following the requirements of GOSTs and OSTs and Technical Specifications for the application of completing elements.

A modular principle of equipment designing satisfies at the maximum most of the requirements. With profound functional design it is also possible to provide satisfactory dimensional characteristics, high maintainability and convenience in service. In this case, heat conditions, protection from noise and minimum intermodular and interblock electrical connections should be taken into consideration.

Requirements for the front panel of CP, i.e. the volume and nature of information presented, type, quantity and arrangement of control devices, are usually preset and approved by the customer. Some control devices, which can cause process violation due to an accidental effect during program fulfilment, should have an automatic or mechanical blocking which can be taken away at any time. It should be noted that application of a more sophisticated element base, the microelectron base in particular, will lead inevitably to increase in requirements both for the developers and for the production management.

After a large number of appropriate tests and pre-flight check-outs, the unit «Isparitel-M» was delivered onboard the OS «Salyut-7». Different experiments, in spite of single failures in operation of hardware, confirmed the validity of the chosen technological solution.

The noise-suppressing diode-choke device (current limiter) at the output of PTI in the anodic circuit was used for the next similar unit «Yantar». In addition, protective inductive-capacitor LC-filters were mounted in CP at the outputs of the analogue information entering from PTI. This allowed the level of noise to be somewhat decreased. But as the communication between the CP and working unit was realized again by a non-screened cable from the space station, no radical improvement was achieved.

As the experiments showed, the tests and operation of the hardware onboard the station, circuits of the primary supply of the power part and control equipment should be separated, i.e, supplied from different feeders, which is, unfortunately, not always possible. Therefore, the LC-filters should be installed obligatorily at the inputs of the primary supply into each part of the equipment. Moreover, the role of the inductance will be played, here, by ferritic circular cores, put on the supply wires in an appropriate way.

This method of prevention of noises in the primary supply circuits showed itself efficient in the hardware with an electron beam versatile welding tool (VHT and then «Universal»). However, as the similar power system (PTI–high-voltage block with filament transformers–electron beam guns) was used here, then the same problems with noise formation and protection from them remained. The problems of control were somewhat simpler, but the logic of protective blockings from irregular situations had to provide high reliability, as this is connected with the safety of the cosmonaut-operator and station itself.

The equipment for the manual EBW consisted of a sealed outboard instrument compartment with a nest for the tool and the tool itself, which represented a high-voltage block with electron beam guns and was connected to the instrument compartment by a cable.

The sealed container included PTI and a block of automatic control, and a small-sized CP was located at its external surface (Figure 6, *below*).

The controllers on the panel provided switching on of the primary supply (with visual indication) and setting of eight levels of beam power with a combination of switching on of three buttons (with a digital light-diode indication of the number of conditions).

The key «Stop» switched off the primary supply at any moment. The tool had a three-position switch of selection of one of two guns and a pressing switch in the tool handle, giving command «Start» for PTI switching on.

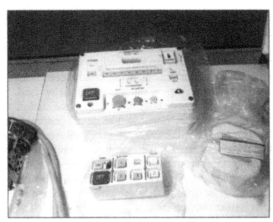

Figure 6 Test-set of hardware «Universal» and its remote control panel.

To prevent useless EVA of the operator the tool serviceability had to be first checked. For this purpose a board CP was developed in designing the hardware «Universal», which made it possible to test equipment located outboard, with a visual control of functioning (test-set is shown in Figure 6, *at the top*). In addition, the tool itself was modified to widen the functioning feasibilities and became more versatile (for example, the variant of welding with a filler metal feeding was added).

The test-set includes a block of information transformation, a TLM block and a secondary power source, whose circuits are similar mainly to those developed for the units «Isparitel-M» and «Yantar».

The block of automatics, located in the instrument compartment, contains a logical block, fulfilling the main role in the control of the tool operation, and also a block of current protections and a block of indication.

All the tests of the tools in on-land and space conditions proved the validity of selection of the element base and principle of design of the information-control equipment of the electron beam hand tool.

The device for the EBW was envisaged to be used in the truss assembly facility (TAF) in construction of large-sized truss space structures [1, 4].

At the stage of an outline project block-diagrams of the technological algorithms of operation of the system for assembly of trusses, elementary electrical diagrams of the system of control using the logic of rigid automation on the basis of the as-checked element base of series 564, were developed and experimental models of separate blocks (e.g. blocks of control of displacement drives, control of ropes tightening) were manufactured.

Even during the statement of the problem it was evident that TAF would be quite a complicated object for automatic control. Therefore, a variant of a microprocessor structure was considered in parallel with the development of diagrams using the principle of rigid automation.

Domestic industry has already mastered the manufacture of microprocessor sets of micro-circuits of an average and high degree of integration, including those on the basis of well-known CMOS technology, which meets strict requirements for onboard equipment. These are series 587 and 588 with a minimum energy consumption, i.e. with the same main characteristics as those in series 564. Series 588 is completely compatible with the series operating on transistor-transistor logic, has a high flexibility at the expense of microprogram control and makes it possible to design both a built-in and self-contained microcomputers [5–7].

The structure of a microcontroller for use as a check-out monitoring set necessary for setting and testing of the control complex, was developed on the basis of series 588. Further, the prospects of realization of the whole task of control of the process of the truss assembly in the automatic conditions were opened up on the base of this controller, but, unfortunately, this direction did not find wide development at that time. Nevertheless, it was quite evident that the wider the task, the more serious the approach to the realization of the information-control equipment, and the more important mastering of advanced achievements in computer engineering for the control of technological processes should be used.

The feasibility of using the onboard computer of the space station was also considered. The unit «Nota-P» for electron beam crucible-free zonal purification of semiconductor materials was developed under the direction of NPO «Research Center» (Zelenograd, Moscow region) for further investigations of electron beam technologies in the semiconductor industry. It was based on the same power link (PTI–high-voltage block with a filament transformer–electron beam gun), creating a circular disc zone of heating. The control system should provide a shock-free smooth heating of the sample until the state of melting, then to switch on the drive of a heater movement at a preset stable rate. The electric drive with a PWM (pulse-width modulation) control and a hinge-helical gearing moved the zone of heating along the sample within a 5–100 mm/h range of speeds. The control system provided the feasibility of presetting such parameters of the program as time of heating, level of anodic current (beam power) and rate of a heater movement from the CP board and display of parameters of PTI. The feasibility of manual correction of the anodic current was provided. In addition, three variants of the unit operation were envisaged: completely manual control, automatic condition, and control from the microcomputer «Oniks» developed at the NPO «Research Center» for some materials science works onboard the space station.

Experiments in this unit proved the necessity of use of full galvanic isolation of the information-control equipment from the circuits of primary supply and power executive circuits for noise protection.

Over recent years, many products appeared on the world market for designing systems of automatic control of technological processes, including those for use in severe conditions. Moreover, the functional characteristics and modularity of these products make it possible to perform the testing of technical solutions, designing and check-out of the software on the basis of the PC.

The works, carried out at the Research Institute of the Aviation Equipment (Zhukovsky city, Moscow region), S.P. Korolyov RSC «Energiya»[2] and other organizations, allowed the «standard» IBM PC to be accepted as the main standard in the design of man-machine interfaces for the modifying variant of spaceship «Soyuz» and ISS [8, 9]. Being based on this, the development of the control system for the electron beam equipment for materials treatment (prototype «Nota-P») was made for conductance of the laboratory-check-out tests to use it further onboard the ISS on the bases of an IBM compatible of Micro PC, designed on CMOS elements [10].

The system highway ISA provided the combination of a sufficiently wide range of modules for realization of control of different technological processes.

Figure 7 presents a schematic diagram of the control system of the equipment, and Figure 8 shows a schematic diagram of the algorithm of its operation.

Module 5025A is used as a processor nucleus. This module can control both an intersystem exchange and the internal means of the control system due to good computational abilities and large resource of the memory. A multicolour electrolumeniscent flat display was used for information imaging.

At present different tendencies are observed in the field of software development, including: the object-oriented approach, which is structuring well both the problem itself and also its solution in the form of an applied system; the use of visual RAD (Rapid Application Development) means based on component architecture; use of compilation, but not interpretation, which increases significantly the rate characteristics of applications and

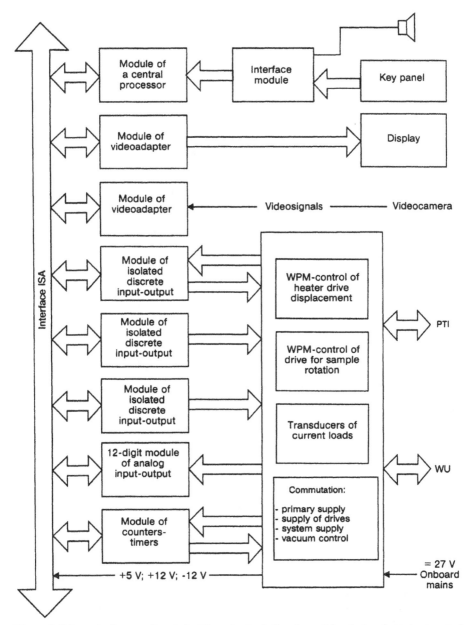

Figure 7 Schematic diagram of control of the technological equipment for electron beam treatment of materials.

decreases the resources content, as there is no need «to carry» an interpreter made in the form of a dynamic library.

The listed tendencies of development and criteria of selection became a basis for the development of a software of the technological equipment in the medium Borland Delphi 3 Professional for Windows 95 [11], which fulfils the following functions: manual control of drives of movement of the heater and a sample rotation; test control in two conditions

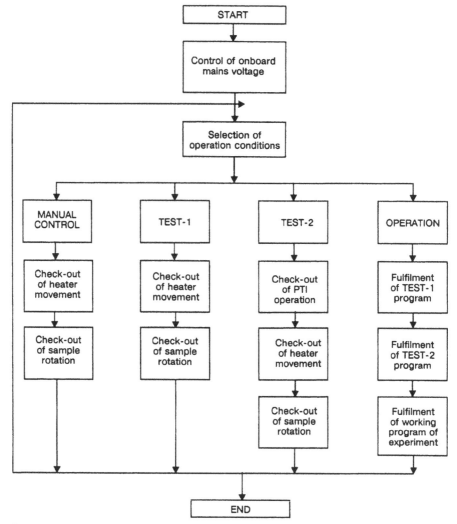

Figure 8 Schematic diagram of algorithm of operation of technological equipment for electron beam treatment of materials.

(TEST-1 and TEST-2); control and monitoring of operation of the equipment in the automatic condition; control and display of parameters of the equipment.

The operator's interface of the equipment is shown in Figure 9. A mimic panel of the equipment is located in the central upper part of the display, where the zone of melt, position of the heater relative to the sample and its displacement and rotation are designated. The dynamics of changing the anodic current is represented in the lower central part under the mimic panel. The left field is assigned for representation of the equipment parameters. The right field contains groups of information about the operating conditions and state; parameters of the program, preset by the operator; selection of the condition, as well as the «Control panel» with indication of the control keyboard.

In the course of fulfilment of the program, the parameters of equipment with a fixation of the real time are recorded into the memory of the processor module with a preset periodicity, that provides the possibility of the next analysis of the technological process and

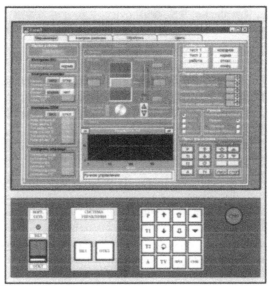

Figure 9 Interface of technological electron beam equipment.

equipment operation. The feasibility of issue of this information by request of the computer of the upper level (onboard computer) via a standard channel of communication is provided.

At present the task of creation of an information-control complex has been studied for the fulfilment of a sufficiently large volume of works onboard the ISS both in the field of materials science and in the construction of various objects outboard the station, including the use of the new versatile and safe electron beam tool. This system should provide simple programming, convenient and sufficient information for the visual inspection, recording of the required quantity of parameters of the operating equipment, control of executive mechanisms and monitoring of their functioning, videorecording, as well as accumulation of information with allowance for real time for the next transfer to the systems of the upper level and telemetry. Here the feasibility of control should be provided not only from the own panel but, also by radio line from the on-land destination through the station system [12].

The way of the development from simple to complex solutions confirms the validity of orientation to the element base CMOS both in structures of rigid automation and also modern programmable automatic control means. However, the problem of formation of noise in power system of the electron beam equipment remains open. The problem of safety of the electron beam tool remains not less real. At present it is solved by a group of specialists.

The development of new technological equipment should be based on the functional and constructive modularity of the equipment using hardware components which make it possible to construct flexible programmable information- control systems with self-diagnostics and high maintainability.

1. Paton, B.E. and Lapchinsky, V.F. (1998) *Welding and related technologies in space*, Kiev, Naukova Dumka, 184 pp.

2. Kogan, A.L., Kolosovsky, A.V. and Sinekaev, V.V. (1979) Micropowerful CMOS of microcircuit of series K564, *Elektron. Promyshl.*, No. 5, pp. 17–19.

3. Shilo, V.L. (1987) Technical specifications KO.347.064. TS, In: *Popular digital microcircuits*, Moscow, Radio i Svyaz, pp. 12–16.

4. (1985) *Systems of assembly of trusses CCT 48*, Kiev, PWI, 56 pp.

5. Dzhkhunyan, V.L., Kovalenko, S.S., Mashevich, P.R. et al. (1978) Micropowerful LSI microprocessors of series K587, *Elektron. Promyshl.*, No. 5, pp. 15–19.

6. Shiller, V.A. (1979) Micropowerful LSI microprocessors of series K588, *Ibid.*, No. 10, pp. 36–38.

7. Shevkoplyas, B.V. (1986) *Microprocessor structures. Engineering solutions*, Moscow, Radio i Svyaz, 210 pp.

8. Tyapchenko, Yu. and Bezrodnov, V. (1997) PC onboard the manned space vehicle, *Sovr. Tekhnol. Avtomatiz.*, No. 1, pp. 34–37.

9. Gobchansky, O. (1997) Application of MICRO PC in special-purpose computer complexes, *Ibid.*, pp. 38–41.

10. Gobchansky, O. (1998) Unified means of onboard computer complexes of space vehicles, *Ibid.*, No. 1, pp. 72–76.

11. Darakhvelidze, P. and Markov, E. (1996) *Delphi — medium of visual programming*, St.-Petersburg, BHV, 482 pp.

12. Grigoryan, O.R., Klinov, S.I., Kaos, Z.T. et al. (1997) Instrument for ecological monitoring at OC «Mir», *Inzh. Ekologiya*, No. 2, pp. 44–50.

Chapter 3.

PROCESSING EXPERIMENTS
UNDER MICROGRAVITY

Chapter 3. Processing Experiments under Microgravity

Experiment is one of the main links in the chain of the research activity.

Problems associated with specific conditions of performing processing experiments in space require thorough earth-based preparation. Development of research in the field of space technologies makes us pay special attention to the peculiarities of training of cosmonauts — operators, optimisation of operations using process equipment under conditions which simulate the space ones, and ensuring safe operation when using a welding tool.

This Chapter includes studies which cover the main directions of development of space technology — electron beam welding, brazing and cutting as methods for heat treatment of structural materials, and fabrication of materials and items possessing unique properties, which are hard or impossible to achieve under earth conditions.

Theoretical investigations and calculations served as a basis for the development of procedures for conducting various processing experiments, design of the electron beam welding hardware and testing it under extreme conditions, as well as analysis of results of the experiments performed using the process equipment.

Studies of technological capabilities of microplasma, laser and combined welding prove that hybrid welding holds high promise for repair operations to be performed on space vehicles.

A number of studies are dedicated to solving one of the problems of space technology, i.e. deposition, restoration and synthesis of various-purpose coatings under space conditions. Thin-film coatings are deposited by the method of thermal evaporation and condensation of materials in vacuum. Comparative analysis of the structure of the metal films confirmed the expediency of continuing research in this field.

Melting and solidification processes used under microgravity make it possible to solve problems associated with production of single crystals, investigate physics of the process, optimise crystallisation process technology under earth conditions and improve the process equipment.

Microgravity involves absolutely unconventional experimental conditions, this adding both pure scientific and applied character to the experiments. The processing experiments may have different tasks:

- *getting an insight into a phenomenon;*

- *obtaining more precise data on properties of a material;*

- *producing a new structure of a material.*

The possibility of using such experiments gives confidence that processing investigations will become an integral part of future space programmes.

GROUND-BASED RETROFITTING

Special Features of Training Cosmonaut-Operators to Perform Processing Operations[*]

G.S. Shonin[3], S.A. Kiselyov[3], V.F. Lapchinsky[1] and A.A. Zagrebelny[1]

Investigations conducted over recent years in the field of space technology, space materials science and production, resulted in a change of the attitude to training cosmonaut-operators. The point is, that in addition to difficulties associated with the specific conditions of space, work of a technical nature will probably require substantial professional training of the operators, as their end product are materials or items, having quite definite properties, that are, as a rule, highly dependent on the cosmonauts' performance. It can be assumed, that with the progress of «space» research, the requirements for professional training of operators will increase continuously.

On the other hand, it is difficult to imagine that crews of the future space vehicles would include only specialists-technologists in a specific narrow area. Evidently, the major scope of research will still be performed by the cosmonauts, whose activity during the flight is extremely broad and multi-faceted, and for whom processing operations are just one of many responsibilities.

There is no doubt, that further progress of research, related to space technology, will require development of a certain procedure of training cosmonaut-operators, that would allow the crew members to develop professional skills necessary for performance of processing operations, with minimal training time.

As a first step in the development of such a procedure, we tried to conditionally divide the entire package of the intended future processing operations in space into a number of complexity categories (in fact, four), differing by the degree of operator's participation in performance of the technological process. The following factors were taken into account:

- conditions of work performance and parameters of the processed material;
- area (working zone) of work performance;
- time of work performance;
- parameters of processing unit operation;
- methods of controlling the quality of the performed work.

Category I includes processing work, where all the above factors are determined in advance, i.e. already on the ground.

Category II covers the work, where the parameters of the processed materials, area (place) and conditions of work performance are known in advance, objective control of the quality of the performed work is possible, but the parameters of the processing unit mode or the time of work performance are not known.

Category III includes operations where the parameters of the processed material are known and objective control of the quality of the performed work is possible, but the area, conditions and time of work performance, or mode of the unit operation, are not known.

Category IV are operations, where practically all the above factors are unknown and objective control of the quality of the performed work, is impossible.

---

[*] (1978) *Technology in space*, Coll. pap. of 7th Gagarin Scientific Readings on Cosmonautics and Aviation, Moscow, 1978. Moscow, pp. 11–20.

It is natural, that the greater the indeterminacy of each of the above factors, the greater is the degree of active participation of the operator in the actual performance of the technological process. Therefore, the operations, belonging to complexity category I, can be performed in stationary automatic units with a preset rigid program. For operations of category II, it is most rational to use stationary automated units with a manually set program or stationary automatic units with feedbacks by one or several parameters of the process cycle. Work of category III can be performed in units of the same type, mounted by the operator in the area of conducting the work just before it is carried out. Extremely scarce information on the parameters and conditions of performance of category IV work makes it necessary to specify manual semi-automatic units, implying the most active involvement of the operator in the technological process.

The operator's functions are also largely dependent on the category of the operations performed.

During performance of work of category I, the operator's functions are reduced to starting the unit, and to acceptance and preservation of the finished products after completion of the process. This category of operations can be illustrated by producing special materials in automatic thermal furnaces with a preset program.

During performance of work of category II, the operator starts the unit, monitors its operating mode and corrects it, if required (in the case of the unit operating without feedbacks), accepts the finished products and performs their preservation and stowing. An illustration of this category of operations can be welding of pipelines or abutted components, using a welding unit, mounted in advance in the place of work performance.

Work of category III requires more active involvement of the operator, consisting primarily in evaluation of the ability and need to perform work in a particular location. As soon as the decision on work performance is taken, the operator performs de-preservation of the unit and mounts it. Further on, operator's actions do not differ from those envisaged for performance of work of category II. Work of category III can be illustrated by the operator's actions on servicing all-purpose processing units or diverse multipurpose automated repair hardware.

Work pertaining to category IV requires the most active participation of the operator. In this case the operator himself conducts the technological process, using manually controlled non-automated or partially automated units. Performance of this kind of work requires a particularly thorough professional preflight training. The complexity of such work performance consists also in that the operator's functions, as a rule, cannot be determined in advance, and are usually defined more precisely by the cosmonaut himself directly before performance of the work. Examples of work of category IV are highly diverse. In particular, they can include various repair or mounting work conducted in planned or unforeseen situations. A more detailed characteristic of the operator's functions for each of the four categories of work is given in the Table.

A special cycle of investigations in a flying laboratory under conditions of short-time zero gravity has been conducted to determine the required level of operators' training for performance of processing operations of different category of complexity in space.

A rather complicated technological process, such as manual welding, was selected for performance of the experiment, as it was assumed that it would be possible to apply the derived results to simpler processes.

Operators with different degrees of professional training took part in the experiments. They were conditionally subdivided into four groups:

• operators having experience of work performance at zero gravity and professional skills of a welder (1st);

• operators having experience of work performance at zero gravity, but no professional skills of a welder (2nd);

Operator's functions during performance of processing operations of various categories.

Work category	Degree of unit automation	Preparatory	Main	Final
I	Stationary automatic units with a rigid program	–	Unit starting	Acceptance and preservation of finished products
II	Stationary units with manually controlled program or feedback	–	1. Unit starting 2. Mode control 3. Mode correction (with manual adjustment)	Same
III	Mobile units with manually-controlled program or feedback	1. Selection of the place of work performance 2. Unit depreservation	1. Unit starting 2. Mode control 3. Mode correction (with manual adjustment) 4. Unit transportation to the place of work performance 5. Unit mounting	1. Acceptance and preservation of finished products 2. Dismantling the unit 3. Unit transportation 4. Unit stowage
IV	Manual non-automated or partially automated units	1. Selection of the place of work performance 2. De-preservation of the unit 3. Unit transportation to the place of work performance 4. Work place preparation 5. Checking professional skills 6. Recovery of professional skills (if required)	1. Performance of the required processing operation with continuous quality control 2. Evaluation of the quality of work performed 3. Correction of processing defects (if required)	1. Finished product acceptance 2. Conservation of the unit work place

• operators having no experience of work performance at zero gravity, but having the required professional skills (3rd);

• operators having no experience of work performance at zero gravity, and having no professional skills (4th).

Such a selection allowed tracing the dynamics of improvement of welder's professional skills in operators for work performance at zero gravity. Testing was performed by experts of the Yu.A. Gagarin CTC[3] and the PWI.

Final results of the operators' activity were evaluated by an expert team by a range of parameters: weld rectilinearity and presence of deviations from the axis (K_1); uniformity of weld width (K_2); penetration uniformity (K_3); presence and number of metal burns-through (K_4).

Each of the above parameters was evaluated using strictly determined objective criteria. For instance, a rectilinear weld without deviation from the axis was given 5 points; that with a deviation of 0.5 mm — 4 points, that with an average deviation of 1 mm — 3 points, etc. Coefficient μ was also introduced, that allowed for the importance of all the estimated parameters.

Thus, the general index K of weld quality was defined as the sum of estimates of individual parameters, taking into account the degree of their criticality:

$$K = \mu_1 K_1 + \mu_2 K_2 + \mu_3 K_3 + \mu_4 K_4.$$

The highest quality index, corresponding to the maximum estimates of all the weld parameters, was equal to 5 points. Prior to performance of zero gravity experiments, testing operators had theoretical training and acquired on the ground the minimal required professional skills.

This paper is not aimed at a detailed analysis of the rather complicated procedure of experimental performance and evaluation of the results. Let us briefly discuss just the main conclusions from the investigations.

Obviously the kind and scope of operator training completely depends on the category of work for the performance of which they are being trained. On the other hand, there exist a number of training stages, required for performance of all the categories of processing operations. Preliminary training of the operator is performed in three stages.

In the first stage of training the operator-cosmonaut should be thoroughly familiarized with the goals and main theoretical principles of the technology of the process to be conducted by him. Making a high-quality product can be expected only in the case, where the operator is well familiar with the final purpose of work performance and has a clear idea of the physical essence of the processes needed to achieve it.

The second stage of training consists in practical familiarization of the operator with the design and principles of operation of the processing unit, and the purpose and functioning of its main components and adjustment and control devices.

In the third stage of training, a qualified specialist should demonstrate the performance of the specified processing operation to the operator on the ground. Then the operator (also on the ground) independently performs the processing operation several times. The number of repetitions depends on the individual qualities of the operator and complexity of the operation.

After that the preliminary training of operators to perform processing operations of complexity category I can be regarded as completed. Directly before performance of the technological process in space, the operator just has to simulate his actions in the unit to recover the acquired skills.

For operators, training for performance of processing work of complexity category II, such a scope of training is insufficient. Since the operator's functions will include adjustment of the unit operating mode, he needs to be familiarized with the performance of the specified processing operations at zero gravity, first with the assistance of a qualified expert, and then try to perform the same operation several times on his own. This is primarily related to the fact that the unit operating mode at zero gravity can be essentially different from that used on the ground.

Operator's training to perform work of complexity category III should be even more extensive. Since the operator should independently perform de-preservation and preservation of processing units, mount them in the specified area, set up and adjust the mode of functioning and, finally, take decisions on the need to perform a particular operation, all the above operations should be mastered during training to the point of the operators developing a steady skill, also at zero gravity. This is due to the fact, that the operator should be able to adapt steadily to the external manifestations of the process unfamiliar to him (heat and light radiation, material heating and melting, gas or vapour evolution, formation of liquid drops, etc.), that have a peculiar nature at zero gravity that cannot be reproduced on the ground.

It should be noted that the only means so far developed, allowing the majority of the technological processes to be conducted at zero gravity with the operator's involvement, is the flying laboratory mission. Experience gained by us shows that this approach is extremely effective, and is one of the main tools in training operators-technologists.

An extremely important element in operators' training turned out to be the method of their independently solving various processing tasks of complexity category IV. This enabled the operator to offer his creative input, thus reducing the risk of emergency situations during flight.

When performing the work of the first two complexity categories, the level of operator's professional training did not have any noticeable effect on the quality of the final product, as the processing operation proper was performed by the automatic device. A much faster adaptation to the conditions and specific features of work performance was found in operators of the first and second groups.

The quality of welds made in the units classified as category III significantly depends on operators' training. The operators of the first group demonstrated the best results, and the operators of the fourth group had the worst results. As regards the second and third groups, preference should be given to those in the third group (operators, having experience of welding performance), who began making quality products faster.

The most interesting results were derived from studying processing operations of complexity category IV. The operators were given the following task: perform four experimental cycles, welding 4–5 welds in each of them. The time interval between the cycles was 1–3 days. In this case also, as could be anticipated, the first group operators had the best results. For instance, from the graph shown in Figure 1a, demonstrating the dynamics of improvement of the grades for the quality of welds, made by a first group operator, it is seen, that already after 6 performed experiments, the weld estimate was higher than 4 points, and after 11 experiments the welds performed by this operator were steadily given 5 points.

Adaptation of third group operators (welders, having no welding experience at zero gravity) to unfamiliar working conditions was a little worse. However, in this case also after 9 experiments their welds were steadily estimated as 4–5 points (Figure 1b). Operators without welding experience, but well accustomed to zero gravity, rather quickly achieved individual grades, higher than 4 points, but they, as a rule, could not steadily maintain the quality of welds, made by them, at the specified level (Figure 1c).

The fourth group of operators had the least stable results. It should be noted, however, that in this case also, the operator achieved grades higher, than 4 points at the end of each cycle, the average grade per cycle increasing from cycle to cycle (Figure 1d).

The obtained regularities allow tracing the dynamics of improvement of professional skills of the tested operators. Analysis of the results indicates, that even untrained operators can rather quickly develop the appropriate stereotype, allowing a technological process of complexity category IV to be conducted at a high level. However, training of this operator category, in addition to the above-named training stages, requires performance of several (at minimum two-three) cycles of practical work at zero gravity. The total number of experiments, performed by operators during this experimental period should be not less than forty. Especially important is the cyclic nature of training, promoting the operators forming steady professional skills. This is confirmed by the fact that the operators who were included in the

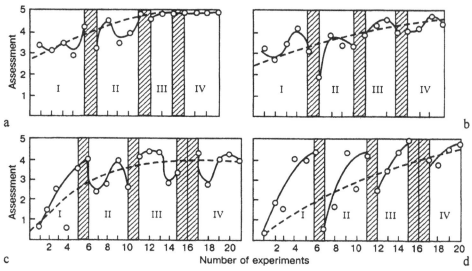

Figure 1 Dynamics of improvement of quality assessment of welds made by operators of the first (a), third (b), second (c) and fourth (d) groups (I–IV are the experimental cycles).

fourth, least qualified group, after several experimental cycles, actually reached the level of training characteristic of the first group.

Further observation of these operators confirmed the earlier conclusions. After an almost two year interval, an operator was able to completely recover high professional skills after 5–6 experiments, during a period typically required for well-trained testing operators.

In conclusion, we would like to emphasise that we are not trying to present the results of our investigations as a comprehensive characterization of the problem of operator training for performance of processing work in space. They do, however, lead to the conclusion of the undoubtedly promising nature of the chosen approach, and of the need to train the cosmonauts in the professions of technologists.

Specific procedures of training cosmonaut-operators, dealing with processing operations in space, can be developed, taking into account the above statements.

The basis for mastering the profession of space technologist, especially for work of category III and IV, should be the ability to perform finely co-ordinated motions at zero gravity, also when wearing outfits that allow EVA performance.

Facility for Studying Technological Processes under Simulated Space Conditions*

B.E. Paton[1], V.E. Paton[1], D.A. Dudko[1], V.N. Bernadsky[1], G.P. Dubenko[1], V.F. Lapchinsky[1], V.V. Stesin[1], A.A. Zagrebelny[1], Yu.N. Lankin[1], Yu.A. Masalov[1], O.S. Tsygankov[1] and V.M. Bojchuk[10]

Steadily expanding space research is bringing forward as one of the goals the performance of various processing operations (metal melting, welding, coating, casting various parts, growing single-crystals, etc.) in space. This problem is particularly urgent now, when the

* (1973) *Space research in Ukraine*, Issue 1, pp. 5–8.

work on creation of long-term orbital scientific laboratories is expanding both in the USSR and abroad.

Any scientific, in particular technological space experiment, requires thorough ground-based training that is the more effective, the more completely can the space conditions be simulated, namely zero gravity, space vacuum, peculiar thermal mode of the object of study and of the experimental units [1]. A facility for performance of technological experiments under conditions simulating those of space, was developed for the first time at the PWI. When the facility was developed, the goal was to simulate the first to conditions, on the assumption that, if required, the thermal mode of the experiment can be achieved by a thermoregulator.

The main difficulties in development of such a facility are primarily related to the fact, that it should meet the following requirements, as a minimum:

• provide, in combination with the flying laboratory, multiple operations in short-time zero gravity mode;

• simulate for 1–2 h the vacuum characteristic of the outboard space environment around the piloted orbital space vehicles;

• have a minimal weight, overall dimensions and maximum reliability, while being self-sufficient;

• provide as high as possible versatility in terms of performance of various technological experiments, in particular, on welding, coating and cutting of metal;

• incorporate a device for reliable recording of the progress of the experiments;

• ensure reliable fixation of the operator relative to the facility with maximum convenience of control and monitoring the experiment.

In addition, the facility should comply with the requirements for test equipment mounted on board the flying laboratories. Solving both these problems can be best analyzed by considering the most important individual functional blocks of the facility.

Working chamber *1* with replaceable processing units is a stainless steel vacuum-tight cylindrical vessel of 500 mm diameter and a capacity of about 100 l with a spherical bottom and a similar flap lid *2* (Figure 1). The chamber body is fitted with neck *8* for connection to high-vacuum pumping unit *7*, vacuum valve *10*, vacuum lead-in *9* for transferring rotation inside the chamber and portholes *6* for observation and filming. The chamber flap lid has neck *5* with bellows corrector for mounting replaceable processing units, portholes *3* for filming and vacuum lead-ins *4* for pressure gauges.

Figure 1 Versatile vacuum facility for studying welding processes under simulated space conditions.

Figure 2 Working chamber with open lid.

The chamber is fitted for mounting various heat sources (electron beam, plasma and consumable-electrode arc) of about 1 kW power, allowing welding, coating and melting of small metal volumes to be performed. Each of the above units is reduced in size and capable of reliable operation under vacuum.

On the chamber bottom (Figure 2) is a sample stage that allows mounting either specimens for welding, or crucibles for metal melting or coating, as desired. The stage can rotate at different speeds around a vertical axis, thus providing displacement of the specimens being welded relative to the heat source.

The required rarefaction was induced and maintained in the working chamber by a vacuum system (Figure 3), consisting of rough pumping unit *1* and special high-vacuum pumping unit *2*. The rough pumping unit includes mechanical pump with 36 V motor, fitted with a nitrogen trap and vacuum valve. It functions only during ground-based preflight training and is powered from the airfield AC mains.

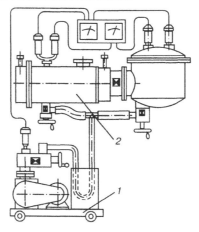

Figure 3 Schematic of the facility vacuum system.

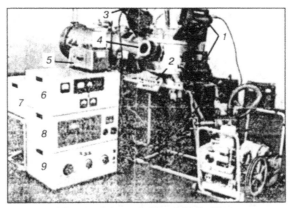

Figure 4 General view of the facility.

Considerable difficulties were encountered in selection of the high-vacuum unit. Steam-jet pumps are not suitable for operation under conditions of overloads and zero gravity. Sorption-ion, molecular and electric discharge high-vacuum pumps cannot be used, as they have considerable weight and overall dimensions. Therefore, B.I. Verkin ILTP[10] developed for the testing facility a special combined adsorption-getter high-vacuum pump that uses activated coal cooled by liquefied nitrogen, for absorption of argon and air, a newly deposited titanium film for absorption of hydrogen, and a nitrogen trap for freezing out the water vapours and carbon dioxide gas.

Ultimate vacuum of the unit is equal to $(1–3) \cdot 10^{-7}$ Torr. In the course of the experiment a pressure of $5 \cdot 10^{-5} – 10^{-6}$ Torr is maintained in the chamber, depending on the conditions of performance of the experiment. This high-vacuum unit does not consume any electric power during operation, this being another quite important advantage under flying laboratory conditions.

For recording progress of the experiments, the testing facility is fitted with a set of recording equipment (Figure 4). Modes of performance of the experiment are registered by a 12-channel loop oscillograph *5* that records the readings of vacuum and overload sensors, as well as voltage, current and a number of other parameters of the process. Filming of the working zone with AKS-2 camera *3* (film width of 35 mm, filming frequency of 24 frames per second, power supply from 27 V DC onboard mains) is conducted during the entire experiment. The most interesting phenomena are recorded with two high-speed cameras SKS-1M *1* (film width of 16 mm, filming frequency of 300–5000 frames per second and power supply from a special source that is part of the testing facility set). The unit set includes a special lighter *4* for filming.

The ability to perform coordinate-time analysis is provided by the presence in the zone of scale and time marks, recorded synchronously on the oscillogram and films. The time marker operates at two fixed frequencies of 1.2 and 10 Hz [2].

Facility controls are mounted in several interchangeable functional blocks and on the control panel. Continuously operating for all the tested technological processes are blocks of general control *9*, measurement *8* and power *7* (Figure 4). Other plug-in modules are blocks 6 of each technological process control. The unit is controlled from special panel *2*. Maximum control automation is provided to facilitate operator's activity during the flight. When an experiment is conducted, the operator, as a rule, performs only the initial switching on of the unit and high-speed filming, all the operations being automatic. The experimental mode can be changed during experiment performance.

The mounting of the testing facilities in the flying laboratory cabin (see Figure 3 on p. 43) depends on the specific conditions of experiment performance. The work places of the operator and his assistant are located in front of the central porthole of the working chamber, under which the control panel is mounted on a frame. The work place is fitted with a board for making records, cassette holder, foot straps and lap belt for fastening the body. For convenience of recharging in flight, SKS-1M cameras are located on the operator's right and AKS-2 camera and the chart recorder are on his left. Overall dimensions of the facility, mounted in the flying laboratory cabin, are $2000 \times 1150 \times 1300$ mm, its weight being 550 kg.

The facility enabled the performance of a number of important studies, related to development of various technologies for space application. The main features of welding and cutting sheet metals were studied, and the fundamental possibility of performing melting of small metal volumes and deposition of metal coatings at zero gravity was demonstrated. Extensive experience of the facility operation fully confirmed the correctness of the main technical decisions taken in its development. The presence of a bellows-type vacuum lead-in for mounting replaceable processing units, allows recommending the facility for studying the most diverse technologies. For instance, with minor design modifications, the facility can be adapted to study the processes of metal cutting, net-shape casting and performance of other operations in vacuum and at zero gravity.

1. Paton, B.E. and Kubasov, V.N. (1970) *See pp. 154–160 in this Book.*

2. Lankin, Yu.N., Masalov, Yu.A. and Shejkovsky, D.A. (1968) Time marker for a high-speed camera and loop oscillograph, *Avtomatich. Svarka*, No. 3, pp. 70–71.

Ergonomic Aspects of Welding Operations Performance in Space*

B.E. Paton[1], D.A. Dudko[1], V.N. Bernadsky[1], V.F. Lapchinsky[1], A.A. Zagrebelny[1], V.V. Stesin[1] and A.A. Moiseenko[3]

The versatility of welding processes opens up broad prospects for their application in the space objects of the future. Welding can be used for production of tight joints on large space objects and their repair; welding equipment can also be the source of heating in technological processes of space metallurgy [1, 2].

As was noted earlier [2] fully automated process units can be used in space in those cases when the area and nature of work performance are known in advance, i.e. only when all the preparation for work performance is done on Earth, and the units are taken to orbit in an already prepared state. Such solutions are not applicable for repair operations, the need for which has already been experienced by cosmonauts more than once, and which in the future will probably be regarded as part of maintenance operations in long-term orbital stations. In performance of a repair, the area of work performance, material thickness, configuration and dimensions of the part being repaired, etc. are usually not known in advance. In such cases direct participation of a specially prepared anpersonined man in this work is required.

Since the time of the first EVA of cosmonaut A. Leonov, Soviet and American cosmonauts have more than once proved that a man wearing a spacesuit can live and work in space. On the other hand, ensuring adequate performance of the cosmonaut under these conditions requires further study of the specific features of the work and solving a number of problems related to creation of all kinds of special tools and fixtures, as well as to professional training of the cosmonauts [3, 4].

* (1976) Paper was presented at the 5th Gagarin Scientific Readings on Cosmonautics and Aviation, Moscow.

The subject of this presentation is consideration of some ergonomic aspects of processing operations performance in raw space and appropriate training of operators. Analysis is made for the case of welding operations performed by the welder wearing a spacesuit. This work was conducted as a preliminary search for ways to solve the posed task. The observations and conclusions made are based, primarily, on the experiments performed in a special facility on Earth and under the conditions of zero gravity in a flying laboratory [5]. We, however, believe that these experiments are of importance for conducting further integrated investigations with a broad involvement of physiologists and specialists on the theory of control, even though they still leave some room for discussion.

The welding process is an especially suitable object for such integrated investigations, as it requires a thorough professional training of the operator. Thus, the conclusions drawn for the welding process can be applied also to other technological processes, without anticipating any additional difficulties in such cases. On the contrary, it will probably be possible even to make considerable simplifications in a number of cases.

When welding is performed manually, the combination of man–working tool–technological process is a closed system of regulation with several adjustable parameters. All the units in this system are non-linear and, therefore, similar to any control system with man's participation, it does not in practice lend itself to any valid mathematical description. The task is further complicated by the fact that each unit in this system should comply with certain requirements whose violation usually results in a low-quality welding process. Still, some objective criterion of the system performance assessment is required to make a judgment on the restrictions imposed on the above system by the space environment.

The performance of man in combination with the working tool and technological process was analyzed based on consideration of a generalized block-diagram (Figure 1) in which the units are characterized by some transfer functions, and the operational conditions and the main comparative characteristics of the units are taken into account through a number of coefficients and time constants of the control circuits. Such an approach is quite acceptable as it allows a more precise determination of the differences between the system functioning under standard conditions (on Earth) and in space.

Let us briefly characterize each unit of the block-diagram under consideration.

Energy W_m, consumed for melting the metal and formation of the weld W_w is supplied to the working tool *1* with transfer function x. Main parameters *2–4* with transfer functions x_1, x_2, x_3 are used by the welder to adjust the welding process. In this case it is assumed that a reliable joining of the parts being welded will be achieved in the case where constant depth of penetration h, weld width b and welding direction φ are maintained within certain limits. These are exactly the parameters the welder should monitor first. In practice, certainly, he simultaneously monitors a whole number of other parameters (uniformity of weld formation,

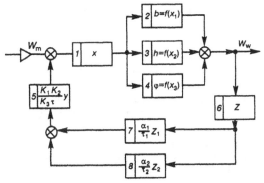

Figure 1 Block-diagram of the system of the technological process organization.

nature of fusion, etc.). However, these parameters were omitted in the first stage for simplification purposes. The physico-chemical processes occurring in weld formation were not taken into account either, and the operation of the systems of stabilization and automatic adjustment of the welding mode was not considered. All the attention was focused on the interaction of man with the working tool and the welding process.

The operator's «end effector» *5* is his hand with transfer function y, time constant τ and coefficients K_1, K_2 and K_3, where coefficient K_1 characterizes the mobility of the end effector, K_2 — the regularity of its motion and K_3 — its fatigue. The operator's information system *6* with transfer function Z includes vision, sense of touch and hearing. Feedbacks *7, 8* with transfer functions Z_1 and Z_2 are used by the operator to control the welding process. In engineering terms it can be assumed that in this case, primarily, two types of feedbacks inherent to man are found, as pointed out in [6]. These are the so-called primary feedbacks Z_1, or information feedbacks and secondary Z_2, or posture feedbacks.

The difference in the nature of these feedbacks consists in that in the case of primary feedbacks the operator's actions are completely determined by his knowledge of the system performance, its behaviour at certain disturbances and the required result of the production process. The speed of response of such feedbacks is as a rule absolutely insufficient, as it is determined by a rather long-time processing of the acquired information and its comparison with theoretical knowledge on the correct functioning of the system.

The secondary feedbacks related to maintenance of muscular tone, are only manifested in well-trained operators in direct contact with the working tool, and feature a much higher response than the primary ones.

In performance of various operations, both these types of feedbacks are practically always in place. However, their ratio can vary in a broad range. The proportion of each of the two types of feedbacks under consideration is taken into account in the block-diagram by α_1 and α_2 coefficients, respectively, and the speed of their response — by time constants τ_1 and τ_2. From the above-said it is clear, that $\tau_1 \gg \tau_2$.

If we take an idealized case, when the operator is fully familiar theoretically with the way welding should be performed, but has not even once done the work in practice, he will conduct the process using mainly the information feedbacks ($\alpha_1 = \max$). From control theory it is known that a complex additive system with several controllable parameters is unstable in operation with one feedback, characterized, in addition, by a greater time constant.

Indeed, an untrained operator using only primary feedbacks, cannot conduct a sound welding process. As practical training proceeds, he gradually acquires certain skills and forms a certain motion stereotype related to the general condition of the muscular system. In this case coefficient α_1 decreases and α_2 grows. In a well-trained, skilled welder, the secondary feedbacks prevail. It can be assumed that the magnitude of α_1 and α_2 coefficients changes in the course of training approximately as shown in Figure 2 (here and further on prime sign indicates a change in the operator's working conditions).

This is the situation if the operator is trained and is working under regular conditions on the ground. The situation changes essentially when the same operations are performed in open space. This is related primarily to the impact of two factors, namely zero gravity and limitations imposed by the spacesuit. These factors influence to a certain degree all the units of the system (see Figure 1). However, when manual welding is performed, the space environment has the major influence on units *5–8*, in other words on the operator himself.

What is the influence of these factors?

Both zero gravity and the spacesuit affect primarily the information system *6* of the operator. In open space his usual coordination referencing is disturbed; the scope of information received through the tactile and hearing organs is markedly decreased, and that

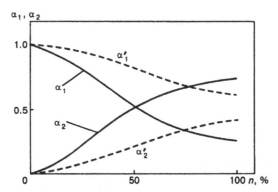

Figure 2 Change of the influence of feedbacks of the technological process control system (hatched lines — zone of satisfactory training of operators, n — level of operator training).

received through the organs of sight is essentially reduced; part of the information can come in a form unusual for ground conditions [7–9].

The influence of these factors on the nature of the operator's feedbacks is not less significant. A drastic reduction of the scope of information perceived by the tactile organs leads to decrease of the proportion of secondary feedbacks ($\alpha_2' < \alpha_2$), and by contrast the proportion of primary, information feedbacks rises ($\alpha_1' > \alpha_1$) (see Figure 2). The changed conditions of information perception and its unusual nature cause an increase of the time constant in this feedback circuit ($\tau_1' > \tau_1$), which is high enough as it is. The latter is manifested in an increase of the time required for processing the information into the form usual for perception and for thinking over one's actions [7]. In addition, a further increase in τ_1 can be anticipated as a result of a psychological strain in the operator during EVA.

Thus, even a quite well-trained operator under these conditions conducts the process mainly using the primary feedbacks, often leading to violation of the system stability and technological rejects during operation.

The difficulties are further aggravated by the fact that the operator's end effector, namely his hand (Figure 1, block *5*) also drastically changes its characteristics under the influence of the stiffness of the spacesuit, which is at a certain excess pressure. The mobility ($K_1' < K_1$) and regularity of motion ($K_2' < K_2$) deteriorate, and fatigue ($K_3' > K_3$) becomes greater. Its time constant ($\tau_1' > \tau_1$) also becomes larger. In this case, the fine movements of the fingers and the wrist are the most affected (Figure 1, block *8*), this being especially important for performance of welding. All this leads to even more frequent disturbances of the system stability and to rejects in welding, which quite often can mean even a burn-through of the metal being welded, this being absolutely inadmissible in the space environment.

Experiments in the flying laboratory using a special training facility were conducted as the first step in the study of the ergonomic features of work performance by a welder wearing a spacesuit under zero gravity conditions (Figure 3) [5]. During the experiments it was found that the above errors are made by both highly skilled welders who had never worked in a spacesuit before, and operators having extensive experience of working in a spacesuit, but no welding skills. As experience of welding performance in a spacesuit and under zero gravity was gained, the number of mistakes made by both in them naturally became smaller, but with the scope of training given to the participants in the tests, errors could never be completely avoided.

The influence of the level of the operators' training on the quality of welding performance at zero gravity when wearing a space suit, was studied by process cinegrams. The coordinate-time analysis of the cinegrams allowed a detailed study of the kinematics of the

Figure 3 General view of the training facility.

tool motions, and their statistical processing permitted determination of several characteristic techniques of work performance by the operators with different levels of training. The most typical cases are shown in Figure 4.

Operators, having an extensive experience of working in a spacesuit, and at zero gravity, familiarized with the welding process in theory, but, not having sufficient practical skills of welding, as a rule conduct the process with abrupt reciprocal motions of the tool with a large amplitude (Figure 4a). This is related to the fact that during work performance they are guided only by the visual perception of the process and use primarily information feedbacks with a large time constant, thus leading to system instability and spontaneous oscillations of extremely great amplitude. This results in an absolutely unsatisfactory formation of the weld with frequent interruptions and repeated deposition of metal in the same part of the weld.

A well-trained welder not having any significant experience of working in a spacesuit, behaves in a different manner at zero gravity. He tries to move the tool along the weld in a strictly uniform, translational manner (Figure 4b), reproducing the working technique which he has mastered well under standard conditions. However, the usual welding technique is made much more complicated by work performance in a spacesuit and at zero gravity. The high rigidity and small mobility of the spacesuit, absence of any direct connection with the tool, and the unusual conditions of observation of the process, lead to fast nervous and muscular fatigue of the operator and, as a result, to periodic disturbances of the stability of the tool displacement. As fatigue increases such disturbances may result in poor quality of welding.

Having gained a certain experience of working in a spacesuit and at zero gravity, the operator works out a more rational technique of work performance when the static loads are replaced by the dynamic loads. Such a working technique is characterized by frequent reciprocal movement of the tool with a small amplitude (Figure 4c). In this case, operator fatigue is minimized, and a quite satisfactory quality of welding is achieved, which is also due to the operator now using not only primary, but also secondary feedbacks, when conducting the process.

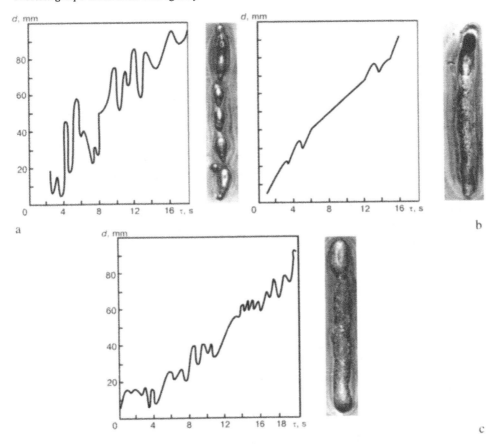

Figure 4 Weld made by the testing operator, having no welding skills (a); by a welder, having no experience of working in a spacesuit (b); welder, wearing a spacesuit, after a period of training (c).

Thus, EVA training of the operators should primarily focus on their acquiring purpose-oriented professional skills of an erector, welder, etc.

Operator's securing in the workplace is absolutely necessary, but a rigid fastening at the waist, used in our experiments, causes additional difficulties when the operator performs the required tool movements, and significantly reduces his working zone.

A preliminary conclusion can also be made that the skills of welding in a spacesuit acquired during training, are lost faster than under standard conditions.

As it is difficult to believe that operators who in the future will perform welding in space, will specialize only in this operation, and will be continuously improving their skills, their insufficient professional qualification should be assumed in advance. Therefore, a system of measures should be developed, which would provide a high reliability of welding performance in space even by insufficiently well-trained operators. The conducted preliminary experiments showed that several methods of improvement of the process reliability can be outlined.

Firstly, this is the application of various kinds of auxiliary devices, allowing automation of the most labour consuming actions of the operator, and facilitating process monitoring. For instance, the tool displacement along the weld lends itself easily to automation even in the most complicated cases.

Another method, essentially improving the reliability and quality of the welding process, is the use of various kinds of systems of automatic regulation of the process, in which the welder only performs the initial setting of the operational mode. Such a division of functions between the man and the working tool will apparently be the optimal one for performance of processing operations in space.

Thus, preliminary experiments showed the principal possibility of performance of manual welding of metals in space. The technique of welding allowing for the difficulties of working in a spacesuit, can be mastered by an operator of medium level of qualifications. In order to achieve a reliable performance of welding for a long time, part of the functions performed by the welder under standard conditions, should be transferred to the tool.

1. Paton, B.E. and Kubasov, V.N. (1970) *See pp. 154–160 in this Book.*

2. Paton, B.E. (1973) Problems of space technology and their influence on science and technology, In: *Coll. pap. of 24ᵗʰ IAF Congr.*, Baku, 1973.

3. Sharp, M.R. (1971) *Man in space*, Moscow, Mir.

4. Umansky, S.P. (1970) *Man in space*, Moscow, Voenizdat.

5. Paton, B.E. (1973) Facilities for investigation of technological processes under space simulation conditions, In: *Coll. pap. of 24ᵗʰ IAF Congr.*, Baku, 1973.

6. Wiener, N. (1948) *Cybernetics or control and communication in the animal and the machine*, New York.

7. Leonov, A.A. and Lebedev, V.N. (1971) *Psychological features of cosmonauts' activity*, Moscow, Nauka.

8. Denisov, V.G. and Onishchenko, V.R. (1972) *Engineering psychology in aircraft engineering and cosmonautics*, Moscow, Mashinostroyeniye.

9. Khorunov, E.V. and Khachaturiants, L.S. (1974) *Man-operator in space flight*, Moscow, Mashinostroyeniye.

Influence of Gravity and Protective Clothing on Cosmonaut-Welder Performance*

B.E. Paton[1], D.A. Dudko[1], V.N. Bernadsky[1], V.F. Lapchinsky[1], A.A. Zagrebelny[1], V.V. Stesin[1], I.F. Chekirda[3] and I.P. Borisenko[3]

The form of interaction in the cosmonaut–space vehicle system is an urgent problem, that needs to be solved in practice, when performing IVA and EVA. Considering special conditions of the cosmonaut's activity, it is necessary to study the influence of the following main aspects on the work performance processes: changed gravity, degree of support, protective clothing, psycho-physiological condition and level of mastering the work skills.

The Yu.A. Gagarin CTC[3] and the PWI, studied the activity of cosmonaut-operator, using different clothing outfits, at normal and zero gravity, to evaluate the influence of gravity and special outfits on the working processes.

In order to solve the problem of optimization of working interaction of the cosmonaut with the welding equipment, several operators were first trained for the purpose of mounting and maintenance of long-term space facilities [1]. They had experience of welding operation performance wearing special outfits and at zero gravity, and had the endurance and capacity for work performance under the conditions of parabolic flight of the flying laboratory. The operators performed welding operations under laboratory conditions, in a horizontal flight, and under the conditions of short-term (30 s) zero gravity in regular overalls, in a spacesuit

---

* (1976) Paper was presented at the 5ᵗʰ Gagarin Scientific Readings on Cosmonautics and Aviation, Moscow.

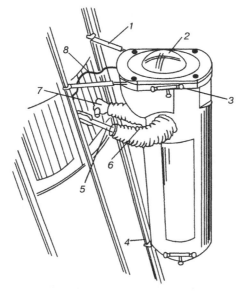

Figure 1 Erection capsule: *1* — mounting stabiliser; *2* — hatch; *3* — manoeuvring nozzles;
4 — outer rail; *5* — special removable tool; *6* — lighter; *7* — removable glove;
8 — power cable, communications line and air hose.

at excess pressure and in a testing facility [2], simulating a special capsule [3], fitted with mobile pressurized sleeves (Figure 1).

The purpose of the operator's activity was application of a model weld on a specimen along a trajectory, marked in advance. Used as a working tool was a light simulator of welding (Figure 2), that is a concentrator with an energy source — a 150 W halogen lamp, placed in its centre. In the second focal point of the ellipse, where the lamp beams are concentrated, a sufficiently high temperature is achieved for low-melting metal welding, and brazing, as well as thermal fracture of non-metallic materials.

The adequacy of the selected procedure was verified in test experiments with the operators performing actual arc welding of stainless steel specimens. These experiments were conducted using the tool shown in Figure 3.

Full comprehensive analysis of the work performance process was conduced using an integrated method of investigation, including:

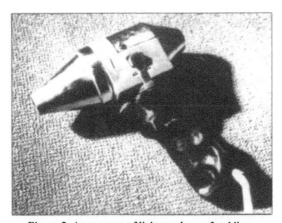

Figure 2 Appearance of light mock-up of welding.

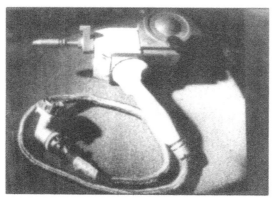

Figure 3 Tool for consumable electrode welding.

- objective procedures of recording welder's performance (accuracy of weld deposition, speed of tool displacement, maintaining the distance to the object of the work);
- physiological-biomechanical methods of investigation of the labour activity (scale cynocyclography, electromyography, goniometry, strain measurement of support reactions);
- operator's subjective impressions.

Evaluation of the welding operator's performance was conducted in a differential manner by the method of statistical processing of the weld trajectory and width, that is the function of welding speed and tool stick-out, i.e the distance to the object of the work. The criterion of tracking accuracy was the mean-root-square value of lateral displacement S, and the criterion of maintaining the stick-out and speed was the mean-root-square value of weld width Δb. Measurements were taken with a 5 mm step. The validity of the results was measured by known methods of mathematical statistics [4].

Physiological-biomechanical assessment of labour activity was performed by recording the bioelectric activity of the muscles, wrist and forearm by the method of global electromiography in four channels, when wearing regular overalls (RO), and in a capsule (C), and in two channels, when wearing a spacesuit (SS). For the forearm muscles quantitative calibration of electromiogram record was further performed at test forces of 10, 20 and 30 kg (Figure 4).

In study [5] it was found that the mean electromiogram amplitude, when applying a force equal to 70 % of the maximal muscle force, was proportional to the force of the voluntary muscle traction. The operator's maximal force of compression of a wrist dynamometer was measured, which was followed by calculation of muscle force in kilograms and its percentage of the maximal force during performance of welding operations under various conditions of gravity and in different outfits. The latter is important, because for long-term and optimal performance of work of a static nature (manual welding has a pronounced static nature), the applied muscle force should not exceed 20 % of the maximal force [6].

The convenience and specifics of welder's activity were also evaluated by analyzing the operators' replies to standard questions, free description of their impressions and estimates of methodologists based on subjective perception of the labour activity at visual observation under the flight experiment conditions.

Relative data on the extent of gravity influence on tracking error $\overline{S}$, using different types of outfit, are given in Figure 5. As is seen from the diagram, zero gravity facilitates tracking, when working in welder's RO, but makes tracking difficult, when working in C. Operators, wearing SS, are indifferent to the absence of gravity with the respect to this parameter.

The influence of different operator's outfits on tracking error under conditions of regular and zero gravity is clearly shown in Figure 6. In this case, an arbitrary unit of tracking

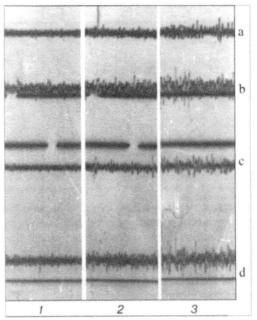

Figure 4 Quantitative calibration of forces in the forearm muscles: a, b — electromiogram of the front and back groups of forearm muscles, respectively; c, d — electromiogram of upper arm bicepts and tricepts, respectively; *1–3* — dynamometer squeezing by the wrist, depending on test forces of 10, 20 and 30 kg.

accuracy is taken to be the error made when working in the welder's RO, for which the error was minimal at zero and regular gravity. When working in a C and in a SS on the ground, the error was essentially higher and practically the same in each case. Zero gravity allowed differentiating the influence of the C and the SS. If the error during work performance in a

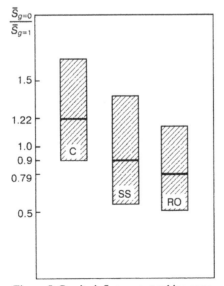

Figure 5 Gravity influence on tracking error
(hatched zone is the interval at confidence probability of 0.95).

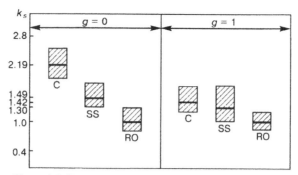

Figure 6 Influence of the welder's outfit on the tracking error
(k_s is the coefficient of outfit impact on tracking error).

SS practically did not increase, when working in a C it became greater compared to the background value under regular conditions.

The influence of gravity and outfit on weld width formation is shown in Figures 7 and 8. Absence of gravity reduced the error during work performance in C, but increased it, when working in RO, and did not change the error during work performance in SS.

The same specimens were subjected to expert evaluation by the same parameters by a five-point scale to determine an integrated performance index [7]. Penetration stability and weld appearance were taken into account in addition to tracking error and weld width. The results are given in Figure 9. As regards the difference in the level of weld estimates in the capsule and the spacesuit, this is probably attributable to the rigidity of shoulder fixation in a capsule that eliminates the involvement of the operator's body and legs during work performance.

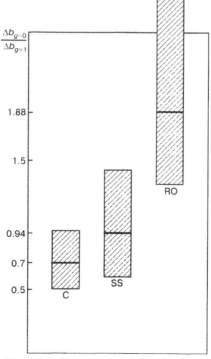

Figure 7 Influence of gravity on weld width.

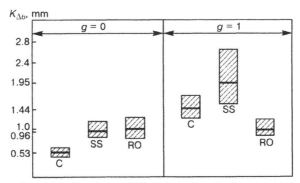

Figure 8 Influence of the welder's outfit on weld width. Averaged width of a weld made by an operator in RO, is taken as a unity.

It should be noted that when working in RO and in C, zero gravity has a favourable influence on the quality of work performance. This is in agreement with the earlier conclusions [8, 9], that zero gravity facilitates man's motion sequence in terms of force and coordination, reduces the applied muscular forces and the number of motion corrections by the higher co-ordination centres.

Application of statistical and expert evaluation procedures demonstrated sufficient agreement and validity of the results on the influence of outfit and gravity on the quality of the performed manual processing operations.

Quite interesting are the results of recording bioelectric activity of the welder's arm muscles, when working under zero gravity. Data are available on the skeletal muscle tone of a man at rest under the conditions of changing gravity. In [10], it is found that the amplitude of muscle biocurrents increases up to 4–5 units, depending on the overload. The bioelectric activity of neck, plain back and hip muscles of a man at rest under conditions of alternating action of overloads and zero gravity was recorded in [11]. It was determined that the activity of neck muscles increased from 130–180 μV in the horizontal flight up to 190–330 μV at twofold overload, whereas at zero gravity it was reduced to 40–50 μV. Sometimes, bioelectric «silence» of the muscles was found at zero gravity. Similar changes occurred in the back and hip muscles.

We found that at dozed muscular traction of 10, 20 and 30 kg, electromiogram amplitude at zero gravity did not change, compared to horizontal flight (see Figure 4). These data allow

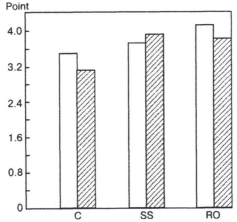

Figure 9 Integrated expert evaluation of weld quality (in the hatched zone $g = 1$).

comparative evaluation of the impact of zero gravity and gear on muscular tension during work activities.

Biomechanical analysis of films established that the work motions of the welder in different outfits, both under the conditions of gravity and zero gravity, require high accuracy, skill of maintaining a constant stick-out of the tool, its comparatively small and constant displacement speed and continuity of movement. Efforts of a static nature prevail during performance of the work.

In all the operators working under zero gravity, the average electromiogram amplitude of the arm and forearm muscles, was the highest at the start of the condition and decreased gradually towards the middle of the condition. By the end of zero gravity condition, further lowering or stabilization of bioelectric activity of the arm muscles was observed during performance of welding, which is indicative of the absence of excess muscular tension or fatigue, and also points to the adaptive processes occurring in central motion regulation (Figure 10).

Muscular tension at gravity during performance of all the welding operations with all kinds of protective gear, was much higher than that during performance of similar work at zero gravity. Figure 11 shows samples of recording the electromiogram of the arm muscle of an insufficiently well-experienced operator during bead deposition in a capsule at gravity [1] and zero gravity [2]. While under the gravity conditions, electromiogram amplitude I_a was higher for all the muscles, and the electromiogram was qualitatively characterized by frequent peaks of activity and instability. At zero gravity, muscular stress decreased by 50 % on the average and was stabilized. Activity peaks were rare and had smaller values. In a more experienced operator, the total electric activity of the arm muscles at zero gravity decreased by 57 %, when working in a capsule (Figure 12).

Such a lowering of muscular tension and facilitation of welding operations performance at zero gravity at a sufficient degree of fixation, agrees well with the evaluation of welding operator performance.

Compared to zero gravity impact, the protective gear was the determinant factor when making work motions in experienced operators. Figure 13 shows measurements of muscular

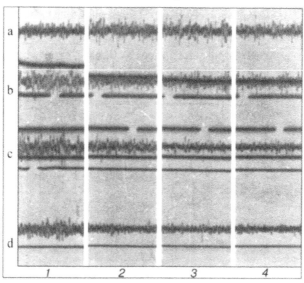

Figure 10 Dynamics of bioelectric activity of the muscles of the operator's guiding arm, when making a weld deposit at zero gravity in a capsule: a–d — same as in Figure 4; *1* — 5th second; *2* — 10th second; *3* — 15th second; *4* — 20th second of the condition.

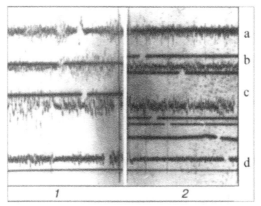

Figure 11 Bioelectric activity of muscles in inexperienced operator, when making a weld deposit in a capsule at regular gravity (*1*) and zero gravity (*2*): a–d — same as in Figure 4.

tension, when working in different outfits. It was established that after the period of adaptation is over (first 8–10 s of zero gravity), forearm muscle tension is equal to 12–14 % of maximal strength when working in RO, 15–16 % when working in a SS, and 17–18 % , when working in a C.

Performance of work in a spacesuit, resulted in redistribution of muscular tension, compared to working in a capsule. Decrease of electrical activity of forearm and upper arm muscles was accompanied by stronger dynamics of the curves of shoulder joint goniograms. The spacesuit enhanced the «rigidity» of the arm kinematic sequence and facilitated joint mobilization for them. The operators also noted subjectively, that arm fixation is easier in a spacesuit. So, the tension of the upper arm bicepts during welding performance in regular overalls, was higher than in a spacesuit.

Figure 14 gives averaged indices of the total electric activity of forearm and upper arm muscles for four operators, working in different outfits under the conditions of regular and zero gravity. For all the kinds of outfit, gravity change from unity to zero did not change the nature of the impact of the outfit on muscular tension. Working in a capsule resulted in the greatest consumption of muscular energy. When working in a spacesuit, muscular tension was lower than that in the capsule, and higher than when working in regular overalls. Absence of gravity smoothed the difference in the level of muscular tension, when working

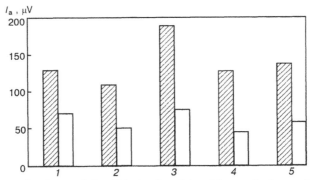

Figure 12 Comparative characteristics of bioelectric activity of the muscles in experienced operator during performance of a weld deposit in a capsule under the conditions of regular (*hatched*) and zero (*light-coloured*) gravity: *1* — front; *2* — back group of forearm muscles; *3* — biceps; *4* — tricepts of the upper arm; *5* — averaged activity of arm muscles.

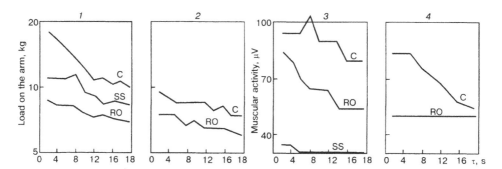

Figure 13 Dynamics of bioelectric activity of the muscles of the operator's guiding arm, when making a weld deposit at zero gravity with different outfits: *1–4* — same as in Figure 12.

in regular overalls, and in the spacesuit, as in the latter, muscle functions of limiting excess degrees of freedom of the arm kinematic sequence were facilitated.

Conclusions

1. Evaluation of the quality of work performance and physiological-biomechanical parameters of the welder's working activity demonstrated the ability of long-term and quality performance of welding in different protective gear under conditions of zero gravity, given sufficient fixation and work experience.

2. Zero gravity conditions only slightly improve the welding quality indices and reduce the applied forces to 50 % of their magnitude at regular gravity.

3. Compared to the impact of zero gravity, the protective gear used is the determinant factor for the welder's work motion sequence and his performance.

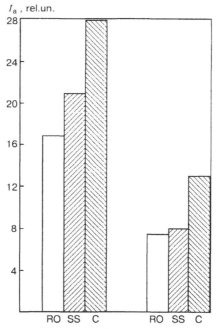

Figure 14 Averaged bioelectric activity of the guiding arm muscles of several operators, when working at regular (*hatched*) and zero (*light-coloured*) gravity.

4. Forearm muscle forces during welding performance at zero gravity were 12–14 % of the maximal force for regular overalls, 15–16 % for the spacesuit and 17–18 % for the capsule, thus meeting the ergonomic requirements for long-term performance of static type of activity.

5. Zero gravity conditions significantly reduce the difference in the magnitude of the total bioelectric activity of the forearm and upper arm muscles during welding in regular overalls and in the spacesuit, compared to that in welding at regular gravity.

1. Khorunov, E.V. and Khachaturiants, L.S. (1974) *Man-operator in space flight*, Moscow, Mashinostroyeniye.

2. Paton, B.E. (1973) Facilities for investigation of technological processes under space simulation conditions, In: *Coll. pap. of 24th IAF Congr.*, Baku, 1973.

3. Denisov, V.G. and Onishchenko, V.F. (1970) *Steps amidst stars*, Moscow, Znaniye.

4. Zajdel, A.N. (1968) *Elementary estimate of measurement errors*, Leningrad, Nauka.

5. Sherrer, Zh. (1973) *Labour physiology (ergonomics)*, Moscow, Mashinostroyeniye.

6. Wudson, W. and Konower, D. (1968) *Reference book on engineering psychology for engineers and artists-designers*, Moscow, Mir.

7. Danilyak, V.I. (1974) *Ergonomics and quality of industrial products*, Moscow, Ergonomika.

8. Chekirda, I.F. (1967) Co-ordination structure and phases of motion skill readjustment at zero gravity and positive overloads, *Kosm. Biologiya i Meditsina*, No. 1, p. 45.

9. Chekirda, I.F. (1967) Co-ordination structure of man's voluntary motions of different complexity under the conditions of flying along Kepler parabola, *Ibid.*, No. 2, p. 46.

10. Babushkin, A.I., Isakov, P.K., Malkin, V.B. et al. (1958) Man's breathing and gas metabolism at radial accelerations, *Fiziol. Zhurnal SSSR*, No. 2, p. 44.

11. Yuganov, E.M., Kasian, I.I. and Asyamolov, B.F. (1963) Bioelectric activity of skeletal muscles under the conditions of alternating impact of overloads and zero gravity, *Izv. AN SSSR*, No. 5, pp. 11–14.

Training Facility for Simulation of Welding Operations Performance in Space*

B.E. Paton[1], D.A. Dudko[1], V.N. Bernadsky[1], V.F. Lapchinsky[1], V.V. Stesin[1], A.A. Zagrebelny[1], A.B. Goldman[1] and O.S. Tsygankov[1]

In 1965 Soviet cosmonaut A. Leonov made the first exit into open space and performed a number of EVA activities, namely filming, hatch opening and closing, removal of cassettes and traps, manoeuvring during movement, etc. This experiment demonstrated, that a man, wearing a special spacesuit, can live and work in open space. However, even then it became clear that achievement of man's full performance under these conditions requires further study of the specific features of work performance and solving a number of problems, related to development of all kinds of special tools [1, 2].

During further experiments, performed outside the flying space vehicle by Soviet and US cosmonauts, spacesuit designs were improved so, that they allow performing practically all the operations that may be required now in open space [3]. Nonetheless, when studying the

---

* (1975) *Space research in Ukraine*, Issue 6, pp. 18–21.

applicability of welding in space [4], without preliminary experiments, it is impossible to reply to the question, whether an operator, wearing a spacesuit, is able to perform such complicated processing operations, as assembly, welding and cutting of various structures [5].

Ground-based experiments demonstrated the possibility, in principle, of manual performance of such work, when wearing a spacesuit, but regular tools turned out to be unsuitable. It also became clear, that when work is performed in a spacesuit, extremely high requirements are made of operation ergonomics and working zone lay-out. In other words, retrofitting of manual welding hardware and technology necessitates a large scope of research and long-term thorough training of the operator. This involves numerous expensive experiments, using the complete package of the spacesuit life support systems.

The problem can be essentially simplified by development of special training facilities [6]. The training facility for optimising the welding units and operator training, shown in Figure 3, p. 126, was developed and manufactured at the PWI.

The facility allows optimisation of the procedure of the operator's performance with manual tools and instruments, when wearing a spacesuit, as well as optimisation of the technology of automatic welding and cutting if required. For better simulation of the space conditions, it can be mounted on board the flying laboratory, providing short-term zero-gravity. Functional layout of the gas-vacuum system of the facility, is presented in Figure 1.

The pressurised working chamber of the facility with portholes for viewing and filming, has a volume of 0.8 m³, thus accommodating large enough objects. The facility design and efficiency of the pumping units, allow the chamber volume to be increased, by adding sections to it. In order to perform manual operations, a spacesuit fragment is mounted on the chamber front wall so, that any required pressure gradient can be provided to simulate the actual working conditions of the cosmonaut. The spacesuit helmet glasses are fitted with a set of replaceable light-filters, allowing operations to be performed with objects of different brightness. All-purpose fasteners are provided inside the chamber for fixation of the objects being tested. One of the walls carries a set of sealed electric lead-ins and leak valves for filling the chamber with air or other working gas, as well as a porthole for specimen

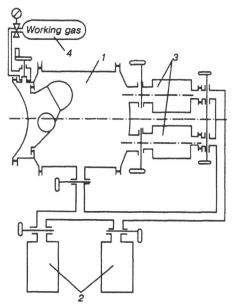

Figure 1 Block-diagram of a gas-vacuum system of the facility: *1* — working chamber; *2* — rough pumping unit; *3* — replaceable high-vacuum unit; *4* — system of chamber filling with working gas.

replacement. If the working chamber accommodates automatic devices, its front wall is closed with a hermetically sealed cover with quick-action clamps.

The required rarefaction is created and maintained in the working chamber by a vacuum system, the principle of operation of which is described in [7]. A feature of the vacuum system in this facility is the use of replaceable high-vacuum units, namely vapour-oil (for ground-based testing) and adsorption-getter (for work performance on board the flying laboratory). The set of each of the units includes two pumps of each type to provide backup and higher efficiency. The system of the chamber filling with the working gas allows achievement of the pressure gradient, required for operation, and filling the chamber with high-purity inert gas, specified by the operating conditions.

The operator and his assistant take part in experiments in the facility. The functions of the assistant include controlling from the facility panel the high-vacuum unit, chamber filling system and recording equipment — regular and high-speed filming cameras, lighting fixtures and multichannel oscillograph. The operator's task is working directly with the object being tested. The operator controls the automatic devices and processes being studied, from a separate panel, and gets into the spacesuit fragment for testing manual devices or tools. The place of the operator's assistant is fitted with a special chair with the leg and waist fastening belts, board for storing the records, and containers for storing the film cassettes and oscillograph.

An essential advantage of the described facility is the capability of unrestrained medical-biological control of the condition of the operator, working in the spacesuit fragment, and convenience of performance of various ergonomic investigations.

Figure 2 Appearance of the training facility, assembled for mounting in the flying laboratory.

Practical experience demonstrated that the facility is a versatile research set-up (Figure 2), allowing performance of a large range of experiments and is an especially effective training facility for the operators.

It should be noted that the capabilities of this facility are not limited to investigation of just welding units. Its design allows placing and testing in the working chamber various manual units, instruments and devices, the operation of which requires thorough earth-based study and training. With the use of the facility, the process of such training is greatly facilitated and shortened.

Use of the above training facility allowed the PWI to develop several ingenious procedures for experimental performance and to conduct a series of investigations.

1. Umansky, S.P. (1970) *Man in space*, Moscow, Voenizdat.

2. Denisov, V.G. and Onishchenko, V.F. (1970) *Steps amidst stars*, Moscow, Znaniye.

3. Sharp, M.R. (1971) *Man in space*, Moscow, Mir.

4. Paton, B.E. and Kubasov, V.N. (1970) *See pp. 154–160 in this Book.*

5. Leonov, A.A. and Lebedev, V.N. (1971) *Psychological features of cosmonauts' activity*, Moscow, Nauka.

6. Denisov, V.G. (1964) *A cosmonaut flies on Earth*, Moscow, Mashinostroyeniye.

7. Paton, B.E., Paton, V.E., Dudko, D.A. *et al.* (1973) *See pp. 118–122 in this Book.*

The issues of safety of cosmonauts during EVA, in particular when they work with hand tools, are given much consideration. The hazard involved in working with process equipment consists in the destructive ability of a heat source (electron beam, arc discharge, exothermal mixtures, etc.). When designing experiments with the electron beam welding hand tool, the hardware developers took into account the probability of a direct impact by the electron beam on the cover of the operator's spacesuit, as well as the space vehicle, outside which the operations were to be conducted, and on its outer structures (aerials, solar batteries, etc.). Protection of the operator or the vehicle in such experiments was ensured through thoroughly training the operator, properly arranging his external work place and strictly following the instructions and procedures for performance of the scheduled welding operations. However, both the hardware developers and specialists who designed those complicated space experiments realised all too well that all the measures taken by them could no more than just partially lower the risk level.

In 1998, at the initiative of Prof. B.E. Paton, the scientific leader of the work on space welding technologies, a competition was announced to advance ideas on how to inexpensively and simply carry out, embody in mock-ups and implement interlocking of the electron beam when it strays outside the working zone. A large team of specialists from both the PWI and other organisations took part in the competition. It resulted in methods for delineating the working zone by a current-conducting frame, or «painting» the working zone with a special dye-sensor, equipping the electron beam tool with a probe to provide an electromechanical contact with the work piece being processed; incorporating into the hardware a device allowing electron beam probing of surfaces on which the electron beam impinges, or placing an infrared «beacon» on the work piece being welded and fitting the hardware with video and radio location systems (Joint Stock Company «Kvant») or «technical vision». Also, it was proposed to consider the cosmonaut space suit as a kind of a multilink manipulator (system of four manipulators, i.e. two arms and two legs, attached to a common base), equipped with angular displacement transducers in the junctions.

Three kinds of systems conditionally called «Probing» (Yu.N. Lankin et al.), «Data beacon» (K.A. Bulatsev et al.), and «Technical vision» (V.V. Tochin et al.) were selected from the presented proposals for further development, making the mock-ups and conducting demonstration tests.

Given are the papers that describe the implementation and results of the tests performed in laboratories of the PWI, Ukraine, and G. Marshall Space Flight Center, NASA, USA, the experts of which M. Vanhooser, C. Russell, M. Terry and S. Clark have made a great contribution to preparation and performance of the tests that is gratefully acknowledged by the authors.

From Compilers

Device for Provision of Safe Operation of Manual Electron Beam Tool in Space[*]

Yu.N. Lankin[1] and S.S. Gavrish[1]

Functioning algorithm. If the electron beam gun is located at the working distance from the surface being processed (less than 150 mm), and the angle of the gun axis inclination is in the specified range ($90\pm30°$), work is performed with a standard sharply-focused beam. Every 0.5 s the electron beam probes the welding zone. Due to the very short probing time (2.25 ms) it does not in any way affect the technological properties of the beam.

If the distance from the gun to the surface, which is hit by the beam, is increased above the admissible level, the protection device will «blur» the electron beam. This will result in an abrupt lowering of the power density effectiveness, as well as of the damaging ability of the beam. The gun displacement into the working zone leads to restoration of the mode with the focused beam.

Principle of operation. A standard electron beam gun *1* (Figure 1) is further fitted with an electromagnetic deflecting system, namely an open magnet core *2* and coil *3*. The electric current flowing through the coil, creates in the magnet core a magnetic field that deflects by angle α electron beam *4* injected by the gun. The coil current changes in time with 0.75 ms period by a saw-tooth law. This results in the beam changing from a point beam of 1.5 mm radius into a divergent linear beam. At 150 mm distance from the gun the beam width is ≈15 mm, i.e. the cross-sectional area is increased by about 13 times, and, therefore, the penetrability is decreased by the same number of times. With the increase of the distance

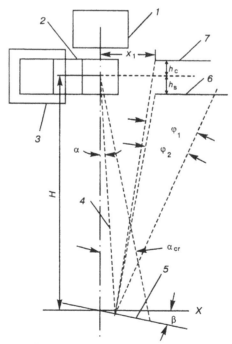

Figure 1 Schematic of beam trajectory in the device (for designations see the text).

[*] (2001) *The Paton Welding Journal*, No. 4, pp. 36–38.

from the gun to the surface, the power density of the periodically deflected beam decreases in inverse proportion to the distance.

Reflecting from the processed surface 5, part of the electron beam hits collector 7. Screen 6 is mounted in front of it, due to which the collector current drops to zero at angle of deflection α_{cr}. As α_{cr} depends on distance H between the gun and the item, its value can be used for interlocking the beam, when it leaves the working zone.

Let us study the dependence of collector current on system parameters and angle of deflection. Let us assume that the electron beam has a normal-circular distribution of current density in the cross-section

$$J = I_m \exp\left(-\frac{r^2}{2r_n^2}\right),\tag{1}$$

where I_m is the current density amplitude; r is the distance from the beam axis in the plane normal to its axis; r_n is the beam «radius». Let us also assume that the electron trajectories are parallel and electron reflection from the surface follows the cosine law. Then, for the value of the electron flow to collector 7 we can write

$$I(\alpha) = I_m \int\limits_0^{2\pi} \int\limits_0^{\infty} \int\limits_{\varphi_1}^{\varphi_2} \exp\left(-\frac{x^2 + y^2}{2r_n^2}\right)\cos\varphi\ d\gamma\ dr\ d\varphi,\tag{2}$$

where

$$\varphi_1 = arctg\left(\frac{x_1 - x(\alpha)}{y_1}\right) - \alpha - 2\beta; \quad \varphi_2 = arctg\left(\frac{x_1 - x(\alpha)}{y_2}\right) - \alpha - 2\beta;$$

$$x(\alpha) = K_1 \cos\alpha + x; \quad K_1 = H\frac{tg\ \alpha\ \sin\ (\pi/2 + \alpha)}{\sin\ (\pi/2 - \alpha - \beta)};$$

$$y_1 = H + h_c + K_1 \sin\alpha; \quad y_2 = H + h_s + K_1 \sin\alpha;$$

$$x = r\frac{\cos\gamma}{\cos\ (\alpha + \beta)}; \quad y = r\sin\alpha,$$

where x_1 is the distance from the gun axis to the collector and screen; r and γ are the beam polar co-ordinates; β is the angle between the gun axis and normal to the item surface; H is the distance from the deflecting system to the item; h_c and h_s is the distance from the deflecting system centre to the collector and screen.

Figure 2a, calculated by (2), gives the dependencies of the collector current on the angle of the beam deflection for different distances from the gun to the reflecting surface. Similar dependencies are also found at the change of the angle of inclination β (Figure 2b). One can see from the Figure that the critical angle of deflection of the electron beam is unambiguously connected with the distance to the reflecting surface and is almost independent of the beam current value. On the other hand, the value of the collector current is directly proportional to the beam current and inversely proportional to the square of the distance. Therefore with large distances from the gun to the reflecting surface, the reflected electron signal is undetectable against the noise background. It is fortunate that this distance is essentially greater than the working distances, and the absence of the collector current can be another signal for the electron beam interlocking.

Control module. The schematic of the control module is given in Figure 3. DD2 microprocessor generates an 8-digit code (port P1) that is converted by DAC (DA1 and DA2)

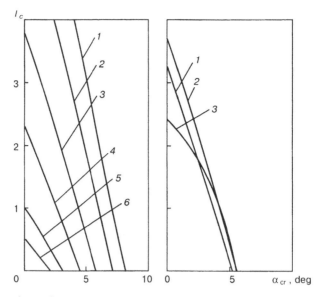

I_c

a   0   5   10 0   5 α_{cr}, deg   b

Figure 2 Dependence of collector current on the angle of electron beam deflection for different distances to the item (a) at H = 35 (*1*), 40 (*2*), 50 (*3*), 65 (*4*), 100 (*5*), 150 (*6*) mm, and for different angles of the gun inclination (b) at H = 50 mm and β = 0 (*1*), 20 (*2*), –20° (*3*).

into saw-tooth voltage. Power amplifier DA3 converts the saw-tooth voltage coming to its input, into the deflecting coil current. The negative current feedback is taken from resistor R6. A signal indicating the specified distance from the gun to the item is formed at the output WR of microprocessor DD2. Green light-emitting diode VD1 is lit on the control module case. A signal of the electron beam interlocking is formed at the output RD. This signal is sent to the hand tool control panel. The signal, coming from the collector, is amplified by operational amplifier DA4. Potentiometer R14 is used to set the value of amplification by direct current, and potentiometer R10 — by alternating current. Comparator DA5, in which the operating level is set by potentiometer R15, generates rectilinear pulses of DA4 output

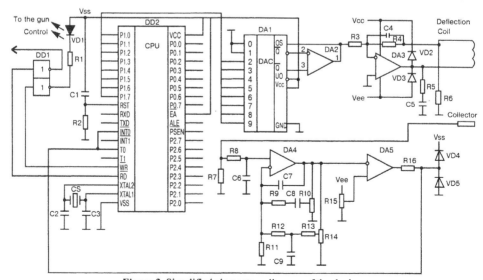

Figure 3 Simplified elementary diagram of the device.

voltage. These pulses are formed by level by diodes VD4, VD5 and come to inputs Tl and INTO of microprocessor DD2. The latter calculates the duration of these pulses during three periods of the electron beam deflection. If the pulse duration is below 50 % of the deflection period, light-emitting diode VD1 is lit, and a logical unity appears at the output of microcircuit DD1.1. Generation of saw-tooth deflection current is interrupted. After 0.5 s three pulses of the electron beam deflection are generated again, the collector current shape is analyzed, etc. Work is performed with the hand tool in the standard mode. The short time of probing the surface with the beam (3 × 0.75 ms = 2.25 ms) and high repetition rate of the probing pulses (512:2.25 = 228) practically do not change the technological properties of the electron beam.

If the pulse duration at output DD5 over three periods of the beam deflection is greater than 50 % of the deflection period, the next three pulses of saw-tooth current are generated and so on. The beam being «blurred» because of periodic deviation at a high frequency, its penetrability drops abruptly, and the greater the distance to the surface onto which it falls, the more it drops. Light-emitting diode VD1 is not lit, after 4 s the signal (logical) for switching off the electron beam of 4 s duration is generated at the output of microcircuit DD1.1. During this time the operator should bring the gun back into the working zone. After 4 s the gun is switched on again. If it is in the working zone, then a standard mode will set

a

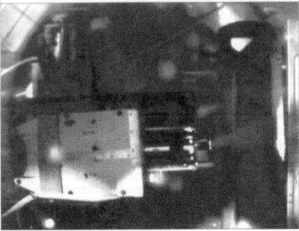

b

Figure 4 Working (a) and blurred (b) electron beam.

in, after three deflection probing pulses. Otherwise, the beam will be «blurred» for 4 s and the beam switching on/off cycle will be repeated.

Experimental verification. The device mock-up was tested together with a set of «Universal» hardware in simulation of manual EBW. The device was set up for operation at 55–65 mm distance from the gun to the stainless steel samples being welded. Beam current was set on the level of 55 mA. With this set-up the influence of the change of the material being welded and the beam current value on the device operation were checked.

It was found that in welding of 304SS stainless steel and Ti-6Al-4V alloy the protection operation range is 55–1000 mm at beam current of 55–75 mA. It is much narrower for aluminium alloy 2219 and for alloy 5456 it is 55–70 mm.

Thus, setting the device up on stainless steel provides guaranteed operation at an acceptable distance in welding of other materials.

With the change of the angle of the gun turning, as was assumed (Figure 2b), protection operates when the electron beam goes beyond the board on which the samples are mounted, i.e. with a significant increase of the distance or at transition to the anodized surface of the board.

The working and «blurred» electron beams at protection device operation are given in Figure 4.

In conclusion, the advantages of the developed protection device should be noted, namely equipment simplicity, small weight and dimensions, reliability, insensitivity to environmental parameters (temperature, vacuum, vapour deposition, lighting, etc.). However, some shortcomings of the device were also found during its operation, namely fitting the gun with the deflecting electromagnetic system with electron collector, which increases the gun overall dimensions; dependence of working zone parameters on the gun current value and processed surface material; and the tested mock-up of the device does not protect the spacesuit material from damage at short range.

The device can be improved by introducing a correction by beam current up to complete elimination of the influence of the latter; improvement of the functioning algorithm and selection of setting up parameters to eliminate the possibility of the space suit burning through at any distance.

In the future the collector current of reflected electrons can be used for automatic correction of the electron beam position relative to the butt being welded, which will greatly improve the welding quality, as well as welding zone visualization.

System of Monitoring the Position of an Electron Beam Hand Tool[*]

K.A. Bulatsev[11] and A.A. Bulatsev[1]

The problem of development of a system of cosmonaut-welder protection during performance of manual welding operations in space, using a hand tool, can be defined as follows. It is the problem of control of the position of the electron beam hand tool in space relative to the structure being welded and interlocking the tool operation when the beam moves outside a certain working zone, or in the case of penetration of the cosmonaut's legs or arms into the working zone.

The objects being welded are structures of arbitrary shape made of ferrous or non-ferrous metals. In keeping with technological requirements the electron beam tool should be located at a distance of 30–100 mm from the surfaces being welded. Deviation of the angle of the

* (2001) *The Paton Welding Journal*, No. 4, pp. 39–41. (Revised and complemented in March, 2002.)

electron beam tool from normal to the surfaces being welded should not be more than ±30°. The working zone, i.e. the area, where welding can be performed, should have an arbitrary shape with linear dimensions of up to 50 cm (limited by the mobility of the cosmonaut's hand in the spacesuit and angle of vision through his view port).

The approach, developed here, is to create a system of monitoring the work tool spatial position, based on a PC, special peripheral devices, applied in virtual reality systems (position/orientation sensors) and appropriate software. Such an approach provides flexibility in determination of the working area. Inside the sensor active zone it is possible to define a working zone of any configuration at the stage of the system start up. A PC with special software is the basis of the system, correlating the tool's current position with the assigned configuration of the working zone and determining the presence of the situation of «leaving the working zone». Use of a PC makes the system flexible, allows the defining of any shape of the working zone, and preserving the specified configuration of the working zone.

The system has two operating modes: setting up and following. In the setting up mode, the user defines (assigns) the working zone configuration and in the following mode the system monitors the tool to determine if it has left the working zone or not. The block-diagram of the monitoring system design is given in Figure 1 and demonstrates the principle of operation of the device.

The above device embodies a variant of system realization, when control is performed using a PC keyboard and mouse. The final embodiment, however, will require adding two keys to the tool control panel, namely «Setting up» and «Following». In addition, the final variant of system realization will lack the capability of representing any information on the PC screen, as there will be no screen or PC in the conventional sense.

The working zone configuration is defined by marking two diagonal limit points of the zone. The sensor with the tool is placed at a point, and by pressing «A» key, or «Setting up» button in the menu, the current point is stored. A similar operation is repeated also for the second point. The created configuration of the working zone can be saved by the user if desired.

A system going over into the following mode is performed by pressing the «Space» key or «Following» button in the menu. When operating in the following mode, the system

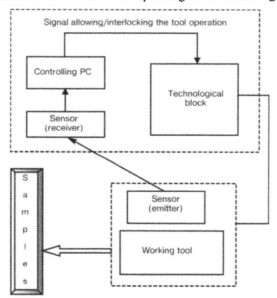

Figure 1 Block-diagram of a device for monitoring the position of the electron beam tool.

periodically (with about 50 Hz frequency) inquires about the position of the sensor. The sensor current position in space is checked for moving outside the specified working zone. The information about the current spatial position of the sensor is displayed. If the sensor (and the tool, respectively) leaves the working zone, the tool operation is blocked. The system is taken out of the following mode by pressing «Esc» key, or pressing a second time the «Following» button in the menu. The algorithm of operation of the system in the general form is given in Figure 2.

Currently there are several types of indirect sensors of position/orientation that can be used as the basis for the following system. The following classes of sensors can be singled out: magnetic (electronic compasses), gyroscopic, ultrasonic, inertia, infrared.

Magnetic sensors have a very high accuracy, up to 6 degrees of freedom, allow determination of the spatial position and orientation, and are not related to the control block, i.e. they are connected with the Earth's magnetic field. The disadvantage of such sensors is the development of deviations when operating near bulk bodies of magnetic materials and the need for correction when the spatial position is changed. To ensure correct operation of the sensor in such situations, it is possible to introduce compensation into calculations or

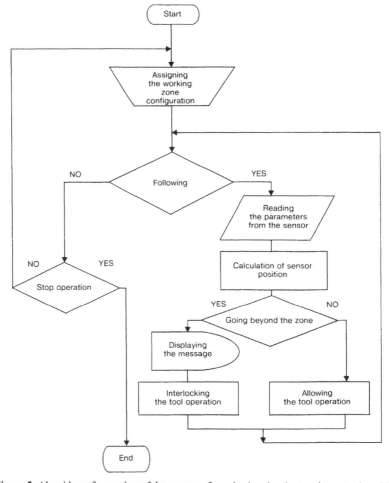

Figure 2 Algorithm of operation of the system of monitoring the electron beam tool position.

perform calibration before each operating session. When these sensors are used in flight, data on flight trajectory should be entered to correct their position.

Gyroscopic sensors allow measuring just the change of orientation, while remaining insensitive to spatial displacements.

Inertia sensors are not tied up to the control block, either. They allow recording the smallest acceleration and converting it into co-ordinates by double integration.

Ultrasonic sensors consist of a signal emitter and receiver. A triangulation method is used to determine the receiver position in space relative to the emitter. The disadvantage is the need to have a couplant, through which the signal should be transmitted, i.e. it will not operate in vacuum.

An infrared sensor consists of a directed emitter and receiver. The receiver, a series of sensory elements, allows determination of the emitter direction, provided the emitter is aimed at the receiver. When the emitter has left the signal receiver working zone, the signal for interlocking the tool operation is still preserved. The possible disadvantages of the system are the insufficient range of the angles of deflection. Possible method of solving it is by the optimal orientation of the emitter–recevier system relative to the working zone.

We were able to get a virtual reality helmet of Union Reality Company and Ultimate 3DZ FreeD device for entering information.

Let us first consider the operation of a system based on the sensor in the Union Reality helmet. The Union Reality sensor consists of a group of sharply-focused infra-red emitters and a receiver. Connection to the PC is via a game port. The receiver module is attached to the work tool and the emitter module — to the fixture with samples. The emitter–receiver system is schematically represented in Figure 3. One projection is considered, whereas the receiver has two dimensions. The sensor allows the distance D to be measured. This distance determines the direction of the source beam.

The working zone is defined by the limit values of distance D and receiver position. To assign the working zone configuration in this case, it is necessary to nominate two limit points for the horizontal and vertical deviation of the beam.

The point position proper is assigned as follows. The tool is placed into the first limit point and «Configure» key is pressed on the panel. The operation is repeated for the second diagonal point.

After zone configuration has been assigned, the system is brought into the following mode by pressing the «Following» key on the panel. After that the system operates in keeping with the algorithm.

Sensor interrogation is performed periodically (about 50 Hz). The derived values are converted into distance D (Figure 3) for the vertical, D_v, and horizontal, D_h, deviation of the beam. Calculated D values arc compared with the specified admissible range of D values. When D value goes beyond the specified range of magnitude, tool operation is interlocked.

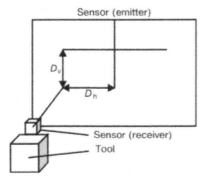

Figure 3 Schematic of Union Reality sensor.

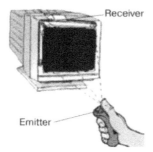

Figure 4 FreeD ultrasonic emitter and receiver.

The system is taken out of the following mode by second pressing of the «Following» key on the panel.

FreeD sensor consists of an ultrasonic emitter and receiver. The emitter is wireless and made as a ring that can be put on a finger, with two control buttons, as on a standard mouse. The receiver is made up by two normally oriented sensor series and filters. The receiving device calculates the emitter position by the triangulation method. The sensor is schematically shown in Figure 4.

Sensor working range is up to 1 m, positioning accuracy being 0.2 mm. Connection to PC is through a serial port.

Fastening the emitter on the work tool, and the receiver on the technological block, we will be able to control both the spatial position of the emitter and of the work tool.

The working zone configuration will be limited by an area of a plane. In our case, as was described above, it is sufficient to assign two points. The working zone height is another parameter, the value of which is assigned at the stage of setting up the system. In the general case, when the number of limiting points is N, we obtain N-polygon, belonging to a plane, determined by the assigned points.

The definition of the points proper is performed as follows. Place the tool into a point and press the «Configure» key on the panel. Repeat the step for all the points.

After the working zone configuration has been assigned, the system is brought into the following mode by pressing the «Following» key on the panel. After that the system functions in keeping with an algorithm.

Sensor interrogation is performed and the obtained values are converted into spatial co-ordinates. Calculated co-ordinate values are compared with the current configuration of the working zone. When the sensor leaves the working zone, tool operation is interlocked. The system is taken out of the following mode by pressing the «Following» key on the panel a second time. On the whole, the system operates similar to Union Reality with practically the same tool control and operator interfaces. The only difference in the principle of operation, namely that FreeD sensor requires a couplant for sound wave transmission and it is not suitable for use in vacuum, respectively, allows using a similar device to control any electrically-driven tool, the functioning of which requires controlling the working zone.

Application of standard position sensors allows the system to be simplified, and the use of a PC with special software makes it flexible and allows experimenting at the setting up stage.

As was noted above, the system, when implemented in the on-board tool, should have two additional keys on the tool control panel, namely «Setting up» and «Following». These keys are required for assigning the working zone points and following the tool operation, respectively. A processor board that enables the co-ordination of the operation of the position sensor and the tool will be located inside the technological block. Before starting the operation after the «Start» key has been pressed and before selecting the welding modes, it is necessary to define the working zone. When assigning the working zone, it will be necessary to bring the tool to the extreme diagonal point of the zone and press «Setting up» key to define the zone, then bring the tool to another diagonal point and again press the same key. After the working zone has been defined, it will be necessary to start operation by pressing the «Following» key.

The system mock-up was tested in a vacuum chamber with the operating tool. Of all the available input devices for using with the tool, Union Reality device was found to be suitable, based on the above considerations. The system of monitoring the position of the electron beam tool was tested in OB-1469 chamber, fitted with a manipulator. A manipulator can provide parallel displacement of the tool, thus permitting processing of four samples, arranged in pairs one above the other. Turning and tilting the tool it is possible to increase the processing zone up to half of the board, but this zone is at the limit of the followed area of Union Reality device. Therefore, it is not very convenient to use the entire zone that can be processed by a manipulator with the tool, in terms of monitoring the position and of the convenience of using the tool proper with additional turning required in the final displacement points.

During testing a working zone was assigned within which the tool should be switched on and when leaving it should be switched off. This area can have an arbitrary rectangular shape, assigned by two diagonal points. It should be assigned every time before starting the work, as described above. In our case of a mock-up embodiment, the zone was assigned by pressing «A» key. The mode of following the process, i.e. interlocking the tool operation, is set by pressing «Space» key. Resetting the assigned working zone after completion of the work is done by pressing «C» key.

The system was tested in the analog joystick mode. Work started with position sensor calibration. Maximum limits of the board working zone were assigned. After that two arbitrary points were assigned in this zone as diagonals of the working zone rectangle. Then work was begun in the following mode with monitoring the tool operation. In the case, when the tool left the working zone, tool operation was interlocked and it was restarted when entering the working zone with the pressed key. A certain incorrectness of the position sensor operation was observed, and sometimes the position sensor «jumped» in the screen at only a small displacement of the tool. Also a somewhat trapezoidal shape was observed instead of the rectangle. This is attributable to low requirements to the game input devices used in this case.

Testing showed the practical applicability of contactless sensors for monitoring the electron beam tool operation. The system has an easily readable interface, i.e. after demonstration of one cycle, including defining the welding zone and conducting welding, we let the operator of our US customer use the tool and she managed to do it without any problems or difficulty. The system is highly flexible, i.e. we will be able to assign any area and the tool will follow it. Another advantage of the system is the high operating speed, reliability and repeatability of the results, inherent to computer systems, guaranteeing the system is switched off outside the working zone. Furthermore, the system is readily incorporated into the existing tools and does not require any retrofitting of the latter.

In addition, we can also consider devising additional means for cosmonaut's protection in the case of an arm being brought into the working zone. This could be a bracelet with a sensor, worn on the arm, and creating a virtual protective zone around the cosmonaut's hand (glove). When the protective zone is brought closer to the operating tool, it will also be interlocked, and, thus provide protection of the operator's hand. We are deliberately not considering here protection of any other parts of the cosmonauts' body, as the working zone is small (about 100 mm at maximum), and it is difficult to imagine the possibility of anything getting into it.

We believe this to be a practicable solution of the problem of cosmonaut's protection from the beam impact in electron beam welding, as well as when working with any electrically-driven tool.

Safety System of Electron Beam Tool*

**V.V. Tochin[9], B.I. Yushchenko[9], P.P. Rusinov[1],
V.V. Demianenko[1] and T.E. Palamarchuk[1]**

A system of electron beam safety (Figure 1) based on processing video information received from the welding zone, includes a video camera, controller and adjustment software-hardware system.

The controller is designed to allow for the specific features of electron beam application, using the hardware of MicroPC series (Octagon Systems) and consists of the central processor unit — MicroPC 5066 (586-133 processor) and MicroPC 5278-RM frame. As MicroPC series lacks a video adapter module, it was additionally developed and made in keeping with the technical requirements of this series. The controller software, consisting of the working program and program of data exchange through the network, is written in Pascal language.

The software–hardware system is based on the IBM PC computer. The software incorporates a user interface and network data exchange program and is developed in Delphi 3 environment. The user interface (Figure 2) is designed for visualizing the video information, setting the control and adjustment window parameters, selection of brightness filtration level, adjustment of video signal amplification, and selection of the video camera position relative to the anode axis (adjustment) during setting up the tool. The parameters derived during setting up are entered into the controller working program, and thereafter it operates in a self-sufficient mode.

The principle of this system operation is based on finding the reference marks in the welding zone. The video camera is mounted on one axis with the tool anode. The video camera lens is aimed into the welding zone through a mirror (video camera optics protection). A control window is created by program means (Figure 3), inside of which the check for reference marks is performed. The control window dimensions and its position in the frame are assigned during setting up.

If three or more reference marks are within the control window when the tool is placed in the welding zone at the working distance from the sample, this is indicative of the tool being within the welding zone and is a sufficient condition for switching off the electron beam interlocking signal. If the number of reference marks in the control window is less than three or the marks are completely absent, a signal for interlocking the electron beam is given.

The algorithm of determination of the presence of reference marks in the control window is as follows:

- digital filtration of video information by brightness;
- creating the data matrix;
- searching for reference marks in the matrix by the specified features;
- if the result is positive, eliminating the electron beam interlock.

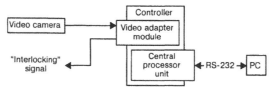

Figure 1 Functional diagram of the electron beam tool safety system.

* (2001) *The Paton Welding Journal*, No. 4, pp. 41–42.

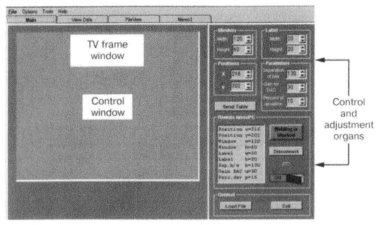

Figure 2 User interface.

Mock-up testing of the above system of electron beam tool interlock as part of «Universal» hardware was conducted in a vacuum chamber fitted with a manipulator. The tool with the video camera was mounted on a manipulator, providing displacement relative to the board with the samples to be welded. In addition to samples, the reference marks were also mounted on the board that defined the working zone in which the welding of samples was to be performed. Observation of switching on and off the electron beam in the working zone and when the beam left the zone, was performed at the tool displacement.

Testing was conducted with two kinds of reference marks, namely passive and active. Good illumination of the welding zone was very important when working with passive marks (non-metal circles painted black against a white background). The light sources were located on the tool and beyond it. When the tool was moved, the lighting of the welding zone was changed and the marks were partially lit up, while the light intervals between the marks were darkened, thus resulting in the change of the mark size and appearance of dark spots. The electron beam interlocking operated in the working zone. Since the work on selection of the correct lighting did not give positive results, an attempt was made to apply active marks (light-emitting diodes defining the working zone), the use of which yielded positive results. Selection of an appropriate level of black and white ratio allowed producing a rather stable image and forming the «allowing-interlocking» signal. When the tool left the welding zone, the interlocking signal switched the electron beam off. The melt zone in the welding area also was a light source. However, correct selection of the control zone eliminates the influence of this light source on taking a correct decision on interlocking.

When the electron beam was switched on during testing, interference appeared, which by distorting the TV signal information led to disappearance of the marks in the frame, as well as unauthorised (erroneous) interlocking of the electron beam. Interference generated

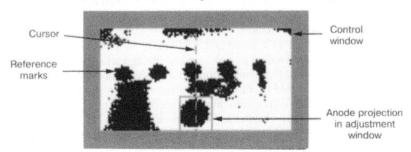

Figure 3 Control window.

by inverter operation and electron beam «noise» confirmed the need for a more thorough investigation of the safety system protection from electromagnetic noise. The faults found and other shortcomings of the above safety system did not permit us (unlike the first two proposals, published in this issue of the journal) to take it to the demonstration testing stage, the terms of which were specified by the competition conditions.

WELDING, CUTTING AND BRAZING OF METALS

Experiment on Metal Welding in Space*

B.E. Paton[1] and V.N. Kubasov[2]

The scope of space research increases every year. Cosmonautics poses ever new problems for researchers and engineers. Performance of various processing operations in space is becoming feasible. This is the reason, why research conducted by welding scientists to develop the fundamentals of space welding and metallurgical technology is so important and critical. Selection of the methods of metal melting and welding that are the most promising in terms of space conditions was of special importance.

Special features of space as an environment for welding operations performance are as follows:

• zero gravity;

• deep environmental vacuum with a very high rate of evacuation (diffusion) of gases and vapours, formed in the metal melting zone or entering it;

• rather broad range of possible temperatures of the molten and solidifying metal.

It is known that space is characterized primarily by deep vacuum. In this connection, the most rational is the application of exactly those processes of joining metals which have been tested and are already used for vacuum welding on Earth. These mainly are diffusion and electron beam welding and cutting. On the other hand, progress of research on space technology of metals led to development of such processes as welding and cutting by a low-pressure plasma arc, consumable electrode low-pressure arc welding, and resistance (spot and seam) welding in vacuum.

Part of these processes are not related to melting or free solidification of the metal being welded, and their use in space will, probably, not run into any particular difficulties. However, the applications of these processes are limited. The other most widely used processes are characterized by melting of the metal being joined and presence of vapours and gases in the place of welding. Therefore, application of these processes in space required conducting the most thorough research.

Simultaneous simulation of vacuum and zero gravity was the greatest challenge. For this purpose, a system of hardware was developed, allowing various welding processes and systems to be tested on board the flying laboratory in vacuum and zero gravity (Figure 1). The hardware system consisted of vacuum chambers, mechanical roughing pumps, sorption-getter pumps, recording instrumentation (high-speed and regular cameras) and controls. The upper cover of each vacuum chamber accommodated units for welding with various processes, namely the electron beam, low-pressure plasma arc and consumable electrode.

* (1970) *Avtomaticheskaya Svarka*, No. 5, pp. 7–12.

Figure 1 Auxiliary device for stabilizing the tool displacement when working in a spacesuit.

When experiments were performed, an oscillograph was used to record the main electrical parameters of the welding mode, pressure in the chambers and gravity in each flight section. Behaviour of the liquid pool and electrode metal drops in consumable electrode arc welding was recorded during the entire process by cameras with a speed of 24 frames per second. Selective filming by SKS-1M cameras at a speed of 1000–5000 frames per second was conducted for recording transient processes. Filming and oscillographing were synchronized; there was also a capability to register a point on the weld line, in which high-speed filming was performed. The flight experiment results were thoroughly analyzed after their completion. Analysis revealed the most characteristic features of each welding process at zero gravity. These features were briefly, as follows.

In electron beam welding and cutting at zero gravity the molten metal is contained in the pool or the cut cavity just by the surface tension force, that in the general case decreases with increase of the metal temperature. The beam pressure and reactive pressure of metal vapours are trying to drive the molten metal out of the melting zone. Therefore, it was extremely important to find out, whether it was possible to provide good weld formation in EBW, characterized by an extremely high energy concentration and, therefore, causing molten metal overheating. In cutting it was primarily necessary to find out, whether the molten metal would be localized along the boundaries with subsequent solidification, or, whether under the beam impact, it will be removed from the cut cavity in the form of drops, which is inadmissible at zero gravity.

Welding and cutting were performed with a low power beam (1 kW at 70 mA current and 30 m/h welding speed). Different types of joints were made, namely on Al-6.2Mg aluminium alloy, overlap joints of VT1 commercial titanium alloy, and flanged joints of 12Kh18N10T stainless steel (C — ≤ 0.12; Cr — 17.0–19.0; Ni — 9.0–11.0; Mn — 1.0–2.0; Si — ≤ 0.8 wt.%).

At short-term stay of the molten metal at zero gravity (up to 25 s), weld shape and penetration depth were found to be identical to those produced under regular conditions (Figure 2). Somewhat higher porosity was observed in aluminium alloy specimens, welded at zero gravity. This can account for difficult evolution of gases from the molten metal in the absence of gravity.

Cutting with the electron beam at zero gravity does not run into any special difficulties, either. Molten metal is not completely removed from the cut cavity, but solidifies along the edges as drops (see Figure 8 on p. 46), or a continuous bead.

Thus, experiments confirmed the ability to perform electron beam welding and cutting of various materials at short-time zero gravity.

Similar results were obtained in low-pressure plasma arc welding and cutting. It was possible to successfully perform quality welding of flanged joints on stainless steel and make butt and overlap joints of titanium alloy at zero gravity. Welded joint micro- and macrostructure

a

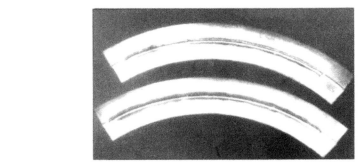

b

Figure 2 Electron beam welded specimens of 12Kh18N10T stainless steel joints: a — at zero gravity on board the flying laboratory; b — under the action of gravity.

were thoroughly studied without observing any significant deviations, associated with zero gravity impact (Figure 3).

Note a minor increase of mechanical strength of titanium alloy welded joints produced at zero gravity.

This welding process is characterized by the influence of vacuum and pumping rate on the arc discharge stabilization. At a high pumping rate the plasma gas coming from the torch nozzle diffuses extremely quickly into the residual atmosphere of the chamber, thus markedly complicating arc stabilization. The performed experiments permitted establishing techniques, that allow enhancing arc excitation reliability.

Investigations of consumable electrode arc welding at zero gravity were focused on the problem of controlling the electrode metal melting and transfer. Welding experiments were conducted in a controlled atmosphere and in vacuum. Investigations demonstrated that the force of surface tension and metal wetting became the most important in formation and transfer of electrode metal drops.

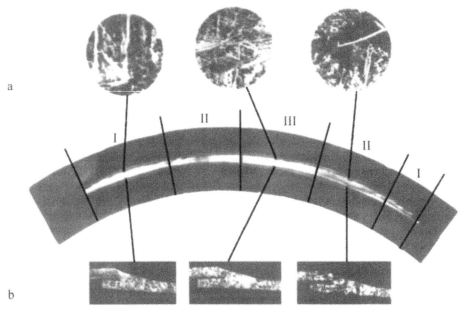

Figure 3 Macro- (a) and microstructure (b) of a welded joint in VT1 alloy, made by low-pressure plasma arc at zero gravity on board the flying laboratory (×20): I — horizontal flight; II — overloads; III — zero gravity.

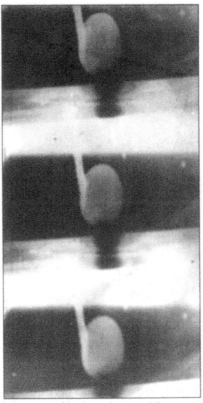

Figure 4 Drop on the electrode in consumable electrode arc welding at zero gravity in a flying laboratory (controlled argon atmosphere, welding current of 50 A, arc voltage of 15 V, feed rate of 180 m/h).

With a long arc, when free growth of the drop is provided, the latter can reach very large dimensions, determined, in principle, by the arc gap (Figure 4). When welding in this mode on the ground, the electrode metal drop is several times smaller.

At zero gravity the drop shape, as a rule, changes continuously, but remains close to spherical, being indicative of the prevailing influence of the surface tension force. In welding some metals the filmograms revealed an intensive rotation of the drops in the meridional direction. In view of continuous change of localizing conditions, the active spots of the arc also continuously and chaotically move over the surface of the drop and the pool. The welding process is unstable and does not provide a sound weld.

Two methods of stabilization of consumable electrode arc welding under zero gravity were studied. With a long arc the process was successfully stabilized by superposition of current pulses. At the moment of the pulse application, the electrodynamic force, providing separation of a small metal drop from the electrode and its falling precisely into the pool, grows abruptly (Figure 5). In the case of a short arc (at a rather high rate of electrode feed and low voltage), welding proceeds with short-circuiting of the arc gap. Stable fine-drop metal transfer is also guaranteed. Weld quality is high in both cases (Figure 6). Welding with short-circuiting turned out to be simpler, and it was used further on when studying the process both in a vacuum and in a controlled atmosphere.

Metal solidification and weld formation at zero gravity also have their special features. Absence of gravity results in the metal being drawn from the weld edges to its axis under the action of surface tension. Weld shape still remains quite satisfactory.

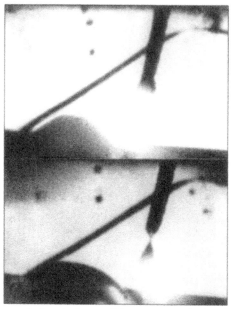

Figure 5 Drop transfer in pulsed-arc welding at zero gravity in a flying laboratory
(welding conditions as in Figure 5).

The final stage of investigations consisted in test welding performed directly in near-earth space under long-term zero gravity and vacuum, that were in place in a depressurized compartment of a space vehicle.

These experiments were conducted using «Vulkan» hardware (Figure 7) — an integrated self-contained system, allowing welding to be performed by several processes, namely electron beam, plasma arc and consumable electrode. The hardware consists of two modules, namely one accommodating the welding units and specimens to be joined, and the other — the power supply system, controls, instrumentation and transducers, as well as automatic devices. The welding control panel is located in the vehicle descent module. Total weight of the hardware is up to 50 kg.

In keeping with the general program of welding research, the first welding experiment was performed on October 16, 1969 in «Soyuz-6» vehicle by cosmonaut-pilots G.S. Shonin and V.N. Kubasov. After depressurizing of the compartment, the cosmonaut-operator, who was in the descent module, switched on the automatic low-pressure plasma-arc welding, in

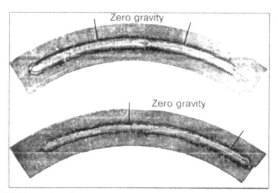

Figure 6 Welds, made with a consumable electrode on board the flying laboratory.

Figure 7 «Vulkan» welding hardware.

keeping with the program. This was followed by switching on automatic devices for electron beam and consumable electrode arc welding. During each test, the cosmonaut monitored the hardware operation by the signal screen on the control panel. All the data on the welding mode and conditions of the experiment performance were transmitted to Earth and recorded by chart recorders.

The performed experiment confirmed the earlier main assumptions and investigation results, obtained in the flying laboratory. The process of electron beam melting and cutting proceeds in a stable manner and provides the required conditions for normal formation of the welded joint or cut (Figure 8).

The main parameters of the mode of consumable electrode arc welding in «Soyuz-6» vehicle, as well the structure of the weld and HAZ were practically the same as with welding on Earth and in the flying laboratory; and the necessary penetration of the metal being joined was achieved. Weld metal was dense, without gas or non-metallic inclusions; gas removal from the molten metal during solidification was satisfactory. No significant deviation from the specified composition of the weld metal or the remelted electrode metal was found.

Study of consumable electrode arc welding demonstrated that a long-term stable arc discharge in the electrode material vapours can be achieved under conditions of long-term zero gravity, despite a high rate of evacuation.

Welding by low-pressure plasma arc in this hardware did not yield the anticipated results. The rate of plasma gas diffusion into the vehicle atmosphere was, probably, higher than the anticipated one. Therefore, its concentration in the arc gap turned out to be insufficient for contraction of the constricted arc.

On the other hand, the high speed of exhaustion of gases through the space vehicle hatch had a positive role in electron beam cutting. The concurrent gas evolution did not affect the reliability of operation of electron beam welding equipment.

Small-sized welding units, incorporated into «Vulkan» system, demonstrated sufficient reliability and performance in space environment. Principal decisions, taken during development of these units, and the data of the experiment proper, can be the basis for designing

a

b

Figure 8 Macrosections of a welded joint (a) and one edge of the cut (b),
made by electron beam in space.

special welding systems for performance of specific processing operations in the space environment.

The experiment on welding in space is a new and important stage in progress of welding science. A technological process involving metal heating and melting, was implemented for the first time in the world.

Note, that the use of high energy density welding heat sources can turn out to be beneficial and needed not only for welding or cutting, but also for processing parts, producing super pure metals and performance of other similar work in space.

Features of Equipment and Processes
of Electron Beam Welding and Cutting in Space*

**B.E. Paton[1], O.K. Nazarenko[1], V.I. Chalov[1], Yu.V. Neporozhny[1], V.K. Lebedev[1],
I.I. Zaruba[1], V.D. Shelyagin[1], D.A. Dudko[1], V.N. Bernadsky[1], G.B. Asoyants[1],
Yu.N. Lankin[1], Yu.A. Masalov[1], G.P. Dubenko[1], Yu.I. Drabovich[12],
N.N. Yurchenko[12], I.V. Afanasiev[2], G.A. Nazarov[2] and N.G. Sidorov[2]**

The high interest shown in the electron beam as a possible source of heating under conditions of space with its inherent vacuum, is determined by a number of important factors:

• high (up to 80 %) efficiency of transformation of electric power into thermal power, consumed for heating and melting of metal;

• as to energy concentration in the heated spot, the electron beam is superior to the other known heat sources (being just slightly inferior to the laser beam). All this allows reducing the power consumed in welding (compared to other heat sources) at equal metal thickness;

• welds made with the electron beam, have small HAZ dimensions, high penetration width to depth ratio and, as a result, good physico-mechanical properties;

* (1970) *Avtomaticheskaya Svarka*, No, 3, pp. 3–8.

- electron beam is the best well-established and versatile tool for metal processing under vacuum. Sharply-focused electron beams are even now widely used for welding, cutting, precise machining, melting, coating and heat treatment of various materials;
- when the electron beam is used, the reactive forces are completely absent.

The need to provide maximum safety when operating hardware for space applications, in addition to the requirements of high reliability and manoeuvrability at minimum weight, makes it necessary to use a special approach to development of this hardware.

The PWI has performed investigations to optimize the processes and develop manoeuvrable, highly reliable and safe hardware for electron beam welding and cutting of sheet material under the conditions of near-earth space.

Selection of hardware parameters. It is known that the main requirements for electron beam welding hardware, i.e. rather high energy density in the focal spot W_f and low angle of convergence α_0 of the beam on the product, are met to a greater extent at a high energy of the electrons [1]:

$$W_f \approx I_b^{1/4} U_{acc}^{7/1}, \tag{1}$$

$$\alpha_0 \approx U_{acc}^{-1/8}, \tag{2}$$

where I_b is the current of the electron beam; U_{acc} is the accelerating voltage.

Use of high accelerating voltages (30–60 and even 100–200 kV) is rational, when it is the case of joining metals $\delta = 10 \div 20$ and $50 \div 100$ mm thick. As shown by our investigations, application of these stresses in welding sheet materials does not offer any advantages in the majority of cases. On the other hand, increase of accelerating voltage makes the hardware more complicated, and the requirements for electrical insulation more stringent. It also markedly increases the integral intensity i of the continuous radiation spectrum, generated in electron braking

$$i \approx U_{acc}^2 I_1 z, \tag{3}$$

where z is the atomic number of the substance braking the electrons. The admissible weekly dose of operator exposure with a 36 h business week, should not exceed 100 mR (at the radiation dose power of 6 µR/s the limit time of working in the radiation zone is not more than 4.5 h per week).

In consideration of the fact that in space the X-ray radiation and the difficulty of providing high-voltage insulation can limit the application of the electron beam, a quite low accelerating voltage of 10 kV was accepted at the PWI. With such a voltage, the power of X-ray bremsstrahlung dose allows the operator to stay in direct vicinity of the welding area, without using biological protection (Figure 1).

It should be noted that at the initial stage of development of the hardware for EBW in space, US researchers selected the accelerating voltage of 80 kV [2]. Lately, however, there was also information about them starting to use $U_{acc} = 5 \div 20$ keV [3].

At ≈ 10 kV accelerating voltage and 0.5–1.5 kW beam power, required for welding and cutting structural materials $\delta = 1 \div 3$ mm thick (Figure 2), as well as effective evaporation of metals (Figure 3), the energy density in the focal spot is not limited by the space charge.

It is known that the beam consisting of particles of the same sign and released into space free of external focusing fields, can expand under the action of the electrostatic forces. However, when certain conditions are met, it is possible to perform ion focusing of the electron beam. Ions, having a positive charge, induce an electric field that practically completely compensates the impact of electrostatic repulsion forces between the electrons in the beam.

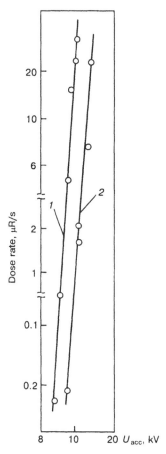

Figure 1 Dependence of X-ray bremsstrahlung dose rate on accelerating voltage in EBW
($I_b = 60$ mA) of tungsten (*1*) and titanium (*2*).

With ion focusing the magnitude of gas or vapour pressure is crucial. It should not be too high, as in a dense medium beam divergence depends not on the electrostatic forces, acting in it, but on collision of the electrons with the atoms of gas or vapour. On the other hand, pressure should not be too low, as in this case, the density of positive ions in the beam area will be insufficient to fully compensate the mutual repulsion of electrons. Ion focusing is optimal at $10^{-4} - 10^{-5}$ mm Hg.

Environmental characteristics, that are of interest to us in the near-earth space [4] are given in the Table.

As can be seen, residual gas concentration at an altitude of 400 km and more, can be insufficient to compensate for the spatial charge in the electron beam.

During material heating and melting, gases evolve and metal vapours form in considerable amounts. Gases and vapours propagate within a quite narrow cone, the axis of which is the electron beam, the particle density in the cone being definitely greater than that of the residual gases in the environment. Therefore, their pressure does not practically affect the beam focusing quality.

Minimal time, during which the beam will be neutralized, i.e. its ion concentration will reach electron concentration,

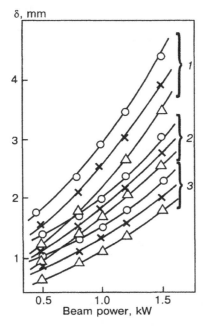

Figure 2 Dependence of thickness of metals being welded on electron beam power (U_{acc} = 10 kV, focal spot diameter at 70 % of total power level is ~ 1 mm): *1* — VT1 commercial titanium alloy; *2* — Al-6.2Mg aluminium alloy; *3* — 304SS stainless steel; $\bigcirc$ — v_w = 10; $\times$ — v_w = 20; $\triangle$ — v_w = 30 m/h.

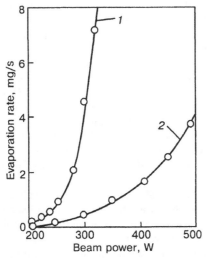

Figure 3 Rate of evaporation of aluminium (*1*) and nichrome (*2*), depending on beam power (U_{acc} = 10 kV).

| Height, km | Pressure, mm Hg | Density, cm⁻³ | | Average molecular weight | Average | |
		electrons	neutral parts		thermal velocity of atoms, m/s	T, K
200	$1.2 \cdot 10^{-6}$	$10^5 - 10^6$	$8.9 \cdot 10^9$	26.0	1063	1227
400	$4.2 \cdot 10^{-8}$	Same	$1.5 \cdot 10^8$	19.6	1266	1436

$$\tau_n \approx \frac{1}{\varepsilon(v)pv} = \frac{0.0169 \cdot 10^{-6}}{p\varepsilon(v) \sqrt{U_{acc}}} \text{ [s]}, \tag{4}$$

where $\varepsilon(v)$ is specific ionization; p is pressure, mm Hg.

Let $U_{acc} \approx 10\,000$ V, averaged pressure along the beam axis $p \approx 10^{-5} \div 10^{-6}$ mm Hg. Then, $\varepsilon \approx 0.9$ and $\tau_n = 20 \div 200$ µs. It is natural that such a short time is not in the way of successful performance of welding and cutting.

Electron-optical system. The electron beam for welding and cutting sheet metal can be effectively formed by the electron optical system with a single-stage electrostatic focusing, incorporating an accelerating electrode that is under the item potential [5]. Such a gun differs from guns with additional magnetic focussing by extreme simplicity of design, small weight and size. So, at unchanged cathode temperature, the focal distance in the single-cascade gun is independent of accelerating voltage, it can be powered by a low-stability or even alternating voltage.

Inner generatrices of the near-cathode and accelerating electrodes of the projector are parts of cones of revolution and are derived proceeding from the known shape of spherical-type projector [6]. The projector was calculated using a solution of Poisson's equation for an electron flow between concentric spheres, limited by a spatial charge. To achieve the maximal distance from the beam smallest cross-section to the accelerating electrode plane and produce a beam of small enough cross-sectional dimensions, an optimal ratio of the cathode and anode spheres of 2.5 has been selected. Proceeding from the required beam current and magnitude of full conductivity of the projector, the half-angle of the beam convergence was assumed to be ~ 3°.

Electrode configuration and interelectrode spacing were determined more precisely in simulation of the electron-optical system. Figure 4 shows the pattern of the potential field in the projector, derived in simulation in EGDA integrator [7]. Electron paths in the beam are plotted allowing for the influence of the spatial charge and ignoring it [8].

Probe measurements showed that in the beam focal spot the specific power is ~1 kW/mm² (beam diameter was measured at the level of 0.7 of full power).

Positive results of development of a small-sized electron beam gun with low accelerating voltage provided a new solution in designing a system of highly reliable maneuverable hardware.

Hardware system. Over the last decade the following classical schematic of designing the electron beam processing hardware has become established: a power source operates from an AC mains and is connected by a flexible high-voltage cable to the electron gun. The high-voltage cable lowers the hardware mobility and its reliability. The operating conditions of the cable and leads are made more complicated by the numerous bends and impact of heat, released in welding. Thorough maintenance and multiple examinations of the high-voltage source and gun lead-ins are required in service.

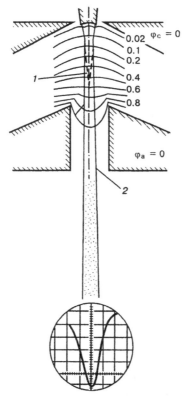

Figure 4 Field pattern, probe characteristic, electron path shape in gun projector without allowing for (*1*) and allowing for (*2*) the spatial charge (U_{acc} = 10 kV, I_b = 60 mA).

These disadvantages were eliminated when developing hardware for operation in space. Figure 5 shows the block diagram of this hardware with self-sufficient power supply from storage battery *1*. Constant voltage is transformed into square-wave variable voltage, using anode *10* and filament *2* transistorized inverters (94 % efficiency). High-voltage transformer *9* and rectifier *8* assembled of semi-conductor valves, generate rectified accelerating voltage. Cathode redundancy is provided by a dual, automatically switched electron-optical system (*3* — filament transformer, *4* — cathode, *5* — near-cathode electrode, *6* — anode, *7* — product).

The block automatically provides the specified modes of welding hardware operation. It incorporates a regulator of anode circuit current, maintaining its required value with an accuracy of ±1 %, as well as devices, providing a temporary delay of switching on the anode

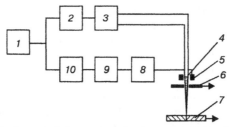

Figure 5 Block-diagram of electron beam hardware (see designations in the text).

inverter, its switching off at short-circuiting of the cathode–accelerating electrode gap and automatic re-switching, when the short-circuiting cause is eliminated.

A higher operating reliability is ensured by element duplicating and deep redundancy, guaranteeing preservation of the nominal parameters of the hardware in case of failure of up to 35 % of circuit elements.

Quite high operating frequency of inverters (1000 Hz) was selected, proceeding from the desire to reduce the weight of the power source and its overall dimensions with acceptable losses in steels. With the increase of frequency, the dielectric losses in the source insulation increased relatively slightly.

The power source and the electron gun are made as one block, based on epoxy compounds, that simultaneously have the function of insulation and structural material [9, 10]. Absence of a high-voltage cable, connectors and leads, allowed a considerable improvement of the hardware reliability and manoeuvrability, reducing its weight and practically eliminating the hazard of electrical shock for the operator.

Application of solid insulation instead of liquid electrically-insulating materials, allowed the hardware overall dimensions to be reduced, thermal load to be lowered due to better heat conductivity of epoxy compounds, and the adaptability to fabrication of the power source to be improved.

The combined power source module, namely the gun for joining 1–3 mm thick metals, is shown in Figure 3a, p. 51.

Discussion of investigation results. Experiments on metal welding and cutting were conducted at constant beam power (1 kW during experiments in the flying laboratory, 0.5 kW under the conditions of a depressurized orbital compartment of «Soyuz-6» vehicle) and speed of specimens displacement relative to a stationary beam of 30 and 18 m/h, respectively.

Flanged joints of stainless steel 304SS and Al-6.2Mg aluminium alloy were welded, overlap joints of commercial VT1 and Ti-3.5Al-1.5Mg titanium alloys were produced and sheet specimens of the same alloys were cut.

Metals of $\delta = 1.5 \div 2.0$ mm thickness were welded without formation of a through-thickness crater in the pool. This allowed the admissible gap between the edges to be increased, and made lower requirements of the stability of mode parameters, namely accelerating voltage, beam current, specific power in the heated spot, angle of beam convergence on the product, welding speed and vacuum.

Investigations demonstrated that the depth and width of penetration are identical at zero gravity and in the earth-based laboratories.

Visual observation and high-speed filming of the process of EBW at simulated zero gravity, did not reveal any noticeable difference, compared to its running in an earth-based laboratory. The small volume of the weld pool and high rates of metal melting and solidification, allow conducting a stable process of sheet material joining under zero gravity conditions.

When welding is performed at zero gravity, certain features of weld shape are observed. A small metal reinforcement is found in the central part of an overlap joint of alloy VT1. The molten metal of the pool, having a sufficiently large surface and volume, tries to take a spherical shape at zero gravity, under the impact of surface tension forces (Figure 6a).

Cutting thin metal with a constant electron beam without modulation by current, produces a cut, the width of which is greater than the plate thickness. The cut is made due to the molten metal drawing to the edges under the impact of the surface tension forces. A considerable part of the initial electron flow passes through the crater and hits a special substrate, mounted on the welding bench.

With good wetting, the molten metal is uniformly distributed along the cut length, for instance, in Al-6.2Mg alloy (Figure 6b), whereas with poor wetting it forms drops (for instance, when processing 304SS steel). To make one of the cut edges smooth, the beam was

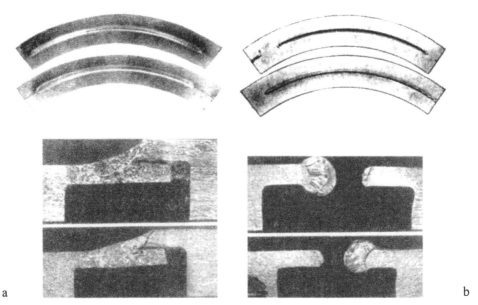

Figure 6 Appearance and macrosections of metal penetration and cuts, made with the electron beam: a — overlap joint of titanium alloy VT1; b — cutting of Al-6.2Mg alloy (*upper* specimens were made on the ground; *lower* specimens at zero gravity).

inclined in the opposite direction, and the metal was ejected in this direction under the impact of the electron flow pressure and counter-pressure of evaporated material (Figure 7).

Influence of zero gravity is readily visible on cut up specimens of aluminium alloy. In the earth-based laboratory, a metal drop is hanging from the cut edge, that is not observed in welding at zero gravity (Figure 6b).

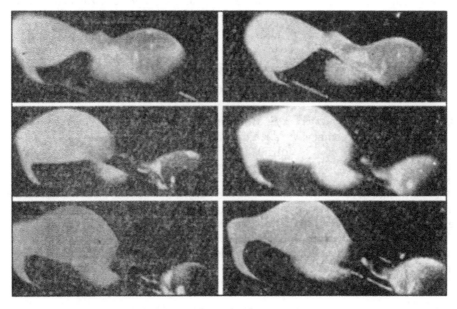

Figure 7 Process of electron beam cutting of sheet stainless steel at zero gravity.

Successful functioning of electron beam hardware in the «Vulkan» system during the experiment conducted on «Soyuz-6» space vehicle confirmed the suitability of the selected electrical parameters and the correctness of the proposed principles of self-sufficiency of the hardware power supply system and application of reliable and effective transistorized converters of direct current into high-frequency alternating current, combining the electron beam gun with the power source, and providing redundancy of critical elements of electron-optical system. The high reliability of the hardware, its small over-all dimensions and weight allow it to be used without changing the block-diagram for solving new specific processing tasks, both under regular conditions on the ground and in space. The developed hardware is already used with success in a number of industrial enterprises for manufacturing critical items.

1. Bas, E.B., Cremosnik, G. and Lerch, H. (1962) Beitrag zum Problem des Erzeugung des Elektronenstrahles für Schmelzen, Verdampfen, Schweißen und Trehnen mit Elektronenstrahlen, *Shweizer Archiv*, März.

2. Schollhammer, F.R. (1963) Electron beam welder for space, In: *Proc. of 6th Nat. Symp. on Space Vehicle Use*, Seatle, 1963. Seatle, United Aircraft, pp. 1–23.

3. (1968) Electron beam welding in space, *Engineer*, **225**, No. 5866, pp. 1024–1025.

4. (1965) *Reference book on geophysics*, Moscow, Nauka.

5. Nazarenko, O.K. (1965) *Electron beam welding*, Kiev, Naukova Dumka.

6. Pirs, D.R. (1956) *Theory and design of electron guns*, Moscow, Sov. Radio.

7. Filchakov, P.F. and Panchishin, V.I. (1961) *EGDA integrators*, Kiev, AN USSR.

8. Strekal, L.P. and Nazarenko, O.K. (1963) Graphoanalytical plotting of electron trajectories in a projector and investigation of magnetic lenses of welding electron guns, *Avtomatich. Svarka*, No. 4, pp. 12–17.

9. Zaruba, I.I. and Shelyagin, V.D. (968) New type of power source for electron beam welding, *Ibid.*, No. 5, pp. 7–9.

10. Shelyagin, V.D., Lebedev, V.K., Zaruba, I.I. et al. *Device for electron beam welding*. USSR author's cert. 206991, Int. Cl. B 23 K, priority 22.02.66, publ. 08.12.67.

Investigation of Melting and Transfer of Electrode Metal in Welding under Conditions of Varying Gravity*

B.E. Paton[1], I.K. Pokhodnya[1], A.E. Marchenko[1], V.G. Fartushny[1], Yu.D. Morozov[1], V.I. Ponomaryov[1], V.F. Lapchinsky[1] and Yu.A. Masalov[1]

Metal electrode welding will undoubtedly be needed for erection work and repair operations in space.

The influence of zero gravity on melting and transfer of electrode metal has not yet been studied, but it became necessary in connection with performance of an experiment on welding in space [1]. The results of the conducted studies are described below.

Experimental procedure. Investigation of arc welding under zero gravity was conducted in the flight along Keppler trajectories in the flying laboratory. Welding was performed in a sealed chamber. The schematics of the unit is given in Figure 1. The arc was powered from silver-zinc type storage battery through a choke. Current polarity was reverse.

---

* (1977) *Space materials science and technology*, Moscow, Nauka, pp. 22–29.

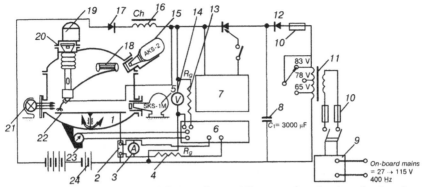

Figure 1 Schematics of the unit for studying metal-arc welding at zero gravity: *1* — chamber; *2* — shunt with ammeter; *3, 5* — high-speed camera SKS-1M; *6* — oscillograph with additional resistors *4* and *13*; *7* — pulse generation block; *8* — capacitor; *9–11* — current converters with a transformer for powering the lighting devices and high-speed cameras; *12, 17* — blanking diodes; *14* — voltmeter; *15* — camera AKS-2; *16* — choke; *18* — coil with electrode wire; *19* — drive for wire feed and welding head oscillations; *20* — axis of welding head oscillation at vibration; *21* — lighting devices; *22* — arc; *23* — accelerometer; *24* — storage battery.

The current was adjusted by changing the wire feed rate, and the arc voltage by changing the number of sections in the storage battery.

Electrode metal melting and transfer were photographed by high-speed SKS-1M camera. The welding mode I_w, U_a and the value of free fall acceleration g_n were recorded by mirror-galvanometer oscillograph. The moment of welding was selected so that weld sections could be produced in the same sample at different values of gravity g_n which was changed in the following sequence in each test: $g_n = g$; $g_n = 2g$; $g_n = 0$; $g_n = 2g$; $g_n = g$.

The influence of zero gravity on melting and transfer of the electrode metal was studied in argon-arc welding with wires of Sv-10Kh18N9T (AISI 304) steel, commercial titanium (VT1) and Al-6.2Mg aluminium alloys of 1 mm diameter. Steel and alloys of the same compositions as that of the wires, were used as the base metal.

Experimental results. The drop of molten electrode metal under zero gravity is mostly exposed to the action of forces of surface tension P_g, reactive pressure of electrode metal vapours, P_v, and electromagnetic force P_{em}. The magnitude and direction of the combination of these forces depend on the welding current value. At small currents the surface tension force holds the drop at the electrode tip, and the drop grows to very large dimensions, exceeding the electrode diameter several times. Under the impact of the gravity field, it is practically impossible to produce drops of such dimensions.

At zero gravity in the absence of any random impacts, the drop stays on the electrode for a very long time. The duration of existence of individual drops in the tests was 4.5–6.0 s. Under the regular conditions of zero gravity, the time between the drops separation was not more than 0.3 s in the same welding mode ($I_w = 40 \div 50$ A).

«Hanging» of the drop and intensive erring of the spot over its surface results in interruption of melting of the metal of the samples being welded.

In a welding arc running under zero gravity, a forced convection of vapours and gases is found, which causes intensive rotation of the drop about its axis normal or inclined at a small angle to the electrode axis.

The angular velocity of the surface layer of the drop was 30–35 cm/s, this being only a little smaller than the electrode wire feed rate.

The nature of heat transfer from the active spot to the electrode tip can be evaluated by the ratio of melting rates and drop volume. If the conditions of heat transfer in the drop deteriorate with increase of its volume, the wire melting rate (and with its constant feed rate also the change of the drop volume in time) would decrease.

In reality only the linear dependence of the drop weight and volume on the time of its existence at the electrode tip is preserved at zero gravity. A similar dependence was also found earlier in welding under the impact of the gravity forces [2].

The conducted investigations showed that in those cases when in welding at zero gravity, the conditions for an unrestrained growth of the drop are in place, it grows up to quite considerable dimensions, and the normal process of weld formation is disturbed.

Transfer of large drops to the workpiece occurs by the drops flowing over under the action of wetting forces at random contact of the drop with the pool.

In order to achieve a regular transfer of electrode metal drops, it was necessary to limit the ability of their free growth. This was achieved by shortening the arc gap. In this case the metal was transferred into the pool at periodic short circuits, and the time of the drop existence on the electrode was reduced. The cinegram of such a process is shown in Figure 2.

Figure 3 gives the dependence of the frequency of drop transfer on welding current for different magnitudes of the gravity force. In all the cases (zero gravity, regular gravity or two time overload) dependence $n = f(I_w)$ is found to be of a parabolic nature.

In the range of 40–50 A currents the frequency of transfers is increased with g_n. Current increase up to 55–60 A achieved by increasing the electrode wire feed rate and accompanied by shortening of the arc gap, neutralized the influence of g_n. The frequency of the drop transitions in this case is independent of g value.

Influence of zero gravity on formation and transfer of electrode metal drops in welding Al-6.2Mg alloy is of the same nature as in stainless steel welding. Welding was performed at 60–80 A current and 14–15 V arc voltage. The metal drops practically all the time took a position coaxial with the electrode, their shape being spherical. The maximum size of free growing drops did not exceed the electrode diameter by more than 3–4 times. The time of such drops existence on the electrode reached 0.5 s.

Welding of VT1 titanium alloy at zero gravity was performed in the following mode: $I_w = 75 \div 80$ A, $U_a = 14 \div 15$ V. The nature of electrode metal transfer under zero gravity hardly changed compared to normal gravity conditions. Drop transfer occurred only through short circuiting. The greatest part of molten metal after each transition remains on the electrode tip. This circumstance, as well as a markedly higher mobility of the weld pool from a titanium alloy, were apparently the cause for more frequent short circuits and practically complete absence of the drop «hanging». The cinegrams of metal transfer in welding Al-6.2Mg and VT1 alloys are shown in Figure 4.

Influence of current pulses on metal transfer. Superposition of current pulses on the arc gap creates an additional axial force which promotes drop separation from the electrode [3]. It turned out to be an effective means of metal transfer control also at zero gravity.

Pulses were formed and superposed on the arc gap using a generator. 10Kh18N9T steel was welded. The results of investigations are given in the Table.

Current pulses stabilize the frequency of drop transfer in all the studied welding modes. The absence of gravity does not influence the transfer frequency, which is entirely determined by the pulse frequency. The volume and weight of the ejected drops depend only on the electrode wire feed rate v_e.

Superposition of current pulses on the arc gap allowed ejection of large drops from the electrode, if they were formed before the pulse generator was switched on. One of the moments of such a transfer is shown in Figure 5a. Figure 5b gives the cinegrams of metal transfer with superposition of current pulses of 50 Hz frequency.

Under conditions of regular gravity of Earth, a drop of molten electrode metal is mainly exposed to the impact of the forces of surface tension P_g, reactive pressure of electrode metal vapours P_v, electromagnetic force P_{em} and gravity g_n. When welding is performed in the

Figure 2 Cinegram of the process of short-circuiting drop transfer (steel electrode).

downhand position, P_g and P_v forces hold the drop at the electrode tip; the gravity force promotes the drop detachment. The direction of the electromagnetic force depends on the ratio of the radii of column r_c and drop r_e at the electrode tip. Deviation of the axis of symmetry and of the drop from the electrode axis gives rise to the tangential component P_d which will twist the drop and promote its separation.

Neutralizing the impact of the gravity field on any mass of liquid of unlimited magnitude, results in its taking an ideal shape under the influence of surface tension, with which a minimal surface and, therefore, also free energy of the liquid, are provided [4–6]. In this case the electromagnetic force remains the only real force which can cause the transfer of a freely hanging drop of electrode metal into the weld pool.

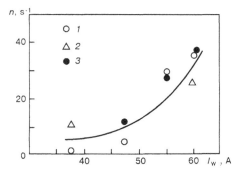

Figure 3 Influence of current on frequency n of electrode metal drop transfer at varying gravity
(steel electrode): $1 — g_n = 0$; $2 — g_n = g$; $3 — g_n = 2g$.

Proceeding from calculations of [7], at 1 mm diameter of the electrode the critical value of current at which the electromagnetic force of current starts promoting the drop separation from a stainless steel electrode, is 32 A. In the absence of gravity, the resultant force required for drops separation can be achieved at a much higher current, than under standard conditions of welding when gravity pull facilitates drop separation from the electrode. The impact of gravity on metal transfer is manifested to the utmost degree at low current when the

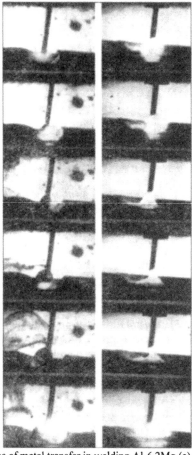

a   b

Figure 4 Cinegrams of metal transfer in welding Al-6.2Mg (a) and VT1 (b) alloys.

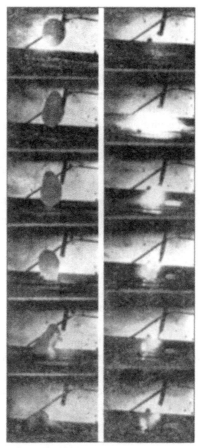

a   b

Figure 5 Cinegrams of electrode metal transfer by current pulses: a — separation of a large drop by means of a pulse; b — drop transfer with a regular superposition of pulses.

electromagnetic force is small. At higher welding currents the impact of gravity is neutralized by the electromagnetic force, and in the case of spray transfer the influence of gravity can be neglected.

Higher welding parameters should be used under zero gravity than at normal gravity, in order to prevent the disturbance of welding and ensure a periodic process of drop transfer.

Influence of current pulses on the parameters of molten metal transfer at zero gravity
(steel electrode of 1 mm diameter, argon-arc welding).

Test number	Background current, A	Arc voltage, V	Transfer parameters			
			without pulse		with pulse	
			τ, ms	n, s^{-1}	τ, ms	n, s^{-1}
8P4G	35–40	16.5	3000	0.3	20.0	50.0
7P2G	45–50	15.0	230	4.5	18.0	57.0[*]
8P2G	55–60	16.3	35	28.2	–	–
7P1G	45–50	15.0	200	4.7	22.5	44.5
8P1G	50–55	16.0	48	21.0	25.0	40.0

[*] Ripple and interruptions of the arc were observed with the switched-on pulse generator.

In pulsed-arc welding or welding with short circuiting the influence of gravity forces on metal transfer is not felt.

Conclusions

1. Experiments on metal-arc welding under zero gravity have been conducted.

2. Absence of gravity forces or their balancing change the ratio of forces acting on the drop. At small welding currents (40–50 A, 1 mm electrode diameter) the electrode metal drops become larger at zero gravity, while their transition into the weld pool becomes irregular and weld formation is disturbed.

3. Deviations caused by zero gravity in the process of welding can be eliminated by shortening the arc gap or superposition of current pulses on the arc gap.

1. Paton, B.E. and Kubasov, V.N. (1970) *See pp. 154–160 in this Book.*

2. Pokhodnya, I.K. and Kostenko, B.A. (1965) Investigation of kinetics of electrode melting in welding, *Avtomatich. Svarka*, No. 4, pp. 2–5.

3. Paton, B.E. and Potapievsky, A.G. (1964) Pulsed-arc welding, *Ibid.*, No. 1, pp. 4–6.

4. Shulejkin, V.V. (1962) Form of wettability of a liquid losing its weight, *Doklady AN SSSR*, **147**, No. 1.

5. Shulejkin, V.V. (1963) Ground-based experiments with weightless liquids, *Ibid.*, **152**, No. 5.

6. Benedikt, E. (1964) General behaviour of liquid at complete or almost complete zero gravity, In: *Zero gravity. Physical phenomena and biological effects*, Moscow, Mir, pp. 35–41.

7. Dyatlov, V.I. (1964) Elementary theories of electrode metal transfer, In: *New problems of welding technology*, Kiev, Tekhnika, pp. 14–18.

Performance of Manual Electron Beam Welding in Space[*]

B.E. Paton[1], D.A. Dudko[1], V.N. Bernadsky[1], V.F. Lapchinsky[1], A.A. Zagrebelny[1], V.V. Stesin[1] and V.I. Chalov[1]

The problem of performance of mounting and welding work in space has been discussed by the experts for more than a decade now. The complexity of the task is primarily attributable to the conditions of work performance making extremely high and sometimes contradictory requirements of the welding process, welding tool and technology. The relative simplicity of EBW performance, high process effectiveness, ability of applying it for all the metals used in space vehicles, make this process, in the opinion of the experts, one of the most promising for the space environment [1].

Industry has gained quite extensive experience of EBW application. On the ground, however, welding is performed solely with mechanized displacement of the specimen being processed or the tool, namely the electron beam gun, that is not always possible and not rational in the space environment. It is sufficient to say that almost all the operations, related to maintenance and repair of space facilities, when neither the area of work performance, nor the parameters of the parts or components being processed are known, will have to be performed either completely in manual mode or with just partial mechanization. This

* (1977) *Space materials science and technology*, Moscow, Nauka, pp. 17–22.

accounts for the greater attention of the experts to studying the possibility of performance of manual EBW in space.

Development of a device for manual EBW in space was reported for the first time in 1965 [2], when a gun with 80 kV accelerating voltage was developed in the USA. A feature of the described welding unit was the design of the electron beam gun, high-voltage power source and welding control system as one module, connected by a low-voltage cable to a storage battery. This engineering approach, as shown by experience, was not promising for manual electron beam tools, as the unit in this case has a high weight, excessive inertia and low maneuverability, that do not allow sound welding to be performed. In addition, with an accelerating voltage of such a magnitude, it is extremely difficult to provide sufficient protection of the welder-operator from hard X-ray radiation.

After some time, another type of electron beam gun [3–5] was introduced, where the inverter, control system and high-voltage power source are made in the form of individual blocks. The gun, fitted with a handle, is connected to the power source by a flexible high-voltage cable. Mounted on the gun case for protection from X-ray radiation is a special protective screen of sheet stainless steel 1.5 mm thick with special portholes for observing the progress of welding and guide rollers for the gun moving over the item. Developers believed such a protective screen to be quite sufficient, as the gun accelerating voltage was reduced to 20 kV at beam power of 1.5 kW.

Overall dimensions of the gun without the handle are ~250 mm length and 75 mm diameter. Weight of the gun with the handle is 4.35 kg. Electromagnetic focusing system of the gun allows beam focusing at 50–70 mm distance from the tool. The gun incorporates an electron projector of Steigervald type with indirectly heated cathode.

Electron beam units with a cold plasma cathode were the next to be developed [6, 7]. Their feature is a low accelerating voltage (about 2 kV) that practically enabled the elimination of any biological protection of the operator. One of the units was developed as a highly specialized system, only allowing welding of tubes of a certain diameter, another one, being somewhat more versatile, was designed for making flat samples. The first of the guns has a constant distance from the circular cathode to the item (tube) of 1/2″ (12.7 mm), thus supporting welding of circular butts of tubes of up to 3/8″ (9.33 mm) diameter. The second unit allows changing the distance from the cathode to the item being welded within 0.5–6.0″ (12–152 mm). The weight of this gun modification is already less than 3 kg.

Electron guns with a cold cathode have the drawback of not being versatile and allowing welding of items of an extremely limited range. In addition, the reliability of welding with such guns is low, and their process characteristics are largely dependent on pressure and plasma gas consumption. This is probably why the design of these units was not pursued extensively enough, despite such an important advantage as there being no need for special biological protection of the operator.

Guns with directly heated cathodes [3–5] are apparently much more promising for manual EBW, even though the biological protection incorporated by the developers is probably insufficient for the selected accelerating voltage of 20 kV [8]. These guns also have another drawback, more important in terms of technology, namely the penetration depth is essentially dependent on the gun-to-item distance (Figure 1), this dependence being the more pronounced the lower the accelerating voltage [9]. This phenomenon, probably attributable to the small focal distance of the gun at a relatively high angle of convergence of the beam, does not allow the anode voltage to be lowered, as, in manual welding, variation of the distance is inevitable.

In addition, the short focal distance, even with the welding zone shielded by a protective cover, is undesirable, because of intensive vapour flow from the weld pool that affects the gun and the welding operator. From this point of view, long-focus electron guns with a low

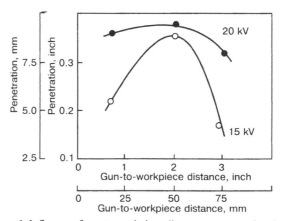

Figure 1 Influence of gun-to-workpiece distance on penetration depth.

angle of convergence of the beam are more promising. Such guns should have much more attractive process characteristics for manual work performance.

The unit for electron beam welding developed at the PWI uses a batch-produced diode electron beam gun (Figure 2) with an indirectly heated cathode of 1.5 kW power and beam working current of 100 mA. The optical system provides stable focusing of the beam at a distance of several dozen or even a hundred millimeters from the gun. The high-voltage power source is a monolithic epoxy block, similar to the block that was incorporated into the «Vulkan» welding system [10]. The block, together with the high-voltage cable, is enclosed in a metal shield connected to the common earthing circuit of the electron beam unit.

To ensure operational safety, the unit accelerating voltage did not exceed 15 kV; and it was not difficult to provide the required biological protection of the operator and service personnel.

The unit for manual EBW was tested in a space simulation test chamber of a large volume, using a special testing facility, also developed at the PWI [11], which allowed various processing operations to be performed under conditions simulating those of space, without man's direct involvement (see Figure 3 on p. 126).

Welding of various welded joints on stainless steel, aluminium and titanium alloys 1.5–3.0 mm thick was performed during the experiments. Butt and lock joints with and without edge flanging were welded. The specimens were placed on a special welding bench at 200–220 mm distance from the gun.

Figure 2 Long-focus gun with directly heated.

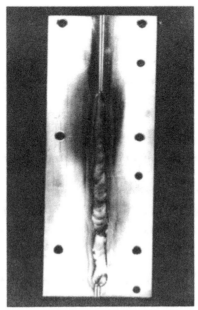

Figure 3 Welded joint on titanium alloy produced by manual EBW.

Experiments confirmed the fundamental ability of manual welding performance, using such a high energy density heat source, as the electron beam. The welding operator could weld around complex contours, strictly maintaining the direction and the required stability of beam displacement over the edges being welded. Welding speed was maintained within 7–10 m/h at beam current of 55–75 mA.

Welding was performed with a slightly defocused electron beam so as to avoid excess energy concentration in the heated spot and have the ability to lower the requirements to the accuracy of samples fit up, butt following and stability of the tool displacement. Change of the gun-to-workpiece distance within ±20 mm (±10 % of the nominal value) does not result in a noticeable change of penetration depth or weld width. If during welding the penetration depth or weld width has to be controlled, this is achieved by changing the gun to sample distance by not more, than 10 %.

Variation of welding speed, always found in manual operations, has a much stronger influence on the stability of penetration. Nonetheless, after a short training the welding operator managed to produce welds of a quite satisfactory quality (Figure 3).

When investigations were performed, studying the influence of gaps on welded joint quality was not a special objective. Manual welding allowed making butt joints with a gap, equal to about 50 % of the sheet thickness. The welder usually had to also apply transverse beam oscillations in addition to longitudinal oscillations. With smaller gaps, welds of an acceptable quality could be made also in the case when the welder imparted just the reciprocal motion to the beam.

Manual EBW requires greater attention and involves considerable loads, thus resulting in fast fatigue of operators. On the other hand, the operator's physiological condition and degree of his fatigue have a significant influence on welding quality. Therefore, special instruction and training of welding operators are required, allowing certain professional skills to be formed under such unusual conditions of work place comfort, as well as novelty of welding process.

1. Paton, B.E., Nazarenko, O.K., Chalov, V.I. et al. (1970) *See pp. 160–168 in this Book.*

2. Hidden, R.D. (1965) USAF. Studying space tool development, *Aviat. Week and Space Technol.*, **13**, No. 9, pp. 58–65.

3. Irwing, R.R. (1966) All systems are go for EBW, *Iron Age*, **17,** No. 9, pp. 59–61.

4. (1966) *Electron News*, No. 14, p. 563.

5. Chvertko, A.I., Nazarenko, O.K., Svyatsky, D.M. et al. (1973) *Equipment for electron beam welding*, Kiev, Naukova Dumka.

6. (1968) Gold-cathode space welder is developed, *Aerospace Technol.*, No. 21, p. 25.

7. (1968) *Amer. Machinist*, No. 2, p. 12.

8. (1969) *Sanitary rules for working with soft X-ray radiation sources*, Moscow, Minzdrav SSSR.

9. (1968) *Welding Design and Fabrication*, No. 5, pp. 55–56.

10. Paton, B.E. and Kubasov, V.N. (1970) *See pp. 154–160 in this Book.*

11. Paton, B.E., Dudko, D.A., Bernadsky, V.N. et al. (1975) *See pp. 137–140 in this Book.*

Testing of Manual Electron Beam Welding Tool in Open Space[*]

V.P. Nikitsky[2], V.F. Lapchinsky[1], A.A. Zagrebelny[1], V.V. Stesin[1] and V.D. Shelyagin[1]

The problem of permanent joining of materials in space has become particularly urgent over recent years. This is attributable to the great success, achieved primarily by Soviet cosmonautics. The flights of Soviet «Salyut» orbital stations, begun in 1973, in combination with «Soyuz» transport spaceships and «Progress» cargo spaceships, led to the establishment of long-term orbital space research facilities. The duration of operation of such facilities is more than five years now, and will steadily increase in the future. The economic effectiveness of their use, naturally, will also be enhanced.

The duration of individual expeditions staying in orbit will also become longer. The most recent expedition that worked in OS «Salyut-7», is known to have been in space for 237 days. In the near future setting up continuously inhabited orbital complexes with a periodical change of the crew, can, probably, be anticipated.

Such a mode of space vehicle operation in the near future requires solution to many problems, including those related to welding hardware. Among the latter the most important are problems of repair of metal structures in these vehicles, restoration of various thin-film coatings on them and erection in space in orbital facilities or near them, of large-sized structures that for some reasons cannot be brought to the service location in their finished form. The majority of these problems can be solved by using permanent joints of materials.

When the team of researchers, dealing with these problems, was given the task to develop hardware and technology for permanent joining of metals in space, the question was raised of how the welding processes should be conducted under these conditions — automatically or with the direct participation of the cosmonaut-operator. For a number of reasons, related chiefly to safety rules and quality of joints, work performance in the automatic mode was extremely attractive. In addition, by this time some experience had already been gained of automatic welding and cutting of metals in space. Experiments, conducted in 1969 in «Soyuz-6» spaceship by cosmonauts G.S. Shonin and V.N. Kubasov,

---

[*] (1985) *Problems of space technology of metals*, Kiev, Naukova Dumka, pp. 7–15.

using «Vulkan» unit, fully confirmed the ability of such welding operations performance in the automatic mode. Such an operating mode, however, will require precise knowledge of where and how the work will be performed, what kinds of permanent joining will be used, and, most importantly, it is related to development of highly sophisticated hardware that it is very difficult to make versatile.

A thorough analysis of the possible of range of work performance led to the conclusion, that, at least at the initial stages of application of permanent joining of materials in space, the most rational approach was to conduct the process with active involvement of the operator. Operations can be performed not completely in the manual mode, but using, in a number of cases, various auxiliary devices, facilitating and simplifying the operator's activity.

These were exactly the concepts that underlay the development, more than ten years ago, of an all-purpose manual welding tool for the cosmonaut, later on named simply «versatile hand tool».

This gave rise to the question of which of the existing welding processes should be selected. It should be noted, that such processes, as EBW, consumable-electrode arc welding, resistance spot welding, plasma arc welding and radiant-energy welding had been tested at different stages of work performance in the vacuum chambers on the ground and on board the flying laboratory.

In the long run, versatility and minimal power consumption of the process had the decisive role — the choice was made in favour of EBW. Indeed, this process covers all the processing operations that may be required for permanent joining of metals in open space, namely heating, brazing, welding, cutting and coating. When going over from one process to another one, in the majority of cases just varying the electron beam focusing can be sufficient, which in the manual mode is achieved by simply changing the distance between the tool and the item being welded. The tool design has to be modified only for the performance of coating, and in especially complicated cases of welding, a filler material feed mechanism may be required. In terms of power consumption, EBW has no competition from the above processes as to effectiveness of the electric power use.

Selection of the welding process certainly did not solve all the problems. As research developed, they became more and more numerous, not only in engineering, but also in psychological terms.

There was an opinion, and not without reason, that such a process as EBW, in view of the presence of a high accelerating voltage, the possibility of X-ray radiation from the weld pool and the need to manipulate a sharply-focused electron beam, would not be able to be implemented in the manual variant. Sceptics' mind could only be changed by an experiment.

Such an experiment was staged for the first time by the experts of the PWI in 1974 in a space simulation test chamber of a large volume, using testing facility. A feature of this facility is that the operator does not have to wear the complete spacesuit, but puts on just a special half-suit (called «cuirass») with a pressure helmet. He is exposed to deep vacuum just down to his waist, so that he can quickly get out of the half-suit in case of an emergency situation. During operation, a pressure, equal to the pressure in the spacesuit, is maintained in the space simulation test chamber and a vacuum, required for the electron beam functioning (10^{-3}–10^{-4} mm Hg) is maintained in the vacuum chamber of facility.

A special indirectly heated triode electron gun with electromagnetic beam focusing was designed and made for conducting these experiments. Gun rated power was equal to 10 kW at the accelerating voltage of 15 kV. The gun generated a focused electron beam at up to 3 m distance from the operator. The power source was made as a separate block, and the gun was connected to it by a special shielded high-voltage cable.

Experimental results turned out to be extremely interesting. First of all it was unambiguously demonstrated, that there were no unsurmountable technical difficulties in implementation of manual EBW. It was also shown, that an operator, wearing a half-suit, can perform such operations as welding and cutting without risk to himself and with good quality. This meant overcoming the psychological barrier in the way of further development of manual EBW in space. On the other hand, experiments performed in 1974, provided a large amount of extremely important information that formed the basis for development of the flight hand tool. It turned out, that a number of engineering solutions, taken in development of the first tool, were flawed. In particular, the level of accelerating voltage was selected incorrectly. At this voltage (15 kV) there was the risk of generation of hard X-ray radiation. Then the accelerating voltage was lowered to 5 kV. At such a voltage hard X-ray radiation is completely absent, while the ability of focusing the electron beam is still quite sufficient.

On the other hand, an indirectly heated gun with accelerating electrode and electromagnetic focusing could not be used to perform the first space experiment, because of its insufficient reliability in the hand tool variant. The simplest and most reliable directly heated gun was selected as the basic variant. However, the gun focal distance was markedly reduced, as long-focus guns run a higher risk of emergency situations developing, and manual control of energy density in the heated spot is more difficult because of defocusing. A number of design modifications were also made, for instance, elimination of high-voltage cable and incorporation of the power source into the tool proper, great reduction of the gun overall dimensions, introduction of a protective shield in front of the handle and of the elements of the tool fixation on the spacesuit glove, etc.

In spite of the positive results of the first ground-based experiments, ten years were required before such an experiment could be conducted in space. This was attributable, on the one hand, to a lack of psychological confidence in the safety of this experiment, and, on the other hand, to the absence of the prerequisites needed to stage it. The majority of processing problems, arising in cosmonautics in that period were solved by, possibly not the best, but earlier proven methods. It was only at the start of the 1980s, that a range of problems were defined that could not be solved in space without using manual welding operations.

The decade certainly was not wasted. During this period, the technological procedures of manual EBW of thin metal were optimized, features of operation of the power source — electron beam gun system at a low consumed power (up to 500 W) were studied, and, finally, the need for conducting another processing operation in space, namely application of metal coatings by the method of material evaporation and condensation was determined and the features of its performance were studied. In the period between 1979 and 1982 this technological process was thoroughly verified in space, using «Isparitel» unit. All this enabled, starting in 1982, the development of the flight variant of versatile electron beam tool.

All the VHT main functional blocks are mounted in a special basic container, designed as a truss structure, the so-called «basket». The fact, that the container is also used for VHT transportation from the pressurized modules of the station to the working area may have had a significant role in giving it this name. The front side of the «basket» carries the fasteners, used to attach VHT to the handrail of the anchoring platform of OS «Salyut» or other similar elements. The set of boards with the samples is mounted on the rear (outer) side in a special holder. The set includes four boards, each of which accommodates samples for conducting one processing operation, namely cutting, welding, brazing or coating deposition. In order to perform each of the operations, the required board is drawn out of the holder, and is moved back in after the operation has been completed. A pressurized compartment with the switching hardware, blocks of the secondary power source (SPS), and the programmer is mounted inside the «basket». The upper cover of the pressurized compartment accommodates the control panel. These blocks are mounted on the left hand side of the operator who conducts the technological processes.

The electron beam hand tool proper is mounted on the right of the pressurized compartment inside the «basket». In the transportation position it is fastened in a special fixture, out of which it is taken to perform the work. Also fastened inside the «basket» in the transportation position, are the cables, used to connect VHT to the station power supply system, and the tool proper to the SPS pressurized compartment. The pressurized compartment and the connecting cables for experiment performance in space were protected by multiplayer thermal insulation. The actual electron beam tool is designed as one block, the basic element of which is a box-shaped case with a special handle made to suit the spacesuit glove. The front wall of the case carries two small-sized electron beam guns, each of which can be assembled both to generate a sharply-focused electron beam (modes of cutting, welding and brazing), and for crucible bombardment with a defocused flow of electrons (evaporation mode). Mounted inside the case is a high-voltage power source, consisting of a high-voltage anode transformer, anode rectifier and two filament transformers, for each of the guns separately. All the components are embedded into epoxy compounds as one block. The toggle switch on the tool case allows power to be selectively connected to the first or the second gun. The block handle is made to rotate, thus allowing the operator to select the most convenient position of the tool relative to the sample during performance of each of the technological processes.

The sample boards for cutting, welding and brazing each carry 6 samples of titanium and stainless steel, having different orientation relative to the operator. This was required so as to verify directly in space the ergonomic recommendations made on the ground, as regards the most rational technique of a particular operation performance. The board for coating carries 2 blackened aluminium samples, on which a silver coating is to be deposited. VHT rated parameters are as follows:

```
Power, W ........................................................ 1000
Accelerating voltage, V ........................................  5000
Maximum duty cycle, s ..........................................  180
Intervals between switching on, s ..............................  180
VHT total weight (without sample board), kg ....................  30
Total weight of sample board, kg ...............................  10
Weight of tool proper, kg ......................................  3.5
```

Performance of space experiments was preceded by thorough ground-based preparation, for which purpose the VHT test sample and its size-weight mock-up were made in advance. Testing was conducted in a special space simulation test chamber, a zero-buoyancy tank and on board the flying laboratory at short-time zero gravity.

Placed into the space simulation test chamber was a part of OS «Salyut-7», including part of the side wall, anchoring platform and handrail. The part is fitted with a weight-neutralizing system, allowing an operator wearing a spacesuit to perform work using the tool for a long time. VHT technological sample was mounted on the handrail.

The main purpose of testing in the space simulation test chamber was optimizing the modes and technique of manual EBW. At the initial testing stages the work was performed by testing operators, having extensive experience of working in spacesuits and having mastered the fundamentals of semi-automatic gas-shielded arc welding of metals. Closer to the end of testing, they were joined by the main and backup crews of «Soyuz T-12» spaceship, who were to perform welding in space. Their participation in testing simultaneously was their preflight training. All together 18 experiments were conducted in the complete programme in the space simulation test chamber, with S.E. Savitskaya and V.A. Dzhanibekov fulfilling this programme twice on the ground. It should be noted that neither the vehicle commander, nor the flight engineer had ever had to deal with welding before that. They started gaining welder's experience right from doing manual EBW in the space simulation test chamber.

Full-scale training mock-up of the facility was set up in the zero-buoyancy tank. The procedures of VHT transportation from the station pressurized modules to the testing location, its attachment to the handrail, taking out and stowing the tool, and the technique of processing operations performance (naturally, without the electron beam switching on) were also optimized in it. This work was conducted by experienced testing operators, using VHT size-weight mock-up, and at the final stages — by both the crews. It should be noted, that the main ergonomic recommendations and the majority of the tool design retrofits were made exactly at this stage.

The work in the flying laboratory was conducted using the part of the OS «Salyut-7» and VHT size-weight mock-up. All the investigation stages that S.E. Savitskaya and V.A. Dzhanibekov were to perform in space were finally optimized.

For VHT testing in space, the following samples were mounted on the boards:

• titanium and stainless steel sheets 0.5 mm and 1.0 mm thick for cutting;

• flanged butt joints of titanium and stainless steel sheets 0.8 and 1.0 mm thick for welding;

• overlap joints of stainless steel sheets 1.0 mm thick and braze alloy strip pre-placed into the joint for brazing;

• blackened aluminium plates 2.0 mm thick and 3 g of silver was loaded into the evaporation crucible for coating deposition.

Material selection and sample thickness were determined by that in the first experiments, the tool full power was not used for safety-related considerations. All the work was to be performed at beam power of 300 W.

Testing of electron beam hand tool in space was conducted on July 25, 1984. After depressurizing of the adapter module and hatch opening at 18.22 Moscow time the vehicle commander took VHT out to the station outer surface and mounted it on the anchoring platform handrail. Svetlana Savitskaya, being in the adapter module, helped the commander to transport the VHT. After that, the cosmonauts changed places. Having fastened herself to the anchoring platform, Svetlana Savitskaya took the tool from the «basket», pulled the first sample board out of the holder and started performing cutting. She cut up the titanium sample, and after that welded, brazed and coated one sample of each kind, successively changing the sample boards. At this time the commander was doing TV reporting and photographing some moments of the flight engineer activity. After completion of the first stage of the work, the cosmonauts again changed places and Vladimir Dzhanibekov completed processing of all the other samples.

At all the stages of VHT testing in space an expert of the PWI was communicating with the cosmonauts, in addition to FCC officers.

After completion of the experiments, VHT was again brought into the transportation position and taken to the station. The open space experiment took 3 h 35 min. Before their return to Earth the cosmonauts removed the samples from the boards and took them with them. The samples are at the PWI and are being prepared for comprehensive testing.

So, what are the first, most general impressions from the conducted experiment?

First of all about welding. Already at the stage of ground-based testing in the space simulation test chamber, it had been demonstrated that manual EBW, similar to other allied processing operations, made with VHT can be quite readily performed by a cosmonaut wearing a spacesuit, and can be recommended for permanent joining of metals in space. Operators that had had just a limited experience of arc welding, performed all the processes using VHT in a quite efficient manner. The samples welded by them met the current requirements, made by industry.

The issue of training the operators who did not have any previous experience of welding, including «Soyuz T-12» crew members to use the technique of manual EBW, was much more complicated.

Unlike arc welding processes, manual EBW allows containment of the weld pool and adjustment of the penetration depth not only by manipulating the tool, but also by changing the beam focusing, thus minimizing the risk of burns-through.

When working at zero gravity the risk of burns-through is even smaller, when the weld pool flowing out under the action of gravity forces is absent, while the role of surface tension forces, preventing burns-through, is markedly increased.

Therefore during ground-based training, the operators were trained to move the tool smoothly, with reciprocal and oscillatory motions of the beam up to achieving complete penetration. On the other hand, their attention was constantly drawn to the inadmissibility of letting the electron beam dwell for a long time in one point of the weld pool, as in this case, burns-through did develop. Therefore, during ground-based training, not a single case of burn-through was observed in welding either the vertical or the horizontal welds. Incomplete penetration was often found at the initial training stages. However, already during the program fulfillment for the second time in the space simulation test chamber, the number of incomplete penetrations was extremely small and the welds made at this stage by S. Savitskaya and V. Dzhanibekov were of good quality. However, the samples welded by the same operators in space differ from those made on the ground by a greater number of incomplete penetrations. The operators apparently were continuously concerned about burning the metal through and thus making an unrepairable defect. The samples naturally do not have a single burn-through.

When evaluating the very preliminary experimental results, we must also take into account the emotional state of the cosmonauts, doing their first EVA and the environment, totally different from that in the ground-based space simulation test chamber and, finally, the very fact that such welding in space was performed for the first time in the world. Therefore, the first samples of manual EBW brought from space should be valued very highly, despite the incomplete penetration found in them.

As was already noted, it is not easy to burn through sheet metal at zero gravity. In the absence of gravity, molten metal after hole formation is again drawn by the surface tension to the cut axis and often in effect heals the already made cut. Therefore, metal cutting under such conditions requires good focusing of the electron beam, uniform movement of the tool and sometimes even going back to poorly cut regions. Nonetheless, in the conducted experiment, the cosmonauts managed to make good cutting of the metal.

Strangely enough the brazing process turned out to be the most complicated. For a reliable performance of this process, the cosmonauts were recommended to preheat the upper of the samples to be joined up to a barely visible reddening and then move the beam along the sample so, that the heated zone moved uniformly together with the beam. In the ground-based space simulation test chamber, where bright illumination was absent, the operators did not have any problems to perform this process. Now in space, where the brightness of solar radiation is extremely high, it turned out to be practically impossible to see the slight reddening of the upper sample. Therefore, during the experiments on brazing the cosmonauts had to change the procedure of their performance and evaluate the degree of samples preheating by the time of the electron beam action on them.

When working under such extreme conditions, it is apparently very difficult to evaluate the time elapsed. Therefore, in the majority of cases, good quality brazing could not be performed. There are, however, several well-brazed places that confirm that brazing under such conditions is certainly possible. The data derived even before the space experiment in the flying laboratory, are indicative of the fact that capillary spreading of the braze alloy under vacuum and zero gravity proceeds in a much more effective manner than on the ground. Therefore it can be anticipated that subsequent experiments on brazing in space will be much more successful. It will, however, be necessary to select precise criteria of sound performance of the process, suitable for the conditions of open space with its bright solar radiation.

The cosmonauts did not have any difficulties in conducting the coating application process, and the samples made in space, at least so far, can meet the requirements of the most stringent standards. This had been probably largely promoted by numerous experiments in «Isparitel» and «Isparitel-M» units, during which the modes and technique of material evaporation and condensation under vacuum were thoroughly verified.

On the whole, cosmonauts' activity during this first in the world test of the electron beam tool in space can be recognized as excellent. By the estimates of the cosmonauts and leading experts, designing flying space vehicles, tools of this type will become widely accepted already in the near future.

From this viewpoint the experimental material, derived by «Soyuz T-12» crew is extremely important. Just the most preliminary conclusions can be made so far. They, however, confirm that all the processes performed by the cosmonauts can be recommended for performance of practical work on board the space vehicles. This will probably require certain changes in the operators' training procedure, technique of operations performance and modes of conducting the processes. In this respect the experimental data derived in July 1984 will also be needed.

Welding Equipment for Space Applications*

**V.A. Dzhanibekov[3], A.A. Zagrebelny[1], S.S. Gavrish[1], V.V. Stesin[1],
V.D. Shelyagin[1], N.N. Yurchenko[12] and A.V. Markov[2]**

Concrete design solutions are considered on the example of several technological units, operating in space, namely «Vulkan», «Isparitel», «Isparitel-M», «Yantar», VHT, «Universal» and the experience of service of these units under space conditions is described.

The space welding equipment is a rather sophisticated hardware system aimed at a solution of a general functional problem. This system in its present form is schematically illustrated in Figure 1. The main its link is the technological equipment *1* representing units for welding, brazing, coating and cutting. All these processes belong to welding technologies. The application of versatile equipment capable to perform several operations is also feasible [1].

Each technological installation needs a specially-equipped work station *2*, which, depending upon the purpose, can be stationary or movable.

The self-link of the space welding equipment system is auxiliary devices *3* designed for widening the capabilities of the equipment units, mechanization of labour-intensive operations or providing additional measures for the operator protection.

Power supply and information processing units are the important link of the system *4*. It is connected to the corresponding systems of the spacecraft using necessary communications *5*.

The difficulty in design of welding equipment for space applications consists in the fact that, on the one hand, it should provide the necessary parameters and quality, typical of welding units, on the other hand, it should satisfy completely the specific requirements for the space vehicles.

The requirements given below are basic and it is necessary to adhere to them strictly:

• conformity to the functional tasks;

_____

* (1991) *Proceedings of Conf. on Welding in Space and the Construction of Space Vehicles by Welding*, USA, Miami, Sept. 24–26, 1991. Miami, pp. 49–58.

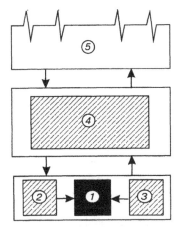

Figure 1 Schematic diagram of space welding equipment: *1* — technological set; *2* — special work station; *3* — kit of auxiliary devices; *4* — system of power supply and information processing; *5* — systems of spacecraft.

- safe service in space vehicles;
- high reliability;
- minimum dimensions, mass, energy consumption;
- possibility of continuous control of parameters of the technological process and periodic diagnostics of the equipment condition;
- compatibility with systems and crews of the spacecrafts;
- repairability.

The fulfilment of above requirements is provided due to the development of a system of design and methodical solutions, whose effectiveness is checked by tests of experimental models of the designed equipment.

Over 25 years the PWI has carried out works on welding in space and many different types of equipment have been developed. In this paper we shall describe only the most interesting models.

In 1969 the automatic welding unit «Vulkan» was delivered into space on the spaceship «Soyuz-6» (see Figure 7 on p. 159). This unit could perform welding using three different methods, such as electron beam, low-pressure constricted plasma arc and consumable electrode welding [2]. The «Vulkan» unit consisted of two independent modules. In one (unpressurized) module the specimens to be welded and devices for the performance of each of above-mentioned methods of welding were located. In the other (pressurized) module the self-contained storage-battery power supply source, converters, control devices, automation means were arranged. The specimens being welded were placed on a special rotatable table. The weight of the unit was about 50 kg. The duration of its continuous operation was determined by the capacity of the storage battery.

«Vulkan» was located in the living bay of the spacecraft, which could be depressurized, whereas crew, consisting of Georgy Shonin and Valery Kubasov, remained in the pressurized landable vehicle.

In October 1969 the first experiments on automatic welding of metals in space were carried out. They confirmed the main technical solutions, which were taken in designing of equipment, and gave valuable information for analyzing the serviceability of different methods of welding under these conditions.

In summer 1979 the technological unit «Isparitel» was delivered to the OS «Salyut-6» (see p. 13, Figure below) [3]. This unit was aimed at the experimental optimizing of processes of thin coating deposition in space on different metallic and non-metallic substrates using the method of thermal evaporation and condensation of materials. The electron beam was used as a heat source. The unit contained two interrelated modules. The first module, arranged in the pressurized bays of «Salyut-6», included the secondary energy supply source (converter), control panel, means of automation and telemetry. The second nodule, which was operating, was arranged in a special airlock chamber of the station. It provided vacuum evaporation of materials used for the coating deposition and the successive injection of specimens to be exposed. Two evaporators, equipped with low-voltage electron beam guns and a drum-type manipulator with substrate samples were its main functional units. The process of coating deposition could start after depressurizing of the airlock chamber and reaching of the overboard vacuum. The heating of crucibles with the evaporating materials was performed by their bombardment with a defocused beam of electrons. The metal being evaporated was deposited on the specimens which were arranged in the manipulator and exposed in succession. The coating thickness was varied over a wide range due to a change in temperature of the material being evaporated (capacity of the evaporator) and duration of the exposure which was preset by the cutoff of the vapour flow by a special curtain.

«Isparitel» unit was working on board the OS «Salyut-6» about three years. During this time more than two hundred samples of different coatings were produced in space. Three crews were working using this unit: Valery Ryumin and Vladimir Lyakhov in 1979, Leonid Popov and Valery Ryumin in 1980, Vladimir Kovalyonok and Viktor Savinykh in 1981.

At the end of 1983 a new experimental semi-industrial unit «Isparitel-M» was delivered to the board of the next OS «Salyut-7» (Figure 2). Its principal design and arrangement on board the station was the same as that of «Isparitel». The mode of operation was also similar. However, the potential of this unit was greater. The working module was equipped with several changeable heads for arrangement of specimens being coated. With the head removed, the working module could be used for coating deposition not on the specimens,

Figure 2 Work module of «Isparitel-M» unit: *1* — changeable head with specimens ; *2* — changeable electron beam evaporators; *3* — high-voltage module.

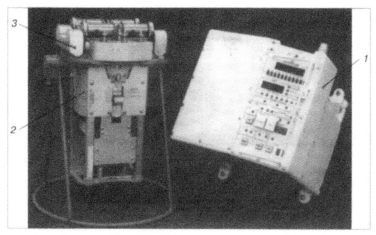

Figure 3 «Yantar» unit: *1* — module of control and power supply; *2* — work module; *3* — changeable head for deposition of coating on strip.

but on the external surfaces of the space vehicles. It was feasible to deposit thick layers of brazing alloys. One of the changeable heads made it possible to perform electron beam melting of metallic materials. The control module of «Isparitel-M» represented a programming device which could realize several dozens of technological procedures of material treatment. The module was equipped with a sophisticated information-monitoring system.

After the equipment delivery to the board of station several crews were involved in the experiments.

In March 1984 cosmonauts Vladimir Soloviov, Oleg Atkov and Leonid Kizim performed a number of fine technological experiments on evaporation of multicomponent systems in space. Layers of tin-silver brazing alloys were deposited on some specimens.

In April 1984 the International crew, consisting of Yuri Malyshev, Gennady Strekalov and Rakesh Sharma (India), performed an experiment «Overcooling» in «Isparitel-M» for melting spheres of eutectic alloys of silver-germanium to study the possibility of producing amorphous metallic materials in space.

In 1987 the technological unit «Yantar» (Figure 3), having a much higher potential, was delivered to the OC «Mir». In addition to the above-mentioned processes, it can deposit coatings on polymeric and metallic conveying strips and also perform welding of metallic specimens of small thickness. Cosmonauts Yuri Romanenko and Aleksandr Lavejkin were working using this unit in 1988–1989.

All the above-mentioned units had a remote or automatic programmed control. Therefore, they could be used in those cases, when the technique of operations was well checked and preset. However, there are many operations, for instance, repair, which are difficult or impossible to optimize preliminary.

To perform these operations a versatile electron beam hand tool (Figure 4) was developed at the PWI [4].

For easy transportation all modules of this tool were arranged in a common wire-mesh container, they were:

• working tool of a gun-type, which included a high-voltage supply unit and two electron beam guns (see p. 16, Figure below);

• pressurized instrument bay consisting of a secondary power source, automation unit and telemetry system;

• control panel;

• cable communications.

Figure 4 Electron beam VHT: *1* — wire-mesh container; *2* — work tool; *3* — pressurized instrument bay; *4* — control panel; *5* — cable communications; *6* — board with specimens.

The container had fasteners with the help of which it could be fixed at the external surface of the space vehicles, the tool being connected to the board mains with a special cable.

The changeable boards with specimens to be treated could be fixed on the container to conduct experiments and training.

The VHT was delivered to the board of the OS «Salyut-7» in 1984. In July this year the first outboard experiments were conducted using manual electron beam welding, cutting, brazing and coating.

Figure 5 Standardized secondary energy supply source (converter).

Figure 6 Standardized high-voltage module (*1*) with an electron beam unit (*2*).

Experiments were carried out by the Soviet pilots-cosmonauts Svetlana Savitskaya and Vladimir Dzhanibekov.

The specially-designed test specimens were treated.

Then, in 1986, cosmonauts Leonid Kizim and Vladimir Soloviov performed the next testing of the tool outboard the «Salyut-7». This time not only specimens were welded and brazed in space, but also separate sub-assemblies and fragment of truss structures. It was shown that cosmonauts can successfully and qualitatively perform such kinds of work using the VHT.

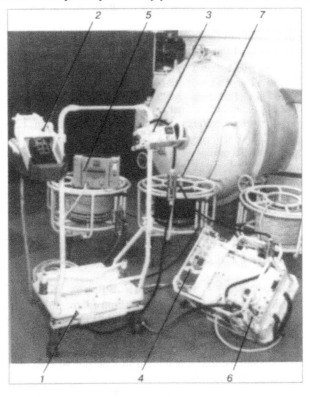

Figure 7 Set of electron beam equipment «Universal»: *1* — positioner of operator; *2* — container with a working tool; *3* — control panel of operator; *4* — control panel of assistant; *5* — system of diagnostics; *6* — pressurized instrument bay; *7* — cable communications.

The development and testing of such a wide range of welding equipment in space made it possible to solve successfully the problem of standardization of its mostly important functional modules, taking into account the maximum possible efficient utilization of the consumed energy at minimum mass-dimensional characteristics of the equipment.

Now, we have already developed standardized modules of the space welding electron beam equipment designed for 1.5 and 3 kW capacity.

As an example, Figure 5 illustrates the standardized secondary power source (converter) of 1.5 kW capacity, and Figure 6 shows the high-voltage module with an electron beam gun of the same capacity. It should be noted that the mentioned values of rated power of these modules can be considerably increased by using forced thermal control.

The design of «Universal» hardware is based on these unified modules (Figure 7). This installation consists of four electron beam tools (see Figure 12b on p. 202) and can perform in space almost all the welding technological operations which can be used in repair of large space stations of the future.

«Universal» is designed for long-term storage in open space. Special cable communications allows work to be carried out on the external surface of the space vehicles within a radius of up to 30 m from the place of connection to the board electric mains.

The set includes a special positioner (work station of the operator which is equipped with auxiliary devices and ensures safe and less tiresome conditions of work for the cosmonaut. The «Universal» hardware passed all ground tests and the preparations are going on for its testing under space conditions. Most of the standardized modules from which it is manufactured can also be used in design of different automated or robotic welding space installations.

This brief paper does not allow detailed descriptions of many technical difficulties which had to be overcome in design of the space welding equipment. It should be noted only that most of the technical solutions which were found will be used in future not only in space, but also in terrestrial industry.

1. Paton, B.E., Dudko, D.A. and Lapchinsky, V.F. (1984) Welding processes in space, In: *Welding and special electrometallurgy*, Kiev, Naukova Dumka, pp. 121–128.

2. Paton, B.E. and Kubasov, V.N. (1970) *See pp. 154–160 in this Book.*

3. Zagrebelny, A.A., Lukash, E.S., Nikitsky, V.P. et al. (1988) Development of coating technique application in flight conditions, In: *Proc. of 22nd Tsiolkovsky Readings*, Kaluga, 1987. Moscow, pp. 134–138.

4. Paton, B.E., Dzhanibekov, V.A. and Savitskaya, S.E. (1986) *See pp. 11–18 in this Book.*

Work on Manual Electron Beam Technology in Space[*]

B.E. Paton[1], S.S. Gavrish[1], V.F. Shulym[1], A.R. Bulatsev[1], V.V. Demianenko[1], V.A. Kryukov[1], B.I. Perepechenko[1], I.G. Lyubomudrov[1], M.A. Strelnikov[1], T.N. Kharkovskaya[1], A.A. Zagrebelny[1], V.P. Nikitsky[2], A.V. Markov[2] and I.V. Churilo[2]

A welding experiment, conducted thirty years ago in a de-pressurized docking module of «Soyuz-6» spaceship marked the start of space environment utilization for solving various research and applied tasks. At that time already, scientists and experts regarded space as man's habitat, realizing, that individual short-time flights are just the first step in mastering

* (1999) *Avtomaticheskaya Svarka*, No. 10, pp. 7–22.

space. Several years have passed and the USSR was the first to start creating long-term large-sized piloted objects, allowing planned systematic research to be conducted. Construction and long-term operation of such facilities envisaged the use of appropriate technologies and tools directly in orbit. Hence the genuine interest of scientists and engineers in the leading countries of the world in this welding experiment. Three «hot» welding processes, namely arc, plasma and electron beam, were tried out in this flight [1, 2]. Postflight experiment analysis showed that the electron beam process was the most promising for the space environment by being relatively simple, highly efficient and applicable for all the metals and alloys, used in aerospace industry.

The «Vulkan» unit, in which the welding experiment was conducted, incorporated a electrostatic sharply-focused directly heated electron beam gun [3].

The first technology experiment in space convinced the experts, that it is both possible and promising to apply automatic welding for producing permanent, also tight joints of metals. However, space hardware developers realized, that almost all the operations on maintenance and repair of long-term flying vehicles, when neither the zone of work performance, not the parameters of the components to be repaired and reconditioned, are known in advance, should be performed manually, with just partial mechanization. This created much higher interest of the specialists in studying the ability of manual welding performance in space. World practice at that time completely lacked the experience of manual use of the electron beam, as such a version of the process was not needed on the ground. Open space is totally different with its «free» vacuum as the necessary environment for the electron beam and constant power deficit, whereas the electron beam is a highly effective heating source, allowing processing of all the metals used in space engineering, even refractory metals. The process generates practically no spatter, and this means that it is safe for performing work in a spacesuit. The advantages of electron beam technology for the conditions of space were so obvious, that a decision was taken to use exactly this process as the basis for the off-board hand tool.

Nonetheless, welding experts had a very dim idea of how a man, especially wearing a spacesuit, would be able to manage such a highly concentrated precise heating source. From foreign publications [4–10] it was known, that a number of US companies (Hamilton Standard, General Electric, Westinghouse Electric, etc.) also started development of a similar tool, but the authors of this paper still have no reliable information on the current status of this work even now. It is known for sure, that the US manual electron beam tool has not been tested in space. There is, however, illustration material, that shows electron beam gun modifications in man's hands (Figure 1).

June 1974. Yu.A. Gagarin CTC[3] conducted mock-up testing of the first sample of the electron beam tool that had the primary objective of evaluation of the ability of manual welding and cutting performance by an operator, wearing a spacesuit [11]. This testing marked the beginning of many years of work on development and improvement of a tool, that is lightweight, small-sized, easy to handle and control — a reliable help for a man living and working in the space environment.

In terms of design, the electron beam unit can be divided into the blocks shown in Figure 2.

The hand tool (Figure 3) was an indirectly heated diode electron beam gun. Its case is fastened to the handle, through which high-voltage cable is passed, providing electrical connection of the gun to the high-voltage power source. The gun optical system ensures electromagnetic focusing of the beam at a distance of several tens or even hundreds millimeters from the end face of its anode. The high-voltage power source is made as a monolithic epoxy block, similar to the one, used as part of «Vulkan» welding unit [12].

For the first mock up testing, a special vacuum facility [13, 14] was created as a horizontally located cylindrical chamber of 2 m^2 volume, one end of which accommodated

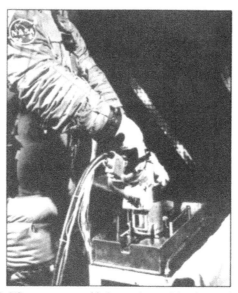

Figure 1 US test operator with the mock-up of the electron beam tool.

a self-sufficient high-vacuum unit, consisting of two sorption-getter pumps (developed by the B.I. Verkin ILTP[10]) and providing a vacuum of up to 10^{-4} Torr in a chamber of such a volume, and a fragment (upper front part) of «Orlan» spacesuit (developed by the «Zvezda» Company) was mounted at its other end. Mounted inside the chamber were the electron beam tool, high-voltage power block and boards with the samples for processing. As the spacesuit shell is designed for a pressure gradient of not less than 0.4 atm (300–350 Torr), the facility was placed into a space simulation test chamber, in which a pressure not higher than 0.4 atm was created (Figure 4). Such a complicated set-up of the first tests was due to an extremely strict approach to the safety of welding operator in case of an authorized contact of the beam with the soft parts of the spacesuit (gloves or sleeves), and, therefore, their thermal destruction or de-pressurizing. Space simulation test chamber volume (60 m^3) was 30 times greater than the volume of the facility chamber. Therefore, emergency lowering of pressure in the space simulation test chamber, where the operator was working, would be just 10 Torr at a rate of pressure drop below 1 Torr/s in the case of an accidental de-pressurizing of the spacesuit fragment, this being not higher than the norms accepted in aviation.

The high requirements for operator's safety dictated the need for a thorough grounding of the hardware (triple system), its protection from current overloads in the power buses and

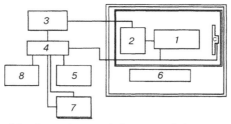

Figure 2 Block-diagram of the electron beam tool: *1* — manual electron beam gun; *2* — high-voltage power block, electrically connected to manual electron beam gun by high-voltage cable; *3* — block-converter of DC voltage into high-frequency AC voltage (inverter); *4* — control and telemetry block; *5* — off-board («ground-based») control panel; *6* — «on-board» control panel; *7* — low-voltage power source; *8* — instrumentation.

Figure 3 Electron beam hand tool.

short-circuiting in case of breakdowns in the gun. In view of the anticipated intensity of secondary X-ray radiation, the spacesuit fragment was fitted with a leaded rubber cover 4 mm thick (equivalent of 1.3 mm of lead) and additional glass 10 mm thick (equivalent of 2.5 mm of lead), suspended in front of the spacesuit viewing window, in addition to the limit value of accelerating voltage (15 kV) set for the tested tool.

The testing procedure envisaged the simultaneous presence of three test operators in the space simulation test chamber. Their functions were distributed as follows: welding operator conducted the scheduled technological operations using the electron beam tool; the first assistant of the operator worked with the onboard panel for welding hardware control and controlled vacuum pumping system of the facility chamber; the second assistant of the operator conducted visual observation of the welding operator work and filming of the testing process. All the test operators were wearing oxygen masks, because of the low pressure in the space simulation test chamber.

Overall supervision was conducted by the leading expert, being outside the vacuum chamber. He operated the «ground-based» control panel, performed monitoring, adjustment and recording of the technological process parameters.

Such a sequence of sending commands to control the hardware and, ultimately, the technological process was selected, that passage of the control signal were provided by pressing appropriate buttons in a certain sequence on both the control panels, namely «ground-based» and onboard. This to a certain extent guaranteed the safety of testing by eliminating random unauthorized starting of the electron beam.

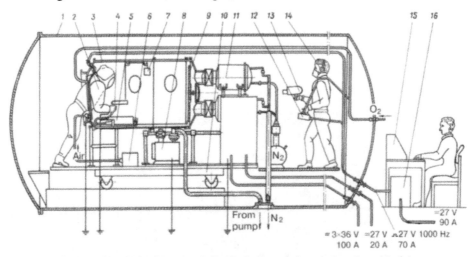

Figure 4 Schematic of conducting manual electron beam gun «hot» testing: *1* — altitude/temperature chamber; *2* — half-suit; *3* — vacuum facility; *4* — manual electron beam gun; *5* — high-voltage block; *6* — compressor for half-suit ventilation; *7* — weight-neutralizing device; *8* — roughing pump; *9* — board with samples; *10* — facility platform; *11* — self-sufficient high-vacuum pump; *12* — control panel; *13* — recording camera; *14* — intercom; *15* — ground-based control panel; *16* — welding converter block.

The entire test cycle consisted of seven experiments (altitude simulations in the space simulation test chamber) with the total time of test operators staying at «altitude», i.e. in the space simulation test chamber, of 28 to 30 h. Direct time of working with the tool during all the seven altitude simulations was just 70 min, this being another proof of the extremely high labour-consumption of such tests.

The following main operations were performed during altitude simulations:

• burning through samples of various materials (distance from tool anode to the sample is 1.5 m);

• cutting up samples of various metals and thicknesses (0.3 m distance);

• welding samples of various metals, thicknesses and weld configuration (0.3 m distance).

Man was holding the electron beam tool in his hands for the first time, so the experts were chiefly interested in the ability to perform an accurate and specified beam displacement required for professional performance of complicated processing operations with a satisfactory quality.

The space simulation chamber testing was followed by a large cycle of aircraft flights (Figure 5), where the arc, plasma and electron beam welding processes were tried out under the conditions of short-time zero gravity in vacuum chambers. Hardware developers received further confirmation that selection of the electron beam process was correct, as under these conditions maintaining a stable arc or plasma is difficult. The testing performed led to the following main conclusions:

• manual electron beam gun can and should be used to perform in open space the mounting and repair operations, involving welding technologies;

• a mandatory condition for safe and skillful handling of the welding hand tool, as on the ground, is professional training of the cosmonaut-operator, who should be reliably fastened during work performance in space;

• in order to increase the level of cosmonaut-operator's safety and spaceship protection from its skin being accidentally hit by the beam, it is rational to incorporate a short-focus gun into the tool;

Figure 5 Manual electron beam gun testing in the flying laboratory has been completed.

• control panel with the keys for starting and switching off the beam, should certainly be handled by the operator, directly conducting the technological process, and the last command for switching the tool on should be given by a key that does not stay put and is located on the tool handle, thus providing automatic switching off of the beam at random loss of the tool;

• to eliminate the use of additional means of operator's protection from adverse influence of hard X-ray radiation, it is necessary to specify the limit accelerating voltage of the high-voltage power source of not more than 10 kV [15];

• to protect the operator's working hand from thermal radiation of the liquid metal and its vapours, it is necessary to fit the tool handle with a protective shield (guard).

Thus, the performed testing allowed the solution of the key technological and design problems and enabled the experts to start development of the flight variant of the onboard tool.

July 1984. An experiment, unprecedented in the complexity of its preparation and organization and having great importance in terms of applied research, was conducted in open space. Technology of manual performance of electron beam processes of welding, cutting, brazing and coating application was retrofitted, using special hardware (Figure 6). The experiment that lasted for about four hours, was conducted by cosmonaut-operators S. Savitskaya and V. Dzhanibekov off board OS «Salyut-7». During this time the operators transported the VHT to the place of experiment performance, fastened it to the hinged handrail of a special platform, fitted with operator's foot restraint, connected the hardware to a power supply system, performed all the scheduled scope of work and returned to the station with the hardware and the processed samples. All together, 20 sample-plates of stainless steel and titanium alloys of different thickness were processed. 6 samples were processed by welding, brazing and cutting and 2 samples were brazed.

Hardware design solutions, technology problems and results of analysis of samples, obtained during this experiment [16–20] are quite well described in various publications, and therefore, this paper presents in greater detail the experiment preparation and schematic of its performance. Nevertheless, we would like to reiterate that the requirements for the hardware and recommendations issued after mock-up testing were implemented in VHT design.

So, specialists decided to develop a new type of secondary power source (inverter), incorporating modern components [21], using ready-made assembly PC boards and output frequency of 20 kHz to reduce its overall dimensions (in «Vulkan» the output frequency was 1 kHz). It was possible to create a small-sized high-voltage block [22, 23], allowing it to be incorporated directly into the tool, thus eliminating the high-voltage cable as the source of increased hazard. The instrument compartment, accommodating the inverter and controls and instrumentation, as well as the operator's control panel and tool seats were designed as

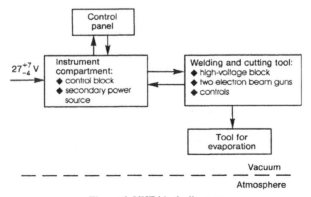

Figure 6 VHT block-diagram.

Figure 7 After completion of the work. S. Savitskaya and V. Dzhanibekov are back to the station
(*extreme right* — L. Kizim; *extreme left* — V. Soloviov).

one small-sized transportation container, that was comparatively easy to take outside the station and fasten to its external handrails.

The entire experiment was conducted at «daytime», i.e. in the «light» flight mode, in the zone of a stable «Station–Earth» radio communication. For this purpose the experiment planning group of the FCC selected the optimal turns of the station orbit, that involve all the ground-based measuring points of the country and several special vessels, that were in the station flight trajectory. In order to be able to issue prompt recommendations to the crew in case of an emergency, a synchronous experiment on the station full-scale mock-up was conducted in the Yu.A. Gagarin CTC[3] zero buoyancy laboratory, with participation of experienced test operators, who had trained the cosmonauts to do this work. In addition to ground-based support from FCC, great assistance to the direct performers who were outside the station, was rendered by crew members L. Kizim, V. Soloviov, O. Atkov, and, especially, I. Volk, who followed V. Dzhanibekov and S. Savitskaya throughout all the stages of their ground-based welding training (Figure 7). They moved up to the portholes around the station, observing what was going on off board, and were ready to provide the required advice-recommendation at any second. This full-scale experiment certainly is the culmination point in the activity of a large team of experts, and so it is necessary to give a brief account also of the preceding period of its preparation.

At the start of 1984 Prof. V.P. Glushko, Designer General, initiated the scheduling of an inspection of the station external surface by «Salyut-7» crew for the second half of July. Members of the primary and backup crews on this expedition, namely V. Dzhanibekov and S. Savitskaya, V. Vasyutin and E. Ivanova were to have the necessary training. The crews included female cosmonauts, who were to make exits for the first time in the world, and so a suggestion was considered that a woman should be given a passive role in this difficult operation, i.e. simply get used to the spacesuit and «walk», while a male cosmonaut should accompany her. The crews, however, and especially, the female crew members strongly objected to such an approach and required to be entrusted with serious flight work.

By this moment, all the required testing of the flight set of VHT welding hardware was being completed, and various stages of operator EVA had to be tried out on a professional level. Test operators were to assess the possibility and convenience of work performance by man in a spacesuit under the conditions of short-time zero gravity in the flying laboratory, at zero buoyancy and in a manned space simulation test chamber. For this purpose, the mock-up of the station's docking module, mounted in the aircraft was used to try taking VHT hardware out through the exit hatchway, VHT transporting to the experiment site, its fastening to the station external surface and preparing for operation.

Ergonomic studies (relative position of the welding operator and the board with the samples for processing and of the controls, etc.) were conducted, optimality of the operator's work place lay out was assessed, and the sequence of auxiliary operations performance (taking the hardware outside, its mounting on the outer skin, laying the electric cable system, dismantling the hardware with the processed samples and its delivery to the station, plotting real-time EVA cyclogram and, finally operators' interaction during performance of all these operations) was determined under zero buoyancy conditions. Such testing was conducted on a full-scale mock-up of the station in a water tank.

The most important element in this experiment was retrofitting the technique of performance of welding operations proper, more precise determination of the process mode parameters, selection, together with the medical experts, of the optimal «work-rest» regime for the welding operator. Testing was conducted in a manned space simulation test chamber of 60 m³ volume, that accommodated a specially designed and manufactured facility. It included a fragment of the station side wall, the handrail on which the welding hardware was mounted, and the welding operator's mobile foot restraints and was fitted with systems of weight-neutralizing (unloading) of the tool and welding operator's spacesuit (tool weight was 5 kg, and that of the spacesuit was above 70 kg). Also evaluated during testing was the need for the spacesuit light-filter that the test operator was using when working with molten metal, and, looking ahead, it should be said, that the crews were recommended to always use the light filter when working with samples of titanium alloys and stainless steel. It goes without saying that used for this specific testing was the welding hardware that by its design and configuration was the «twin» of the VHT flight set (dimension-weight, training and zero buoyancy mock-ups of the hardware were used in the flying laboratory and the water tank). And, certainly, test operators were using EVA suits at excess pressure.

It should be also added that at the request of the primary and backup crews of the short-term expedition, despite the busy schedule of their own flight training, they were allowed to follow all the operations with VHT, and participate in discussion of the results and analysis of test operators' mistakes. It is not difficult to guess that after completion of the test cycle, the crews approached the Designer General with the suggestion to entrust them with conducting the space welding experiment. Taking into account the opinion of the experts, who staged the experiment and test operators' recommendations, the Designer General agreed to include into the short-term crew training schedule this work, which was critical and quite complicated even for professionals.

Considering that none of the crew members had ever used any kind of welding tool, or was familiar with any manual welding technique or skills, the experts had in an extremely short time to not only train the cosmonauts, but also give them a short theoretical course, familiarize them with the objectives of the experiment and the hardware they would have to operate. Further difficulties were encountered because of the need to prepare protective gear for female operators that require smaller spacesuit sizes because of their anthropometry. Such spacesuits were not available on board the station and it did not seem possible to deliver them there, as the volume and weight resources of «Soyuz» transport spaceship, carrying three members of the short-term crew and welding hardware of about 45 kg weight, turned out to be exhausted. Assistance was rendered by experts of the spacesuit manufacturer, headed by G.I. Severin, Designer General. They provided the outfit for the female contingent during ground-based training, and all the necessary recommendations for fitting the spacesuits available in the station. This was quite «simple», especially, if we compare the height of S. Savitskaya (165 cm) with that of V. Ryumin (almost 190 cm), whose spacesuit she had to wear, when doing the work. To the great surprise of welders, doctors and instructors, the «pupils» coped with the task in a excellent manner: they mastered the techniques and skills of manual welding, learned to control the behaviour of the molten metal pool, and to operate the welding hardware controls, etc., each primary and backup crew member requiring just two altitude simulation cycles in the space simulation test chamber. Purely psychological barriers certainly had an important role — the fervent desire to master an unusual profession and the ability to instantly evaluate the results of their work. However, credit should be given to the perfection of the tool that enables performance of one of the most complicated technological processes in an easy, simple and safe manner.

Yu.A. Gagarin CTC[3] decided to approve an unprecedented case of parallel training. Altitude simulation in the space simulation test chamber (about 5 h), a flight in the flying laboratory (2.5–3 h) and work in a water tank (4–5 h) had to be performed on the same day. All the kinds of training were concentrated in one testing centre, fitted with training and mock-up equipment; the team of test operators and service personnel started working two shifts, with thorough planning for training two crews. 18 altitude simulations in the space simulation test chamber, 20 flights in the flying laboratory and 8 immersions in the water tank were performed in one month, from May 4 to June 4. During each training, specialists conducted a final, individual (for each crew member), fitting of the hardware, tool handles, controls, catch locks, hooks safety tethers, latch hooks and other «small items», hindering or being in the way of quality work performance (Figure 8).

The general «run» of the crews in the zero buoyancy laboratory, where the flight experiment scenario was implemented by the full time cyclogram, was conducted in the presence of the authorities — the organization staging the experiment, the flight operations officer, members of the State Board and the Minister of the appropriate industry. Approval was given to this experiment, exotic at that time, the degree of its criticality being indicated even by the fact that just two days before the expedition left for the station, Yu.P. Semyonov,

Figure 8 Operations leaflet, devoted to completion of preparation to the flight of V. Dzhanibekov, S. Savitskaya and I. Volk.

Figure 9 Afterflight meeting of the teacher with his favourite pupils.

Designer General, conducted «dry» training in Bajkonur, actually being a discrete examination to check the knowledge of the hardware and allocation of responsibilities during preparation and implementation of the flight experiment.

It is difficult to single out the primary factor that brought about such stunning success of this experiment, conducted on July 25, 1984, so that all its participants have every right to feel proud (Figure 9).

The results of this work — the welded samples, were brought to Earth by the short-term crew and handed over to the researchers, while the welding hardware remained in the station for subsequent operations.

May 1986. In this year the Soviet Union began construction in orbit of station of the new generation «Mir», designed for operation for 5 to 7 years. The plan was to conduct its construction in stages, with an annual addition of one or two special-purpose modules, namely astrophysical, technological, biological, etc. It was intended to further fit the modules with various-purpose structures, mounted directly in orbit by the crews that continuously kept watch in the station [24]. This station, as the stations of the next generation, had to be fitted with diverse mounting and repair tools, namely bench tools, welding tools, etc. Therefore, already back in May 1985, when the mainframe of «Mir» was in the building berth, and «Salyut-7» was completing its stay in orbit, it was decided to conduct one more experiment, using VHT hardware that was on board the «Salyut». This time it was necessary to try out welding, brazing and cutting of three-dimensional fragments of actual structures, that were planned to be built and dismantled in the stations in the future.

For this purpose, VHT developers designed and made, instead of the board for flat samples, a cassette (Figure 10) that carried samples of a rod truss structure, hinged joints of foldable structures, tubular connections with a specially developed braze alloy, applied onto their surface [16, 20, 25]. Using the procedure, already tried out in 1984, two crews — L. Kizim and V. Soloviov, A. Viktorenko and A. Aleksandrov, had a full course of instruction and training. One of the operators, A.Viktorenko, had the skills of using the welding tool and one altitude simulation in the space simulation test chamber turned out to be enough for him: the Board and the developers, having examined and analyzed the quality of specimens, processed by him, recognized his training to be excellent. The other crew members went though two altitude simulations for «hot» training, while the scope of other kinds of training practically corresponded to the scope of training of the first space manual welders.

It should be borne in mind, that shortly before this new experiment, OS «Salyut-7» was brought into the unmanned mode, and became uncontrollable. As a result, the station lost its orientation, power and life support systems failed and it «froze». Cosmonauts V. Dzhanibekov and V. Savinykh «reanimated» the station, but temperature increase in the modules led to excess condensate, and that, certainly, could affect the standard and experimental

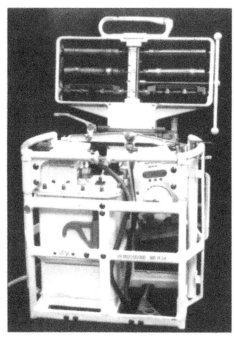

Figure 10 VHT with the board and samples of future space structures.

onboard hardware, in particular, the VHT. Therefore, it was decided, that the crew going up to replace the rescue crew, would conduct trial «idle» switching of the VHT, i.e. with the switched off tool, before taking the hardware outside. Switching on confirmed the usability of the hardware and removed the doubts which had arisen on its integrity after being stowed for several months under extreme conditions in a disabled state. This unforeseen situation led the researchers to the idea of mandatory inclusion of special testing instrumentation into the future tool models.

As in 1984, the indubitable success of this experiment was promoted by the desire of the crews and all the experts to perform the assigned task perfectly. The cosmonauts had so well prepared for the job, that FCC management decided not to conduct the synchronous «ground-based» experiment (in the water tank) and cleared the crew for performance of the flight assignment outside the zone of radio communication with Earth. One can imagine the joy felt by the experts and the degree of their confidence in the correct selection of welding as one of the construction-mounting and repair techniques in space, when, entering the radio communication zone, the crew reported the successful fulfillment of the task.

Welding operations, using VHT, conducted in 1984 and 1986, confirmed the need to modify the tool, introduce new engineering solutions, taking into account some essential corrections of the technological parameters, claimed by the customer, as well as comments and wishes of the test operators and the crews, expressed by them during testing, training and flight experiments. The developers had the following main objectives:

  • raising the tool output power to 1 kW for working with aluminium and its alloys;

  • fitting the tool with a filler wire feed mechanism;

  • incorporating into the welding hardware a system of on-line information and monitoring the technical and process parameters for its transmission to the station recorders and FCC monitors.

July 1998. The primary and backup crews (G. Padalko and S. Avdeev, S. Zalyotin and A. Kaleri) completed their preparation to conducting «Flagman» experiment in OC «Mir»

[26]. In the first stage of this experiment it was planned to test a new, more powerful model of EBW hardware [27–30], and optimize the manual technologies of welding, cutting and brazing on samples, simulating the most vulnerable sections of the future ISS. By this time the flight set of the welding hardware proper and the samples to be processed, of a total weight of 100 kg, were delivered by «Progress» cargo spaceship to the station, where the trained crew was to arrive in August, and in September–October also delivered was the remaining part of the hardware, including the electrical and information cable systems, the rotating holder for samples to be processed and the foldable platform with the welding operator's work place, of a total weight of 143 kg. For a number of reasons, however, the main one being the indeterminate situation of the extension of the complex operation period, the range of necessary payloads scheduled for delivery on board the station, was radically changed, and by the time of this paper preparation the further fate of the OC «Mir», and, naturally, of the experiments planned to be performed in it, is still unclear [31, 32].

Therefore, we will now consider just the preparation of the welding experiment, using «Universal» hardware, which is the electron beam tool modification, coming after VHT. The design of «Universal» welding hardware, begun in 1987, was based on a statement of work, put together, taking into account the results of experiments, conducted in OS «Salyut» in 1984 and 1986. Flight retrofitting of this modification of electron beam unit was intended to be conducted in the OC «Mir», the construction of which began in 1986. Similar to previous developments («Vulkan», «Isparitel», «Yantar», VHT hardware, etc.), «Universal» was created, allowing for the specific features of working with molten metal in space, the unique nature and high cost of the hardware. It was also necessary to provide the maximum safety under special service conditions (zero gravity, space vacuum, operator's outfit, that completely limits his motion and visual capability, high temperature gradient, presence of sharp light and shadow boundaries, etc.). The draft design was also based on the main requirements for the space hardware, specified in the codes on development of experimental and scientific onboard hardware and described in Reference [20] in sufficient detail.

Hardware block diagram and appearance are given in Figures 11 and 12. Compared to the previous VHT model, the following essential changes were made in the «Universal»:

• hardware output power was considerably increased (2–2.5 times);

• tools proper were made functionally purpose-oriented, single-gun, with a standby cathode, thus allowing transition from one processing operation to another one, by just changing the tool (VHT is fitted with two guns, having diverse processing functions);

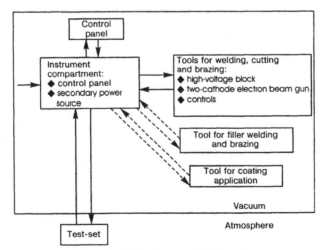

Figure 11 Block-diagram of «Universal».

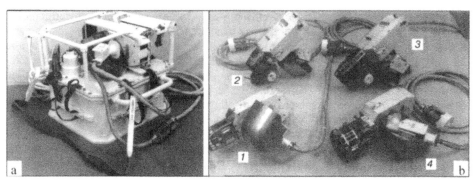

Figure 12 «Universal» hardware: a — instrument compartment; b — set of specialized tools; *1* — base tool; *2* — tool for brazing with feeding of brazing alloy; *3* — tool for welding with filler wire feeding; *4* — tool for coating deposition.

• «Universal» is fitted with the basic tool for welding, brazing and cutting; a tool with filler feed mechanism for welding and brazing; a tool with a turret extension, incorporating four crucibles with evaporation materials, for application of special optical, thermo-regulating and other coatings.

Tool architecture remained practically unchanged, but the tool starter button was made as a two-position push button, thus allowing in the basic and turret tools a discrete power increment by pressing the button a second time (by almost 15 % from that initially set by the operator on the control panel and reached by the first pushing of the button). In tools with filler wire feed mechanism, the second pressing of the button sends a command for starting this mechanism drive. In addition, the hardware is fitted with the so-called test-set. This instrumentation is placed into the station habitable module, is connected to the «Universal» instrument compartment by the cable and information systems and has an output connection to the station telemetry system. The test-set is used for preliminary and in-process diagnostics of welding hardware, monitoring the unit electrical parameters during its operation and «downloading» the measured and monitored parameters to the «board» and to FCC.

At the end of 1990 the first sample of «Universal» hardware [28, 29] after its setting up and factory testing for compliance to the specification, was turned over to the experts for adjustment, dynamic and thermal testing, «runs» in the limit power modes and other testing as part of the so-called laboratory and design retrofitting, in keeping with the accepted standards for space research hardware. After receiving positive results of all the kinds of tests and correction of technical documentation, manufacture of the first flight set of the hardware was begun, the one to be used to perform the space experiment. Testing in the manned chamber was conducted in parallel, to optimize the techniques of handling the tool in its envisaged manual variant. Also checked was the lay-out of the workplace of the welding operator that was to be mounted on the outer surface of the OC «Mir» module under construction. Testing was also intended to alleviate the concerns of hardware developers as to the need to devise additional means of spacesuit protection from the by-products generated during the welding processes (soot and intense thermal radiation from the weld pool leading to thermal destruction of the spacesuit upper cover, molten metal drops, etc.). Considering the higher power of the hardware, planned operations with the aluminium-magnesium alloys, and incorporation of tools with filler material feed, all this was new and required thorough verification.

The scope of this testing was equal to six 5-hour altitude simulations, performed by several test operators with «hot» processing of many tens of various samples, mounted on the flap boards. The obtained results finally determined the scenario of the space experiment. It was decided to:

• mount the samples on a rotating multiface sample holder;

Figure 13 First model of the work platform for OC «Mir».

• place the welding operator on a rotary mobile platform, fitting it with a foot restraint («anchor»);

• mount the control panel of the welding hardware on the mobile platform handrail;

• further fit the instrument compartment of «Universal» hardware with seats for manual tools and place it to the right of the operator, so that he could easily change the tools, when going over from one operation to another;

• envisage additional easily removable protective clothing, shielding the spacesuit front part, its sleeves and glass, from by-product contamination when conducting the technological processes.

As soon as testing was over, an operating full-scale mock-up of the platform of frame design was fabricated. It carried the mobile work place, rotating sample holder with the samples and «Universal» hardware (Figure 13). Fit demonstration testing confirmed the correctness of the selected procedure of the flight experiment performance.

When a delegation of specialists from NASA and such major US aerospace companies as Martin Marietta, McDonnel Douglas Aerospace and Rockwell International visited our country at the start of 1991, the work on «Universal» and construction of «Mir» was in full swing. After familiarization with the program of the scheduled research, the Americans suggested to S.P. Korolyov RSC «Energiya»[2] (customer for «Universal» hardware) and the PWI (welding experiment organizer) that the first phase of the experiment be conducted in one of the US shuttles, and then the subsequent phases in «Mir».

In 1991 and 1992 several series of demonstration familiarization flights of KC-135 flying laboratory (L. Johnson Space Flight Center, NASA, Houston) and zero buoyancy testing in the water tank were performed involving US astronauts and specialists (Figure 14a and b). The first operating sample and electric and hydraulic mock-ups of «Universal» welding hardware, as well as the mock-up of the work station with a mobile work place of welding operator were used in testing [33–36]. The outcome of this testing was an agreement on conducting in 1997 the International Space Welding Experiment on board the US «Columbia» vehicle (STS-87 mission). It envisaged joint development of the experimental procedure, where Teledyne Brown, US company, a manufacturer and supplier of various-purpose load-carrying platforms for the Shuttle cargo bay, was to design and manufacture a specialized platform, and our specialists were to develop, manufacture and fit to it all the required hardware (rotating sample holder, carrying the US samples, mobile work place, designed to fit the US spacesuit, container for replaceable hand tools, as well as the flight variant of the «Universal» hardware proper).

In 1995 the first fit demonstration testing at zero buoyancy was conducted at G. Marshall Space Flight Center, NASA, Huntsville, during which ergonomic evaluation of the newly developed mock-up hardware was performed, with the hardware mounted on the load-carrying

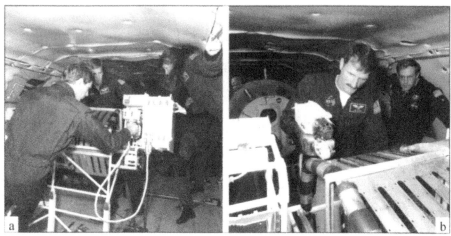

Figure 14 US astronauts are mastering unfamiliar hardware under conditions of zero gravity.

platform (Figure 15), installed in the zero buoyancy mock-up of the Shuttle cargo bay (Figure 16). Preliminary procedure of the flight experiment was optimised simultaneously, namely sequence of commands and operators' actions and their relative position during the welding experiment. Some minor omissions and imperfections, revealed during testing, were eliminated by the designers; technical documentation was finally corrected and the manufacture of the flight set began.

Special attention was given to protection of the operators and the vehicle from the possible direct impact of the electron beam, that might lead to the spacesuit de-pressurising, damage to the vehicle skin, its systems and experimental hardware located in the cargo bay. It was necessary to localize the working zone and prevent the by-products from penetrating beyond this zone. Unfortunately, no radical solution of this difficult issue was found, and just a protective shroud covering the rotating sample holder from three sides, had to be used; the front side was left open. This is the side where the operator and a mobile fork for the working hand with the tool, preventing the operator from moving his hand beyond the working zone, are located. The fork was further fitted with the seat for the work tool that

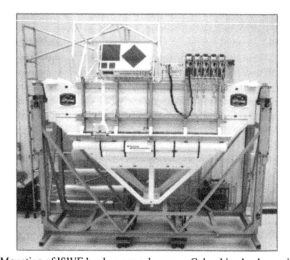

Figure 15 Mounting of ISWE hardware mock-up on «Columbia» load-carrying platform in the building berth.

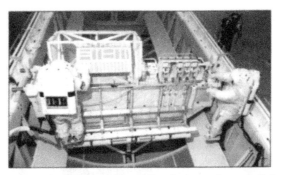

Figure 16 Zero buoyancy testing (load-carrying platform in the mock-up of «Columbia» cargo bay, special welder's protective clothing worn over the spacesuits).

can be used by the operator in the intervals between the operations, for instance, if he has to bring the rotating sample holder into a new position. It was also finally decided to create easy-to-remove protective clothing, put over the spacesuit by the operator for contamination protection (Hamilton Standard Company). These new features had to be evaluated by the US astronauts during additional zero buoyancy tests. Looking ahead, we can say, that testing, conducted in the same year 1995, confirmed the correctness of the incorporated solutions, and a special ISWE Safety Board assessed the measures as being sufficient.

In parallel, during «hot» testing, welding technologists determined a set of samples for in-flight processing, their configuration, materials, thicknesses, the most suitable for placing on the rotating holder.

During the entire second half of 1995, endless qualification and special testing of the flight welding hardware were conducted, which are required by the appropriate NASA standards. The most important and labour-intensive were:

• testing for electromagnetic interference, i.e. electromagnetic compatibility of welding hardware with the vehicle systems. Electromagnetic interference, affecting the power, control, telemetry and radio communications systems (Figure 17) during the welding hardware operation, as well as the impact of interference, generated during operation of various vehicle instruments and systems, on the service and technical parameters of the welding hardware, its control and information system were measured and assessed;

• evaluation of preservation of welding hardware serviceability in a disabled state at long-term (up to two days) impact of low (−113 °C) and high (+80 °C) temperatures on it;

Figure 17 Evaluation of the level of radio interference, created by «Universal» hardware.

Figure 18 Evaluation of the thermal resistance of materials for the spacesuit protective shroud.

• thermal cycling tests (–40 – +60 °C; cycle number being 8);

• thermal testing, i.e. evaluation of the thermal life of the instrument compartment and tools in the planned flight experiment mode (5 min work–5 min rest) of processing all the samples, envisaged by the ISWE program;

• testing for toxicity, i.e. hazardous evolutions from the hardware, mounted in the vehicle habitable module (test-set, cable system);

• quantitative estimate of spattering, when working with various metals;

• qualitative evaluation of the nature of the molten metal spatter impact on the materials, suggested by the developer companies for the upper clothing of the spacesuit and the protective shroud;

• evaluation testing with direct impact of the electron beam on the same materials (Figure 18);

• testing for vibration strength and vibration resistance of the hardware (Figure 19);

• testing for the possibility of corona discharge initiation;

• X-ray radiation measurement;

• quantitative estimation of the strength of light radiation in processing various metals, that was conducted with the purpose of issuing the necessary recommendations to welding operators on the use of a standard light filter of the spacesuit.

Figure 19 Determination of resonance frequencies of the mobile work place of welding operator.

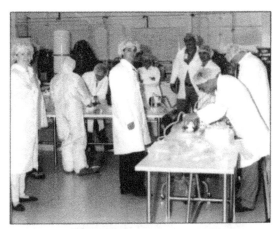

Figure 20 Control acceptance of flight equipment («clean» laboratory, Huntsville, USA).

Unfortunately, not all the testing results were encouraging. Testing for electromagnetic compatibility demonstrated, that a situation of an adverse impact of simultaneous operation of several power systems of the vehicle on «Universal» control block is theoretically possible. On the other hand, when work was performed using the filler, it was found, that the tool «makes a noise» and creates an increased background noise in the flying armchair receiver.

Joint efforts of our and US specialists allowed these faults to be eliminated.

Totally unexpected, and very bitter for us, was the failure of several high-voltage blocks in the tools. Analysis showed that the technology of impregnation with epoxy was violated when the blocks were manufactured. A second complete check of all the manufactured blocks and their stringent selection allowed this problem to be solved.

Qualification of support equipment (rotating sample holder, mobile work place, tool container, etc.) was finally begun by the start of 1996, following its double joint acceptance in the manufacturing plant (Kiev) and special «clean» laboratory (Huntsville) (Figure 20). Acceptance consisted in external examination, taking the necessary measurements, weighing, functional checks of various mechanisms, laboratory checks of the used lubrication and optical properties of special coatings. After that, the hardware elements «dispersed», following their itineraries to the respective G. Marshall Space Flight Center testing departments and laboratories, where they would get flight clearance.

In December 1996, when qualification testing was almost completed, astronaut-operators were preparing for the final «hot» training in the manned space simulation test chamber (L. Johnson Space Flight Center) and specialists were getting ready to conduct pre-start retrofitting of the hardware at Cape Canaveral (J. Kennedy Space Flight Center, NASA), fatal news, as we may say, was received. All the work in connection with this complex and expensive experiment was interrupted for reasons unrelated to ISWE in any way. The point is that the program for the next scheduled mission of the «Columbia» vehicle in October 1996 was completely disrupted, because of failure of the opening mechanism of the exit hatch in the cargo bay. The mission, the program of which envisaged retrofitting the procedure of the initial stage of ISS construction and was of paramount importance, was, naturally, interrupted. ISWE was to be one of the key activities during the next «Columbia» mission, and this was exactly the vehicle for which the special platform for the welding hardware had been developed, but NASA administration decided to repeat the unfulfilled program in the next «Columbia» mission and put our experiment into a «deep» reserve. The experiment was postponed for an indeterminate period, while the guarantee period of the welding hardware had already begun. Therefore, the management of the S.P. Korolyov RSC

Figure 21 Check assembly on the ground of the work platform for «Flagman» experiment.

«Energiya»[3] and the PWI, in their turn, decided to withdraw the hardware from the USA and start preparation for experiment performance in the OC «Mir».

All of 1997 was spent in design and manufacture of the working platform, that was a foldable frame structure, which carried the mobile work place, rotating sample holder and «Universal» with the changeable working tools (Figure 21). Unlike the US lay out procedure, in which everything was to be mounted on the ground on the load-carrying platform in the Shuttle cargo bay, in the «Mir» the hardware, as well as the power and information cable systems were to be taken out by the cosmonauts, block by block, through the docking module hatch of 1000 mm diameter and mounted on the station outer surface, so that it was necessary to cardinally modify the entire project and make all new support hardware.

In January 1998 specialists from Russia and Ukraine started certification testing of the newly made support hardware. Fortunately, the qualification of «Universal» welding hardware proper was not done, as similar testing had been conducted in the USA, while comparison of the US and Russian standards demonstrated their practically complete coincidence.

In April–June 1998 training of the primary and backup crews was conducted, the crews having to implement the first stage of the complex «Flagman» experiment, i.e. mount outside the station the continuously operational work platform and try out the welding technologies. The operators trained in the manned space simulation test chamber, steadily gaining professional experience of the welding hardware operation (Figure 22) and at zero buoyancy (Figure 23a–c), where they retrofitted the mounting operations for assembling the work platform on the «Mir» mock-up and finally checked the procedure and cyclogram of the crew EVA. The first stage of the integrated experiment fitted into two EVA sessions of 4.5–5 h each. During the first exit the operators took the hardware stowages outside the station, transported them to the experiment site, using the station's cargo boom, and completely assembled all the hardware. The second EVA was devoted to conducting the welding experiment, dismantling the rotating holder with the processed samples and the hardware and their transportation to the station. In the intervals between the EVA the operators, using the test-set, mounted in one of the modules, were to test the «Universal» and its state, making test switching on from the test-set panel and reading its working parameters.

Figure 22 Defining some final details of the assignment before closing the manned space simulation test chamber.

It should be noted, that training and testing were performed under the same conditions in the space simulation test chamber and the water tank. During testing all the techniques, technology and procedure of training were preliminarily retrofitted, in particular the design of the upper protective clothing for the Russian EVA suit «Orlan-DM» was finally determined. During training the operator's mobile work place was further fitted with a special light for working in the shadow part of the orbit. Testing and training allowed the responsibilities of each operator participating in the hardware installation, and in the experiment proper, to be finally determined.

By July 1998 preparation of the hardware and the crews to perform the welding experiment was completed, and in August the primary crew (G. Padalko and S. Avdeev) was

Figure 23 Zero buoyancy testing: a — station modules immersion; b — transporting «Flagman» component blocks from the docking module to the zone of work performance; c — operator's lifting after the end of training.

going to go up to the station, the experiment being scheduled for November. Already in May the station took on board the flight set of «Universal» welding hardware and cassettes with samples for welding, cutting and brazing. The plan was to deliver the second part of the hardware on «Progress» cargo ship in October...

In July the issue of quick sinking of «Mir» suddenly became very urgent. Regular research work in the station was discontinued. Starting from August work mostly related to keeping the station in the operating mode, began to be performed, while experiments were conducted using just the available hardware and consumable materials. The stream of supply to the station was disrupted, and the second half of the hardware, required for our experiment, is still on Earth [31]. The station fate is yet undetermined, and the prospects for conducting such a costly and important experiment are very vague.

Time passes, however, and, despite the fatal turn of events, we still have grounds to feel optimistic. The need for modern welding hardware, specifically adapted to building and maintaining in the operating condition large-sized space structures for 20 to 30 years, is becoming more and more obvious. Prospects are opened up for using laser tools to produce permanent joints in space and other «earthly» methods for doing work inside the station modules.

Parameters	«Vulkan» 1968	Mock-up sample of manual EB gun, 1974	VHT 1984–1986	«Universal» 1991–...
Supply voltage, V	26–28	27–32	23–34	23–34
Consumed power, W	800	Up to 2000	Up to 500	Up to 1500
Accelerating voltage, kV	10	15	5	8–10
Electron beam current, mA	60	Up to 100	Up to 70	Up to 110
Cathode type	Directly heated, Ta, 1×1 mm, $\delta = 0.06$ mm	Indirectly heated, LaB_6, $d = 3$–4 mm	Directly heated, Ta, 2×2 mm, $\delta = 0.06$ mm	Directly heated, Ta, 2×2 mm, $\delta = 0.06$ mm
Focusing	Short-focused, electrostatic	Long-focused, electrostatic	Short-focused, electrostatic	Short-focused, electrostatic
Operating mode	Continuous	Adjustable	4 fixed modes	9 fixed modes
Duty cycle, min	3	Long-time	5	5
Number of electron beam guns in the unit, pcs	2	1	1	1 (two-cathode)
Processing operations	Welding, cutting	Welding, cutting	Welding, cutting, coating deposition from crucible	Welding, cutting, brazing, filler welding, coating deposition from crucible and using filler wire
Control	Semi-automatic	Manual	Manual	Manual, semi-automatic testing
Monitoring	Instrumentation block, telemetry	Instrumentation block	No monitoring	Telemetry, test-set

That is why we purposefully moved away from the traditional description of the scientific aspect of the work, done over the last thirty years and told the readers about the specific features of staging and conducting in space complex experiments with man's participation, that, in our opinion, is well illustrated by the summary Table of technical data of all the above-mentioned space welding units. An interested reader will be able to find extensive information on the hardware, technological and scientific solutions of the considered issues in earlier published articles.

1. Paton, B.E. and Kubasov, V.N. (1970) *See pp. 154–160 in this Book.*

2. Bogdanov, V.V. (1970) Three welding processes, tested in space, *Khimiya i Zhizn*, No. 7, pp. 17–20.

3. Paton, B.E., Nazarenko, O.K., Chalov, V.I. et al. (1970) *See pp. 160–168 in this Book.*

4. Hidden, R.D. (1965) USAF. Studying space tool development, *Aviat. Week and Space Technol.*, **13**, No. 9, pp. 58–65.

5. Schollhammer, F.R. (1963) Electron beam welder for space, In: *Proc. of 6th Nat. Symp. on Space Vehicle Use*, Seatle, 1963. Seatle, United Aircraft, pp. 1–23.

6. Schollhammer, F.R. *Portable beam generator.* Pat. 3392261, USA, publ. 06.07.68.

7. Schollhammer, F.R. *Apparatus for welding large diameter pipes with a beam of charger particles.* Pat. 3483352, USA, publ. 09.12.69.

8. Irwing, R.R. (1966) All systems are go for EBW, *Iron Age*, **17**, No. 9, pp. 59–61.

9. Lienan, H.G., Lowry, J.F. and Hassan, C.B. (1966) Electron beam welder for use in space, *Westinghouse Eng.*, No. 14, pp. 41–45.

10. (1968) Gold-cathode space welder is developed, *Aerospace Technol.*, No. 21, p. 25.

11. Paton, B.E., Dudko, D.A., Bernadsky, V.N. et al. (1977) *See pp. 174–178 in this Book.*

12. Nazarenko, O.K., Bondarev, A.A., Kajdalov, A.A. et al. (1987) *Electron beam welding.* Ed. by B.E. Paton, Kiev, Naukova Dumka, 255 pp.

13. Zagrebelny, A.A., Lapchinsky, V.F., Stesin, V.V. et al. *A method of simulation of the cosmonaut's working conditions in space and a device for its implementation.* USSR author's cert. 563783, Int. Cl. B 64 G 7/00, priority 09.06.75, publ. 09.03.77.

14. Paton, B.E., Dudko, D.A., Bernadsky, V.N. et al. (1975) *See pp. 137–140 in this Book.*

15. (1981) *OSP-72/80*, Main sanitary rules for working with radioactive substances and other sources of ionizing radiation, Moscow, Energoizdat, 57 pp.

16. Paton, B., Dzhanibekov, V., Savitskaya, S. et al. (1989) The test of the versatile hand electron beam tool in space, In: *Proc. of IIW Conf. on Welding Under Extreme Conditions*, Helsinki, 1989. Helsinki, Pergamon Press, pp. 189–196.

17. Nikitsky, V.P., Lapchinsky, V.F., Zagrebelny, A.A. et al. (1985) *See pp. 178–184 in this Book.*

18. Paton, B.E., Lapchinsky, V.F., Zagrebelny, A.A. et al. (1986) *See pp. 283–288 in this Book.*

19. Paton, B.E., Semyonov, Yu.P., Arkov, P.F. et al. *Versatile electron beam hand tool.* USSR author's cert. 222960, Int. Cl. B 23 K 15/00, B 64 G 9/00, priority 31.07.84, publ. 01.08.85.

20. Paton, B.E. and Lapchinsky, V.F. (1998) *Welding and related technologies in space*, Kiev, Naukova Dumka, 184 pp.

21. Drabovich, Yu.I., Yurchenko, N.N., Shevchenko, P.N. et al. (1985) Power supply system of the versatile electron beam hand tool, In: *Problems of space technology of metals*, Kiev, Naukova Dumka, pp. 65–67.

22. Shelyagin, V.D., Mokhnach, V.K., Koritsky, V.A. et al. *A device for electron beam welding.* USSR author's cert. 1287419, Int. Cl. B 23 K 15/00, publ. 03.01.85.

23. Shelyagin, V.D., Mokhnach, V.K., Stesin, V.V. et al. *A device for electron beam welding.* USSR author's cert. 1626542, Int. Cl. B 23 K 15/00, publ. 08.10.90.

24. Paton, B.E., Bulatsev, A.R., Mikhajlovskaya, E.S. et al. (1991) *See pp. 484–495 in this Book.*

25. Zagrebelny, A.A., Gavrish, S.S., Bulatsev, A.R. et al. (1986) *See pp. 258–264 in this Book.*

26. (1998) Preparation for conducting an experiment in space, *Avtomatich. Svarka*, No. 9, p. 73.

27. Zagrebelny, A.A., Nikitsky, V.P., Gavrish, S.S. et al. *A tool for electron beam welding.* USSR author's cert. 1655722, Int. Cl. B 23 K 15/00, priority 05.06.89, publ. 15.02.91.

28. Paton, B.E., Semyonov, Yu.P., Nikitsky, V.P. et al. *A system of manual electron beam hardware for EVA.* USSR author's cert. 1745617, Int. Cl. D 64 G 9/00, priority 03.06.89, publ. 08.03.92.

29. Paton, B.E., Dzhanibekov, V.A., Gavrish, S.S. et al. *A method of electron beam cutting.* USSR author's cert. 1811462, Int. Cl. B 23 K 15/08, priority 30.07.90, publ. 10.10.92.

30. Paton, B.E., Lapchinsky, V.F., Stesin, V.V. et al. *Device for manual electron beam processing of materials in space.* Pat. 5.869.801, USA, priority 14.02.96, publ. 09.02.99.

31. Devyatiarev, E. (1999) Hundreds of arguments in favour of «Mir», *Novosti Kosmonavtiki*, No. 2, pp. 10–11.

32. Devyatiarev, E. (1999) «Mir» for the world. State Duma: «Mir» should fly. «Energiya» project is on Primakov's table, *Ibid.*, No. 1, pp. 38–39.

33. Bulatsev, A.R., Zagrebelny, A.A., Nikitsky, V.P. et al. *A foot restraint device.* USSR author's cert. 280769, Int. Cl. B 64 G 9/00, priority 05.10.87, publ. 01.08.88.

34. Bulatsev, A.R., Zagrebelny, A.A., Lapchinsky, V.F. et al. *A device for operator fastening.* USSR author's cert. 313349, Int. Cl. B 64 G 9/00, priority 26.06.89, publ. 03.05.90.

35. Saprykin, Yu.I., Linevich, Yu.B., Zagrebelny, A.A. et al. *Anchor.* USSR author's cert. 322133, Int. Cl. B 64 G 9/00, priority 04.08.89, publ. 02.01.91.

36. Paton, B.E., Kryukov, V.A., Gavrish, S.S. et al. *Astronaut's work station device.* Pat. 5.779.002, USA, priority 14.02.96, publ. 14.07.98.

Some Features of Formation of Brazed Joints with Radiant Heating under Conditions of Gravity and Zero Gravity*

I.E. Kasich-Pilipenko[13], V.S. Dvernyakov[13], V.V. Pasichny[13], V.F. Lapchinsky[1], V.S. Novosadov[13], L.B. Beloborodova[13] and A.A. Zagrebelny[1]

The radiant energy of the Sun is attracting ever greater attention of researchers. Especially great prospects are opened up when it is used for performing various technological processes in space. Works [1, 2] show the feasibility of conducting the processes of welding, brazing and cutting in stationary ground-based solar power units. While insufficiently localized energy input into the item sometimes has its impact in welding and cutting processes, in brazing the density of radiant flows, concentrated in the solar power units is sufficient for production of braze welds with a high productivity, using even brazing filler metals.

---

* (1980) *Materials and processes of space technology*, Moscow, Nauka, pp. 62–68.

It is known that natural deep vacuum and zero gravity are inherent to space, they having an essential influence on running of the processes of braze weld formation. In order to reveal the features of wetting with the braze alloy and filling of the butts being brazed with it, optimize the process modes and reduce the cost of the above investigations, preliminary experiments were conducted under conditions of artificial short-time zero gravity. The work was carried out in cooperation with the PWI and the Yu.A. Gagarin CTC[3].

The testing facility with the process chamber (Figure 1), which accommodates stage *1* and radiation unit *2* for simulation of concentrated solar flow, was mounted on board the flying laboratory which can create a weightlessness state for 24–28 s. Rarefaction of up to $5 \cdot 10^{-3}$ mm Hg can be created in the chamber using vacuum pump. In order to conduct experiments in deeper vacuum, the facility was further fitted with an attached steam-jet vacuum unit, which in the flight experiments was replaced by an oil-free adsorption vacuum unit, providing and maintaining in the chamber a rarefaction of up to $5 \cdot 10^{-5}$ mm Hg for 3 h. The optical system of the unit Figure 2 consisted of an elliptical glass concentrator *3* of 358 mm diameter with the front aluminized coating and angle of coverage of 200°. Xenon arc lamp *4* was used as the solar radiation simulator. The lamp was mounted so that its arc was in the first focal point of the ellipsoid. The beam flow being reflected from the ellipsoid, was concentrated in the second focal point into which the object under study *5* was introduced. The lamp could be displaced along the ellipsoid axis by means of electric drive *6*, and the density of the radiant flow on the sample varied in keeping with its arc displacement relative to the ellipsoid focal point. The flow intensity was adjusted through the mode of the radiator arc running.

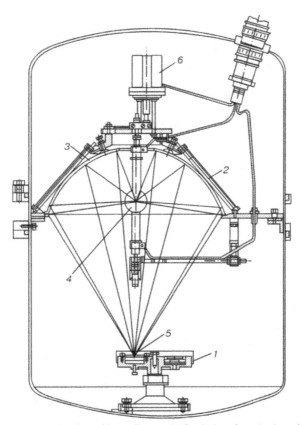

Figure 1 Process chamber with a radiator unit (for designations *1–6* see the text).

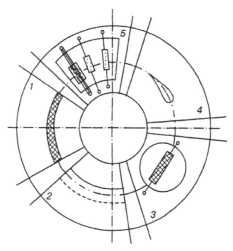

Figure 2 Types of samples being tested: *1* — butt joint with flanges; *2* — overlap joint; *3* — coupling of braze alloy on rods, placed with a gap; *4* — hole with a patch; *5* — single- and multistrand conductors with braze alloy couplings.

Experiments were conducted with the aim of studying the processes of surface wetting by braze alloys, formation of drops, melt and solidified structures at zero gravity, overlap and flanged butt brazing of plates, brazing of single- and multistrand conductors, and placing patches on holes. Commercial titanium alloy VT1, steel 12Kh18N10T (C — ≤ 0.12; Cr — 17.0–19.0; Ni — 9.0–11.0; Mn — 1.0–2.0; Si — ≤ 0.8; S — up to 0.02; P — up to 0.035 wt.%), copper and braze alloys POS-40 (St — 39.0–40.0; Sb — 1.5–2.0 wt.%; Pb — balance), POS-61 (St — 59.0–61.0; Sb — ≤ 0.8 wt.%; Pb — balance), PSr-72 (Ag – 72.0; Cu — 28.0 wt.%) were the materials used. The samples whose types are shown in Figure 2, were fastened in the composite cassette placed on the stage so that the heating zones were located along the circumferences of 80 mm diameter. The samples were moved through the focal zone by a motor drive in the range of speeds from 0.2 up to 16 m/h. In order to generate comparative data, the flight experiment was always preceded by a similar ground experiment.

Special features of braze weld formation were studied on samples, cut out of plates of thickness δ = 0.3÷0.8 mm in the form of sectors with 1.0÷2.5 mm high flanges, between which braze alloy plates were placed into 0.3÷1.5 mm gap. In steel samples with copper used as the braze alloy on the ground, at more than 0.8 mm gap, copper flowed down from the flanges, solidifying on the lower part of the butt. At zero gravity a uniform filling with copper melt without sagging was found in a gap of up to 1.5 mm. Formation of braze alloy structures in the gap is similar in the ground-based experiments and at zero gravity. In the braze weld, formation of a composite structure was observed due to the plate dispersion in the braze alloy. At zero gravity the dispersed particles are uniformly distributed across the weld section, and their gradual increase is found at the interface (Figure 3a). Grain blocks pass into the copper melt, which then disintegrate, and partially dissolve in the central part of the weld.

During copper melting in the gap between the flanges of the butt of VT1 plates, incomplete melting of the filler in the lower part is observed, and, therefore, inclusions in the form of needles, oriented in the heating direction, are found in the structure. This is a manifestation of a purely surface impact of the sample heating by the radiant flow. Intensive dissolution of titanium in copper results in the braze weld having a pronounced metal lustre. The weld structure mainly consists of Ti-base Ti–Cu system. The amount of the eutectoid is the greater the closer to the heating surface.

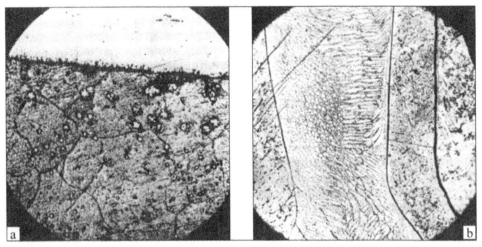

Figure 3 Microstructures of brazed joints at sample surface (a) and of braze weld, plate and roll (b) (×20).

Copper flowing over the outer part of the sample was observed at zero gravity. It is obvious that in the absence of the gravitation forces, the copper melt under the impact of the surface tension forces moved over the flange and solidified in the corner between the flange and the outer surface of the plate. Investigation of the roll microstructure (Figure 3b) showed the absence of directional solidification in it. This, probably, is a feature of solidification at zero gravity. An intermetallic interlayer with a microhardness of 300–400 kg/mm^2 is found on the interphase.

In brazing of steel plates ($\delta = 0.5$ mm), using POS-40 braze alloy, no sagging of the braze alloy is found in the butt root in the samples, made at zero gravity unlike those made on the ground. No structural differences were detected.

In order to study the features of wetting and formation of a melt drop at zero gravity, experiments were conducted with the following samples. A coupling of foil of POS-61 braze alloy was wound on two coaxially located copper rods of 1 mm diameter, the edges of which were spaced at 10 mm. When the rods were displaced along their axis through the focal spot, the braze alloy coupling melted. In the ground experiment the braze alloy wetted the rods and fell off under gravity pull. At zero gravity the coupling being melted, wetted the rod end, and with further melting the melt shifted to the second rod, having formed a sphere. After that, the sphere, probably, due to change of surface energy with dissolution of copper in the braze alloy, shifted to the rod end and solidified at its sharpened cone. Microstructural analysis showed the presence of an intermetallic compound on the boundary of the copper rod and the solidified braze alloy. White needles which are chemical compounds of Cu–Sn of 0.2–0.4 mm length, oriented relative to the cooling surface, are uniformly distributed through the entire volume of the sphere. The needle microhardness is ≈ 118 kg/mm^2, that of the matrix is 8 kg/mm^2. The uniformity of dissolved copper distribution through the volume of POS-61 sphere, can be attributed to the features of structure formation at zero gravity (Figure 4).

Experiments were conducted on brazing butts of copper conductors and bundles of single- and multistrand conductors. A butt of copper rod of 1 mm diameter was wrapped into a coupling of braze alloy (POS-61; PSr-72). The coupling melted while passing through the focal spot of radiation, wetted the rods and solidified, forming a reinforcement on the butt, without sagging at zero gravity, and with a certain sagging in the ground-based experiment. In multistrand bundles POS-61 melt spread uniformly between the conductor

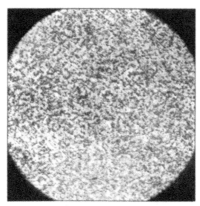

Figure 4 Microstructure of impurities distribution through the sphere volume (×50).

wires at zero gravity, whereas flowing down of the braze alloy occurred in the ground-based experiment.

Special samples were prepared to study the process of melts flowing into capillaries of a variable cross-section and the ability to apply patches on holes by brazing. A braze alloy plate was placed on a plate with a variable gap of 0 to 5 mm on a 20 mm length, and a continuous plate (patch) was placed on top. The package assembly was fastened by rivets. When the package passed through the focal spot along the variable gap, the braze alloy plate melted, and the plates were bonded over their wetted areas by the solidified interlayer of the braze alloy. The gap was filled with the braze alloy only up to the width of 1.5 mm, as after that it was drawn into the gap between the plates. The patch was not bonded to the hole around the entire perimeter as a result of the plates buckling.

Braze welds on 12Kh18N10T steel, Ti-3.5Al-1.5Mg and VT1 titanium alloys, as well as copper were made, using PSr-72 braze alloy, in the solar power units of the I.M. Frantsevich IMSP[13]. Braze welds of plates and cylindrical samples of up to 2 mm thickness demonstrated good results in terms of strength and tightness. This provided the data for development of a special small-sized solar unit which would be used for retrofitting the elements of systems and mechanisms for further recommending them for application in the orbital vehicle [3].

Thus, the preliminary investigations conducted to study the features of surface wetting by melts, formation of braze welds and weld structures at zero gravity, using a simulator of concentrated solar light as the heat source, showed that:

• at zero gravity (in contrast to ground conditions) the gaps in the butts are filled with the braze alloy melt uniformly without sagging;

• wetting and spreading of braze alloys over the surface occurs under the action of the capillary and interatomic forces of adhesion;

• distribution of dispersed particles of the base metal in the braze alloy melt is uniform through the volume, and the structures demonstrate a proneness to fine-grain texture;

• it is possible to perform brazing of butt and overlap joints, as well as butts and bundles of conductors with sound weld formation;

• it is possible to repair holes by applying brazed patches.

The performed research showed the good prospects for using concentrated radiant energy of the Sun in space to make braze welds.

1. Fransevich, I.N., Dvernyakov, V.S., Passichny, V.V. et al. (1972) In: *Coll. pap. 23rd IAF Congr.*, Vienna. Vienna, pp. 160–161.

2. Shiganov, N.V., Bareskov, N.A., Frantsevich, I.N. et al. (1972) In: *Coll. pap. of 1ˢᵗ All-Union Sci.-Techn. Conf. on Regenerative Energy Sources*, Moscow, 1972. Moscow, Energiya, pp. 246–253.

3. Frantsevich, I.N., Dvernyakov, V.S., Kasich-Pilipenko, I.E. et al. (1975) In: *Space research in Ukraine*, Issue 6, pp. 15–18.

Welding Aluminium Alloy 1201 by Radiant Energy of the Sun*

A.A. Bondarev[1], V.S. Dvernyakov[13], I.E. Kasich-Pilipenko[13], A.A. Korol[13], V.F. Lapchinsky[1], A.V. Lozovskaya[1] and E.G. Ternovoj[1]

Practical use of solar energy is attracting greater attention of researchers every year. The Sun being an inexhaustible source of energy, with no additional expense involved in its generation, use of this energy in the space environment looks highly attractive even now [1–3].

Welding of metals is one of the technologies requiring large amounts of concentrated thermal energy. Therefore, the attention, given by welding scientists, dealing with permanent joining of metals in space, to the development of practical application of solar energy, is quite natural. From this point of view, of special interest is the direct use of solar energy as a concentrated thermal energy, that is achieved by applying all kinds of concentrators. Development of this area, however, runs into a number of serious design and technological difficulties, their overcoming requiring numerous experiments.

Particular technological difficulties can be anticipated in welding materials, having a higher reflectivity, high heat conductivity and heat capacity and forming high-melting oxide films on their surface (aluminium and magnesium alloys). On the other hand, these are exactly the materials that currently are the main structural material in space engineering. Therefore, the problem of their permanent joining in space is highly urgent.

The PWI and the I.M. Frantsevich IMSP[13] conducted joint experiments to study the principal possibility of welding aluminium alloys by the concentrated radiant energy of the Sun.

Welding was performed in a solar welding unit with radiant energy concentrator of 2000 mm diameter and focal distance of 864 mm and coverage angle of 120° [4]. The cylindrical working chamber of the unit is fitted with a mechanism of samples fastening and displacement and allows work to be performed both in vacuum (at up to $133 \cdot 10^{-2}$ Pa pressure), and in a controlled atmosphere of various gases. For this purpose the chamber is fitted with a vacuum unit, vacuum meter, gas analyser, leak valve and mixer.

The radiant flow, focused by the reflector on the sample being welded, is brought into the chamber, through a quartz viewing window and water-cooled aperture with changeable inserts of various diameter, selected depending on the conducted technological process.

Flanged cylindrical and flat samples of Al–Cu alloy 1201 (Figure 1) were welded during experimental studies. Sample shape, their weight and configuration closely enough simulated the conditions of heat removal from the weld in unsupported welding of items. Flat samples were welded in the annealed condition with base metal hardness of *HRB* 800 MPa, cylindrical samples — after quenching and artificial ageing with base metal hardness of *HRB* 1000–1020 MPa. Prior to welding the edges were scraped to a depth of 0.1 mm. Samples were welded, using an aperture with 8.0 mm hole and light flow density of 1200–1300 W/cm². Prior to welding the stationary sample was first pre-heated for 5 s. Welding was performed in a controlled argon atmosphere.

* (1981) *Space research in Ukraine*, Issue 15, pp. 34–37.

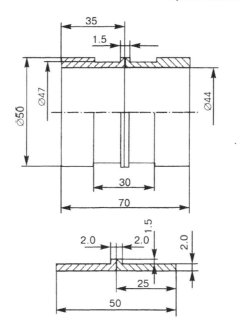

a

b

Figure 1 Schematic of welded cylindrical (a) and flat (b) sample.

Alloying element content in different points of the welded joint was determined by local spectral analysis.

Optical microscope MIM-8 was used to conduct metallographic studies. Hardness was determined in a Rockwell instrument, using a steel sphere of 1.6 mm diameter at 600 N load by scale B. Measurements were taken in the direction, normal to the weld. Measurement steps were 2.0 mm. The results of the conducted experiments showed, that this thermal energy density allows welding flanged aluminium samples up to 2.0 mm thick without special reduction of their reflectivity at a speed of not more than 4 m/h. Further increase of welding speed leads to lack-of-penetration and discontinuities in the weld pool.

Formation of short welds is satisfactory with sufficiently uniform reinforcement of the face and the root sides (Figure 2). The quality of the longer welds is non-uniform. Lack-of-fusion, undercuts and burns-through are periodically observed. A large number of oxide film inclusions is even visually seen. Tight joints cannot be obtained. The main cause for such a

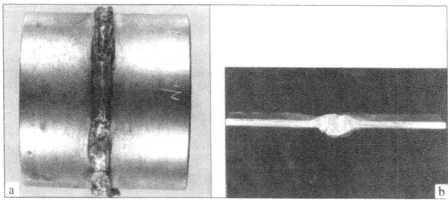

Figure 2 Appearance of welded joint on a cylindrical sample (a) and its macrosection (b).

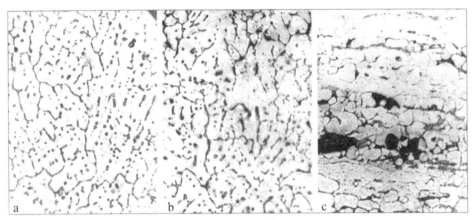

Figure 3 Microstructure of 1201 alloy welded joint: a — weld metal; b — fusion zone; c — defects in base metal near the fusion zone (×150).

low quality of extended welds is, probably, related to unsatisfactory cleaning of the surface of the material being joined from the oxide film.

Welded joint properties were studied in greater detail.

The data of spectral analysis did not reveal any significant differences between the base metal and the weld metal in terms of the content of alloying elements of copper, manganese, iron and silicon. No loss of alloying components was observed during welding.

The results of metallographic examination of the structure of different sections of the joint demonstrated, that the weld metal has a cast coarse-grained structure with a multitude of coarse globular precipitations of $CuAl_2$ phase (Figure 3a). Closer to the fusion line the weld metal has a pronounced dendritic structure (Figure 3b). The dendrites mainly grow out of several partially-melted base metal grains. The fusion zone boundary is wide and not well-defined.

Metal structure in the HAZ differs noticeably from the base metal before welding. Metal adjacent to the weld has been recrystallized with grain coarsening. Formation of discontinuities in the form of pores and looseness is found in this section (Figure 3c). Such phenomena usually occur in the base metal at its long-term overheating up to high temperatures, associated with a low welding speed, and are accompanied by a considerable loss of strength and ductility.

Weld metal hardness is equal to *HRB* 52 on average, that of the HAZ being *HRB* 67.

Ultimate tensile strength of flat samples is in the range of 160–170 MPa, this being a little more than 50 % of base metal strength in the annealed condition, while bend angle of the joints is approximately 2 times lower than that of the base metal (not more than 40°).

Thus, the results of welded sample analysis indicate, that today we can only state the principal possibility of performing heliowelding of aluminium and magnesium alloys, at least heat-hardenable ones. This method can be used only to make short non-critical welds.

Use of heliowelding for joining more critical elements of aluminium and magnesium alloys is only possible if a method can be found for reliable cleaning of the surfaces being joined to remove the oxide film and considerable (up to 3–4 time) increase of welding speed.

It can be anticipated that when working under the conditions of space an increase of welding speed can be achieved due to increased level of solar radiation beyond the atmosphere (about 0.14, compared to 0.07 W/cm^2 in testing site of solar welding unit) and significant heat removal from the weld pool, when working in vacuum. As regards cleaning techniques, systematic purpose-oriented studies are required to establish them.

1. Frantsevich, I.N., Trefilov, V.I., Dvernyakov, V.S. et al. (1977) Some results of investigation of the technological processes, implemented using solar energy under conditions, simulating those of space, In: *Abstr. of pap. of 28th IAF Cong.*, Prague, 1977. Prague, p. 6.

2. Belyakov, I.T. and Borisov, Yu.D. (1974) *Technology in space*, Moscow, Mashinostroyeniye, 290 pp.

3. Kasich-Pilipenko, I.E., Dvernaykov, V.S., Pasichny, V.V. et al. (1980) *See pp. 212–217 in this Book.*

4. Dvernyakov, V.S. (1975) Use of heliowelding units of IMSP of Acad. of Sci. of Ukr. SSR for radiant technology, investigation of the properties and development of new materials, In: *Investigation of materials under radiant heating conditions*, Kiev, Naukova Dumka, pp. 3–12.

Features of Chemical Element Migration in the Subsurface Layers of Metals and Alloys During Thermal Cycling*

O.D. Smiyan[1], G.M. Grigorenko[1], E.S. Mikhajlovskaya[1],
L.M. Kapitanchuk[1] and S.O. Antonov[1]

Under orbital flight conditions the space vehicle structural material is exposed to multiple action of the heating–cooling cycle, both as a result of its following a near-earth orbit and revolution about its axis. The temperature difference between the sunlit and the shadowed surface in the absence of the station axial rotation can be up to 600–650 °C and with its axial rotation this difference reduces to 150–200 °C. This circumstance allows reducing thermal stresses in the vehicle structure, but substantially increases the number of thermal cycles of the structural materials. The process is made more complicated by the fact, that simultaneously with the thermal cycles, the surface of these materials is also exposed to the impact of the environment, namely the space gas plasma at $P \leq 10^{-4}$ Pa, flows of ultraviolet, charged particles with different kinetic energy and a number of other factors.

The issues of thermal cycling influence on various properties of metallic materials and their structure have already been discussed in publications [1–3]. However, there is no data on the variation of the chemical or elemental composition in the metal subsurface layers directly during thermal cycling under the conditions of a high or ultrahigh vacuum. This is the subject of this study.

Materials and methods of investigation. Investigations were conducted on samples of steel 12Kh18N10T of $9 \times 9 \times 1$ mm size and commercial titanium alloy VT1-0 of $9 \times 9 \times 0.6$ mm size. These materials are widely used as structural materials in space construction. Samples were mounted in a special holder, having five degrees of freedom. A heater — an archimedean spiral of tungsten, fastened in a holder, was placed in the centre of an 8.0 mm hole. Samples were heated from below, i.e. by radiation heating. The upper surface of the sample was irradiated by flows of electrons ($E = 3$ keV in electron Auger-spectrometry (EAS)), photons (chromatic X-ray radiation with a fixed quantum energy in roentgen-electron spectrometry) or ions ($E = 4$ keV at ionic cleaning of the surface by Ar$^+$ ions). Changes of the elemental composition of the samples surface during their heating and subsequent cooling in the analysis chamber were recorded by the methods of EAS and X-ray photoelectron spectroscopy (electron spectroscopy for chemical analysis — ESCA) in an instrument system LAS-2000 (Riber Company, France) for studying the surface. The data obtained were

* (1996) *Proceedings of Conf. of the Int. Society for Optical Engineering on Space Processing of Materials*, Denver, Aug. 4–5, 1996. Denver, Vol. 2809, pp. 311–315.

processed in a computer. Heating temperature was controlled by a chromel-alumel thermo-couple. Sample heating (up to 453 K) and their cooling (up to 123 K) were performed directly in the analysis chamber of this unit at a pressure of $10^{-7} - 10^{-8}$ Pa. Measurements by EAS method were the basic measurements, as their results are not influenced by the sample temperature. The ESCA method was used as the reference one.

Before the start of investigations, the samples were subjected to desorption treatment by the thermoflash method, i.e. heated to 673 K, soaked at this temperature for 10 min in the collision chamber of LAS-2000 unit at $P = 10^{-6}$ Pa, and then cooled to 123 K, then again heated up to 300 K and bombarded with Ar^+ ions at this temperature to remove the surface layer of adsorbates and surface contamination, not removed by the thermoflash. Chemical element distribution in-depth from the surface was studied by the method of Auger-pro-filometry in steel, and by ESCA and EAS methods in the titanium alloy (since Auger-peaks of titanium and nitrogen coincide). The rate of etching the surface of the metal samples by argon ions with $E = 4$ keV was 1.5 nm/min over the entire surface of the sample. This was enough to avoid the influence of surface diffusion of oxygen, carbon, chlorine, sulphur and other elements on the validity of the obtained results, as the rate of this diffusion may exceed that of bulk diffusion of the same elements by 4–6 orders of magnitude at the same temperatures and pressure. The kinetics of oxygen penetration into the metal bulk was studied by concentration profiles of oxygen, derived by Auger-profilemetry method.

The fracture surface of a sample after its breaking up directly in the unit at ultradeep vacuum (10^{-8} Pa) was studied to examine and compare the processes of oxygen penetration into the metal depth through the grain bulk and along their boundaries. At such a vacuum the time of formation of monomolecular film of adsorbates on the atomically pure fracture surface is 17 h. This is sufficient for conducting appropriate research.

In order to clarify the influence of temperature on the migration ability and corrosion activity of oxygen, the sample surface was first cleaned by the thermoflash method (heating up to 673 K for 2 min, soaking for 10 min, cooling up to 123 K, soaking for 2 h) and using ionic cleaning with subsequent heating up to 423 K (soaking for 2 h) and cooling to 300 K. At each of the above processing stages, Auger-spectrum or ESCA was taken from the sample surface. Samples of 12Kh18N10T steel were given similar treatment.

In a number of studies [4–6], it was determined that the principal changes of the properties of metals and alloys during their thermal cycling are observed at not more than 10 thermal cycles. With further increase of the number of thermal cycles, just accumulation of defects formed in the solid during the first five thermal cycles, takes place, this being exactly what leads to material failure. Since we were looking for changes in the elemental composition in the surface zone at thermal cycling, just 5 to 8 thermal cycles were sufficient to trace these changes down.

Investigation results and their discussion. *12Kh18N10T steel.* Elemental composition of the surface layers of the metal after its cleaning by the thermoflash method and ion cleaning in LAS-2000 system is quite close to the specification (C — ≤ 0.12; Cr — 17.0–19.0; Ni — 9.0–11.0; Mn — 1.0–2.0; Si — ≤ 0.8; Ti — 0.4–0.5; S — up to 0.02; P — up to 0.035 wt.%). During sample heating and subsequent thermal cycling oxides form on its surface as a result of interaction of the elements present in the alloy with residual gases. Already during the first heating a complex compound of the type of $(Fe, Cr)_2O_3$ forms on the steel sample surface. At the stage of the 1st and 2nd thermal cycles no nickel oxide is formed, as it is not present on the surface. Beginning from the 4th thermal cycle, however, when nickel, as a result of thermal «play» of the elements [7], starts coming to the sample surface, also its oxide formation becomes possible. Titanium «comes» to the surface after the 3rd–4th thermal cycle, so that titanium-con-taining compounds, including its oxide, can appear on the surface only after the 3rd thermal cycle.

During thermal cycling each new heating cycle leads to increased content of carbon on the surface (Figure 1). This process is especially abruptly accelerated after the 4th thermal

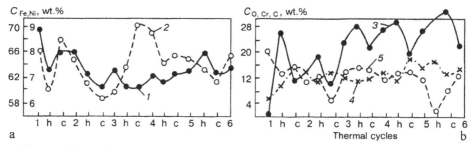

Figure 1 Change of the content of iron (*1*), nickel (*2*), oxygen (*3*), carbon (*4*), chromium (*5*) on the surface of 12Kh18N10T steel sample at the vacuum of 10^{-8} Pa during thermal cycling in the temperature range of 123–423 K (on abscissa axis *1–5* — start numbers of the next cycles; h — heating up to 423 K; c — cooling to 123 K).

cycle: while during the first four thermal cycles the content of this element on the surface is increased by 1 % (from 11 up to 12 wt.%), and just in one 5th cycle it at once rises by 6 % (up to 18 wt.%). This is attributable to the fact, that thermal «play» of the elements results in a sufficient number of carbide-forming elements appearing on the steel surface that form a chemical compound with carbon. Thus, part of the elements, rising from the metal depth, become fixed on the surface.

It is characteristic that, in heating, the highest content of chlorine and nitrogen on the metal surface is achieved after the 2nd thermal cycle, and that of sulphur — after the 2nd and 4th thermal cycles. On the other hand, oxygen content on the surface becomes minimum after the 2nd thermal cycle (Figure 2).

At samples cooling after heating, formation of other compounds is also observed (for instance, chromium carbide in the 4th thermal cycle, etc.), in addition to the processes of chemosorption and oxide formation.

Observation of the nature of variation of various elements concentration during thermal cycling showed that, for the majority of them, these changes follow the periodic law (this is more clearly traceable for nickel, chlorine, sulphur, chromium and is less obvious — over five thermal cycles — for oxygen, titanium and phosphorus). At more than ten thermal cycles, such a regularity of content variation is characteristic for all the elements present in the metal. This is the consequence of wave mass-transfer of elements, discovered by one of the authors of this paper as far back as in 1969 [8]. This process is characterised by the concentration wave length and period, inherent to each element (sulphur has peaks in the 2nd and 4th cycles, nickel in 3rd and 5th cycles, etc.) (Figure 2).

Titanium alloy VT1-0. Oxygen and carbon form stable chemical compounds with titanium in the form of oxide or carbo-oxide films and can be removed from the metal surface only by ion etching. The thickness of carbo-oxide layer on the studied sample of alloy is about 100 nm in the initial condition. When the composition of the initial alloy VT1-0 (commercially pure titanium) was determined, a considerable amount of impurity elements, namely carbon, oxygen, calcium, sodium, sulphur and phosphorus, was found in the thin subsurface layer 2 to 3 nm thick.

As was already noted, in Auger spectra of titanium alloys one of the titanium peaks and nitrogen peak coincide. This hinders the determination of these elements content by EAS method. ESCA method is more suitable in this case. In order to equalise the analysis conditions (EAS method is a microprobe method, and ESCA analyses over 2×2 mm^2 area), analysis by EAS method was conducted with scanning by the microprobe (of 10 μm diameter) over 2×2 mm^2 area with accumulation, summing up and averaging of the result in the computer. The results of these investigations are given in Table 1. It is seen, that in

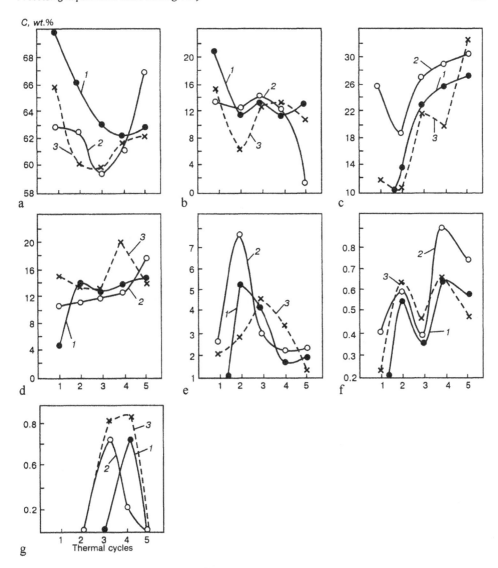

Figure 2 Influence of the number of thermal cycles on the change of the content of iron (a), chromium (b), oxygen (c), carbon (d), chlorine (e), sulphur (f) and titanium (g) at the start of the thermal cycle (*1*), after heating to 423 K (*2*) and cooling to 123 K (*3*) in 12Kh18N10T steel.

the initial condition both the methods yield practically the same results. Deviations are up to 2 %, this being within the measurement errors.

From Figure 2 one can see, that after four thermal cycles carbon content in the metal surface layer rises by 1.5 times, and that of oxygen by 4 times. Nitrogen and calcium on the surface disappear already after the first heating. Thus, under the conditions of this experiment, nitrogen and calcium are only present in the adsorbate. Moreover, nitrogen does not form any chemical compound with the titanium alloy, and, therefore, in the thermal cycling temperature range (123–423 K), EAS can be used, as a more convenient and independent of temperature method of investigation. The ESCA method, as a more complex one for application, calculation, and interpretation, was used in further experiments as the reference one.

Table 1 Change of elemental composition of the surface of VT1-0 alloy during thermal cycling, determined by ESCA method.

Element	Magnitudes in initial condition at $T = 300$ K		Cooling to 123 K	Heating to 423 K	Cooling to 300 K, soaking for 2 h in LAS-2000 chamber
	EAS method	ESCA method			
C	7.73	7.56	10.76	8.77	11.63
O	10.07	9.84	21.02	27.96	39.93
N	0	1.30	0	1.89	0
S	0	0	0	3.85	0
Ti	82.08	81.29	68.21	57.63	48.88
Ca	0.27	0	0	0	0

Results of studying element distribution in the subsurface zone of VT1-0 alloy under the conditions of thermal cycling at comparatively low temperatures (cooling to 123 K, heating up to 423 K) are given in Figure 3. It is seen, that the main changes at thermal impact are observed during the first two cycles. Already during the first heating up to 300 K, carbon and chlorine are partially «thrown off» the metal surface and oxygen sorption begins. At 423 K temperature chlorine is mostly removed from the metal, and carbon sorption begins at the same temperature. In the 2nd thermal cycle the cleaning processes are absent, and just the enhancement of the adsorption processes for oxygen and carbon is found. Repeated heating up to 423 K leads to complete removal of chlorine from the metal surface. In the 3rd thermal cycle all the processes are stabilised, and metal oxidation goes on.

It is of interest to study the concentration profile of various elements on the surface of the fracture, obtained directly at ultrahigh vacuum on a metal sample after eight thermal cycles in the temperature range of 123–423 K. These data are given in Table 2.

It turned out, that the metal in the intergranular spaces (the fracture being brittle) contains a considerable amount of various impurity elements: sodium (in up to 3 nm layer) and calcium (in up to 20 nm layer) directly on the fracture surface. Carbon, oxygen, sulphur and chlorine are present across the entire thickness of the intergranular interlayer metal (80 nm). The nature of concentration profiles of chlorine and sulphur indicates, that these are not just «relic» (charge) impurities. Probably they partly penetrated into the metal during its plastic

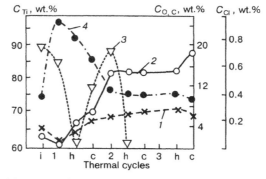

Figure 3 Change of the content of carbon (*1*), oxygen (*2*), chlorine (*3*), titanium (*4*) during thermal cycling of alloy VT1-0 in the temperature range of 123–423 K (designations are the same as in Figure 1): i is elemental composition on the initial sample surface.

Table 2 Change of elemental composition of alloy VT1-0 in wt.% after eight thermal cycles at different duration of the layer-by-layer ion etching of fracture surface et v_{et} = 2 nm/min, determined by EAS method.

Ele-ment	Etching time, min												
	0.0	0.3	0.6	1.0	1.5	2.0	3.0	4.0	5.0	7.0	10.0	16.0	40.0
C	67.63	42.26	30.91	21.69	16.82	14.64	9.62	8.33	7.10	5.47	4.98	4.56	4.50
O	12.25	15.33	15.02	12.14	9.96	7.87	4.87	3.66	3.06	1.87	2.16	2.06	3.20
Ti	6.70	26.75	41.70	54.42	66.06	72.46	81.75	86.44	86.50	90.02	90.72	91.89	90.94
Ca	3.71	2.67	2.16	1.73	1.56	0.48	0.32	0.61	0.60	0.59	0.38	0	0
Na	4.49	5.38	2.36	3.77	0.51	0	0	0	0	0	0	0	0
S	2.98	3.21	3.92	3.41	2.46	1.83	1.42	1.29	0.95	0.74	0.80	0.49	0.31
Cl	2.24	2.90	2.33	1.53	1.33	1.62	1.43	1.11	1.14	0.96	0.63	0.62	0.62
Ar	0	1.50	1.56	1.31	1.30	1.10	0.58	0.55	0.59	0.31	0.34	0.38	0.42

deformation. The same goes for concentration profiles of carbon and oxygen with just the difference, that their absolute content is much higher, especially at 80 nm depth from the fracture surface. The zone of fracture surface analysis was located at 0.35 and 0.25 mm distance from the outer surface of the sheet metal. The above impurities could not penetrate into the intergranular spaces and moving from them saturate the metal to 80 nm depth by a purely diffusion mechanism at these temperature (up to 423 K) and times (up to 10 min.). Therefore, another mechanism of impurity elements sorption by the metal from the environment is active here. One of these possible mechanisms of impurities absorption by the metal is described below.

During heating–cooling the metal sections, located at the grain boundaries and between them, are the most exposed to alternating loads of compression (heating)–tension (cooling). As a result, these metal sections act as a kind of sorption pump: now absorbing the impurity from the environment in the tension phase, now releasing it in the compression phase, if it has not formed a chemical compound in the metal. In a certain sense, this process is similar to impurity sorption from the environment at repeated-alternating loading [9] or contact loading [10]. At each step of the thermal cycle, the intergranular space metal absorbs a certain «portion» of the impurities. This pertains (under our conditions) primarily to oxygen and carbon, but is also found for other impurities. Oxygen penetration into the metal (and it is the first of the above impurities to migrate) induces additional stresses in the metal of intergranular spaces (as a result of internal oxidation) and other impurity elements are moving there, until they form chemical compounds there. By this moment, formation of a continuous oxide film is completed on the metal outer surface. Thus, active penetration in-depth of the metal along the intergranular boundaries is taking place during thermal cycling, primarily of oxygen (hydrogen was not studied or considered in this work). In principle, hydrogen is the first to migrate into the metal, inducing a stressed state in this zone and promoting penetration into these areas of other impurities, namely carbon, chlorine, sulphur, calcium, sodium (listed by the decrease of their activity). As a result, defects pile up and chemical compounds form in the intergranular spaces, promoting metal embrittlement.

Thus, during thermal cycling in the range of temperatures of 123–300–423 K not only formation of the oxide layer and its thickening is found, but also its «growing» along the grain boundaries in-depth of the metal, accompanied by inner oxidation and promoting active migration into this zone, after oxygen, also of a number of other impurity elements (chlorine, sulphur, phosphorus, nitrogen, carbon), some of which form their compounds there.

Conclusions

1. At thermal cycling in the temperature range of 123–423 K oxides of elements, present in the material composition, form on the surface of the metal (12Kh18N10T steel and VT1-0 alloy) and grow through its thickness.

2. It is found, that during thermal cycling, not only new chemical compounds form, but also some of the already present ones dissociate. It is shown, that the most resistant under the thermal cycling conditions are the compounds more depleted in oxygen (Me_2O type), that are preserved even to comparatively high temperatures. On the other hand, compounds of $MeO–Me_2O_3–MeO_2$ type, comparatively rich in oxygen, fail at heating, even up to relatively low temperatures (up to 300–423 K).

3. It is found, that thermal cycling promotes interstitial element migration along the intergranular spaces into the metal bulk, forming chemical compounds, that lower the metal strength and intergranular bond strength. It ultimately results in metal failure. The greatest activity in this respect is demonstrated (ranged from higher to lower) by oxygen, carbon, sulphur, chlorine and nitrogen.

4. The optimal amount of thermal cycles for 12K18N10T steel and VT1-0 alloy is three, thus resulting in the highest homogenising of the metal, minimising of the heating-cooling process influence on the surface elemental composition (it remains practically the same for the 3rd thermal cycle in all the stages). At other numbers of thermal cycles, the difference in the surface elemental composition at the heating–cooling stages can be quite considerable (up to 2–3 times).

5. A new mechanism is proposed of impurity element sorption by the metal during its thermal cycling in the temperature range of 123–423 K.

1. Lebedev, A.A., Kovalchuk, B.I. and Zajtseva, L.V. (1995) Influence of the mode of temperature-time impact on the kinetics of phase transformations in metastable austenitic steel, *Problemy Prochnosti*, No. 3, pp. 28–34.

2. Polyansky, V.I., Pirog, O.I. and Silis, M.I. (1993) Interrelation of mechanical properties with quantitative characteristics of microstructure for alloys VT3-1 and VT23, *Fizika Metallov i Metalloved.*, **75**, No. 4, pp. 176–181.

3. (1980) *Heat treatment in mechanical engineering. A handbook.* Ed. by Yu.M. Lakhtin and A.G. Rakhshtadt, Moscow, Mashinostroyeniye, 783 pp.

4. Dzyadikevich, Yu.V., Smiyan, O.D. and Gorbatyuk, R.M. (1996) Refractory metal refining from interstitial impurities, *Dopovidi NAN Ukrainy*, No. 8, pp. 98–104.

5. Smiyan, O.D., Kasatkin, V.S., Musiyachenko, V.F. et al. (1973) Diffusion displacement of gaseous impurities in metals during their heating, In: *Methods for determination and investigation of gases condition in metals*, Moscow, Nauka, pp. 61–63.

6. Kasatkin, B.S., Musiyachenko, V.F., Smiyan, O.D. et al. (1973) Heating influence on hydrogen distribution in a welded joint of 14Kh2GMR steel, *Avtomatich. Svarka*, No. 12, pp. 63–64.

7. Smiyan, O.D. and Antonov, S.O. (1990) *Mechanism of chemical inhomogeneity formation in solids in welding and spraying*, Reprints of sci. pap. of the NAS of Ukraine, Kiev, PWI, 12 pp.

8. Smiyan, O.D. and Kruzhkov, A.G. (1972) On some features of movement of the diffusion flow of gases in metals, *Doklady AN SSSR*, **202**, No. 6, pp. 1311–1313.

9. Smiyan, O.D. (1983) Hydrogen mass transfer in the metal under the impact of repeated-alternating loading, *Fiziko-Khim. Mekhanika Materialov*, No. 3, pp. 29–32.

10. Smiyan, O.D. and Chajkovsky, B.P. (1983) Redistribution of impurities during contact loading of ShKh-15 steel, *Ibid.*, No. 6, pp. 106–107.

Design, Based on EBW Application under Extreme Conditions*

A.A. Bondarev[1] and E.G. Ternovoj[1]

Repeated inclusion of welding operations into space flight programmes indicates that this technology is given not the least important role in mounting large-sized orbital complexes directly in orbit in the future. Moreover, as shown by the results of the most recent experiments that involved S.E. Savitskaya and V.A. Dzhanibekov performing welding operations in open space [1], out of the many welding processes, tested already in the first experiment by V.N. Kubasov [2], the electron beam turned out to be the most flexible heat source, allowing it to be used in multi-faceted experiments.

The main problem in implementing the processes of assembly and welding, where the electron beam is used as the heat source, is provision of the required accuracy of geometrical parameters, guaranteeing the absence of gaps or edges misalignment in the butt area, maintaining the alignment of the assembled blanks and ensuring a sound welded joint. It is known, that both in automatic welding using «Vulkan» hardware [2], and in manual welding in open space [1], the samples to be welded were in the restrained condition in stationary fixtures, prepared and assembled on the ground. Under such conditions it was not difficult to meet high requirements to assembly.

Now, when assembly is performed without man's involvement by mechanisms or when they are remotely-controlled, this process turns out to be more complex and labour-intensive. Provision of the required accuracy and quality of assembly should be envisaged both in the design of the elements to be welded, and in the mechanisms and fixtures, making the performance of this operation possible. It is difficult to select such a process from the known processes of assembly and welding of circumferential butts of joints [3–8], performed under regular conditions on the ground, that would have sufficient reliability under the conditions of remotely-controlled fit-up of the butt for welding, without man's participation. Even minor deviations from the abutted elements alignment or axes refraction, result in large gaps in the butt and make achievement of the required quality of the welded joints much more difficult. The more so, since none of the known processes had been developed for the application of multiple re-welding of defects that may arise in case of welding operations performance under extreme conditions.

The objective of this study was to develop the geometry and procedures for welding aligned ring-shaped blanks under extreme conditions, as well as ensure repairability of the welded joint in case defects are detected after the first weld has been made.

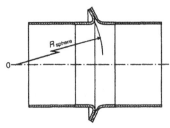

Figure 1 Schematic of preparation and assembly of a joint with spherical surfaces of the edges welded, forming the flanges.

* (1985) *Problems of space technology of metals*, Kiev, Naukova Dumka, pp. 51–53.

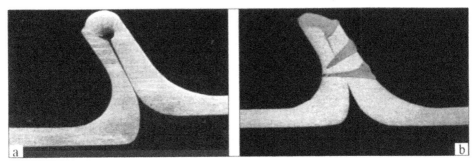

Figure 2 Macrosections of joints made by EBW in one pass over the edges flanging (a) and after application of the second and third repair slot welds (b).

Ground-based retrofitting of welding technology and technique using the samples on board the flying laboratory [9, 10] already demonstrated that regular formation of a joint at zero gravity with large gaps in the butt, requires a sufficient amount of the liquid phase to form the weld metal, that may be provided by melting off flanged edges. The butts in the gap should be minimal, so that the liquid metal is not ousted beyond the edges been welded by vapour pressure.

Geometry of a circumferential butt joint with flanged edges (Figure 1) was developed [11], proceeding from the already gained experience, as well as high requirements to fit-up, that are made of joints, produced by EBW. A feature of such a butt joint geometry is making the flange elements by plastic deformation process in the form of parts of spherical surface of a sphere of the same radius (see Figure 1). The shape of the abutted edges in the form of congruent parts of the surface of a body of revolution forms parts of a ball support during assembly. This, on the one hand, provides self-centring of the adjacent elements, and, on the other hand, this being especially important, does not result in gaps forming between the abutted edges in the butt area, even in the case of refraction of the assembled element axes. Such a shape of the abutted elements is all-purpose, as it can be used also in assembly of structural elements, the axes of which form any angle in the abutment point. Its advantages of preventing gaps between the abutted edges are also preserved in this case. The abutted elements are sometimes subjected to heat fixing so that they consistently preserve the specified shape after plastic deformation.

Butt welding is performed by melting off two flanged edges simultaneously (Figure 2a). Welding can be performed both in the automatic and manual mode. In automatic welding the flanged edges can have the role of tracers, along which the welding gun can move. After welding the joint is controlled for tightness, and repair welds are made in case of a lack of tightness. The first repair weld can be made over the earlier deposited one. This is also an advantage of a joint, made with flanged edges. If the joint is still not tight after that, repair slot welds are made on the side surface of the flange (Figure 2b). In this case, two slot welds are made, in addition to the main weld, with a shift closer to the cylindrical surface of the shells. It is seen in Figure 2, that the gap in the butt is practically absent, and this despite the fact, that the shell diameter is 200 mm, and the flanging height was up to 25 mm. Mechanical properties of such a joint are close to those of the base metal. Changing of the flanged edges thickness relative to the shell side wall thickness allows making an equivalent welded joint.

Conclusions

1. Welded joint geometry has been developed that in case of alignment of the assembled elements (cylindrical or conical), allows assembly operations without gaps in the butt to be performed with remote control without man's involvement.

2. A process of welding circumferential butt joints has been developed, that allows improvement of the reliability and producing a sound welded joint with remotely-controlled assembly and welding of structures under extreme conditions. In case of a lack of tightness, the method allows implementing multiple operations of joint repair.

3. The process can be also used is assembly-welding of cylindrical and conical elements of structures with refracted axes in the butt zone.

1. (1984) Woman-cosmonaut in open space for the first time, *Trud*, July 27.

2. Paton, B.E. and Kubasov, V.N. (1970) *See pp. 154–160 in this Book*.

3. (1976) USSR author's cert. 519305, *Inf. Bull.*, No. 24.

4. (1978) USSR author's cert. 625872, *Inf. Bull.*, No. 36.

5. (1974) USSR author's cert. 434597, *Inf. Bull.*, No. 45.

6. Pat. 3268248, USA, publ. 23.08.66.

7. Pat. 1453477, France, publ. 11.08.65.

8. Pat. 1372793, Great Britain, publ. 09.02.72.

9. Bondarev, A.A., Lapchinsky, V.F., Lozovskaya, A.V. et al. (1978) *See pp. 240–245 in this Book*.

10. Ternovoj, E.G., Bondarev, A.A., Lapchinsky, V.F. et al. (1976) Investigation of some aspects of weldability of aluminium alloys by the electron beam at zero gravity, *Space research in Ukraine*, Issue 9, pp. 5–11.

11. Bondarev, A.A. and Ternovoj, E.G. (1982) *A method of producing circumferential butt joints*. USSR author's cert. 963767, *Inf. Bull.*, No. 37.

Investigation of the Properties and Structure of 1201 Alloy Joints Electron Beam Welded at Different Gravity Levels and Low Temperatures*

D.M. Rabkin[1], V.F. Lapchinsky[1], E.G. Ternovoj[1], A.V. Lozovskaya[1], S.V. Mnyshenko[1], A.A. Bondarev[1] and I.Ya. Dzykovich[1]

Welded structures of aluminium and its alloys have become widely accepted in different industries (structures subjected to complex loading, to elevated and low temperatures and aggressive media). Small weight, quite high specific strength and corrosion resistance of aluminium alloys enabled development of structures capable of long-term operation in the space environment. Aluminium alloys currently are the main structural materials in aerospace engineering. It may become necessary to weld these alloys directly in space in the near future.

It is known that in fusion welding of heat-hardenable aluminium alloys, the weld and the HAZ develop various kinds of macro- and microdefects, not detectable by X-ray transmission [1]. Such defects sometimes lead to reduction of the strength, ductility, and corrosion resistance, as well as loss of the joint tightness [2, 3]. Appearance of such defects can be anticipated in particular in welding under such conditions as the impact on the weld pool of low temperatures, deep vacuum and different gravity levels, i.e. accelerations essentially differing from the free fall acceleration on the Earth surface [4].

In order to study the features of electron beam welded joints of heat-hardenable aluminium alloys under these conditions, experiments were conducted in TU-104A flying

* (1985) *Problems of space technology of metals*, Kiev, Naukova Dumka, pp. 94–101.

laboratory. A-1084 type unit with a power system consisting of a high-voltage power source and U-729 electron beam gun of 1.5 kW power was installed on board the laboratory [5–7]. The samples to be welded were rigidly fastened on a stationary table. Samples were cooled by liquid nitrogen. The absolute pressure in the chamber did not exceed $1.3 \cdot 10^{-1}$ Pa. The following parameters were recorded during welding: beam current, focusing current, supply voltage for power system from on-board DC mains, total load current, welding speed, acceleration affecting the weld pool, sample temperature and absolute pressure in the chamber. Welding was performed in the following mode: accelerating voltage of 15 kV, beam current of 100 mA, welding speed of 26–32 m/h, beam diameter of up to 2.5 mm at gun edge to sample surface distance of 120 mm.

During investigations through-penetration welding of 1201 (2219Al) alloy plates of 180×50 mm size 2.0 mm thick was performed. The plate surface in the places of the beam entering and exit was scraped to the depth of 0.05 mm from both sides prior to welding. Beads were deposited with one weld along the plate axis or with two parallel welds at 10 to 16 mm distance with an opposite direction of welding. Welding was performed at the following magnitudes of accelerations, acting on the weld pool (g — acting acceleration; g_0 — free fall acceleration):

• $g/g_0 \leq 10^{-2}$ (practically zero gravity);
• $g/g_0 \sim 1/6$ (free fall acceleration on lunar surface);
• $g/g_0 = 1$ (free fall acceleration on terrestrial surface);
• $g/g_0 \geq 2$ (more than 2 times overload).

The samples temperature was set to be as follows: +20, –100, –120 and –183 °C. The error of temperature measurement is ±0.5 °C for the positive and ±3 °C for the negative temperatures.

Analysis of the derived results was performed proceeding from:

• 100 % control of the tightness of joints by X-ray radioscopy;
• determination of the composition by X-ray microprobe analysis;
• performance of mechanical tests for static rupture (sections without any defects visible on the X-ray film were selected for mechanical testing and macro- and microstructural investigations);
• investigation of the distribution of phases and microdiscontinuities;
• element analysis in X-ray microprobe analyzer «Cameca»;
• macro- and microstructural investigations in MIM-8 microscope.

Processing and analysis of the investigation results have demonstrated the following. The welds made under the conditions close to zero gravity ($g/g_0 \leq 10^{-2}$) are characterized by an increased number of pores, which was noted earlier by the Soviet and foreign researchers [8–11].

Composition of the weld metal with the studied variants of welding technology does not practically depend on the accelerations acting on the weld pool, or the initial temperature of the sample being welded (Table 1).

The produced joints were used to prepare special small-sized samples for mechanical tests, and their ultimate tensile strength was determined (Table 2). Analysis of the derived data shows that the welded joints produced under the conditions close to zero gravity ($g/g_0 \leq \leq 10^{-2}$), are characterized by the lowest strength values. With increased accelerations the welded joint strength rises from 230 up to 250 MPa. In welding with samples cooling to the temperature of –183 °C, the strength of the joints made at $g/g_0 \leq 10^{-2}$, also increases from 230 up to 270 MPa, and at $g/g_0 \geq 2$ it increases considerably from 250 up to 320 MPa. Comparing the derived values of strength for the welded joints directly after welding and after heat treatment, we can see than the application of artificial ageing allows achievement of a higher ultimate strength for each of the studied variants of welding.

Table 1 Composition of electron beam welded joints made on 1201 alloy 2.0 mm thick under different conditions of welding.

Analysis location	Elemental content, %							
	Cu	Mn	Ti	Zr	V	Mg	Fe	Si
Base metal	6.20	0.29	0.06	0.13	0.07	0.01	0.03	0.03
Weld central part	6.17	0.27	–	–	–	0.01	0.03	0.03

In terms of their structure the produced joints are close to argon-arc welded joints, this, apparently, being attributable to the fact that welding was performed by a defocused beam. Similar to argon-arc welding, some joints demonstrate an increased number of micropores located in the area of the strengthening phases surface melting. It should be noted that the dimensions of these pores are almost by an order of magnitude smaller, than in the weld, and they are not detected by X-ray inspection. The weld is characterized by a directional

Table 2 Values of ultimate tensile strength, MPa, of welded joints produced at different conditions of welding 1201 alloy.

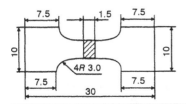

Base metal in the initial condition

$$\sigma_t = \frac{448 - 460}{454}$$

Artificially aged metal

$$\sigma_t = \frac{397 - 401}{399}$$

Initial temperature of samples, °C	Post-weld heat treatment	Acceleration acting on the weld pool, g			
		$\leq 1 \cdot 10^{-2}$	$\sim 1/6$	1	≥ 2
+20	Without heat treatment	212–244 / 237	204–257 / 228	233–262 / 245	228–269 / 248
	Artificial ageing	230–255 / 243	252–274 / 261	280–310 / 292	294–315 / 301
–100	Without heat treatment	244–272 / 259	223–287 / 254	246–297 / 269	295–314 / 303
	Artificial ageing	252–278 / 264	267–292 / 281	284–315 / 297	298–324 / 308
–120	Without heat treatment	255–286 / 271	273–291 / 279	262–303 / 282	305–309 / 307
	Artificial ageing	268–280 / 273	276–302 / 289	281–318 / 291	300–328 / 312
–183	Without heat treatment	244–274 / 266	283–300 / 291	315–326 / 310	303–325 / 314
	Artificial ageing	264–290 / 275	288–314 / 296	320–338 / 328	325–336 / 329

Note. The numerator gives the minimum and maximum values, and the demoninator — the average values.

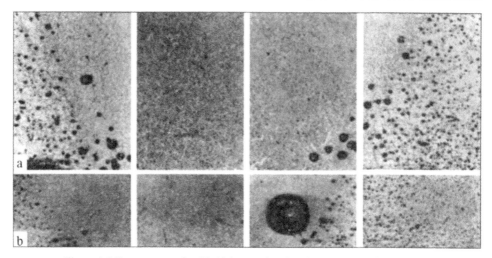

Figure 1 Microstructure of welded joints produced under conditions of $g/g_0 \geq 2$ (a)
and $g/g_0 \leq 10^{-2}$ (b) (×50) (reduced by 3).

cellular-dendrite structure near the fusion zone and a more equilibrium one in the weld center
(Figure 1), and is a solid solution of copper in aluminium and finely dispersed inclusions of
$CuAl_2$ phase. The conducted metallographic investigations did not reveal any particular
differences in the structure, amount or nature of distribution of the phase components in the
joints, welded at different temperatures and accelerations.

Quantitative evaluation of the micropores and of the ratio of phase components in the
HAZ and in the welds, proper produced at different variants of the welding technology, was
of considerable interest. These investigations, as was mentioned, were conducted in «Quan-
timet-720» instrument and «Cameca» microanalyser. As the pores in the welds were
individual, of comparatively large dimensions (1 mm and higher) and their number is
determined with certain accuracy by X-ray inspection, «Quantimet-720» was used to
determine only the number of micropores in the HAZ and of the phase components in the
HAZ and in the weld.

Samples for investigation of the distribution of alloying elements and determination of
the number of pores and phases, were prepared so that the pores were dark, and the phase
inclusions were clearly defined, weakly etched, of gray colour and were easy to distinguish
from pores and other discontinuities. With deeper etching, the phases acquire a darker colour
and it is difficult to distinguish them from the pores.

The percentage of phase components and micropores was assessed using «Quantimet-
720» computational image analyzer at a magnification of Epi 32, PK ×8. Qualitative
estimation was performed by the instrument automatically with up to 1 % accuracy for
objects taking up not less than 5 % of the area of the field of vision. The area of the field of
vision was 0.036 mm^2. For transverse microsections the number of pores and phase
inclusions was recorded every 0.16 mm in different zones of the welded joint, including the
base metal, the HAZ and the weld. The derived results are presented in the form of graphic
dependencies of the change of the number of pores and excess phases at different distances
from the fusion line in the direction of the weld and the HAZ (Figure 2). Analyzing the
obtained data, it is possible to draw the conclusion that for the welded joints, produced at
different accelerations ($g/g_0 \leq 10^{-2}$; $g/g_0 \sim 1/6$; $g/g_0 \geq 2$), the degree of porosity development
in the HAZ is practically the same. It should be noted that the pores are found, as a rule, in
the base metal, in the sites of phase component partial melting, their number and size being
the larger the closer to the fusion line (Figures 3 and 4). It was also noted that development

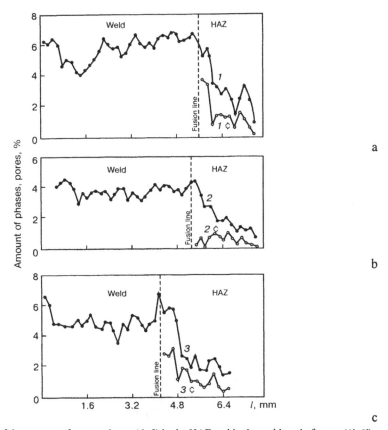

a

b

c

Figure 2 Change of the amount of excess phases (*1–3*) in the HAZ and in the weld; and of pores (*1'–3'*) in the HAZ for welded joints, produced at $g/g_0 \leq 10^{-2}$ (a), $g/g_0 \sim 1/6$ (b) and $g/g_0 \geq 2$ (c).

of porosity in the HAZ at the start of welding is somewhat smaller, than at the end, both for one weld and for the case of welding two parallel welds, located close to each other. Comparison of the derived data on porosity for welds made under overloading conditions ($g/g_0 \geq 2$) shows that the maximal number of pores for one weld does not exceed 3 % and reaches 6–8 % for two parallel welds at 6–8 mm distance from each other (see Figures 2 and 4).

Dependencies of the change of the amount of excess phases in the HAZ with the studied variants of welding, were derived in a similar manner. It was found that in the direction from the base metal towards the fusion line, the amount of excess phases increases through decomposition of the solid solution, partial melting of intermetallic inclusions and formation

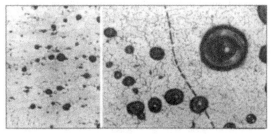

Figure 3 Microstructure of a welded joint, produced under conditions of $g/g_0 \leq 10^{-2}$ (×150) (dashed line is the fusion line).

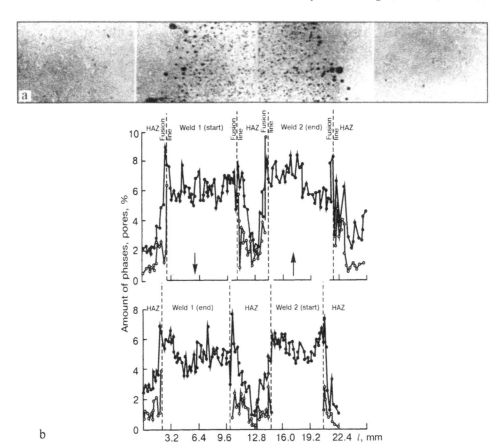

Figure 4 Microstructure (a) and change of the amount of excess phases and pores (b) for two closely
located welds, produced under conditions of $g/g_0 \geq 2$ (welding start: weld 1 — *top*;
weld 2 — *bottom*) and $g/g_0 \leq 10^{-2}$ (end of welding: weld 1 — *bottom*; weld 2 — *top*):
● — amount of phases; ○ — amount of pores.

of eutectic interlayers. The maximum amount of phases is 5–6 % and does not practically
depend upon the conditions of welding. Somewhat contradictory results were derived for
the weld metal, namely for welds made at $g/g_0 \leq 10^{-2}$, average phase content is 6 %, at $g/g_0 =$
$= 1$–4 %, and under the conditions of $g/g_0 \geq 2$ overloading it is about 5 %. As variations of
the phase amount from 4 to 7 % can be found even within one weld, variation of the amount
of phases was studied on longitudinal sections cut out of the welded joints, produced at the
moment of a drastic change of the gravity conditions from $g/g_0 \leq 10^{-2}$ up to $g/g_0 \geq 2$.
Investigations showed that the amount of phases, i.e. features of weld solidification, does
not practically depend on the accelerations and is equal to 6–8 % for these conditions,
whereas the boundary of transition from overloads to zero gravity practically cannot be
detected (Figure 5).

A 2 μm diameter microprobe was used to study the distribution of the main alloying
element (copper), as well as impurities of iron, silicon and manganese in different sections
of welded joints produced at $g/g_0 \geq 2$ and $g/g_0 \sim 1/6$ on transverse and longitudinal
microsections. Distributions were derived in the form of continuous diagrams recording the
content of elements in the solid solution and in individual phase components. For two closely

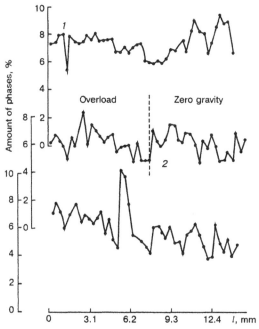

Figure 5 Change of the amount of phases in welded joints produced under conditions of $g/g_0 \geq 2$ (*1*) and $g/g_0 \leq 10^{-2}$ (*2*).

located welds, one of which is made at $g/g_0 \geq 2$, and the second at $g/g_0 \sim 1/6$, copper distribution is approximately the same, namely 2.2–2.5 % in the solid solution and up to 40 % in the intermetallic inclusions.

As was found earlier [12] some copper-rich phases are also enriched in iron and silicon, this being accompanied by peaks in the distribution diagram, of up to 1 % for iron and up to 0.65 % for silicon. Wider and higher peaks are found in the HAZ, both for one and for the other weld, corresponding to phases rich in copper, this being related to surface melting of intermetallic inclusions due to welding heat. The content of copper in the solid solution in this case varies from 5 % (for the base metal) to 2.7 to 3.0 %, becoming close to copper content in the weld metal solid solution. Analysis of the copper distribution diagram produced at different recording rates for the longitudinal microsection of a weld made under the conditions of transition from $g/g_0 \geq 2$ to $g/g_0 \leq 10^{-2}$ shows that copper content in the solid solution does not depend on the conditions of welding.

Distribution of elements for welded joints produced at $g/g_0 = 1$ and $g/g_0 \leq 10^{-2}$ accelerations and −120 and −183 °C temperatures was studied in order to determine the influence of low temperatures on the conditions of weld solidification and structure formation in the HAZ. When analyzing the inhomogeneity of the HAZ and the weld, one can see a certain increase in the copper content in the solid solution of the weld with lowering of the initial temperature of the sample prior to welding. A lower degree of the solid solution decomposition and, accordingly, softening in the HAZ zone metal points, equidistant from the weld, are also found in this case. Intensive cooling during welding promotes preservation of a higher content of copper in the solid solution and, probably, is the main factor, responsible for a higher strength of welded joints, made under conditions of lower gravity and low temperatures (see Table 2).

Analysis of the results of the conducted investigations leads to the following conclusions:

• 1201 alloy can be welded by the electron beam at zero gravity;

• gravity levels and low temperatures do not have any noticeable influence on the composition, element distribution and structure of the metal of the weld on 1201 alloy;

• low initial temperature of the sample in EBW of 1201 alloy promotes higher values of ultimate tensile strength and a narrower HAZ;

• superposition of the temperature fields from two welds close to each other in EBW of 1201 alloy, causes an increase in the amount of micropores in the base metal, located between these welds, phase precipitations being the sites of discontinuity nucleation.

1. Nikiforov, G.D. (1972) *Metallurgy of fusion welding of aluminium alloys*, Moscow, Mashinostroyeniye, 264 pp.

2. Andrew, R.C. and Waring, I. (1974) Effect of porosity on transverse weld fatigue behavior, *Welding J.*, **53**, No. 2, pp. 85–90.

3. Roll, W. (1974) Einfluß von Poren in Schweißnähten an $AlMg_{4.5}Mn$ und $AlMgSi_{0.8}$ auf die Ergebnisse des Zugversuchs, *Schweißen und Schneiden*, **23**, No. 4, pp. 150–151.

4. (1980) Schweißen im Weltraum. Erstarrungsbedingungen beim Elektronen-strahlschweißen, *Techn. Rundsch.*, **72**, No. 18, pp. 7–11.

5. Paton, B.E., Paton, V.E., Dudko, D.A. et al. (1973) *See pp. 118–122 in this Book.*

6. Shelyagin, V.D., Lebedev, V.K., Zaruba, I.I., et al. *Device for electron beam welding.* USSR author's cert. 206991, Int. Cl. B 23 K, priority 22.02.66, publ. 08.12.67.

7. Paton, B.E., Dudko, D.A., Bernadsky, V.N. et al. (1977) *See pp. 174–178 in this Book.*

8. Paton, B.E. and Kubasov, V.N. (1970) *See pp. 154–160 in this Book.*

9. Ternovoj, E.G., Bondarev, A.A., Lapchinsky, V.F. et al. (1976) Investigation of some aspects of weldability of aluminium alloys by the electron beam at zero gravity, *Space research in Ukraine*, Issue 9, pp. 5–11.

10. (1973) Skylab tests space welding, *Engineer*, **236**, No. 6114, p. 9.

11. (1973) Skylab crew tests electron beam welding in space, *Iron Age*, **211**, No. 26, p. 21.

12. Bondarev, A.A., Lapchinsky, V.F., Lozovskaya, A.V. et al. (1978) *See pp. 240–245 in this Book.*

Influence of Gravity Forces, Dissolved Hydrogen and Initial Temperature on the Properties and Tightness of Joints in EBW of Light Structural Alloys*

E.G. Ternovoj[1], A.A. Bondarev[1], V.F. Lapchinsky[1] and A.V. Lozovskaya[1]

Successful performance of long-term space flights and construction of multi-purpose long-term piloted complexes in the near future [1] make scientists and designers address the task of development of various structures and constructions directly in the space orbit [2]. When such work is performed, one of the urgent problems certainly is permanent joining of materials in the space environment [3]. Significant features of this environment, compared to the terrestrial one, is change of the gravitation forces, namely microgravity [4, 5], as well as a special temperature mode [6], i.e. temperature gradient from 100–120 °C (at higher solar radiation) to $-100 - -120$ °C and lower (when the object is in shadow).

* (1986) *Problems of space technology of metals*, Kiev, PWI, pp. 56–60.

Figure 1 Appearance of a weld on alloy AMg3 made when going over from $g/g_0 = 2$ to $g/g_0 = 10^{-2}$ and sample temperature of -183 °C.

A quantitative measure of long-term zero gravity condition is the ratio of acceleration g in the system of co-ordinates, referenced to the space object and acceleration g_0 in the terrestrial gravity field, and under the actual conditions it can be $g/g_0 = 10^{-2} \div 10^{-5}$, that in the case of technological processes, related to material melting, often leads to undesirable side effects [7–9].

Above-zero temperature of the object (up to 120 °C) in technological processes, involving material heating and melting, has mainly a positive role, for instance, in welding aluminium alloys [10]. On the other hand, during performance of these operations, the low temperatures can make fundamental changes in the melt solidification process, its fast cooling, and, as a result of these impacts, achievement of comparable properties of these joints.

The purpose of the work consisted in studying the influence of gravity forces, below-zero temperatures and initial gas content of the metal on the properties and tightness of welded joints in non heat-hardenable structural aluminium and magnesium alloys in their EBW.

Experimental work was conducted in TU-104A flying laboratory and on the ground by the procedure, described in [11].

The following alloy grades were used during investigations: ADOO — commercial purity aluminium; AMg6 — Al–Mg alloy with weight fraction of 5.8–6.8 % Mg and of dissolved hydrogen of 0.2, 0.3, 0.5 and 0.6 cm³/100 g; AMg3 — Al–Mg alloy with weight fraction of 3.2–3.8 % Mg; IMV-2 — magnesium alloy with weight fractions of 8 % Li, 5 % Al, 4 % Gd.

Visual monitoring by the operator of process dynamics in welding of AMg3 alloy showed, that despite a low content of magnesium, the nature of molten metal melting and transfer from the weld pool front part to its tail part was unstable, with liquid metal ejection in all the ranges of gravity forces variation and temperature modes. The evidence of that is the extent of ripple and its non-uniform distribution on the weld surface (Figure 1).

In welding IMV-2, containing about 80 % Mg, the process was highly stable, weld formation was good, but the instruments demonstrated an abrupt drop of beam current (to 50–30 mA on 10–130 mm of weld). As was established later, an intensive flow of magnesium vapours causes a fast contamination of the gun cathode, resulting in the loss of its emissive ability.

The nature of melting and behaviour of the molten metal pool was recognised to be the most stable of all the process variants of welding alloys ADOO and AMg6 with the content of $H_2 = 0.2$ cm³/100 g Me (Figure 2a and b). These data are in agreement with earlier work [11].

Processing of results after welding is performed by:

• determination of weld composition by local spectral analysis;

a

b

Figure 2 Appearance of welds on alloys ADOO (a) and AMg6 (b) with $H_2 = 0.2$ cm³/100 g Me, made when going over from $g/g_0 = 2$ to $g/g_0 = 10^{-2}$ and samples temperature of -120 °C.

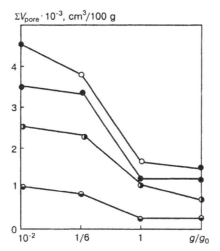

Figure 3 Porosity in welds on aluminium and magnesium alloys at different magnitudes of gravity forces: AMg3 (○), ADOO (●), IMV-2 (◐), AMg6 (◒) at $H_2 = 0.2$ cm³/100 g Me.

• 100 % X-ray control of joint tightness;

• control of tightness in the most interesting sections in the longitudinal and transverse cuts and longitudinal surface cuts of macrosections;

• performance of static breaking test in P-05 machine.

Analysis of investigation results shows that the weld metal composition in the studied alloys in different welding process variants does not practically depend on the gravity forces acting during welding, initial temperatures and metal gas content.

Study and analysis of filmograms and macrosections of longitudinal cuts and layer-by-layer longitudinal cuts of the welds of the joints indicated that in all the variants of welding the porosity was minimal in the welds of alloys AMg3, IMV-2, ADOO and AMg6 with $H_2 =$ = 0.2 cm³/100 g Me (Figure 3). Welds of alloy AMg6 with hydrogen content of 0.3, 0.5 and 0.6 cm³/100g Me have a tendency to an abrupt increase of porosity, especially at $g/g_0 \sim 1/6$ and $g/g_0 = 10^{-2}$, the amount of which sometimes reaches 0.442 cm³/100 g Me (Figure 4).

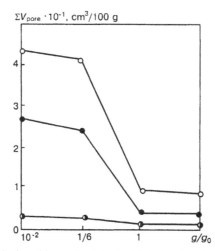

Figure 4 Change of porosity in welds on alloy AMg6 with different content of hydrogen, depending on gravity conditions with $H_2 = 0.6$ (○), 0.5 (●) and 0.3 (◐) cm³/100 g Me.

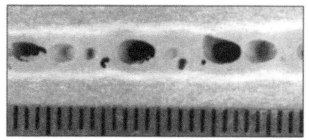

Figure 5 Longitudinal cut in the upper part of the weld, made under conditions of $g/g_0 = 10^{-2}$ on alloy AMg6 with $H_2 = 0.6$ cm^3/100 g Me, dissolved in the base metal at sample temperature of -100 °C.

Pore size increased considerably to 3.0–3.5 mm in diameter (Figure 5). Low temperatures do not have any impact on pore number and size under the above conditions.

Special small-sized flat samples were cut out of the produced joints for mechanical testing, and ultimate strength of welded joints was determined for all the variants of welding technology.

The results of mechanical testing of samples are given in Figure 6a and b. Comparing the derived strength values for welded joints, made at different values of g/g_0 and temperature of 20 °C, a certain lowering of strength can be noted at $g/g_0 \sim 1/6$ and $g/g_0 = 10^{-2}$. This is, probably, attributable to a lower tightness of welds, because of defects in the form of fine pores, that were observed at lower magnitude of gravity forces (Figure 6b).

Conclusions

1. Alloys ADOO, IMV-2 and AMg6 with dissolved hydrogen content of 0.3 cm^3/100 g Me and less are readily weldable by the electron beam at lower gravity and low temperature.

2. Level of gravity and low temperature do not have any noticeable influence on weld metal composition in electron beam welded joints on light structural alloys.

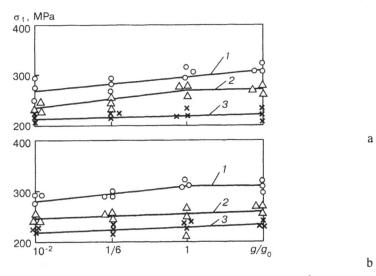

Figure 6 Dependence of σ_t on g/g_0 for welded joints on alloys AMg6 with $H_2 = 0.2$ cm^3/100 g Me in the base metal (*1*), IMV-2 (*2*) and AMg3 (*3*) at sample temperatures of 20 (a) and -183 (b) °C.

3. Alloys AMg6 and AMg3 with dissolved hydrogen content of more than 0.3 cm^3/100 g Me are prone to higher porosity in EBW under conditions of lower gravity forces.

4. Lower gravity leads to a 3–5 % decrease in mechanical properties of joints on light structural alloys, compared to values, obtained on the ground.

5. In welding of non heat-hardenable alloys ADOO, AMg6, IMV-2 the low temperature of samples prior to welding does not have any influence on welded joint strength values.

1. (1986) In space — a new research station, *Izvestiya*, Febr. 21.

2. Paton, B.E. and Semyonov, Yu.P. (1983) To future orbits, *Pravda*, Nov. 29.

3. Paton, B.E. and Kubasov, V.N. (1970) *See pp. 154–160 in this Book.*

4. Belyakov, I.T. and Borisov, Yu.D. (1977) *Technology in space*, Moscow, Mashinostroyeniye, 290 pp.

5. Benedikt, E. (1964) In: *Zero gravity. Phisical phenomena and biological effects*, Moscow, Mir, 275 pp.

6. Ganiev, R.F. and Lapchinsky, V.F. (1978) *Problems of mechanics and space technologies*, Moscow, Mashinostroyeniye, 34–36 pp.

7. (1980) Schweißen im Weltraum. Erstarrungsbedingungen beim Elektronenstrahlschweißen, *Techn. Rundsch.*, **72**, No. 18, pp. 7–11.

8. (1973) Schweiß-versuche im Skylab Programm, *Schweißtechnik*, **63**, No. 6, pp. 195–199.

9. Ternovoj, E.G., Lapchinsky, V.F., Bondarev, A.A. et al. (1977) Investigation of the influence of zero gravity on weldability and quality of aluminium alloy joints, welded by the electron beam, In: *Space materials science and technology*, Moscow, Nauka, pp. 29–34.

10. Kiselyov, S.N., Khavanov, V.A., Roshchin, V.V. et al. (1972) *Gas electric welding of aluminium alloys*, Moscow, Mashinostroyeniye, 176 pp.

11. Ternovoj, E.G., Bondarev, A.A., Lapchinsky, V.F. et al. (1976) Investigation of some aspects of weldability of aluminium alloys by the electron beam at zero gravity, *Space research in Ukraine*, Issue 9, pp. 5–11.

Investigation of the Structure and Element Distribution in Welded Joints, Made by the Electron Beam on Alloys 1201 and AMg6 at Zero Gravity[*]

A.A. Bondarev[1], V.F. Lapchinsky[1], A.V. Lozovskaya[1] and E.G. Ternovoj[1]

The electron beam as a possible heat source in welding and cutting of metals in space attracted interest both in the USSR and the USA [1–3]. The electron beam is the best mastered and versatile tool for performance of materials processing operation in vacuum. The welds, made with the electron beam, are characterized by high physico-mechanical properties.

In order to determine the ability to apply the electron beam to make vacuum-tight joints of alloys 1201 (used in as-annealed condition with base metal hardness of *HRB* 80 kgf/mm^2) and AMg6, experiments were conducted at zero gravity in the flying laboratory TU-104A, with a facility for studying the technological processes [4], incorporating 1.5 kW electron

---

* (1978) *Production and behaviour of materials in space*, Moscow, Nauka, pp. 21–29.

beam gun, mounted on board. Welding of samples was performed in the modes, providing edge penetration to the depth of 2.0–4.0 mm.

Check experiments were conducted on Earth under the same assembly conditions and welding mode parameters to compare the results.

The power to the welding hardware and recording instrumentation was supplied from the on-board 27 V power system. The following parameters: beam current, input voltage, total load current, welding speed and degree of rarefaction in the chamber were recorded during welding.

185 × 50 × 5 mm samples were prepared for welding without grooving and assembly was performed with an overlap. Fillets of 1.5 × 2.0 mm cross-section were machined out along the edges to be welded. Plate edges were scraped prior to welding in the butt zone.

Element composition in different points of the welded joint was determined by local and layer-by-layer spectral analysis. Metallographic examination was conducted in an optical microscope MIM-8. Hardness was determined in a Rockwell instrument with a steel sphere of 1/16 inch diameter at the load of 60 kgf by scale B. Measurements were taken in the direction, normal to the weld. The measurement point spacing was 2 mm. Element distribution in various sections of the joint was determined by X-ray microprobe analysis in «Cameca» instrument. Joint tightness was controlled by radiography of welded samples, as well as metallographic examinations of macro- and microstructure in the longitudinal and transverse sections. The data of spectral analysis of the base metal, central part of the weld metal and weld reinforcement did not reveal any significant differences in the content of the main alloying elements, namely copper for alloy 1201 (2219Al), magnesium for alloy AMg6 (Al–6.2Mg), as well as manganese, iron and silicon. The obtained results of layer-by-layer analysis every 0.03 mm demonstrate, that the content of copper and manganese in the welds is the same, as in the base metal, and is independent of the welding conditions. The content of other elements is practically unchanged, either in alloy 1201 or AMg6.

Macrostructure of welded joints, made in the zero gravity condition, is characterised by the presence of somewhat coarser crystallites in the weld metal, compared to those produced on the ground. Crystallite coarsening in the weld metal in zero gravity welding is also visible on longitudinal macrosections, cut out along the butt axis (Figure 1). Microstructure of welded joints of alloys 1201 and AMg6 was studied both on the transverse and the longitudinal sections. Base metal of alloy 1201 consists of a solid solution of copper in aluminium and coarse and fine precipitations of intermetallic phases of $CuAl_2$ type. Weld metal microhardness in alloy 1201 is fine-grained in the weld lower part and coarser in its upper part and in the reinforcement area. Hardening phase precipitates are finely dispersed and uniformly distributed across the weld. No differences in the distribution density or hardening phase size were found in the welds, made at zero gravity.

Base metal of alloy AMg6 consists of elongated grains of the solid solution of magnesium and other elements in aluminium and intermetallic phase fringes along the grain boundaries along the direction of rolling. No significant differences were found in the microstructure of the joints, made at zero gravity and on the ground, except for shrinkage type defects, namely pores or microlooseness, located along the fusion zone (Figure 2). In addition, a considerable amount of porosity is observed in welds made at zero gravity (see Figure 1d).

Hardness measurements of joints on alloy 1201 demonstrated, that the lowest hardness value was recorded in the weld metal (75 kgf/mm^2), while in the partial-hardening zone of the HAZ, located at 1.5–2.0 mm distance from the fusion zone, it was maximum (90 kgf/mm^2). In other measurement points, the hardness values corresponded to those earlier derived for the base metal in the annealed condition, i.e. 80 kgf/mm^2. It was found, that the nature of hardness distribution in the joints of alloys 1201 and AMg6 is identical in welding on the ground and at zero gravity.

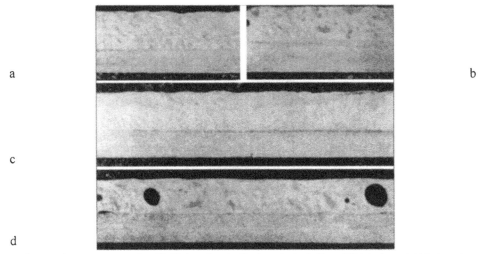

a

b

c

d

Figure 1 Macrostructure of welded joints (longitudinal section) of alloys 1201 (a, b) and AMg6 (c, d), made on the ground (a, c) and at zero gravity (b, d) (×50).

Investigation of alloying and impurity elements distribution in «Cameca» microanalyzer was conducted across the fusion zone in three sections of the joints, namely base metal, HAZ and weld metal. Diameter of electron probe for irradiation was ~ 2.0 μm. Element distribution in different zones of the welded joints is given in Figure 3.

As indicated by the data of X-ray microprobe analysis, copper content in the base metal solid solution in alloy 1201 varies between 1.2–1.5 %, and is about 43 % in the intermetallic phases (Figure 3a and b). (Copper content in the solid solution in different sections of the welded joint was determined by the lower levels in the copper distribution diagram across the fusion zone.) The low copper content in the base metal solid solution is attributable to the metal in as-annealed condition.

Distribution diagrams show a partial hardening zone 250–300 μm wide in the welded annealed metal, both for joints produced on the ground and at zero gravity. Copper content in the solid solution increases closer to the fusion zone and is 2.7 % right at this zone. Eutectic inclusions of Al–CuAl$_2$ with up to 28–30 % copper content, form in the zone of partial melting of intermetallic phases.

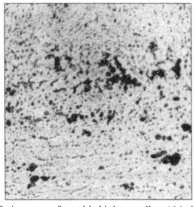

Figure 2 Microstructure of fusion zone of a welded joint on alloy AMg6, made at zero gravity (×150).

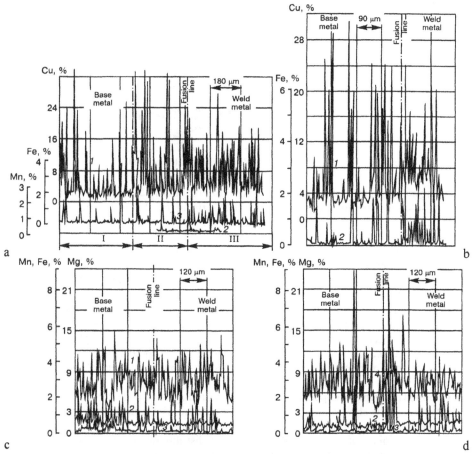

Figure 3 Diagrams of distribution of alloying and impurity elements in welded joints
on alloys 1201 (a, b) and AMg6 (c, d), made with the electron beam on the ground (a, c)
and at zero gravity (b, d): *1* — copper; *2* — manganese; *3* — iron; *4* — magnesium; I — base metal;
II — partial hardening zone; III — weld metal.

In the weld metal, phase precipitates are commensurate with the probe size, so that copper content in them is not determined.

In the joints, welded at zero gravity, a slight difference in copper content distribution (5.0–5.8 %) is observed in the weld metal solid solution; in the section, adjacent to the fusion line, it is somewhat higher than in the more remote weld sections and in similar sections of a weld (3–4 %) made on the ground (Figure 3a).

Diagrams of iron distribution across the fusion zone in different sections of welded joints, produced on the ground and at zero gravity for alloy 1201, are practically similar. This diagram precisely determines the boundary of the weld metal fusion with the base metal, and no particular differences are found in iron distribution in the HAZ base metal. Despite a low content of iron in the alloy (0.23 %), its distribution by individual phase components in the base metal and weld metal is highly non-uniform (Figure 3a and b) and in some Al–Cu inclusions it abruptly rises up to 5–6 % (isolated peaks in the distribution diagrams). When going over to the weld metal, the distribution diagrams record iron peaks of a smaller magnitude, compared to the base metal, their frequency, however, rising, which is probably due to a higher density and dispersion of the Al–Cu phase containing iron.

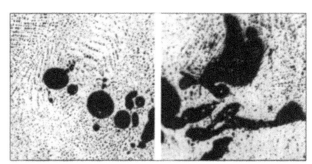

a   b

Figure 4 Porosity (a) and oxide films (b) in AMg6 weld metal, produced at zero gravity (×150).

Manganese is non-uniformly distributed in different phase constituents of the alloy: in the solid solution manganese content is 0.15–0.3 %, rising to 1.0 % in individual phases. No differences in manganese distribution were found depending on welding conditions.

Copper distribution in welded joints in alloy 1201 was also recorded in the form of scanegrams, obtained in X-rays (K_αCu) and absorbed electrons. X-ray radiation and absorbed electron scanegrams were taken from the same sections.

Sections of weld metal and HAZ demonstrate, that in the weld metal the dispersed intermetallic phases, rich in copper, are uniformly distributed across the weld, while fine and coarse inclusions of Al–Cu eutectic (Al + $CuAl_2$) are found in the HAZ. In addition, an interlayer, dividing the base metal and the weld metal, is observed in some sections of welded joints in the fusion zone, irrespective of the welding conditions. In X-rays this interlayer is light-coloured and in absorbed electrons, it consists of fine light and dark-coloured components.

No particular differences were found in copper distribution in welded joints produced at zero gravity and on the ground.

Obtained diagrams of copper and iron distribution (Figure 3a and b) in different sections of the welded joints showed, that the majority of eutectic Al–Cu inclusions are enriched in iron, while no iron was found in the other, finer inclusions.

In welded joints of alloy AMg6, magnesium distribution across the fusion zone in samples produced at zero gravity and on the ground, does not have any significant differences, except that in the solid solution of the weld metal adjacent to the fusion line, magnesium content is higher for welds made at zero gravity, compared to those, produced on the ground (Figure 3c and d).

Diagrams of distribution of iron and manganese demonstrate that these elements are non-uniformly distributed, either in the base metal or in the weld metal. Alongside with the low content of iron and manganese, individual phases are found in the solid solution characterised by a high content of iron and manganese, this probably corresponding to $(FeMn)Al_6$ phase.

Analysis of weld filmograms showed that the welds made on board the flying laboratory, have a greater quantity of pores of different size [5].

Metallographic analysis revealed individual pores in alloy AMg6 welds, made on the ground. However, the welds, produced at zero gravity, demonstrated a considerable amount of coarse and fine porosity, and oxide film clusters (Figure 4). Weld metal porosity on alloy 1201 was somewhat less increased (see Figure 1b).

Thus, the above-stated leads to the conclusion, that investigations of the macro- and microstructure and element distribution did not demonstrate any significant differences between welded joints of alloys 1201 and AMg6, made by the electron beam at zero gravity, compared to those, made on the ground.

Results of X-ray and metallographic analyses showed, that the welds, made at zero gravity, are characterised by a large number of pores of different size.

1. Paton, B.E. and Kubasov, V.N. (1970) *See pp. 154–160 in this Book.*

2. Lienan, H.G., Lowry, J.F. and Hassan, C.B. (1968) Electron beam welder for use in space, *Westinghouse Eng.*, No. 14, pp. 41–45.

3. (1968) Electron beam welding in space, *Engineer*, **225**, No. 5866, pp. 1024–1025.

4. Paton, B.E., Paton, V.E., Dudko, D.A. et al. (1973) *See pp. 118–122 in this Book.*

5. Ternovoj, E.G., Bondarev, A.A., Lapchinsky, V.F. et al. (1976) Investigation of some aspects of weldability of aluminium alloys by the electron beam at zero gravity, *Space research in Ukraine*, Issue 9, pp. 5–11.

Calculation of a Multilayer Tubular Item Heating in Brazing*

A.T. Zelnichenko[1], S.A. Vakulenko[1], A.R. Bulatsev[1] and V.F. Lapchinsky[1]

Design of advanced structures, where the main element are multilayer brazed tubular structures, requires a number of features of the technology of their fabrication, especially, the brazed joint quality, HAZ dimensions, etc. Setting up an effective technological process of fabrication of multilayer tubular billets (MTB) with layers of different thickness requires an optimal combination of the following factors, namely heater power, brazing duration, braze alloy melting temperature, and heater position (outside and inside the billet).

This work is devoted to studying the methods of mathematical modeling of thermal processes in MTB brazing, in particular in evaluation of the influence of the heating process and mode on MTB temperature and HAZ length.

MTB are heated by a thermoelectric heater of a constant power. Two variants of the heater location relative to the multi-layer tubular billet are possible, namely external and internal. The braze alloy is applied on boundaries $r = r_3$, $r = r_4$ (Figure 1).

Mathematical models. For evaluation of temperature distribution in MTB in the brazing zone, let us assume that temperature gradients in the axial direction are negligible; then in view of azimuth symmetry the equation of heat conductivity, describing the process of non-stationary heat transfer in each of regions D_n of a multi-layer item, can be written in the form of

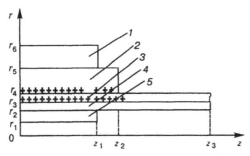

Figure 1 Schematic, illustrating the mathematical description: *1, 5* — heater, located internally and externally, respectively; *2* — washer; *3, 4* — pipe layers; ✚ — limits of braze alloy application area.

* (1988) *Avtomaticheskaya Svarka*, No. 2, pp. 30–32.

$$c_n \gamma_n \frac{\partial T_n}{\partial t} = \frac{1}{r} \frac{\partial}{\partial r} \left(r \lambda_n \frac{\partial T_n}{\partial r} \right) \quad r \in D_n, \quad t > 0;$$

$$D_n = \left\{ r : r_{n+1} < r < r_{n+2}, \quad n = \overline{1.3} \right\}, \tag{1}$$

where $T_n(r, t)$ is the absolute temperature; c_n, γ_n, λ_n is the specific heat capacity, density and heat conductivity, respectively.

On outer boundaries of region

$$D = \bigcup_{n=1}^{3} D_n \quad \text{at} \quad r = r_2 \quad \text{and} \quad r = r_5,$$

let us assign the conditions of heat exchange in keeping with the Stefan–Boltzmann law (or a combination of the above condition and symmetry condition):
external location of the heater:

$$\frac{\partial T_1}{\partial r} \Big|_{r=r_2} = 0;$$

$$\lambda_3 \frac{\partial T_3}{\partial r} \Big|_{r=r_5} = \varepsilon \sigma_0 \left[T_3^4 (r_5, t) - T_h^4 \right];$$

internal location of the heater:

$$\lambda_1 \frac{\partial T_1}{\partial r} \Big|_{r=r_2} = \varepsilon \sigma_0 \left[T_1^4(r_2, t) - T_h^4 \right];$$

$$\lambda_3 \frac{\partial T_3}{\partial r} \Big|_{r=r_5} = \varepsilon \sigma_0 \left[T_3^4(r_5, t) - T_a^4 \right],$$

where ε is the coefficient of the reduced degree of blackness; σ_0 is the Stefan–Boltzmann constant; $T_h(t)$ is the heater temperature (a priori unknown and to be determined); T_a is the ambient temperature.

In view of the presence of gaps, transparent for thermal radiation, conjugate conditions, associated with radiant heat exchange, are satisfied on inner surfaces of region D (at $r = r_3$ and $r = r_4$):

$$\lambda_n \frac{\partial T_n(r_{n+2} - 0, t)}{\partial r} = \varepsilon \sigma_0 \left[T_{n+1}^4(r_{n+2} + 0, t) - T_n^4(r_{n+2} - 0, t) \right];$$

$$\lambda_n \frac{\partial T_n(r_{n+2} - 0, t)}{\partial r} = \lambda_{n+1} \frac{\partial T_{n+1}(r_{n+2} + 0, t)}{\partial r}, \quad n = 1, 2. \tag{2}$$

At the contacting layers temperature above 570 K a non-ideal radiation contact changes into an ideal thermal contact, which is due to melting of the braze alloy, present in the gaps at $r = r_3, r = r_4$.

To determine heater temperature $T_h(t)$, let us draw the equation of energy balance

$$m_h c_h \frac{\partial T_h}{\partial t} = W_h - S_1 \varepsilon \sigma_0 (T_h^4 - T_a^4) - S_2 \varepsilon \sigma_0 (T_h^4 - T_i^4), \quad i = 1, 3, \tag{3}$$

where m_h, c_h is the heater mass and the specific heat capacity of the heater materials; W_h is the heater power; S_1, S_2 is the heater surface area through which its heat exchange with the environment and the item proceeds, respectively.

Equation (3) at $i = 1$ corresponds to the internal position of the heater and at $i = 3$ — to its external position.

The system of equations (1) with conjugate conditions (2) can be written in terms of generalized functions in the form of one equation, meaningful in the entire region D, including boundaries $r = r_n$, $n = 3, 4$.

As was shown in study [1], a generalized thermal resistivity is given by

$$p = \frac{1}{\lambda(r)} + \sum_{n=1}^{2} \frac{\delta(r - r_{n+2})}{\alpha_n},$$

where $\delta(r - \xi)$ is the delta function, concentrated in point $r = \xi$, while coefficient α_n is

$$\alpha_n = \varepsilon\sigma_0\psi[T_{n+1}(r_{n+1} + 0, t), \quad T_n(r_{n+2} - 0, t)];$$

$$\psi[a, b] = (a + b)(a^2 + b^2).$$

In order to determine the HAZ in brazing a multi-layer tubular item, let us use a one-dimensional equation that describes the process of heat propagation in the axial direction and is derived by averaging the two-dimensional equation of heat conductivity along the radial co-ordinate (in view of axial symmetry let us ignore the heat flows in the azimuth direction, as for equation (1)):

$$c\frac{\partial T}{\partial t} = \frac{\partial}{\partial z}\left(\lambda\frac{\partial T}{\partial z}\right) + \varepsilon\sigma_0\,(T_h^4 - T^4)\eta(z - z_1) - \varepsilon\sigma_0\,(T^4 - T_a^4), \tag{4}$$

where $\eta(z - z_1)$ is the unit Heaviside function;

$$c = \begin{cases} c_1\gamma_1 h_1 + c_2\gamma_2 h_2 + c_3\gamma_3 h_3, & 0 < z < z_2, \\ c_1\gamma_1 h_1 + c_2\gamma_2 h_2 + 0.5c_3\gamma_3 h_3, & z = z_2, \\ c_1\gamma_1 h_1 + c_2\gamma_2 h_2, & z_2 < z < z_3; \end{cases}$$

$$\lambda = \begin{cases} \lambda_1 h_1 + \lambda_2 h_2 + \lambda_3 h_3, & 0 < z < z_2, \\ \lambda_1 h_1 + \lambda_2 h_2, & z_2 \leq z < z_3; \end{cases}$$

z_i, $i = \overline{1, 3}$ are the geometrical characteristics of the item (Figure 1); $h_i = r_{i+2} - r_{i+1}$, $i = \overline{1, 3}$.

Let us solve equation (4) at the following boundary and initial conditions:

$$\frac{\partial T}{\partial z}\Big|_{z=0} = 0; \quad \frac{\partial T}{\partial z}\Big|_{z=z_3} = 0; \quad T(z, 0) = T_a. \tag{5}$$

Heater temperature $T_h(t)$ is found from equation (3).

Investigation results. The principal goal of this study is selection of the method and mode of MTB heating. Investigations were conducted at the following geometrical characteristics of the item and the heater: $r_1 = 0.4$ cm, $r_2 = 1.27$ cm, $r_3 = 1.29$ cm, $r_4 = 1.3$ cm, $r_5 = 1.4$ cm, $r_6 = 3$ cm, $z_1 = 1.3$ cm, $z_2 = 1.3$ cm, $z_3 = 15$ cm. Thermophysical properties of steels were selected by the data of [2]. Weight of heater, located externally was 88 g, and that of heater, located internally, was 21 g. Values T_a, $T_n^{(0)}$ ($n = 1, 3$) were taken to be 293 K. Braze

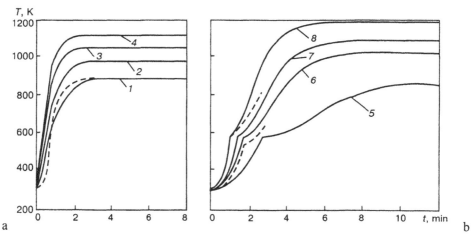

a   b

Figure 2 Dependence of MTB temperature on time with the heater, located internally (a) and externally
(b), respectively, at W_h = 220 (*1*), 280 (*2*), 380 (*3*), 480 (*4*), 40 (*5*), 80 (*6*), 120 (*7*), 140 (*8*) W
(dashed lines are experimental data).

alloy melting temperature was 573 K, brazing temperature being 873 K. HAZ was determined by 873 K isotherm. Different heater power was used.

In order to identify the mathematical model parameters, several full-scale experiments
were conducted, the results of which were compared with the calculation data. A satisfactory
coincidence of the design and experimental data (Figure 2) confirms the adequacy of the
mathematical model to the actual technological process. Thus, brazing conditions can be
determined by the computational experiment method.

Figure 2a gives the dependencies of MTB temperature in the brazing zone, calculated
by equation (1) for the case of an external location of the heater. At heater power below
240 W, no brazing occurs even at unlimited time. With the internal location of the heater
(Figure 2b), the power level, required for brazing, is essentially reduced to 60 W. Figure 3
gives the dependencies of brazing duration on heater power for different variants of its
location.

It is found, that in the case of external location of the heater, softening occurs only when
a heater of more than 480 W is used, HAZ length being 1.7 cm for W_h = 480 W. Evaluation
of the HAZ size with the internal location of the heater, is given in Figure 4.

Comparing the derived results leads to the conclusion, that internal location of the heater
is more cost-effective in terms of the time, required for brazing, and the heater power.

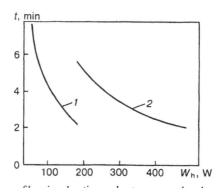

Figure 3 Dependence of brazing duration on heater power: *1* — heater located internally;
2 — heater located externally.

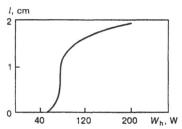

Figure 4 Dependence of HAZ length on heater power.

Proceeding from the admissible HAZ dimensions, the dependence in Figure 4 can be used to select the optimal heater power, and Figure 3 — to determine the required brazing duration.

1. Demchenko, V.F., Misyura, R.S. and Tarasevich, N.I. (1978) Investigation of contact heat exchange in production of a sheet EBM ingot, *Problemy Spets. Elektrometallurgii*, No. 8, pp. 71–78.
2. Kazanstev, E.I. (1975) *Industrial furnaces*, Moscow, Metallurgiya, 367 pp.

Modelling of Thermal Process in Piping Repair in Space*

Yu.L. Vasenin[1], A.A. Zagrebelny[1], A.T. Zelnichenko[1], I.V. Krivtsun[1] and V.F. Shulym[1]

One of the major problems facing aerospace experts is ensuring the service life of the space vehicle and its service systems, providing normal operation of a quite expensive item for several years (satellites for various purposes and interplanetary vehicles), or even decades (orbital stations and production facilities on other planets). This problem is being currently solved in two ways [1]:

• development of both advanced materials and new components, incorporating them into the designs of the promising space vehicles;

• provision of space vehicle repairability under service conditions.

It is understandable that the optimal solution of the repair problem is replacement of failing modules or complete systems. This, however, is not always feasible and is always expensive. Therefore, well-established methods of individual component repair should not be discarded, being especially effective in the case of an emergency failure of any component, for instance in the space vehicle life support system. In this case it is proper to make a comparison with orbital stations and modern nuclear submarines that are at sea for months on their own and are fitted with workshops having machining and welding equipment as well as a wide range of tools.

The urgency of the problem of ensuring repair of long-term space vehicles became particularly obvious over the last years of the operation of OC «Mir». From data provided by experts, responsible for maintenance of station operability, ≈ 50 % of the work, performed in the station was related to repair and restoration of its various system functions.

One of the vulnerable, but also the most vital life support sections in such complexes, is the many kilometres of piping (steel, titanium, aluminium) for various purposes. The ability to repair a failing pipe by its simple replacement seems improbable (especially in an emergency). In this case it is almost impossible to do without the traditional methods of joining materials by welding and brazing processes.

* (2001) *The Paton Welding Journal*, No. 4, pp. 18–23.

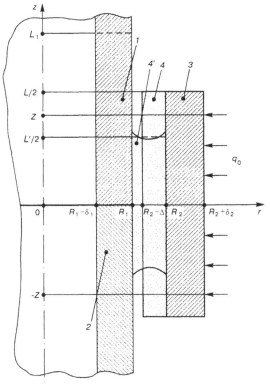

Figure 1 Schematic of a brazed joint in repair of the space vehicle piping: *1* — piping; *2* — insert; *3* — coupling; *4* — initial layer of the braze alloy; *4'* — braze alloy melt (for other designations see the text).

The subject of this study is an investigation of thermal processes involved in the repair of space vehicle piping by brazing in space. Solving this task will permit determination of the most effective process parameters of future standard on-board hardware to implement the above process, allowing long-term vehicle piping system repair to be performed by crew who have limited or no skills in brazing.

The schematic suggested by Prof. B.E. Paton is the simplest one for making a joint by brazing piping in space (Figure 1). According to this schematic, the damaged piping section is replaced by a tubular insert of the same material, having appropriate length, outer and inner diameters. The insert is centred relative to the piping axis using two couplings (just one of them is shown in the drawing) that are moved off the insert by half of their length towards the solid sections of the piping after the insert has been placed into the piping rupture. Then a heat source distributed over the outer surface of the coupling, as shown in Figure 1, is used to heat the item up to complete melting of the braze alloy earlier deposited onto the coupling inner surface, its filling the part of the gap between the piping and the coupling and, finally, reaching the brazing temperature in the zone of contact of the braze alloy with the outer surface of the insert and the piping. (The given schematic of brazing the joint is conditional, the actual design variant of the above component should provide complete filling of the gap with the molten braze alloy and fillet formation.) This results in formation of the brazed joint.

This paper is devoted to development of a mathematical model and detailed computer modelling of thermal processes in brazing of stainless steel piping in space. The aim of modelling is determination of the optimal geometrical dimensions of the coupling, selection of its material, thickness of the layer of the braze alloy required for deposition on the coupling inner surface, as well as optimal conditions and time of the item heating to provide a sound brazed joint.

Mathematical model. A non-stationary quasi linear equation of heat conductivity will be used for calculation of the temperature field in all the components of the joint (areas *1–4* (*4′*) in Figure 1) in the considered method of repair of space vehicle piping. Taking into account the azimuthal symmetry of the system, this equation for each of the above areas $D_i(i = 1, ..., 4, 4′)$ in the cylindrical system of co-ordinates (see Figure 1) can be written in the form of

$$\rho_i C_i \frac{\partial T_i}{\partial t} = \frac{1}{r}\frac{\partial}{\partial r}\left(r\lambda_i \frac{\partial T_i}{\partial r}\right) + \frac{\partial}{\partial z}\left(\lambda_i \frac{\partial T_i}{\partial z}\right), \quad r, z \in D_i, \quad t > 0. \tag{1}$$

Here, $T_i (r, z, t)$ is the space-time distribution of temperature in i-th region ($i = 1, ..., 4, 4′$); $\rho_i(T)$ is the density; $\lambda_i(T)$ is the heat conductivity coefficient; $C_i(T)$ is the effective heat capacity of the appropriate material that is determined allowing for the heat of its melting:

$$C_i(T) = c_i(T) + W_{mi}\, \delta(T - T_{mi}), \tag{2}$$

where $c_i(T)$ is the specific heat; W_{mi} is the latent heat of melting; T_{mi} is the material melting temperature in i-th region.

Assuming the insert length to be large enough and the coupling position symmetrical relative to the plane of contact of the insert with the main piping (plane $z = 0$ in Figure 1), the temperature field can also be assumed to be symmetrical relative to the above plane. In this case in the section from the start of heating up to complete melting of the braze alloy, equation (1) can be solved only in three of the four considered areas, for instance, in region *1* and in the upper halves of regions *3* and *4* (see Figure 1).

$$D_1 = \{r : R_1 - \delta_1 < r < R_1, \quad 0 < z < L_1\},$$
$$D_3 = \{r : R_2 < r < R_2 + \delta_2, \quad z : 0 < z < L/2\},$$
$$D_4 = \{r : R_2 - \Delta < r < R_2, \quad z : 0 < z < L/2\}, \tag{3}$$

where R_i is the outer radius of the piping and the insert; δ_1 is their wall thickness; R_2 is the inner radius of the coupling; δ_2 is its wall thickness; L is the coupling length; Δ is the thickness of the initial layer of the braze alloy. Let us assume that after assembly of the joint under consideration, there is a gap of a finite width between the piping external surface and inner surface of the braze alloy layer. Considering that brazing is performed under zero gravity, i.e. in the absence of the convective component of heat exchange between the above surfaces, the heat exchange between the piping and the coupling at the initial stages of the joint heating is considered as just radiation heat exchange [2]. In the case when brazing is performed inside the space vehicle living compartments, i.e. at atmospheric pressure, the conductive component of heat exchange, related to heat conductivity of the air present in the gap, should be also taken into account in addition to radiation heat exchange.

When solving the heat conductivity equations, we assume that after the braze alloy melting temperature (liquidus) has been reached in all the points of region D_4, it changes its shape under the impact of surface tension forces and fills the gap between the coupling inner surface and the outer surface of the piping, as shown in Figure 1. When solving the system of equations (1) further, let us approximately consider area $D_4′$ as an area having a rectangular cross-section that can be defined as follows under the condition of preservation of the braze alloy volume (see Figure 1):

$$D_4' = \left\{ r : R_1 < R_2, \quad z : 0 < z < L'/2 \right\} \tag{4}$$

where, provided condition $\Delta \ll R_2$ is satisfied for the length of this region L', we have

$$L' \approx L \frac{2\Delta/R_2}{1 - (R_1/R_2)^2}. \tag{5}$$

It is further assumed that after the braze alloy contact with the piping surface at $z < L'/2$, an ideal thermal contact between regions D_1 and D_4 is established.

Let us assign the initial and boundary conditions for solving equations (1) in the considered areas. Let us select the obvious condition as the initial one

$$T_i \big|_{t=0} = T_a , \tag{6}$$

where T_a is the ambient temperature. The boundary conditions for equations (1) are assigned, proceeding from the following considerations. As it is assumed that the temperature field is symmetrical relative to plane $z = 0$, boundary conditions for temperature in all regions at $z = = 0$ are selected in the form of conditions of solution symmetry. On the inner boundary of region D_1 it can be assumed that

$$\frac{\partial T_1}{\partial r} \big|_{r=R_1-\delta_1} = 0, \tag{7}$$

as under zero gravity conditions, convective heat exchange is absent, and radiation heat exchange between the regions of the inner surface at slow change of its temperature in the longitudinal direction can be neglected. On the boundary of region D_1 opposite to plane $z = = 0$, we assume that

$$T_1 \big|_{z=L_1} = T_a . \tag{8}$$

Here, L_1 is the length of region D_1, selected to be so large that with its further increase, the change of the temperature field in the zone of joint heating becomes negligible. On the piping outer surface beyond the coupling ($z < L/2$) conditions of radiation heat exchange with the environment are assigned:

$$\lambda_1 \frac{\partial T_1}{\partial r} \big|_{r=R_1} = \varepsilon_1 \sigma_0 \left(T_a^4 - T_1^4 \big|_{r=R_1} \right), \tag{9}$$

where ε_1 is the degree of the piping material blackness; σ_0 is Stefan–Boltzmann constant. As it is assumed that assembly of the joint results in a gap of a finite width between the coupling and the piping surface, transparent for thermal radiation, at the initial stages of heating (up to the braze alloy melting) the boundary conditions on the outer boundary of region D_1 at $0 < z < L/2$ are taken in the form similar to (9) with replacement of T_a by $T_4 \big|_{r=R_2-\Delta}$ and of ε_1 by the reduced degree of blackness $\overline{\varepsilon}_{1,4}$, calculated by the equation from [2]:

$$\overline{\varepsilon}_{1,4} = \frac{\varepsilon_1 \varepsilon_4}{\varepsilon_1 + \varepsilon_4 - \varepsilon_1 \varepsilon_4}, \tag{10}$$

where ε_4 is the degree of blackness of the braze alloy material. After part of the gap has been filled with the braze alloy, the boundary condition in the zone of the braze alloy contact with the piping surface, i.e. at $0 < z < L'/2$ is replaced by the condition of an ideal thermal contact.

It is assumed that on the remaining part of the piping surface under the coupling, the conditions for radiation heat exchange with the coupling inner surface are still satisfied.

Let us assign boundary conditions for region D_3 (see Figure 1) as follows. On the coupling outer surface in the region of the heat source impact on it, i.e. at $0 < z < Z$, we assume

$$\lambda_3 \frac{\partial T_r}{\partial r}\bigg|_{r=R_2+\delta_2} = q_0, \tag{11}$$

where q_0 is the density of the heat flow introduced through the coupling surface. For the remaining part of the coupling external surface, similar to (9), let us assume conditions of radiation heat exchange with the environment. As regards the inner boundary of region D_3, conditions of ideal thermal contact with the braze alloy layer are assigned that are satisfied at $z < L/2$ up to the moment of complete melting of the braze alloy initial layer, and at $z < L'/2$ after change of the shape of region D_4, whereas at $L'/2 < z < L/2$ conditions of radiation heat exchange between the coupling inner surface and the piping surface are assumed.

Let us define the boundary conditions for the region taken up by the braze alloy (region 4 ($4'$) in Figure 1). On the boundary of this region contacting the coupling metal, let us assign the conditions for ideal thermal contact. On the inner boundary up to the moment of complete melting of the braze alloy, let us assume the conditions of radiation heat exchange with the piping surface that are replaced by the conditions of an ideal thermal contact after complete melting of the braze alloy. Finally, on the upper boundary of the considered region, i.e. at $z = L/2$, ($z = L'/2$ after the braze alloy melting), conditions of radiation exchange with the environment are assigned.

This completes the definition of boundary problems for equations of heat conductivity (1) for the case of the method of piping repair in space considered in this paper. Note that the described model of thermal processes in brazing of piping is a two-dimensional one unlike, for instance, the model suggested in [3], where unidimensional equations of heat conductivity were used for calculation of the temperature fields in brazing of thin-walled tubular items.

Modelling results. Equations (1) with the respective initial and boundary conditions were solved numerically by the finite difference methods. The derived differential equations were solved by the method of changing directions [4]. All calculations were made with the following geometrical parameters of the joint: $R_1 = 6$ mm; $\delta_1 = 1$ mm; $L = 20$ mm; $R_2 = 6.5–6.7$ mm. In this case thickness Δ of the braze alloy initial layer was varied in the range from 0.3 up to 0.5 mm, so that the gap width between the coupling and the piping surface remained constant and equal to 0.2 mm[*]; $\delta_2 = 1–2$ mm. The piping and insert were of 10Kh18N9T (AISI 304) stainless steel; the coupling was of copper or the same steel. Considered as the braze alloy was Sn–Ni–Ge alloy proposed in [5] with a high content of tin and, hence, close to tin values of thermophysical properties. All the required thermophysical properties of the above materials were taken from [6–8]. Temperature of stainless steel brazing with the above braze alloy was assumed to be equal to 900 K.

First the optimal length of the zone of the coupling heating by the external heat source (heater) was determined. It was assumed that the heat flow was uniformly distributed over the external surface of the coupling within the region of $-Z < z < Z$ (see Figure 1), and its density q_0 was given by the following relationship:

---

[*] The above size of the gap is the maximal admissible one for brazing and it was selected, allowing for the ability to perform manual assembly of the brazed joint during extravehicular activity.

$$q_0 = \frac{P}{4\pi(R_2 + \delta_2)Z}, \qquad (12)$$

where P is the total power applied to the item by the heater.

Modelling results demonstrated that the temperature fields that are optimal for brazing the considered joint, in the zone of the braze alloy contact with the piping surface, are in place when condition $Z_1 \approx L'/2$ is satisfied. Shortening of the heating area length is accompanied by overheating of the central regions of the coupling surface (up to melting of its material). Increase of Z_1 leads to overheating of the coupling outer edges after the braze alloy melting and formation of an ideal thermal contact between the coupling and piping in $z < L'/2$ region. As a result, the temperature profile in the zone of the braze alloy contact with the surface of the piping and the insert is rising towards the coupling edges, this being especially significant in the case when a stainless steel coupling is used, which has a much lower heat conductivity than copper. Such a temperature profile in the brazing zone is inadmissible, as decrease of the contact angle of wetting of the surface being processed with the braze alloy can lead to the braze alloy spreading to the coupling edges with increase of this surface temperature, which, in its turn, does not allow a sound brazed joint to be made.

With the selected length of the heating zone (for the considered geometrical parameters of the joint, values $Z_1 = L'/2$ are in the range of 6.2–7.5 mm), calculations were performed of the dependence of temperature in different points of the item on the duration of its heating at the initial temperature $T_a = 300$ K (brazing in living compartments of the space vehicle). Figure 2 gives the appropriate dependencies for the central points of the piping external surface ($r = R_1$, $z = 0$) and central points of the coupling inner surface ($r = R_2$, $z = 0$). As follows from the presented curves, the piping surface temperature remains practically

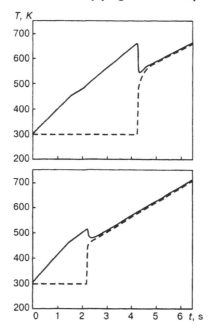

a

b

Figure 2 Dependence of temperature T in different points of the item on heating time t when a coupling of stainless steel 10Khl8N9T (a) and copper (b) is used at $R_1 = 6.0$ mm; $\delta_1 = 1.0$ mm; $R_2 = 6.6$ mm; $\delta_2 = 1.5$ mm; $\Delta = 0.4$ mm; $P = 500$ W (solid curves — for a point with co-ordinates $r = R_2$, $z = 0$; dashed curves — $r = R_1$, $z = 0$).

constant, equal to the item initial temperature, up to complete melting of the braze alloy, change of its shape under the impact of the surface tension forces and formation of an ideal thermal contact between the braze alloy and the piping surface. The temperature of the coupling inner surface, equal to the braze alloy temperature in appropriate points, at the initial stage of heating grows practically linearly in time (except for the process of braze alloy melting). Then, after contact of the molten braze alloy with the relatively cold surface, it abruptly cools down, while the piping surface is heated, so that their temperatures are equalized, and further heating of the item leads to a linear growth of temperature in the brazing zone. Note that in the case, when a copper coupling is used, complete melting of the braze alloy proceeds almost 2 times faster, than in the case of a stainless steel coupling (Figure 2), this being attributable to a higher heat conductivity of copper and its lower heat capacity.

As noted already, contact of the molten braze alloy with the cold surface of the piping and the insert leads to considerable cooling down of the braze alloy (up to partial solidification) that is followed by further heating and fusion. Figure 3 shows the evolution of the thermal condition of the braze alloy after its contact with the piping surface; time was counted since the moment the braze alloy touched the surface being brazed.

In order to determine the energy characteristics of the brazing process, the time of the joint heating up to the specified thermal condition and the energy required for it, depending on the power applied to the item, was calculated. The appropriate calculation data derived for the stainless steel coupling, are given in Figure 4, and those for the copper coupling — in Figure 5. The lower curves in Figure 4a and Figure 5a correspond to the time of the joint heating up to complete melting of the braze alloy and establishment of an ideal thermal contact between the braze alloy and the piping surface. The lower boundary of the grey area

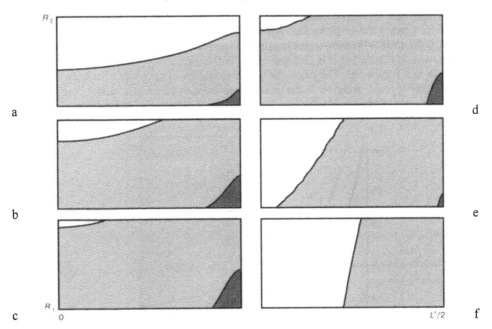

Figure 3 Evolution of the thermal condition of the braze alloy after its contact with the piping surface when a stainless steel coupling is used under the conditions, mentioned in Figure 2 and $P = 250$ W (unhatched areas correspond to the liquid condition of the braze alloy ($T \geq T_L = 493$ K); dark areas — to the solid state ($T \leq T_S = 456$ K); grey areas — to the intermediate state): a — $t = 0.01$; b — 0.04; c — 0.10; d — 0.20; e — 0.30; f — 0.40 s.

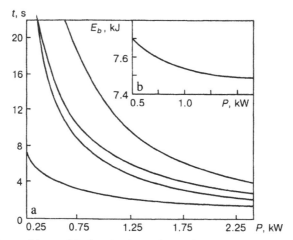

Figure 4 Dependence of time t of the item heating to the specified thermal condition (a) and energy E_b, required for the brazing process (b), on the input power P when a coupling of stainless steel 10Khl8N9T is used (R_1 — 6.0 mm, R_2 = 6.6 mm, δ_1 = 1.0 mm, δ_2 = 1.0 mm, Δ = 0.4 mm, T_a = 300 K).

corresponds to the time of achievement of the brazing temperature at least in one point of contact of the braze alloy with the piping surface, and the upper boundary — to the time of heating of the entire surface being brazed up to the specified temperature. Finally, the upper curves in these Figures correspond to the time of the start of melting of the coupling surface, i.e. limit the maximal admissible time of the item heating at the specified applied power.

As follows from the curves, given in Figures 4 and 5 the time of heating up to the specified thermal condition is naturally reduced with the increase of P. As regards the energy, required for heating the entire surface being brazed up to the brazing temperature, $E_b = Pt_b$, where t_b is the time of heating to the brazing temperature (upper boundary of the grey area in Figure 4a and Figure 5a), it is also reduced with the increase of power applied to the item (Figure 4b and Figure 5b). The cause for that is reduction of the share of energy losses for heat conductivity and radiation in the total energy balance of the joint under consideration with the increase of the speed of energy application, determined by value P.

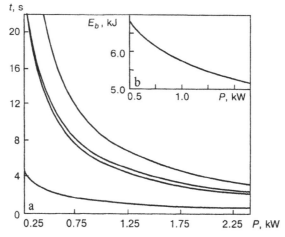

Figure 5 Dependence of time t of the item heating up to the specified thermal condition (a) and energy required for the brazing process (b) on the input power when a copper coupling is used (see parameters and designations in Figure 4).

Calculations showed that at other conditions being equal, the brazing time t_b and, therefore, also energy E_b, required for making the brazed joint, increase with the coupling wall thickness, practically in proportion to the increase of its material volume. In addition, the energy, required to provide the brazing process in the case a copper coupling is used, turns out to be ≈ 20 % smaller than when a steel coupling is used (Figure 4b and Figure 5b). Thus, the method of making the brazed joint, that is the most energy effective in repair of stainless steel piping, is to use a copper coupling of a minimal admissible thickness.

The optimal range of powers applied to the item by the heater, that are required for making the brazed joint according to the schematic under consideration, is between 250 and 500 W, this amounting to not more than 1 kW of the consumed electric power, allowing for the efficiency of the heater proper and the process of heat transfer to the item. Such power levels, on the one hand, can be easily provided by the on-board power supply system of the space vehicle, and, on the other hand, allow brazing to be performed in a relatively short time (about 10–30 s). So, for instance, the time required to achieve the brazing temperature in all the points of the surface being brazed at $P = 500$ W and $T_a = 300$ K, is equal to 12.3 s in the case of a copper coupling (Figure 4a) and 15.5 s, if the coupling is of 10Khl8N9T steel (Figure 5a). In addition, as follows from the same plots, applying such power to the item provides a sufficient (more than 6 s) time interval between the completion of the brazing process and the start of the coupling material melting.

Finally, if brazing is performed in raw space, the time for the joint heating up to the brazing temperature and the energy required for it, respectively, can essentially depend on the initial temperature of the item. In particular, in repair of piping on the sunlit side of the space vehicle, where the initial temperature is 400 K, the time of brazing the considered joint, when a copper coupling is used, is 10.4 s ($E_b = 5.2$ kJ), whereas on the shadow side at $T_a = 180$ K — 14.4 s ($E_b = 7.2$ kJ).

In conclusion it should be noted that this work is of a theoretical nature and its results require experimental verification, including flight testing.

Conclusions

1. A mathematical model of thermal processes in brazing of piping in space, an appropriate computational algorithm and software for implementation of this model in the computer have been developed.

2. Modelling results have shown that use of a copper coupling in brazing of the considered joint, provides an energy saving up to 20 %, compared to the steel coupling, as well as allows achieving a more uniform distribution of temperature in the brazing zone.

3. In brazing joints of such geometrical dimensions, the optimal powers are 250–500 W, applied to the item by the heater, thus allowing the zone of braze alloy contact with the piping surface to be heated up to the brazing temperature in 10–30 s.

1. (2000) *Space: technologies, materials science, structures*. Ed. by B.E. Paton, Kiev, PWI, 528 pp.

2. Kutateladze, S.S. (1979) *Fundamentals of heat exchange theory*, Moscow, Atomizdat.

3. Zelnichenko, A.T., Vakulenko, S.A., Bulatsev, A.R. et al. (1988) *See pp. 245–249 in this Book.*

4. Anderson, D., Tannekhil, J. and Pletcher, R. (1990) *Computational hydromechanics and heat exchange*, Moscow, Mir.

5. Zagrebelny, A.A., Gavrish, S.S., Bulatsev, A.R. et al. (1986) *See pp. 258–264 in this Book.*

6. (1984) *A handbook on brazing*. Ed. by I.E. Petrunin, Moscow, Mashinostroyeniye.

7. Larikov, L.N. and Yurchenko, Yu.F. (1985) Structure and properties of metals and alloys, In: *Thermal properties of metals and alloys*. A handbook, Kiev, Naukova Dumka, 367 pp.

8. (1976) *Tables of physical values*. A handbook. Ed. by I.K. Kikoin, Moscow, Atomizdat.

Analysis of Tubular Element Permanent Joints of Braze-Welding and Brazing Type, Made Using the VHT[*]

A.A. Zagrebelny[1], S.S. Gavrish[1], A.R. Bulatsev[1], V.F. Khorunov[1], V.I. Shvets[1] and S.A. Skorobogatov[1]

The work, performed over recent years in connection with space exploration, has laid the foundations for a new stage in the mastering of space by man for peaceful purposes. Solving practical problems of the national economy, using units put into near-earth orbit, already is an urgent issue now. In this connection, erection of various large-sized constructions, using truss structures, in open space will be required in the near future. Specific features of such installation conditions are microgravity, high vacuum ($10^{-2} - 10^{-6}$ Pa), limited heat sink and higher radiation. The first of the above factors makes it rational to apply in the load-carrying structures thin-walled elements, one of which can be a cylindrical elastic open shaped section [1]. The initial billet for it, is a cold-rolled highly-deformed strip of dispersion-hardened 36NKhTYu alloy. Shaped sections are produced from the strip by its cold forming in the zone of elasto-plastic deformation and subsequent heat treatment (tempering) in the fixed condition (thermofixing). Shaped section wall thickness is 0.15 mm [2].

The three-face truss structure of such shaped sections requires making joints during assembly in space, that are required in certain points between the longitudinal girths (ribs) of the truss of elastic shaped sections and the transverse elements, namely triangular frames with node bushings, located in their apices (Figure 1). The node bushings are made of steel 30 (C — 0.27–0.35; Ni — 0.25; Cr — up to 0.25 wt.%), their thickness being equal to 1 mm in the joint area.

As shown by investigations, the most acceptable technology for joining the truss elements in space is brazing or braze-welding.

Brazing is attractive by high adaptability to fabrication. It does not require sophisticated hardware, as just one parameter, namely the heating temperature, is to be maintained. However, erection of large-sized load-carrying structures based on brazed joints is a fundamentally new approach and requires solving a number of problems.

First of all it is necessary to develop such a braze alloy, that would meet the following requirements:

• providing fluxless brazing in space;

• easy and simple application onto the surface of the node bushing on the ground;

• high corrosion resistance (preliminary application of the braze alloy implies subsequent long-term storage).

There is also the problem, related to process hardware, in that brazing performance in space requires a heater capable of long-term reliable operation under these conditions.

Development of braze alloy composition is closely related to the problem of preliminary preparation of the elastic shaped section surface. It should be noted, that 36NKhTYu (C — ≤ 0.05; Si — 0.3–0.7; Mn — 0.8–1.2; Ni — 35.0–37.0; Cr — 11.5–13.0; Ti — 2.7–3.2; Al — 0.9–1.2; S, P — up to 0.2 wt.%) alloy presents special difficulties for brazing, as it contains elements with a high affinity to oxygen [3]. In any operations in air and even in

---

[*] (1986) *Problems of space technology of metals*, Kiev, PWI, pp. 13–22.

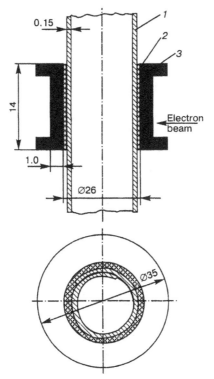

Figure 1 Schematic of a truss connection elements: *1* — elastic shaped section; *2* — braze alloy; *3* — node bushing.

commercial vacuum, stable Al_2O_3, Cr_2O_3 and other oxides form on its surface. Application of a metal coating, having less stable oxides, is proposed to ensure braze alloy wetting the surface being brazed. Analysis of constitutional diagrams of various alloy systems reveals, that application of a nickel coating is the most rational. In addition, such a coating provides a high enough reflectivity of the truss elements.

Nickel coating was applied by two methods: electroplating and electron beam vacuum deposition, preceded by surface cleaning by an abrasive mechanical method to remove scale and oxides. Braze alloy spreading does not depend on the coating application method. However, nickel adhesion to 36NKhTYu alloy is maximal in electron beam deposition. The optimal thickness of the nickel layer was selected from the condition of achievement of the best spreading of the braze alloy and was equal to 2 μm.

The braze alloy used was Sn–2Ni–4Ge alloy, developed specially for low-temperature vacuum brazing of a shaped section with nickel coating and node bushing. During investigations, contact, radiation and electron beam methods of heating the truss elements were considered, resulting in selection of the electron beam process as the most readily adaptable to fabrication. Development of special hardware, based on the electron beam gun, is envisaged for truss construction in space. However, in the preparatory stage, VHT was used during the preliminary experiment. The tool had already demonstrated its high reliability and serviceability in open space during performance of the experiment in the OS «Salyut-7» by cosmonauts S. Savitskaya and V. Dzhanibekov [4]. In keeping with the new research programme, operator training in brazing the truss assemblies, using VHT, was conducted in a special space simulation test chamber. The tool operation mode was set as follows: 5 kV voltage; about 40 mA current. The distance from the electron beam gun to the bushing surface was about 150 mm, which provided heating by a defocused beam. In such a mode, the bushing was heated for 10–20 s (from two sides), resulting in the braze alloy melting and formation of a brazed joint. It should be noted, that in view of the local nature of heating, the temperature field induced by the electron beam, is non-uniform and is characterized by a high temperature gradient. In this connection, the braze alloy should meet still one more requirement, namely it should be able to perform in a broad temperature range to produce a brazed joint over the entire area.

The results of experiments in the space simulation test chamber demonstrated that Sn–2Ni–4Ge meets practically all the requirements.

It should be noted, that during experiments at terrestrial gravity, part of the braze alloy flows out and does participate in the joint formation. At zero gravity this phenomenon will probably be absent, which will have a positive effect on the joint strength. Nonetheless, evaluation of the already produced joints indicates, that their strength reaches 2000 N and can be much more in a number of cases. This is higher than the strength of individual truss elements.

In addition to vacuum brazing, the ability to produce a complex joint was studied. It was a slot weld made by placing individual plug lap joints in combination with simultaneous brazing of the gap between the surfaces being joined. In this case the VHT operates in the welding mode, resulting in melting of not only the braze alloy, but also the base metal of the truss elements. A spot welded joint is formed in that part of the bushing, which is exposed to the direct impact of the beam, and a considerable overheating of the shaped section occurs in the other part, resulting in breaking up of the oxides on its surface, improvement of wettability, and production of a brazed joint. This method has the following advantages:

• substantial improvement of the joining reliability due to the fact that, even if welding defects develop (burn-through, lack-of-penetration, etc.) the strength is still sufficient, owing to a brazed joint formation;

• higher efficiency of processing hardware due to excess heat, removed into the item through heat conductivity, being used for braze alloy melting.

In this case VHT operation modes have the following parameters: voltage of 5 kV; current of 50–55 mA. Distance from the gun to bushing surface is 70 to 100 mm. It should be noted, that in this case heating of the bushing is possible, both by a sharply focused and a somewhat defocused beam.

Figure 2 General view of a joint sample.

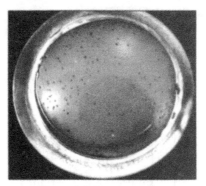

Figure 3 Joint cross-section.

When experiments were conducted in the space simulation test chamber, braze alloy of the same composition as in brazing Ni-plated sections, was used in such a joint of node bushings with cleaned shaped sections.

Obtained samples of these joints were studied, using metallographic and X-ray microprobe analyses. Figure 2 is a general view of the sample and the produced joint in the shaped section, passing through the bushing middle. On one side the bushing is partially melted to a small depth, and on the opposite side base metal is melted through with formation of a hole (Figure 3). The observed difference is attributable to the fact that, in the first case, the bushing was heated by a defocused electron beam, and in the second by a focused beam. Thus, the joint forms under the conditions of a non-uniform heating, and the joint microstructure heterogeneity is observed accordingly. In the low-temperature zone, the weld has the structure of a cast braze alloy and the interdiffusion of the braze alloy and the base metal elements is insignificant (Figure 4). Germanium is an exception, which is present in the brazed joint zone in an amount much higher than its content in the braze alloy (weight fraction of 12 %). Such an anomalous behaviour of germanium, obviously, brings about the observed

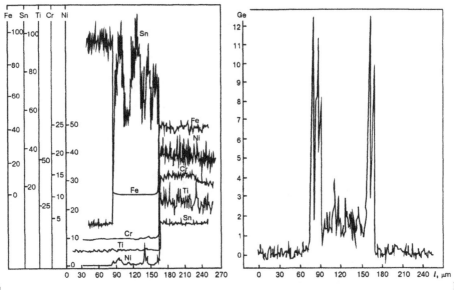

a   b

Figure 4 Distribution of element content, wt.%, in the low-temperature joint zone: a — iron, tin, titanium, chromium, nickel, b — germanium.

Figure 5 Microstructure of a joint, produced in heating with a defocused beam
(*A* — zone of practically complete dissolution of the profile).

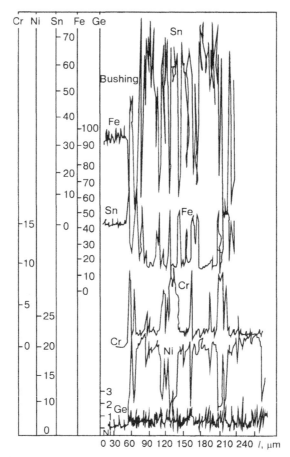

Figure 6 Distribution of element content, wt.%, in the zone of practically complete dissolution of the shaped section in the joint, produced with a defocused electron beam.

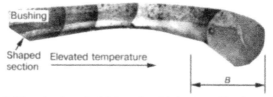

Figure 7 Microstructure of a joint, produced in heating with a focused beam
(*B* is the zone, adjacent to the partially-melted zone of the joint).

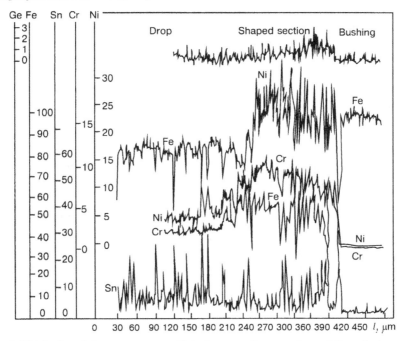

Figure 8 Distribution of element content, wt.%, in the zone, adjacent to the partially-melted zone of the joint, produced by a focused electron beam.

strengthening of the joint. Additional studies are required, however, to gain a deeper insight into this phenomenon.

With the increase of temperature in the brazed joint area, interaction of the liquid braze alloy with the base metal results in formation of a structural component, the amount of which increases with temperature that intensifies the diffusion processes, up to filling the entire gap. In case of heating with a defocused beam, however, this component has not become highly developed. Further temperature rise is accompanied by intensive dissolution of 36NKhTYu alloy.

In heating by a defocused beam, this process proceeds up to complete dissolution of the shaped section (Figure 5). Intercrystalline erosion of the alloy is observed, with formation of austenite grains, separating from the matrix. Element distribution in the weld zone at practically complete dissolution of the shaped section is shown in Figure 6. Curves of element distribution show that the average phase composition in this area is as follows, wt.%: 60Sn–17Ni–15Ge–2Cr.

A feature of the process of heating by a concentrated beam is that the shaped section interacts with the braze alloy from the inner side, and from the outside it interacts with the drop, formed in the base metal melting (Figure 7). The time of interaction of overheated braze alloy with the shaped section becomes much shorter, and no complete dissolution is observed, despite a certain alloying with the braze alloy elements. In this case, this area has the following phase composition, wt.%: 50Fe–25Ni–8Cr–12Sn–1Ge (Figure 8). Evaluation showed the strength to be much higher than the strength of the joint produced in the bushing heating by a defocused electron beam.

In this connection, it is recommended to use focused electron beams for joining the truss elements during its construction in space. Conducted studies demonstrated that already now reliable hardware is available to perform such work, that allows making strong joints.

1. Trishevsky, I.S., Shugaenko, V.K., Voronstov, N.M. et al. (1985) State-of-the-art and prospects for development of fabrication and application of shaped sections for deploy-

able space metal structures, In: *Proc. of 1ˢᵗ All-Union Seminar on Problems of Space Technology of Metals*, Kiev, 1985. Kiev, Naukova Dumka, pp. 16–21.

2. *TU 14-2-333–78*, Shaped sections with high elastic properties for antennae and rods. Effective from 01.05.79 till 01.05.84.

3. Rikhshtadt, A.G. (1982) *Spring steels and alloys*, Moscow, Metallurgiya, 253 pp.

4. Nikitsky, V.P., Lapchinsky, V.F., Zagrebelny, A.A. et al. (1985) *See pp. 178–184 in this Book*.

Technology of Vacuum Soldering of Connections in Truss Structures of Aluminium Alloys*

V.F. Khorunov[1], V.F. Lapchinsky[1], V.I. Shvets[1] and V.F. Shylum[1]

The conquest of space requires the development of technologies for making permanent joints in orbit, in particular at high vacuum. Selection of the optimal technology is associated with hardware capabilities and should envisage performance of the maximum number of operations in the preparatory stage, low power consumption and simplicity of conducting the process.

The method of heating with an electron beam is the most developed and most practiced in space. Let us consider the technology of soldering connections of aluminium alloy truss structures, developed from this method.

The parts being joined are massive parts of a complex configuration (Figure 1). Therefore, when the electron beam is used, which is a concentrated heat source, a non-uniform temperature field is induced on the item surface. In addition, the initial temperature of the item in orbital conditions depends on its spatial position relative to the Sun. It should be also noted that it is difficult to assemble structures with capillary gaps in orbit.

In order to produce brazed joints directly in orbit, it was necessary, first, to develop solders which ensure joint formation over a broad range of temperatures, and, secondly, provide a forced feed of the molten solder into the gap.

Analysis of the published data on the currently available solders and the results of our studies led to the identification of such a material, an earlier one developed by us [1] a solder of the Sn–Ni–Ge system that does not have any elements with a high vapour pressure and is recommended for application at high vacuum. The solidus temperature of the solder, equal

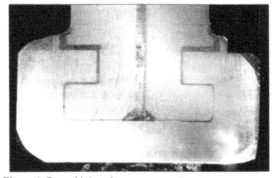

Figure 1 Brazed joint of a truss connection element (section).

* (1992) *Avtomaticheskaya Svarka*, No. 2, pp. 52–53.

to 235 °C, is higher, than the maximum possible heating temperature of the structure in space (150 °C).

Vacuum soldering is a little studied process. It is found, that thermal dissociation of oxides at low temperatures is possible in a limited number of metals [2]. Therefore, surface preparation for soldering requires developing a special operation.

Different techniques are known of removing chemically stable aluminium oxides from the surface of the parts. Each of them has its special features, limiting its use under orbital conditions. The operation of mechanical breaking up of aluminium oxide directly during soldering requires additional equipment. Use of special fluxes involves additional technological operations for their removal; otherwise the corrosion resistance of the joints decreases. In addition, the volatile components of the flux adversely affect the hardware.

Fluxless vacuum soldering, using reactive metals in the form of vapours, as well as in the composition of cladding, is made difficult by the need to create the vapour phase in the first case, and a cladding coating on a complex-shaped item in the second case.

An interesting method is soldering aluminium alloys with preliminary application on the surfaces to be joined of layers of a material lending itself easily to soldering, and preserving this property over time. We studied as such materials nickel, nickel-copper and nickel-tin layers (nickel and copper oxides with a low dissociation temperature form in air in heating up to temperatures above 520 and 180 °C, respectively [3]).

The obtained data demonstrate that spreading of the solder over nickel coating is satisfactory at a temperature above 500 °C, and over nickel-copper at above 400 °C. On the other hand, good spreading over nickel-tin coating is found at a temperature, close to that of the solder melting.

The developed technology and the proposed solder were used for laboratory soldering of elements of a truss structure assembly over nickel-copper and nickel-tin coatings. Pressure in the chamber was $1 \cdot 10^{-2}$ MPa. Solder in the form of a tablet was placed into a cylindrical cavity of the inner part (Figure 2). Solder melt penetrated into the soldering gap through the functional hole under the pressure of an expanding spring. Heating was carried out by a defocused electron beam. The beam direction is indicated by an arrow in the schematic.

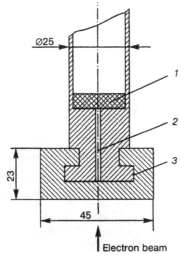

Figure 2 Schematic of assembly of a truss connection for soldering: *1* — solder; *2* — functional hole; *3* — surfaces to be joined.

The time of the electron beam impact was limited by the moment of the solder filling the gap. The temperature on the item with a nickel-tin coating was up to 250–350 °C, and on that with a nickel-copper coating — 350–450 °C.

It should be noted that soldering is accompanied by evolution of gas that is contained in the coatings applied by electroplating. Gas evolution is observed as braze alloy «boiling».

The results of studying the produced items demonstrate the good quality of the soldered joint. Despite the gap width reaching several millimetres and the gas evolution processes running in it, the joint has no dry areas or gas pores.

Examination of the microstructure and element distribution in the soldered joint was performed in «Camebax» system and scanning microscope JSM-840 with «Link Systems» microanalyser. Characteristic regions of the microstructure are shown in Figure 3.

It was found that the joint microstructure is characterised by inhomogeneity, corresponding to the non-uniform degree of heating of different sections of the soldered surface. Structural inhomogeneity is manifested on the joint contact boundary and results from interaction of the solder melt with the base metal. The soldered joint proper in all the cases is a tin-based matrix, containing inclusions.

a

b

c

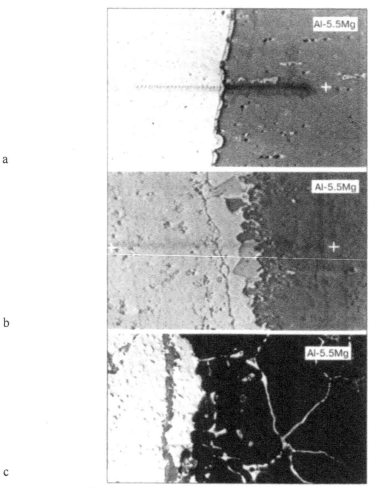

Figure 3 Microstructure of a soldered joint: a, b — over nickel-tin coating (×500);
c — over nickel-copper coating (×400) (reduced by 2/3).

Comparative analysis of the microstructure of the joints soldered over nickel-tin or nickel-copper coatings, showed, that in the first case, the transition zone forms diffusion layers (Al–Ni, Ni–Sn) (Figure 3a). A section adjacent to the area of the electron beam impact is an exception, where spalling and breaking up of the coating layer is found (Figure 3b). In the second case, the joint microstructure is characterised by spalling of the coating layer and formation of a soldered joint of aluminium with tin (Figure 3c). A pronounced intergranular erosion of the aluminium alloy is also found.

The microstructure is responsible for the strength properties of soldered joints. Additional research is required to study the interaction of the soldered joint properties and their microstructure. The results of such research can be used to adjust the soldering mode and select the coating to be applied prior to soldering.

1. Khorunov, V.F., Shvets, V.I., Lapchinsky, V.F. et al. *Solder for nickel soldering*. USSR author's cert. 1606295, Int. Cl. B K 35/26, publ. 15.11.90.

2. (1985) *Element properties*. A handbook. Ed. by M.E. Drits, Moscow, Metallurgiya, 672 pp.

3. Nikitinsky, A.I. (1983) *Soldering aluminium and its alloys*, Moscow, Mashinostroyeniye, 192 pp.

Formation of Brazed Joints of Thin-Walled Structures in Space[*]

V.F. Khorunov[1], V.I. Shvets[1], A.R. Bulatsev[1] and S.S. Gavrish[1]

Methods and means for producing permanent joints should be developed for construction of long-term large-sized space facilities, as well as repair of metal structures in open space.

The electron beam currently is the optimal source of thermal energy in space [1]. VHT, designed for performance of repair and assembly work, has been developed at the PWI.

The paper presents the results of investigation of joints in the connections of the truss structure elements, made of thin-walled open shaped sections (36NKhTYu alloy) (Figures 1 and 2), that were produced using VHT on the ground and in space by technology developed by the authors [2].

In assembly of a three-face truss structure it was necessary to make a joint of an elastic shaped section 0.15 mm thick with transverse elements, the apices of which had node bushings, made of steel 30 (C — 0.27–0.35; Ni — 0.25; Cr — up to 0.25 wt.%) (see Figure 1 on p. 259). EBW is actually unacceptable for making joints of this type, because it is impossible to provide accurate assembly of parts with a minimal gap. The optimal method in this case seems to be brazing with general heating that does not involve any difficulties on the ground, but requires developing and taking into orbit specialized hardware for use in space. VHT application made it necessary to look for non-traditional solutions when developing the brazing technology. The issue of formation of brazed joints under conditions of non-isothermal heating by an electron beam, not previously considered, became urgent.

The heating mode in VHT operation is determined by current, degree of beam focusing and heating time. VHT has three degrees of mode adjustment by current. The focusing degree depends on the distance from the electron beam gun to the object.

When work is performed manually under extreme conditions, provision of a stable focusing degree, or evaluation of the time intervals, is highly problematic. To produce joints

* (1999) *Avtomaticheskaya Svarka*, No. 10, pp. 31–38.

Figure 1 Truss structure with girths of 25 mm diameter tubes during assembly.

in a broad range of heating modes taking into account the thermal instability of alloy 36NKhTYu (C — ≤ 0.05; Si — 0.3–0.7; Mn — 0.8–1.2; Ni — 35.0–37.0; Cr — 11.5–13.0; Ti — 2.7–3.2; Al — 0.9–1.2; S, P — up to 0.2 wt.%) and allowing heating up to 650 °C without loss of ductile properties, brazing was performed using a low-melting Sn-2Ni-4Ge braze alloy developed by the authors for the conditions of vacuum (232 °C solidus temperature) [3].

It should be noted that the technological process envisaged conducting on Earth the preparatory operations for brazing. The braze alloy in the form of a tinning layer was applied onto a bushing inner surface. After magnetic-abrasive cleaning, the surface of the shaped section was coated with a layer of nickel ≈ 10 μm thick by thermovacuum spraying.

During preliminary laboratory experiments it was established that the first stage of VHT operation by current (I = 60÷65 mA) was not suitable for brazing thin-walled objects, because of the high energy concentration. The brazed joint can be produced in the second and third stages.

In order to determine the influence of the heating modes on joint formation, samples were analyzed that were produced using the limit parameters of heat input, namely a defocused beam in the third stage (I = 40 mA, distance from the electron beam gun to the bushing L = 150 mm) and a sharply-focused beam, melting the assembly through (Figure 3) with formation of a through-thickness hole in the second stage (I = 50÷55 mA, L = = 70÷100 mm).

In reality any intermediate variant can be implemented by manual operation.

Figure 2 A fragment of a truss structure with girths made of 45 mm diameter.

Figure 3 General view of a truss connection during heating by a focused electron beam.

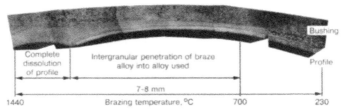

Figure 4 Macrostructure of a brazed joint, produced with electron beam heating (×25).

Ground-based experiments and operator training were performed in a special space simulation test chamber in a vacuum of $1 \cdot 10^{-3}$ Torr. The truss connection elements were heated from the bushing outer side in four points successively.

Optical microscopes MIM-8 and «Neophot-2» were used for metallographic examination of the joints. It was found, that in bushing melting by a defocused beam, a brazed joint forms (Figure 4) in the assembly at 7–8 mm distance from the zone of electron beam impact in the braze alloy melting section (in the temperature range of 232–1440 °C). No brazing defects were found in the joint structure. Structural inhomogeneity along the weld length was a characteristic feature.

A band of austenite grains that separated from the matrix was found in the high-temperature region along the boundary with 36NKhTYu alloy. The band grew wider closer to the zone of the electron beam impact, and this could go on up to complete dissolution of the shaped section.

At shorter heating in the case of preservation of the shaped section integrity, braze alloy «rolls» that deformed it were found on the assembly inner surface in the zone of the electron beam impact (Figure 5).

When the assembly is melted through by a sharply focused beam, the shaped section and the bushing of the brazed joint on the boundary with the hole, are connected to the solidified melt, the composition of which consists of the metal of the bushing, shaped section and braze alloy (Figure 6). A combined brazed-welded joint forms in the considered section, and adjacent to it is a section of the brazed joint proper, similar to that produced with a defocused beam, being, however, somewhat shorter.

Joint formation in time can be represented as follows (Figure 7). At the initial stage the electron beam partially melts the bushing and melts the braze alloy. The conditions of the assembly heating induce directed mass transfer of the braze alloy into the zone of the electron beam impact, resulting in the shaped section deformation and formation of «rolls» on the assembly inner part (Figure 7a).

The next step is complete dissolution of the shaped section in the braze alloy (Figure 7b) that is followed by melting of the bushing and formation of a liquid bridge in the assembly (Figure 7c). The liquid bridge, contained by the surface tension forces, alongside the hydrostatic forces, is also exposed to the forces, induced by the electron beam (electron beam pressure, back pressure of the flow of evaporating atoms, thermal electrons, as well as pressure, induced by the thermal and reactive radiation) [4]. The liquid bridge breaks up, the

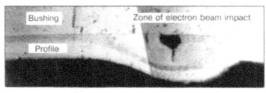

Figure 5 Joint macrostructure in the region of the electron beam impact at the stage of the braze alloy «roll» formation (×100).

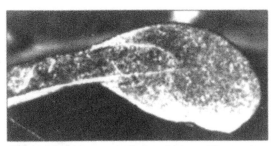

Figure 6 Macrostructure of a joint in the welding section.

melt enclosing the edges of the bushing and the shaped section (Figure 7d). Increase of the melt volume due to braze alloy mass transfer improves joint formation in the welding area.

Analysis of the theoretical and experimental work on non-isothermal flow in the capillary gaps [5, 6], demonstrated, that the directed mass transfer of the braze alloy can be caused by the surface tension gradient on the interface (Marangoni effect), as well as thermostatic sliding.

Considering, however, that the actual process is made more complicated by the braze alloy interaction with the base metal and the presence of electric and magnetic fields, and that the theoretical issues are insufficiently developed, the phenomenon of directed mass transfer of the braze alloy requires further studies.

Intergranular penetration of the braze alloy into 36NKhTYu alloy in the high temperature region is attributable to the change of the grain boundary conditions as a result of dissolution of the strengthening γ-phase in austenite, and grain boundary enrichment by such surfactants as titanium and aluminium. According to the data of [7], phase transformations start at the temperature of 650 °C.

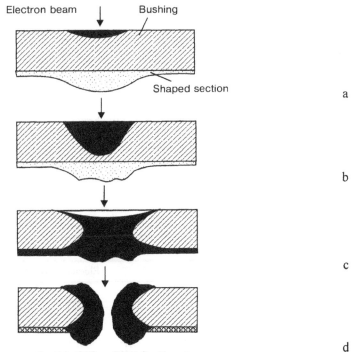

Figure 7 Schematic of joint formation.

Table 1 Results of X-ray microprobe analysis of structural components in the range of brazing temperatures, close to the braze alloy melting temperature.

Structural component	Element content, wt.%					
	Fe	Ni	Cr	Al	Ti	Sn
Boundary of steel 30	89.6	8.9	0.20	0.80	–	0.5
Dark-coloured layer	25.5	72.3	0.20	1.40	–	0.6
Light-coloured crystallites	3.1	36.2	0.10	1.30	–	59.3
Light-coloured weld sections	0.3	0.3	0.03	1.26	–	98.0
Dark-coloured weld sections	1.0	3.3	0.34	1.49	–	93.8
Boundary of 36NKhTYu alloy	46.3	37.5	12.60	1.20	2.4	0.1

The performed testing showed that the combination of the brazed joint with the welded joint greatly improves the structural strength (up to 10000 N). In this connection during truss assembly in space, it is recommended that its elements be joined by using a focused electron beam. Furthermore, the moment of melting through the assembly can be a sign of process completion for the operator, which is important as the bright illumination makes visual control difficult.

Chemical and structural inhomogeneity of the brazed joint along its length was studied. Investigation and analysis of the characteristic weld zones have been performed for this purpose.

Information is limited on the constitutional diagram of Sn–Ni–Cr–Fe–Cr system, as well as on ternary Sn–Ni–Fe and Sn–Ni–Cr constitutional diagrams. Therefore analysis of the joint structure and phase composition was conducted only on the basis of the derived experimental data, using a raster scanning electron microscope SET-15 and microscope-microanalyser MS-46.

In the region of low brazing temperatures, close to the braze alloy melting temperature, the weld, according to the data of X-ray microprobe analysis, is a Sn-based solid solution (Table 1). The revealed non-uniformity of the solid solution etching is related to slight enrichment of the dark-coloured regions with iron, nickel and chromium. In the transition zone on the boundary with the bushing a residual layer of nickel coating is found with Ni_3Sn_2 crystallites adjacent to it, judging from the chemical composition. The technological nickel coating in the shaped section is dissolved in the braze alloy.

With increase of brazing temperature, the nickel coating of the bushing also dissolves in the braze alloy (Table 2). Inclusions of the Fe–Ni–Sn based phase, containing 12–13 wt.% Fe; 24–32 wt.% Ni and 54–62 wt.% Sn are found in the weld. The volume fraction of the inclusions increases closer to the zone of the electron beam impact, and this phase ultimately fills the weld.

Table 2 Results of X-ray microprobe analysis of structural components of a joint in the low brazing temperature region.

Structural component	Element content, wt.%					
	Fe	Ni	Cr	Al	Ti	Sn
36NKhTYu alloy	48.1	36.2	12.6	0.90	2.1	0.1
Inclusions in the weld	12.0	32.0	1.4	0.01	–	54.6
Steel 30 interface	88.2	7.1	0.4	0.01	–	0.3

a

b

c

Figure 8 Microstructure of a brazed joint in the region of high brazing temperatures:
a — T_{br1} in section *1*; b — T_{br2} in section *2*; c — T_{rb3} in section *3* (×810).

In the region of high brazing temperatures, the joint microstructure changes significantly. Comparative metallographic analysis has been performed of three sections located at three different distances from the zone of the electron beam impact, and having different brazing parameters, respectively, with $T_{br1} < T_{br2} < T_{br3}$ (Figure 9). The joint microstructure is similar in all three sections. The difference of the thermal modes of brazing is manifested in the intensity of the braze alloy interaction with 36NKhTYu alloy. While intergranular penetration of the braze alloy in section *1* leads to formation of diffusion wedges (Figure 8a), a narrow band of austenite grains, separated from the matrix, is found in sections *2* and *3* (Figure 8b and c), being wider in section *3*.

The Fe–Ni–Cr based primary phase solidifies along the boundary with the bushing in the form of the continuous layer and dendrites growing from the layer. Isolated primary phase crystallites are observed inside the weld. The primary phase has a similar composition in all the sections. Iron, nickel and chromium contents are 50, 28–30 and 7–10 wt.%, respectively.

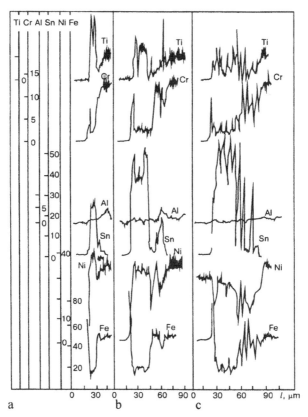

Figure 9 Element distribution, wt.%, in the brazed joint in the region of high brazing temperatures: a — T_{br1}; b — T_{br2}; c — T_{br3}.

The weld middle part is formed by a lower melting Sn–Ni–Fe based structural component, containing fine dark-coloured inclusions.

According to the element distribution in the weld (Figure 9) and data of X-ray microprobe analysis (Table 3), tin content in the weld middle part grows from section *1* to section *3*, while nickel content becomes smaller.

The curve of tin distribution in section *2* shows an abrupt change of concentration, namely tin content in the central part corresponds to its content in section *1*, whereas in the adjacent sections it corresponds to its content in section *3*. Tin distribution correlates with iron distribution at a constant content of nickel, in section *3*. This implies that, depending

Table 3 Results of X-ray microprobe analysis of Sn-containing structural component of a joint in the region of high brazing temperatures.

Object	Element content, wt.%					
	Sn	Ni	Fe	Cr	Al	Ti
Section *1*	32–35	39–38	16	2	0.25	6.5–7
Section *2*, centre	36–37	31–32	19	3	0.2	3–4
Section *2*, periphery	48–51	31–32	16	3	3	3–4
Section *3*, periphery	48–52	28–29	16–17	2–3	2.5–3	2–3

on brazing parameters, the Sn-containing structural component can consist of one or two phases, supposedly based on Sn–Fe compounds.

A typical feature of base metal element distribution in the weld is titanium localized in the Sn-containing structural component, and chromium localized in the primary crystallites. In view of that, as well as the fact that concentration minima on the curve of tin distribution in the weld in section *3* coincide with the concentration minima on titanium distribution curves and maxima on iron and chromium distribution curves, the fine dark-coloured inclusions, observed in the weld middle part, can be assumed to be particles of Fe–Cr–Ni based phase.

Increase of tin concentration and decrease of the concentration of the base metal elements (titanium, nickel) in the Sn-containing structural component with increase of brazing temperature are caused by the brazing gap becoming larger under the impact of mass transfer in the high-temperature area closer to the electron beam impact zone. In addition to the indirect influence, mass transfer can also have a direct impact on the composition of the melt in the gap.

Therefore, while in the low-temperature region the braze-weld structure forms under the conditions of prevailing solution-diffusion processes, intensifying with temperature increase, in the high-temperature region, in keeping with the brazing laws, it forms during melt solidification in cooling.

The brazed joint microstructure in each section is determined by brazing parameters and gap size. Under non-isothermal conditions of heating by the electron beam, also the influence of the possible convective flows should be taken into account.

An experiment on brazing truss assemblies, made with VHT by cosmonauts L. Kizim and V. Soloviov in May 1986 in OS «Salyut-7», confirmed the good prospects for brazing to be used to perform external work in space. Unfortunately, the information on its results was then restricted because of the classified nature of the work, and therefore the possibility of brazing application for assembly in space can only be mentioned now. Analysis of space samples showed that during heating in orbit a deviation occurred from point energy input, namely the contour of surface-melted metal was elongated along the bushing surface. The brazed-welded joint was not produced, while a joint in the brazing mode formed over a large area.

As was anticipated, brazed joint formation controlled by surface tension forces improved at microgravity. It should be noted, that the tinning layer melt on the bushing does not flow out of the gap; the braze alloy fills the gaps, non-capillary under ground conditions; and a fillet is formed on the boundary of the bushing and the shaped section in the area, adjacent to the zone of electron beam impact.

It is interesting that braze alloy mass transfer into the zone of the electron beam impact is enhanced. This confirms the fact that directed melt flows initiate under the influence of the surface tension gradient on the interfaces.

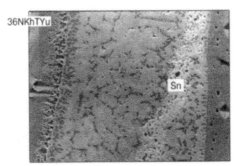

Figure 10 Microstructure of a brazed joint produced in space in a section with a greater width of the gap (×422).

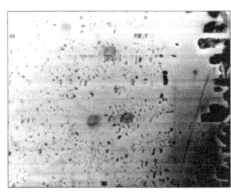

Figure 11 Eutectic colonies in the middle part of a weld in joints produced in space (×1490).

The features of brazed joint microstructure formation found on Earth are preserved under the conditions of space.

When space samples were analyzed, data were derived additionally on the weld microstructure in the region of high brazing temperatures with wide gaps (0.2–0.5 mm). In this case, the braze alloy melt is characterized by a lower concentration of elements of the materials being brazed, and a band of a Sn-based phase forms in the weld middle part in solidification (Figure 10).

Characteristically, very pronounced eutectic colonies form in the middle part of the weld in the high-temperature region, one of the components of which are dark-coloured crystal-lites of Fe–Ni–Cr based phase (Figure 11). In the microstructure of the joints made on the ground, this phase is chaotically distributed in the matrix, and the issue of eutectic transformation during braze alloy melt solidification remained open.

In the high-temperature region, sections having developed columnar dendrites of Fe–Ni–Cr based primary phase oriented across the weld are found with a predominant frontal dissolution of 36NKhTYu alloy in addition to the microstructure similar to that produced on the ground (Figure 12).

It is known, that brazing with copper braze alloys of steels containing different contents of carbon, is accompanied by dissolution of the low-carbon component and growth of the columnar phase of Fe–Cu–C system from the high-carbon steel boundary [8]. The amount of the phase becomes greater with time and the steel boundaries are ultimately joined. The joint strength reaches the strength of steel.

The mechanism of the process of dissolution–deposition of the metal being brazed is so far insufficiently studied. Interaction of the solid and liquid phases in brazing of 36NKhTYu

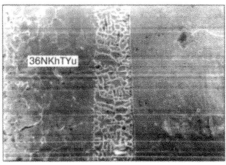

Figure 12 Microstructure of a brazed joint produced in space in a section with well-developed columnar dendrites (×406).

alloy containing no carbon to steel 30 using tin braze alloy possibly proceeds in a similar manner.

Formation of the observed microstructure in this case is due to the thermal cycle, which led to development of the process of dissolution–deposition of the metal being brazed, i.e. long-term soaking at a low brazing temperature (probably, as a result of electron beam scanning).

It is seen that optimizing the thermal cycle of brazing can produce a structural strengthening of the weld. Application of a dosed heat input in the optimal mode, will allow high-strength joints to be made.

It should be noted, that one of the goals in staging the space experiment was to clarify the possibility of shaped section wetting by the braze alloy without the nickel coating after magneto-abrasive cleaning. The experimental results demonstrated that no dissociation of the oxides proceeds on the surface of 36NKhTYu alloy in the low-temperature region, namely the joint forms only in the region of high brazing temperatures.

Conclusions

1. Use of a low-melting braze alloy in electron beam brazing of thin-walled structures allows joints to be produced in space in a broad range of heating modes, which is important when the operators perform manual operations in extreme conditions. A brazed, or a brazed-welded joint, forms.

2. A brazed-welded joint improves the structural strength of the connection, allows visual monitoring of the process and is suitable for construction of thin-walled structures in space.

3. Conditions of microgravity improve brazed joint formation which is controlled by the surface tension forces. No influence of the space environment on the weld microstructure was found.

4. In the future it will be necessary to carry on with the investigation of surface phenomena at microgravity, taking into account the earlier derived data [9–11], as well as to study the influence of such external factors, as non-isothermal conditions, gravity level, electric and magnetic fields, etc. on the nature of the convective flows in the brazing gap.

1. Nikitsky, V.P., Lapchinsky, V.F., Zagrebelny, A.A. et al. (1985) S*ee pp. 178–184 in this Book.* Testing of electron beam hand tool in space.

2. Zagrebelny, A.A., Gavrish, S.S., Bulatsev, A.R. et al. (1986) *See pp. 258–264 in this Book.*

3. Khorunov, V.F., Shvets, V.I., Lapchinsky, V.F. et al. *Solder for nickel soldering.* USSR author's cert. 1606295, Int. Cl. B K 35/26, publ. 15.11.90.

4. Bashenko, V.V. and Vanshtejn, V.I. (1970) Analysis of forces, applied to the weld pool, in electron beam welding, *Svarochn. Proizvodstvo*, No. 8, pp. 1–2.

5. (1989) *Space materials science. Introduction to the scientific fundamentals of space technology.* Ed. by B. Foerbaher, G. Hamaher and R.J. Nauman, Moscow, Mir, 478 pp.

6. Deryagin, B.V., Shutor, Yu., Nerpin, S.V. et al. (1965) Investigation of thermo-osmotic effect in glass capillaries, *Doklady AN SSSR*, **161**, No. 1, pp. 147–150.

7. Chernyakova, L.E., Shugaenko, V.K. and Vorontsov, N.I. (1973) Electron microscopy examination of the process of excess phase precipitation in the wrought 36NKhTYu alloy, *Metalloved. i Term. Obrab. Metallov*, No. 8, pp. 16–19.

8. Yoshida, T. and Ohmura, H. (1980) Dissolution and deposit of base metal in dissimilar carbon steel brazing, *Welding J.*, **59**, No. 10, pp. 278–282.

9. Siewert, T.A., Heine, R.W., Adams, C.M. et al. (1977) The Skylab brazing experiment, *Ibid.*, **56**, No. 10, pp. 291–300.

10. Sasakibe, Yu. (1982) Brazing in space, *Yosetsu Gijutsu*, **30**, No. 4, pp. 15–18.

11. Favicz, J.J. (1984) Results of Spacelab-1, In: *Proc. of 5th Eur. Symp. on Mater. of Sci. Microgravity*, Schloss Elmau, 1984. Paris, pp. 435–436.

Some Problems of Welding Sheet Metal in Space*

V.F. Lapchinsky[1]

One of the main requirements for the space vehicle structures is minimum weight. As a result the range of materials applied in space vehicle construction is mostly limited to aluminium, titanium, high temperature and refractory alloys, the thickness of which is, as a rule, minimal. Usually, metal 0.1 to 3.0 mm thick has to be processed.

Assembly and preparation for welding of the parts to be joined (especially, when repair work is performed) is very complicated. Therefore, the technology of welding in space should provide the ability to join sheet metal with greater and in a number of cases highly non-uniform gaps. Welding, as a rule, will have to be performed without heat removal or clamping fixtures with greater deformation of the parts being joined. The latter circumstance in combination with the high cost of power for welding in space vehicles makes it necessary to use heat sources with maximum thermal efficiency and high power density, namely electron beam, microplasma, etc. [1, 2].

It is obvious that the above features are contradictory, as welding of sheet metal should be performed without backing with large and non-uniform gaps, using highly concentrated heat sources. Experience of work performance on the ground demonstrates that such a combination of contradictory requirements will inevitably lead to a great number of burns-through in the metal being welded, defects which are difficult to remove and, therefore, inadmissible during work performance in space.

Thus, the problem of burn-through elimination is one of the main problems in development of the technology of welding in space. The causes leading to burn-through, specifically in sheet metal welding, have not yet been thoroughly established. The following definition is considered to be universally accepted [3]: «Burns-through are cavities in the weld formed as a result of the weld pool flowing out... Welding sheet metal sometimes leads to formation of specific defects, looking like burns-through, but not associated with the weld pool flowing out. The nature of their formation is not clear so far».

Development of a reliable technology of welding sheet metal and looking for effective measures of burn-through elimination required a more thorough study of the causes for their development. A simple set-up was developed for investigations, that is based on a special cinematographic recorder (Figure 1). The experimental procedure consisted in filming the process of penetration and subsequent burning through the metal sheet by a unit rectangular heat pulse. A special microplasma torch was used as the heat source. Variation of heating power was simultaneously recorded by an oscillograph. The capability of adjustment of power density of the penetrating thermal pulse was provided. Burn-through of a continuous sheet was studied first, to eliminate many factors associated with the quality of the welded joint fit up. Experiments were conducted in a flying laboratory that allowed working in gravity fields with accelerations of $2.5–10^{-2}$ g_0 ($g_0 = 9.8$ m/s^2).

When the filmograms were interpreted, it was noted that complete penetration of the sheet is in place after a certain period of time after feeding the heating pulse (complete penetration time is t_{pen}. Then, for a certain time interval (pool existence time, t_p) the molten pool grows larger, which is followed by a burn-through — a phenomenon, accompanied by

* (1984) *Space research in Ukraine*, Issue 18, pp. 9–15.

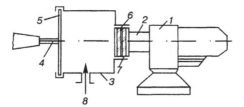

Figure 1 Schematic of a set up to study the process of sheet metal burn-through: *1* — cinematographic recorder; *2* — lens; *3* — inert gas chamber; *4* — heat source; *5* — sample burned through; *6* — protective glass; *7* — monochromatic light filter; *8* — shielding gas feed.

breaking up of the pool surface and formation of a through-thickness hole in the sheet. The time of burning through t_{br} the sheet is equal to the sum of t_{pen} and t_p.

Indeed, formation of a through-thickness hole is not accompanied by metal flowing out of the weld pool. The metal formed during the pool surface breaking up, deposits along the hole edges in the form of a torus. When the sheet thickness and the metal thermophysical properties are changed, t_{pen}, t_p and t_{br} values and ratios also change, all the three time intervals becoming smaller with smaller thickness of the sheets.

Dependence of t_{pen}, t_p and t_{br} on power density in the heated spot is of specific interest (an example of such a dependence for a titanium sheet, melted through with microplasma, is given in Figure 2). At high power density the time of the pool existence becomes extremely small (up to 10^{-3} s). At low concentration at zero gravity the following phenomenon is observed: when a molten pool is formed, its diameter is gradually increased, while sheet burn-through does not occur (in the studied range of pulse duration of up to 6 s).

Check experiments at different levels of acceleration (in the range of 10^{-2}–$2g_0$) demonstrated that t_p essentially depends on overload only at a large volume or great thickness of the molten pool (for instance, in vertical welding). With a small volume of the pool the overload influence is negligible.

Filmogram analysis in a microphotometer demonstrated that liquid metal temperature distribution around the pool perimeter has a pronounced maximum located, as a rule, near

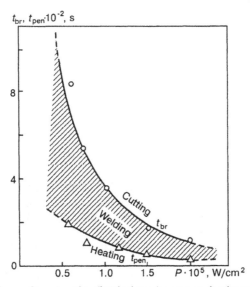

Figure 2 Dependence of t_{pen}, t_p and t_{br} (hatched zone) on power density on in the heated spot (heat source is a constricted microplasma arc in vacuum, titanium is 0.35 mm thick).

the pool centre [4]. On the other hand intensive drifting of the most heated zone over the pool surface was often found, related probably, to the arc erring or metal stirring by the convective flows and reactive pressure of the vapours. In all the cases, however, a well localized zone was found in the filmograms, overheated to a much higher temperature, than the metal melting temperature. The presence of such a zone is attributable to radial temperature distribution by the Gauss law [5], inherent in concentrated welding heat sources and specific heat removal conditions in welding sheet metal without a backing strip.

The process of burning through the sheet can be explained, proceeding from the existence of such a zone. As surface tension of metal melts, σ, decreases with temperature rise (temperature coefficient $d\sigma/dT$ is negative) [6–8], the magnitude of surface tension force F_{σ_1}, applied around the overheated zone perimeter, is much smaller than the magnitude of force F_{σ_0}, applied around the perimeter of the molten pool. This results in a radial temperature gradient of the surface tension force dF_σ/dT, creating force vector $\overline{\Delta F_\sigma} = F_{\sigma_0} - F_{\sigma_1}$, directed along the radius from the centre to the periphery and tending to break up the molten metal surface in the most overheated zone. As this is the exactly the zone where the magnitude of the metal surface tension responsible for its surface continuity is minimum, this is the very region where the most favourable conditions for breaking up are in place.

An approximate estimate of the maximum admissible metal overheating when breaking up of the pool surface does not yet take place, leads to the following conclusions.

If D_0 denotes the molten zone diameter, and D_1 is the conditional diameter of the localized high-temperature zone, forces F_{σ_0} and F_{σ_1} will be equal to

$$F_{\sigma_0} = \sigma_0 \pi D_0, \tag{1}$$

$$F_{\sigma_1} = \left(\sigma_0 - \frac{d\sigma}{dT} \Delta T \right) x D_1, \tag{2}$$

as

$$\sigma_t = \sigma_0 - \frac{\Delta \sigma}{dT} \Delta T, $$

where ΔT is the metal overheating above the melting temperature.

The condition of the pool surface continuity in the most overheated zone, can be written as

$$F_{\sigma_1} > \Delta F_\sigma = F_{\sigma_0} - F_{\sigma_1} \tag{3}$$

or from (1), (2):

$$\frac{2D_1}{D_0} \left(\sigma_0 - \frac{d\sigma}{dT} \Delta T \right) > \sigma_0. \tag{4}$$

Analyzing expression (4), we can conclude that the continuity of the pool surface depends on value ΔT (overheating) and D_1/D_0 ratio. At D_1/D_0 values of about 0.6–0.8 a high (two- or even three-fold) overheating of molten metal above its melting temperature is admissible, which is practically never observed. On the contrary, if D_1/D_0 is small, for instance 0.2–0.3, overheating by 30–60 % can cause breaking up of the pool surface, i.e. a burn-through.

It is assumed that the main cause for burns-through in thin metal welding is formation of a local overheated zone in the molten metal, its dimensions being much smaller than the molten pool diameter. The probability of such a zone formation becomes higher with increase of power density in the heated spot. Direct evidence of formation of a small-diameter local overheated zone in the molten pool prior to the burn-through has also been obtained.

The research performed allowed development of effective measures of eliminating thin metal burns-through, which should, essentially be aimed at prevention of local overheated zones formation in the weld pool and shortening the total time of the pool existence. Experiments showed that a very effective measure for eliminating burns-through is adding filler metal to the weld pool, preventing metal overheating. It should however be noted, that in welding on the ground, this measure is less effective than during work performance at zero gravity. This is associated with increase of molten metal mass and of hydrostatic pressure in the pool, this being especially noticeable in vertical welding. At zero gravity filler material feeding practically completely eliminates the risk of burn-through.

Therefore in melting through a sheet with a consumable electrode arc on the ground, burn-through occurred in 4 to 5 s, despite absence of a dangerous overheating zone. The molten metal flowed out of the pool under gravity impact. In melting through a sheet in a similar mode at zero gravity, the burn-through did not occur even after 22–23 s. This time interval was limited by the duration of zero gravity condition in a flying laboratory.

A still more effective measure of burn-through elimination turned out to be dosing of the thermal energy applied to the metal being processed. In this case, heating, welding and cutting of metal (i.e. all the modes, shown in Figure 2), can be performed with a heat source with the same power density that is modulated with a certain frequency and repetition rate.

With the duration of heating pulses shown on t_{pen} curve in Figure 2, the mode of dosed heating of the metal is implemented without its melting. Pause duration is selected, depending on the specified thermal cycle, but so as to provide a significant cooling of the metal between the pulses. If pulse duration falls between curves t_{pen} and t_{br} (see Figure 2), a periodic melting of the processed material is observed. With a duration of pauses resulting in complete solidification of metal, metal can be melted through without shifting the heat source for a practically unlimited time. If the duration of the heating pulses is above curve t_{br}, the sheet is burnt-through already under the impact of an isolated pulse.

The above observations were made in heating and melting through a continuous sheet, while in welding a gap between the sheets being joined is always present in the spot of the molten pool formation. The following can be noted in this case.

A sheet metal butt joint behaves as a continuous sheet at a gap size of not more than 0.2δ (sheet thickness) and the heated spot located symmetrically relative to the butt. Increase of the gap size and, to an even greater extent, shifting of the heated spot to one of the edges, leads to an abrupt drop of t_{pen} and t_{br}. For each metal thickness and gap size, there exists a certain maximum admissible value of displacement, when the time of the pool existence is still sufficient for a weld nugget to form. Exceeding this displacement results in burning through one of the abutted edges, to which the heated spot is shifted. Burning through time is so small that it is lower, than the accuracy of measurement during the experiment. This is related to non-uniform heat removal because of the heated spot shifting relative to the butt.

Such a strong dependence of the abutted edge heat mode on the heated spot location requires the development of special measures providing the maximum possible accuracy of following the butt. It should be specified, that when filler material is fed, as was already noted, the required accuracy of finding the butt can be much lower at zero gravity. An overlap joint is even more sensitive to the size of the gap between the sheets being joined. In this case, a reliable contact should be provided between the sheets practically without a gap, as otherwise burn-through of the upper sheet occurs before that of the lower sheet.

All the above applies to stationary heat sources and covers spot welding or, more precisely, electric riveting. In the case of a tight joint or joint of equivalent strength, a continuous weld is performed by a uniformly moving heat source.

No.	Power density $\cdot 10^8$, W/m^2	Minimum welding speed $\cdot 10^{-3}$, m/s	Maximum welding speed $\cdot 10^{-3}$, m/s	Range of welding speeds, %	Maximum admissible tool dwelling time $\cdot 10^{-3}$, s
1	5	3	12	75	64
2	10	7	25	72	26
3	15	12	41	70	13
4	20	20	57	67	5

In this case, it is possible to experimentally determine the minimum possible welding speeds below which cutting starts, and the maximum possible speeds, above which complete penetration of the joined sheets cannot be achieved. The requirements to the stability of the heat sources displacement were determined from the ratio of these speeds.

The time of the weld pool existence was used to evaluate the maximum admissible time of tool dwelling and not leading to burn-through. Calculations were performed for different metals and thicknesses under conditions of minimum gap in the butt. An example of such calculation for titanium sheets $\delta = 0.35$ mm thick, experimentally confirmed at zero gravity, is given in the Table. It is seen from the Table, that the admissible limits of variation of the heat source displacement speed in welding sheet metal are large enough (about 70 %) and are almost independent of power density. Maximum admissible time of the tool staying in one point, by contrast, is drastically reduced with increase of energy concentration, thus increasing the risk of burns-through at random dwelling of the tool.

Summing up it should be noted, that sheet metal welding by concentrated heat sources in space is feasible. To ensure the reliability of this process, the following set of measures should be envisaged: steady displacement of the tool along the weld, without dwelling in a point; use of dosed thermal energy pulses, where the frequency and repetition rate are adjusted depending on the required task; use of filler material in a certain form; assembly of the parts being joined with minimum gaps as far as possible.

1. Paton, B.E. and Kubasov, V.N. (1970) *See pp. 154–160 in this Book.*

2. (1980) *Space technology.* Ed. by L. Steg, Moscow, Mir, 160 pp.

3. (1974) *Technology of electric welding of metals and alloys.* Ed. by B.E. Paton, Moscow, Mashinostroyeniye, 275 pp.

4. Kustanovich, I.M. (1972) *Spectral analysis*, Moscow, Vysshaya Shkola, pp. 169–184.

5. (1979) *Microplasma welding.* Ed. by B.E. Paton, Kiev, Naukova Dumka, 248 pp.

Analysis of the Results of Space Experiments, Conducted Using VHT*

B.E. Paton[1], V.F. Lapchinsky[1], A.A. Zagrebelny[1], S.S. Gavrish[1], G.M. Grigorenko[1], Yu.B. Malevsky[1], V.F. Grabin[1] and I.Ya. Dzykovich[1]

When evaluating the results of experiments conducted in space [1] using the VHT, attention was focused on hardware performance, operator's actions and properties of the made joints and cuts. Such data will further allow reliable forecasting of the ability to perform work of various degrees of complexity using VHT.

Experiments showed that the wide scope of ground-based retrofitting of hardware conducted and the changes in the design made as a result, allowed the development of a sufficiently reliable and convenient tool which raised no criticism during operation, either from the operators or the technical specialists [2]. As was anticipated the operators' orientation relative to the samples being processed and of the work place relative to solar illumination, are highly important for sound performance of the technological processes. Selection of the optimal orientation conducted during the ground-based tests and operator training, turned out on the whole to be correct. It became clear that it is much easier for the operators to work in space as the zones of reach and optimal performance become larger, due to the absence of gravity and weight neutralizing devices. The issue of selection of operator orientation relative to natural illumination turned out to be rather complicated. It was established, that the best illumination is uniform solar illumination without any sharp light-shadow boundaries. The tentative angle of the position of the Sun relative to the plane of the surface being processed is 45 to 60°. Under such conditions when the standard light filter of the space suit was used, the processes of cutting, welding and light-coloured coating application on dark surfaces did not present any difficulties for the operators.

As regards the brazing process, its performance was made much more complicated compared to the trial conditions because of the difference in illumination levels. During ground-based retrofitting the operators were instructed to preheat the sample up to a slight reddening to ensure sound performance of the brazing process. At a relatively low level of illumination this criterion was quite sufficient. In space, however, it is extremely difficult to see the reddening of the samples at a high level of illumination. As a result, the braze alloy turned out to be unmolten in the majority of the samples. In the future, higher melting braze alloys should be used which allow mixing with partially molten base material, when going over to braze-welding.

When making rectilinear cuts, the operators noticed that it is more difficult to do it in space with the electron beam than on the ground, as at zero gravity the surface tension forces in the molten metal lead to the cut self-closing. Nonetheless, both operators were able to make continuous cuts by lowering the speed and periodically going back to uncut sections. Coating application did not present any difficulties for the operators, while their quality was essentially higher, than in the ground analogues. Values of radiation thermal characteristics A_s and ε (coefficient of solar radiation absorption and total emissivity factor) were 0.12–0.14 and 0.05–0.60 in the ground-based samples, respectively, and 0.07 and 0.03 in the space samples, this being higher than the requirements of the industrial standard. The average speed of silver deposition is about 8 nm/s at coating thickness of 0.7–0.8 μm. In the ground-based sample, the coating thickness was about 0.5 μm and the average deposition rate was about 5.5 nm/s with the same exposure time. The consistency of the coating thickness (scatter is

---

* (1986) *Problems of space technology of metals*, Kiev, PWI, pp. 5–13.

not more than 0.1 μm) shows that the operators are able to make a good visual estimate of the uniformity of the applied coating and successfully homogenize it by moving the tool.

Samples were studied to determine the tracking accuracy, when operations on metals cutting and welding were preformed. The accuracy of operations performance for each sample was estimated by the following formula:

$$S_n = \left(\frac{1}{n-1} \sum_{i=1}^{n} h_i^2 \right)^{\frac{1}{2}},$$

where S_n is the mean root square displacement of the beam; n is the number of measurements on samples; h is the beam displacement in each measured point, equal to $h = (B/2 - A)$, where B is the width of the cut (weld) and A is the distance from one of the edges of the cut (weld) to its axis.

Samples from operators with varying degrees of training were evaluated and operators' performance on the ground and in space were compared. The data obtained led to the following conclusions:

• design of VHT work tool allows the processing operations to be performed with a sufficient degree of precision;

• an operator required not less than 2 to 3 training sessions in space simulation test chambers to be capable of high-quality performance of processing operations;

• skills acquired during training allow high-quality performance of work in space. Data on tracking accuracy do not practically differ from the ground-based data (about 0.5 mm in both the cases).

The set task did not include tightness testing of the welds. Considering, however, their quite satisfactory quality, it is rational to assess their tightness. As the welded joint samples were made of non-magnetic materials, liquid-penetrant testing turned out to be the only acceptable method for tightness check. Defect detection in welded joints made in space was conducted for 18 h. No defect traces were detected. This is indicative of a high tightness of the joints, despite partial lack-of-penetration found in several places.

Investigation of the structure, phase composition and gas saturation of metal, remelted in space, were of great interest. However, in the case of welded joints that as a rule form at high rates of liquid metal solidification, significant differences in the weld structure can hardly be anticipated, as was confirmed by the first experiments on automatic welding in space in the «Vulkan» unit. Nevertheless, each new welded sample made in space is highly interesting to study, due to its unique nature. In appearance and penetration shape welded joints, made on commercial titanium in space are actually identical to their ground analogues. Weld microstructure has several differences, which consist of the following:

• the prevailing form of α'-phase are thin-needle precipitates easily revealed by etching (Figure 1a) growing inside from serrated grain boundaries;

• precipitates inside the grains are intercrossing α'-phase plates and coarse-plate formations with inner twining of α'-phase plates (Figure 1b).

For comparison Figure 1c and d show the microstructure of reference welds on commercial titanium, welded using VHT in a manned space simulation test chamber on the ground.

The above morphological features of the weld metal structure are close to that of the cast metal cooled at very high rates under conditions of a highly stressed state. The microhardness of weld metal made in space was H_μ 248–262. The microhardness of weld metal made on the ground was H_μ 248–258. In electron-microscopic examination of weld metal on commercial titanium, a cellular substructure of non-uniform size and distribution

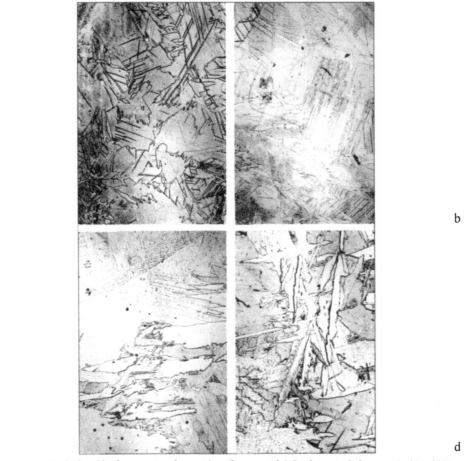

a

b

c

d

Figure 1 Typical weld microstructure in samples of commercial titanium, made in space (a, b) (×200) and on the ground (c) (×200), (d) (×400).

is observed. Cell dimensions are 0.5 to 7.0 μm. Cell walls are wider than in the ground analogue. No cells were found in some areas. In this case the following structural features are observed on the interface of the type of «cell-free area–fine cells»:

• needles or bands separating the two areas. Dislocation density in the cell-free area is quite high (Figure 2a);

• long plates (Figure 2b). Microdiffraction investigations show them to be an fcc-phase, similar to the base metal.

For comparison Figure 2c and d give the microstructure of a weld on reference samples of commercial titanium obtained by electron microscope.

Small amounts of particles of two types were found in addition to the above features. The first type is similar to particles observed in the base metal with fcc-lattice and $a \cong$ $\cong 0.24$ nm. The second type has bcc-lattice and $a \cong 0.32$ nm. Welded joints of stainless steel of 12Kh18N10T (304SS) type were also studied. The metal structure of welds on this steel made on the ground (Figure 3a) is somewhat coarser, while the amount of α'-phase in it is greater than in the weld made in space (Figure 3b). The intensity of austenite decomposition in the welded joint HAZ is higher than in welding in space (Figure 4). The amount of ferrite, determined by the point method, is 1.8 to 2.0 % in welding in space, and 2.1–2.2 % in the

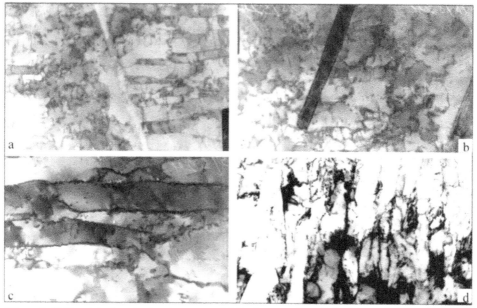

Figure 2 Weld metal microstructure in commercial titanium samples, obtained in an electron microscope for space (a, b) (×10000) and ground (c) (×10000), (d) (×6000) conditions.

ground analogues. The microhardness of the ground and space samples is approximately the same at H_μ 268–278. Electron microscopy examination indicates that the metal structure of 12Kh18N10T type steel welds made in space is mainly similar to the metal structure in welds made on the ground. Differences are found in the α-phase dispersity and number of dislocations. The α-phase in the space samples is more disperse, while their density is higher (Figure 5). From the above results of metallographic and electron-microscopy analysis it is obvious, that in both cases the structure of welds (commercial titanium and stainless steel) made in space, features higher dispersity and dislocation density. This is apparently attributable to the high rate of cooling (solidification) of the weld metal and thus induced thermal stresses. Chemical microinhomogeneity was studied in a «Cameca» microanalyzer with an

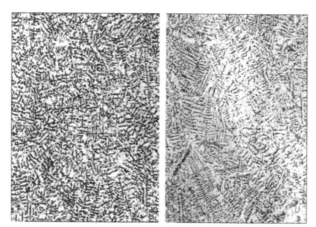

Figure 3 Typical microstructure of welds on samples of 12Kh18N10T type steel, welded on the ground (a) and in space (b) (×200).

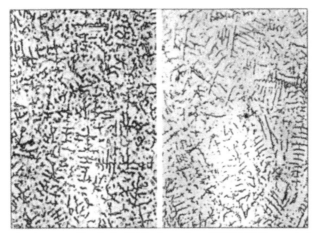

a

b

Figure 4 Austenite decomposition in the HAZ of samples of 12Kh1810T type steel, welded on the ground (a) and in space (b) (×400).

electron probe of MS-46 model. Pure elements, namely nickel, manganese, chromium, titanium, silicon, were used as standard samples.

The structure of the metal of a weld on 12Kh18N10T type steel in space samples and their ground analogs is cast (dendritic) with α-phase precipitates along the primary boundaries. Investigations showed that the structure of welds, made in space, has finer boundaries with a smaller amount of α-phase. This, as was noted above, is probably, associated with the higher cooling rate. Element content (in percent), in the weld metal structural components, averaged from 8 to 10 measurements in an area of 1 μm diameter, is given in the Table for both variants.

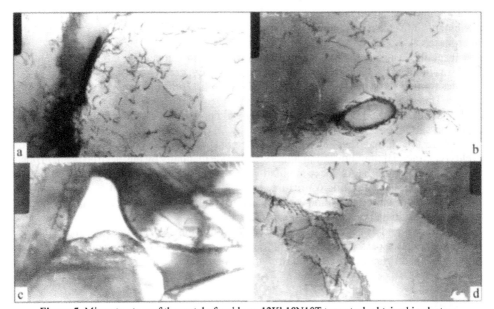

Figure 5 Microstructure of the metal of welds on 12Kh18N10T type steel, obtained in electron microscope for space (a, b) and ground (c, d) conditions (×20000).

Element content (wt.%) in structural components of welds on 12Kh18N10T type steel, made in space and on the ground.

Conditions of making	Ni	Mn	Cr	Ti	Si
Earth	11.0 / 7.0	0.1 / 0.1	16.8 / 22.5	0.31 / 0.45	0.25 / 0.35
Space	12.0 / 8.5	–	17.8 / 23.7	0.36 / 0.50	0.30 / 0.45

Note. Numerator is γ-phase (grain), denominator is α-phase (boundary).

Therefore, stainless steel welded joints produced on the ground and in space are identical in terms of the extent of chemical microinhomogeneity. Also no significant differences have been found in the element content distribution in titanium welded joints.

Attention was also given to the investigation of gas saturation of welds, especially in commercial titanium samples.

The method of hot extraction in a neutral gas flow was used to determine oxygen and hydrogen content in samples of VT1-00 commercial titanium. LECO instruments PO-16 and PH-2 were used. In EBW on the ground a certain increase of oxygen content in the welds and 1.5 to 2 times lowering of hydrogen content are found compared to the base metal.

The increase of oxygen content is also found in samples welded in open space (oil-free vacuum with pumping rate close to the infinite rate). Oxygen weight fraction was 0.058–0.061 % in the initial metal and 0.071–0.084 % in the weld.

It should be noted that oxygen concentration in the weld is rather low, being in the range specified by the GOST standard for VT1-00 grade titanium.

The hydrogen weight fraction is 0.0056–0.0072 % in the initial metal and 0.0010–0.0015 % in the weld. In other words, a significant cleaning of the metal from hydrogen occurred that resulted in the content of the latter dropping by approximately 5 times. This is attributable to intensive removal of hydrogen, evolving from the metal away from the interphase.

It should be also noted that isolated pores were found at the weld start, where the liquid pool was induced by a stationary heat source. Under zero gravity conditions removal of gas bubbles formed in the molten metal is probably difficult, as the buoyancy force is practically absent. This should be taken into account for melting of a relatively large mass of the metal. If gas bubble evolution occurs, the solidifying weld pool metal can become porous. Special measures should be taken to avoid it.

The results of materials science studies demonstrate that the properties of the metal of welds (commercial titanium and stainless steel) made manually by the electron beam in open space and in ground-based space simulation test chambers, are quite close. The observed minor differences in the microstructure, in the distribution of impurities and secondary phases and in the gas composition can be attributed to differences in the external conditions, mainly the residual atmosphere and heat removal.

The above-said leads to the conclusion, that application of VHT type tools in open space guarantees reliable performance of operations of processing preheated and molten metal and making welded structures with good strength properties.

1. Nikitsky, V.P., Lapchinsky, V.F., Zagrebelny, A.A. et al. (1985) *See pp. 178–184 in this Book.*

2. Paton, B.E., Dzhanibekov, V.A. and Savitskaya, S.E. (1986) *See pp. 11–18 in this Book.*

Manual Electron Beam Welding of Sheet Metal in Open Space*

B.E. Paton[1], V.F. Lapchinsky[1], E.S. Mikhajlovskaya[1], V.F. Shulym[1], S.S. Gavrish[1], G.N. Gordan[1] and V.P. Nikitsky[2]

The concept of practical mastery of space in the interests of mankind is manifested in the development and implementation of international projects for the creation of large-sized long-term orbital stations of a new generation «Freedom», «Alpha», «Mir-2». The service life of these space vehicles should be not less than 15 to 20 years, and their operation will naturally require preventive maintenance and repair-restoration work. Assembly of orbital complexes which should be performed directly in orbit from individual elements or modules, raises the issue of application of permanent joints of various structural materials. In this connection the interest of experts in welding as one of the most reliable methods for making permanent joints, is understandable.

Extensive experience of welding ideology has been accumulated by the PWI, which, while being at the origins of space technology, still remains a world leader in this area.

From the first welding experiment in space [1] and development of a versatile hand tool [2] up to co-operation with NASA and S.P. Korolyov «Energiya»[3] — such is the creative path covered by one of the teams of the PWI. However, the published materials [3–7] mostly pertain to the hardware and technology issues of welding in space. The main scientific information on the results of the conducted experiments became generally accessible only lately, which is related to the earlier existing classified nature of space-related work.

This paper analyses in terms of material science, the earlier published [8], as well as additional investigations by the results of the experiment conducted by S.E. Savitskaya and V.A. Dzhanibekov [3], with the aim of comparison of the quality of the welds made in space and on the ground, as well as possible assessment of the influence of space environment on welding metallurgy.

Flanged butt joints of VT1-0 titanium alloy (δ = 0.8 mm) and 12Kh18N10T steel (δ = 1.0 mm) sheets of 47×180 mm size were welded by electron beam using the VHT in raw space and on Earth in the manned space simulation test chamber.

In orbit, welding was conducted at a pressure of $10^{-3} - 10^{-4}$ Pa and microgravity level $g = 1 \cdot 10^{-5} g_0$, where g_0 is the gravitation on the Earth surface, and the pressure in the vacuum chamber was $5 \cdot 10^{-2}$ Pa. The speed of welding on Earth was $\approx$ 5–7 mm/s. The welded joints made in space do not practically differ by their appearance or shape of penetration from their ground analogs. Checking of the continuity of welded joints produced in space was conducted for 18 h by the method of liquid-penetrant testing [8]. No traces of defects were found in this case. Space samples have lack-of-penetration in several locations. These defects result not from any design or technological imperfections, but from a lack of experience of performance of such work in space. Without detracting from the uniqueness of each of the samples, the lack-of-penetration has lowered the statistical estimate of the derived results. So, after defectoscopy of the available welded joints it was possible to cut out the samples for mechanical tests, whose working zone had no detectable internal defects.

X-ray testing of each space sample was conducted in parallel with its ground analog in RAN 150-300 unit (1.2 mm focus; F = 1 mm; U = 100 kV; I = 8 mA; RT-5 film). The control method sensitivity was 0.05 mm.

* (1998) *Avtomaticheskaya Svarka*, No. 10, pp. 3–8.

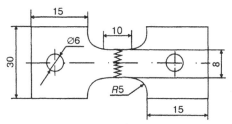

Figure 1 Shape and dimensions of samples for mechanical tests.

Mechanical tensile tests of the samples (Figure 1) were conducted in the tensile testing machine ZD-4 of VEB Werkstoffprüfmaschinen Company (Leipzig) (samples for testing were selected by I.G. Lyubomudrov).

It can be seen from Table 1 that in welded joints produced in space and on Earth, σ_t values are close for the appropriate materials. By the data of [8], microhardness H_μ of weld metal produced in space and on the ground is practically the same and is in the range of 248–262 for commercial titanium alloy VT1-0 and 268–278 for 12Kh18N10T steel (C — ≤ 0.12; Cr — 17.0–19.0; Ni — 9.0–11.0; Mn — 1.0–2.0; Si — ≤ 0.8; Ti — 0.4–0.5; S — up to 0.02; P — up to 0.035 wt.%).

Fracture surface in the rupture zone after mechanical testing of the samples was studied in the scanning electron microscope ISM-T 200 with an X-ray energy spectrum analyser of JEOL Company, Japan. Samples No.1–4 (VT1-0) and No.5–7 (12Kh18N10T) were selected for this purpose (see Table 1).

From Figure 2 it can be seen that the fractures of the rupture surfaces after mechanical testing of welded joints of VT1-0 sheets (0.8 mm), produced by manual EBW in space and on Earth, are identical, their nature being that of a tough coarse-crystalline fracture.

Figure 3 shows the appearance of fractures in the rupture zone after mechanical tests of welded joints from 12Kh18N10T sheet. Comparison of Figures 3a and b shows that fractures of rupture surfaces in the HAZ are heterogeneous in their structure. Sections of tough and brittle fracture are adjacent. However, the width of the brittle fracture region is much greater in the welded joint produced in space, than in its ground analog. Judging by the spectrograms, this region in the first case is enriched in sulphur, compared to the ground analog (Figure 4a).

Fracture of the surface of rupture along the weld of 12Kh18N10T welded joint, made in space, is of a coarse-crystalline nature (Figure 3c). Different sections of one and the same surface somewhat differ in silicon content (Figure 4b). Comparison of Figures 4a and b shows that the content of the main alloying elements (chromium, nickel, manganese) is somewhat higher in the weld zone, than in the HAZ. It can be assumed that, on the whole, the process of welding in orbit proceeds similar to that on the ground, namely during the period of co-existence of the liquid

Table 1 Results of tensile mechanical tests of welded joints.

Sample No.	Material	σ_t, MPa	Fracture location	Conditions of welded joint production
1	VT1-0	472.9	In HAZ	In space
2	VT1-0	435.8	In the grip	Same
3	VT1-0	397.2	Same	On the ground
4	VT1-0	435.5	In HAZ	Same
5	12Kh18N10T	619.0	Same	In space
6	12Kh18N10T	600.6	»	On the ground
7	12Kh18N10T	404.5	In weld	In space

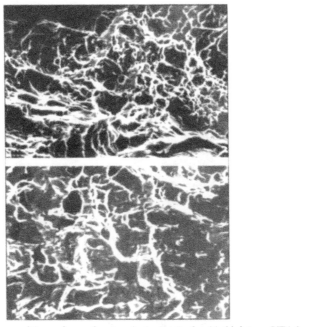

a

b

Figure 2 Fractures of the surfaces of rupture in the HAZ of welded joints on VT1-0,
produced on Earth (a) and in space (b) (×750).

and solid phases in welding, those alloying elements and impurities diffuse from the solid
metal into the liquid metal, which dissolve better in the latter. During solidification, the
solidification front «drives» these elements to the weld center.

The microstructure of the cross-section of welded joints was studied in a «Neophot-2»
optical microscope. After etching of 12Kh18N10T welded joints, the diffusion zone of
co-existence of the solid and liquid metal during welding is clearly seen both in the space
and ground samples. The microstructure of the weld metal (Figure 5) is cast (dendritic) with
α-phase precipitation along the primary boundaries. Comparison of Figures 5a and b shows
that the microstructure of the weld metal produced in space has grain boundaries with a
smaller amount of α-phase, whereas γ-phase (white sections) is more finely dispersed.
Microstructure of space samples is also characterised by dark areas without clearly defined
boundaries, located predominantly in the HAZ. Electron microscopy investigations of weld
metal in 12Kh18N10T welded joints conducted in [8] showed that α-phase in the space
welds is more disperse and the dislocation density in them is higher. Microstructures of the
cross-sections of welds on commercial titanium are given in Figure 6. It can be seen from
the Figure that the grains with characteristic fine-needle precipitates of α-phase are smaller
in welds on commercial titanium, produced in space (Figure 6b), than in their ground
analogs (Figure 6a). This is probably evidence of the fact that the rate of weld solidification
in space was higher than on Earth. There is published information both on grain refinement
in space welds [9] and on its coarsening [10], compared to the ground analogs. Nonetheless,
it does not seem possible to speak unambiguously about any regularities, not having the data
on the temperature parameters of the process of welding in the space.

The substructure of the metal of welds on commercial titanium is also different in the
ground and space samples, proceeding from the data of [8]. The latter are characterised by
a cellular substructure (0.5–7.0 μm cell sizes) non-uniform in size and distribution. In the
areas free from the cells, the dislocation density is quite high.

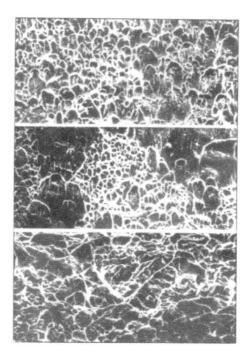

a

b

c

Figure 3 Fractures of surfaces of rupture of welded joints on 12Kh18N10T, produced on Earth (a) and in space (b, c) with fracture in the HAZ (a, b) and across the weld (c) (×750).

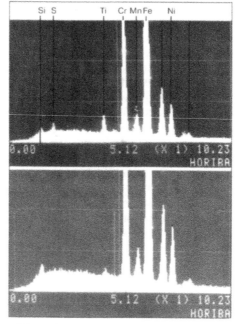

a

b

Figure 4 Spectrograms of the surfaces of rupture of 12Kh18N10T welded joints in the HAZ (a) and in the weld (b), produced in open space.

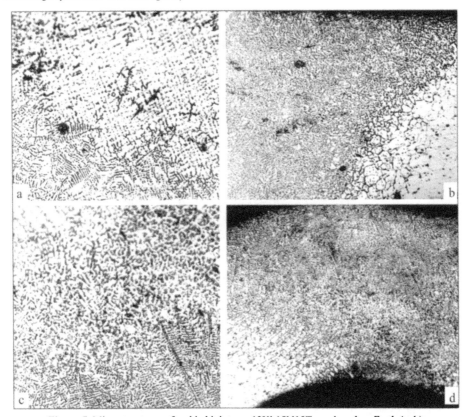

Figure 5 Microstructures of welded joints on 12Kh18N10T, produced on Earth (a, b)
and in space (c, d) (a,c — ×100; b, d — ×50).

Element distribution over the weld metal cross-section was studied on transverse microsections, using a scanning electron microscope-analyser Camebax Sx-50 of «Cameca» Company, France. In welded joints of commercial titanium no significant differences were found between the space and ground samples. The produced elemental spectra (of titanium, silicon, iron) correspond to their average content in the initial alloy and demonstrate a uniform distribution in the studied region. The pattern of peak distribution in the spectra of the space and ground samples is somewhat different, this being due to the different grain size in their structure.

The content of such elements as chromium, nickel, manganese, titanium, sulphur in the weld and near-weld zone of 12Kh18N10T welded joints produced in space and on Earth corresponds to their content in the initial material (Figure 7). A chemical inhomogeneity is found, which is characterised by enrichment of the grain boundaries (α-phase) in chromium, titanium, silicon and of the grains proper (γ-phase) in nickel. Such a regularity is observed both in the space and in the ground samples. Dark areas in the HAZ of 12Kh18N10T joints, produced in space (revealed in the microstructure in etching), are enriched in sulphur. It is seen that the increase of sulphur content is accompanied by the increase in titanium content. This is obviously related to formation of titanium sulphides under the influence of the thermal cycle of welding.

Welding in open space was conducted at an altitude of about 300 km. In this orbit the atmosphere contains Ar, He, O_2, H_2, N_2, H_2O, CO_2, etc., as well as ions of CO_2^+, O^+, H_2O^+,

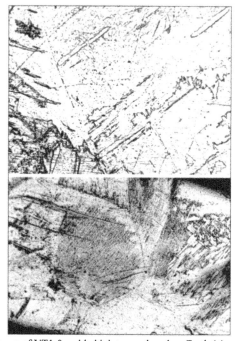

a

b

Figure 6 Microstructures of VT1-0 welded joints, produced on Earth (a) and in space (b) (×100).

H_3O^+, N^+, N_2^+, NO^+, O_2^+, OH^+. The gas background in the vicinity of the orbital station depends on surface temperature, orbit geometry, cargo bay contents, but the main component of the environment at the mentioned altitude is atomic oxygen [11]. Analysis of the content of oxygen, hydrogen, nitrogen and carbon in the welding zone of the studied samples was performed on transverse sections for assessment of the possible influence of the external atmosphere on the process of weld solidification. EKhO-4M unit with a laser probe, as well as LAS-2000 system of Riber Company for surface studies were used to conduct X-ray microanalysis.

The elemental composition was determined by the Auger spectrometry method in scanning of the surface of a transverse microsection. Investigations showed that the content of oxygen and carbon varies along the weld axis from 4–8 at.% on the surface to 0.1–1.0 at.% in-depth to a distance of 50–100 μm. In the HAZ and the weld center the content of oxygen and carbon is in the range of 0.1–1.0 at.%. Such a distribution of oxygen and carbon in the welding zone is characteristic of all the studied samples.

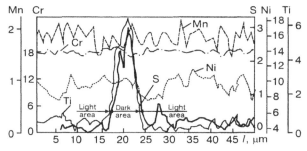

Figure 7 Distribution of elements, wt.%, in the HAZ of 12Kh18N10T welded joint, produced in space (microsection).

Table 2 Content of impurities, wt.%, in different regions of welded joints.

Material	Element	Earth				Space			
		Base metal	HAZ	Fusion line	Weld	Base metal	HAZ	Fusion line	Weld
VT1-0	$H_2 \cdot 10^{-2}$	0.0770	0.051	0.088	0.0610	0.074	0.051	0.085	0.069
	C	0.0900	–	–	0.0500	0.070	–	–	0.090
	O	0.1200	–	–	0.1300	0.080	–	–	0.110
	N	0.0260	0.028	0.028	0.0160	0.017	0.012	0.018	0.012
12Kh18N10T	$H_2 \cdot 10^{-2}$	0.0140	0.079	0.012	0.0150	0.014	0.011	0.011	0.016
	C	0.1100	–	–	0.0600	0.100	–	–	0.100
	O	0.0190	–	–	0.0190	0.019	–	–	0.013
	N	0.0066	0.009	0.009	0.0087	0.011	0.025	0.021	0.022

Table 2 gives the results of determination of the content of hydrogen, oxygen, carbon, and nitrogen in different regions of the welding zone, using a laser probe of 300 μm diameter. From the Table it is seen that for VT1-0 the content of hydrogen, carbon, and nitrogen does not exceed the critically admissible values, leading to material embrittlement [12].

Generalising the data, derived by various methods, allows the conclusion to be made that the content of oxygen, hydrogen, carbon and nitrogen in the welding zone of all the studied materials does not exceed the average values allowed by the appropriate GOST standards. The data of [8] where the content of hydrogen and oxygen in the welds on VT1-0 produced in space and on Earth, was determined by the method of hot extraction in a flow neutral gas, coincide with the results given in Table 2. As regards oxygen, a certain increase in its content in the weld is found, both in the space and in the ground samples, compared to the base metal. A significant decrease in the content of hydrogen in the welds (by 2 times on Earth and by 5 times in space) [8] can be related to the errors made, when preparing special samples for investigations (with the hot extraction method it essentially influences the result).

Despite the fact that the derived results are not of a statistical nature (this probably being one of the features of the unique experiments), the performed investigations lead to the following conclusions.

Conclusions

1. The process of manual EBW of VT1-0 ($\delta = 0.8$ mm) and 12Kh18N10T ($\delta = 1.0$ mm) sheets in open space at about 300 km altitude can be performed and on the whole proceeds similar to that on the ground.

2. Mechanical properties of VT1-0 and 12Kh18N10T welded joints, produced in space and on Earth are identical. So, the ultimate tensile stress of VT1-0 welded joints is 453.5 MPa (ground samples) and 472 MPa (space samples). For 12Kh18N10T welded joints similar values of σ_t are 600.6 and 619.0 MPa, respectively.

Microhardness ($H\mu$) of the weld metal, produced in space and on Earth, is in the range of 248–262 for VT1-0 and 268–278 for 12Kh18N10T.

3. Structural characteristics of VT1-0 and 12Kh18N10T welded joints, produced in space and on Earth, are different. For VT1-0 they are manifested in that the grains with the characteristic α-phase precipitates are smaller in the welds produced in space than of their ground analogs. The substructure of space welds is cellular and non-uniform in size. Dislocation density is higher than that of their ground analogs. For 12Kh18N10T the

differences in the microstructure are manifested in that the grain boundaries of the metal of the weld, produced in space, contain less α-phase, and γ-phase is more finely dispersed. The substructure of space welds is characterised by higher dislocation density than that of the ground welds.

4. The content of the main elements of VT1-0 alloy and 12Kh18N10T steel in the welds, produced in space and on Earth, corresponds to their average content in the initial material. The microstructures of the space and ground welds on 12Kh18N10T are characterised by chemical inhomogeneity at which the grain boundaries are enriched in chromium, titanium, silicon, whereas the grains proper are enriched in nickel.

5. The content of oxygen, hydrogen, carbon and nitrogen in the VT1-0 and 12Kh18N10T welds, produced in space and on Earth, does not exceed the average values, admissible by the appropriate GOST standards. There is every ground to believe that the welds on VT1-0 and 12Kh18N10T sheets produced by manual EBW in orbit, will not be inferior to their ground analogs as to their quality.

1. Paton, B.E. and Kubasov, V.N. (1970) *See pp. 154–160 in this Book.*

2. Nikitsky, V.P., Lapchinsky, V.F., Zagrebelny, A.A. et al. (1985) *See pp. 178–184 in this Book.*

3. Paton, B.E., Dzhanibekov, V.A. and Savitskaya, S.E. (1986) *See pp. 11–18 in this Book.*

4. Paton, B.E. (1995) Advanced trends in improvement of welded structures, In: *Proc. of Int. Conf. on Welded Structures*, Kiev, 1995. Kiev, PWI, pp. 1–18.

5. Dzhanibekov, V.A., Zagrebelny, A.A., Gavrish, S.S. et al. (1991) *See pp. 184–190 in this Book.*

6. Paton, B.E., Lapchinsky, V.F., Bulatsev, A.R. et al. (1989) *See pp. 468–473 in this Book.*

7. Lapchinsky, V.F., Rieck, U., Sobisch, G. at al. (1993) *See pp. 297–304 in this Book.*

8. Paton, B.E., Lapchinsky, V.F., Zagrebelny, A.A. et al. (1986) *See pp. 283–288 in this Book.*

9. (1996) Lost in Space, *New Scientist*, March 16.

10. Masubushi, K. (1992) Laser microscopy opens up new opportunities for welding, *Welding J.*, September, pp. 69–71.

11. Zimcik, D.G. and Maar, K.P. (1988) Consequences of reactions between atomic oxygen and materials being on board of the space Shuttle in STS-41G mission, *J. Spacecraft and Rockets*, No. 2, pp. 162–168.

12. Grabin, V.F. (1975) *Fundamentals of metal science and heat treatment of titanium alloy welded joints*, Kiev, Naukova Dumka, 252 pp.

The results of the welding experiment which was scheduled for performance in «Mir» station in 1998, would have allowed to discuss with a greater degree of certainty the features of the process of welding sheet metal in raw space, as well as the life of welded joints under these conditions (Editor's remark).

Bases of Orbital Joining Technologies.
Requirements for Equipment for Welding in Space*

V.F. Lapchinsky[1], U. Rieck[14], G. Sobisch[15] and S. Keitel[15]

Introduction. With the development of space engineering the welders faced the following problems:

• space stations, functioning now and designed for a long-term operation, for example, OC «Mir» and OL «Spacelab», are subjected to wear and emergency situations typical of orbital conditions. Strong and reliable joints can be produced using welding or brazing and other methods (for example, adhesion). These are main technologies which ensure technical functioning of the space station and guarantee the safety of the crew;

• to create large structures, demanded, for examples, for the space stations energy-supplied solar batteries or withstanding high pressures of hollow elements (space object compartments), the technologies of manual and automatic joining of materials with parameters meeting requirements for the strength of the structure, are required. The quality assurance for these works should be guaranteed by a safe and simple method of joining, as the cosmonaut can follow instruction on welding only partially. The welded joints are characterized by a number of special features to meet space requirements. In particular, these are serviceability and leak-tightness.

Unfortunately, experience gained on Earth cannot be used in space completely. The reason for that are the physical conditions of space differing from those on Earth (Table 1).

The selection of the welding method for use in space is limited with allowance for the following peculiar features:

Spattering. Due to limited heat dissipation in vacuum the welding spatters preserve their heat energy for a long time and because of their uncontrollable movement, they are hazardous for a welder. Spatters are undesirable also from the metallurgical point of view and therefore should be avoided.

Energy supply. Electrical energy required for welding is provided by solar batteries, and thus its amount is limited. The method of welding should have a high efficiency factor.

Manipulation. The physical load on the welder in terrestrial conditions becomes higher in space because of the protective spacesuit. The advantages of the negligible weight of the welding equipment in weightlessness are rather conditional as it must be taken into account at acceleration of the object during welding. Therefore, the process should possess good feasibility of mechanization and automation.

Table 1 Specific extreme conditions in space.

Physical conditions	Positive effect	Negative effect
Weightlessness	Melt stabilizing, no «deflection»	Gas pores have no buoyancy in melt Inclusions There is no stable position of «welder»
Vacuum	Perfect metallurgical protection	No gas protection of molten pool No cooling from convection

* (1993) *Schweisstechnik im Luft- und Raumfahrzeugbau*, Duesseldorf, DVS, pp. 19–24.

Table 2 The first investigations of welding methods for space applications.

Year	Unit	Material	Method	Purpose
1969	«Vulkan»	Al	Electron beam	Welding
USSR		Ti	Plasma	
		Cr–Ni steel	Arc	
1973		Al	EB	Welding
USA		Cr–Ni steel		Brazing
		Ta		
		Ni		
1979	«Isparitel»	Al	Electron beam	Coating
USSR		Ti		Condensation
1984	VHT	Al	Electron beam	Welding
USSR		Ti		Coating
USA		Steel	Laser	Welding

Materials. In accordance with requirements for mass in space, aluminium and titanium, which are not simple as to the weldability and strength properties produced, justified themselves well.

In these special conditions, however, such methods as shielded-gas arc or flash-butt welding are hardly suitable.

The first investigations of welding methods for space applications were carried out in 1969, however, some of them were also performed earlier (Table 2).

In the former USSR, EBW was recognized to be a suitable method at the PWI, Kiev. The need for vacuum, the greatest disadvantage of the method under Earth conditions, is transformed into a significant advantage in space.

Owing to the good relations existing for many years between the PWI and SLV Halle[15], the project work was started, at the request of DARA, in cooperation with DASA/ERNO Company[14] for the development of an effective technology and proper equipment for welding in space. This work was aimed at evaluation of the technological capabilities of already existing welding equipment and creation of new equipment taking into account the requirements of space.

Prerequisites. Within the scope of the project the following welding methods and equipment were considered:

Method of welding	*Hardware*
EBW	VHT, «Universal» (the PWI)
EBW	60/15 (SLV)
CO_2-laser	RS1700 SM
Nd:YAG laser	RSY500P

The EBW VHT unit, developed at the PWI, was the only equipment which had been already used in space. It was used for welding titanium and Cr–Ni steel samples in the conditions of a real experiment in orbit.

The comparative examination envisaged for determination of the requirements for the electron beam, was performed using electron beam units of SLV Halle[15], as it included an advanced system of deflection of the beam, which satisfied the existing requirements.

Laser welding was regarded as an alternative to EBW to achieve the required properties of the welded joint. It was necessary to clarify in principle what laser systems and what laser capacities are required for similar conditions to space. The project was aimed at the evaluation of the possibility of using in space the advantages of both methods obtained in atmospheric conditions. The main emphasis was on laser welding realization in space.

For the experiments a sheet of aluminium 2219 and tubes of titanium 3.7191 were used. Aluminium 2219 is an American alloy of the following chemical composition, wt.%: Cu — 6.3; Mn — 0.3; Ti — 0.06; V — 0.1; Zr — 0.18; Al — the rest.

The butt, overlap and butt joints with edges flanging were examined. Before welding the oxide film was removed from the aluminium plates by mechanical treatment.

Alongside the above-described factors, the result of welding depends greatly on the distance between the tool and the workpiece being treated. In selection of a manual or mechanized method the length of focusing and depth of focus are important factors. Even before the investigation it was established that EBW is predominant in terms of better depth of focus in manual welding.

Welding unit VHT. The VHT was designed taking into account requirements specified from the practical experience of Soviet cosmonauts, leading to continuous improvement of the equipment from the time of the first experiments in orbital conditions. Therefore, the welding unit is very manoeuvrable, and for users who are not as a rule welders, it is simple to operate. A reserve feasibility of beam formation was provided in the unit. In case of failure it is possible to use another electron gun without work interruption. Other aspects concern the reliability of the electron gun design. The necessary vibrational tests were successful on a diode system based on a tantalum cathode. The unit design is such that a supply system and manual welding gun are connected only with a cable of 42 V. (Variable voltage on a low-voltage cable, connecting electron beam manual tool with an instrument compartment, was 27 V for VHT and up to 100 V in hardware «Universal» at 20 kHz frequency.) The high voltage was 5 kV in the first version, and 8 kV — in the next version («Universal»), being generated in the EBW gun.

The supply voltage in station OC «Mir» is varied between 25 and 35 V. A maximum 30 m length of cable is recommended. The available X-ray radiation in aluminium welding at 8 kV accelerating voltage is within the admissible ranges. (see p. 16, Figure below) shows a manual welding gun, and (see Figure 12a on p. 202) presents a welding gun and supply connector in combination. All the elements of maintenance are reduced to minimum for simple operation of VHT under space conditions. To meet the known requirements, the unit was manufactured for fulfilment of the following tasks:

- welding without filler material;
- welding with filler material;
- thermal cutting;
- coating deposition by evaporation.

All the complex of works was easily realized using auxiliary devices in the system «Universal» (see Figure 12b on p. 202). A melting crucible is used for coating deposition, whose special design provides the evaporation of material.

During investigations only welding without filler material using different modifications of equipment by power was performed.

Investigations with use of welding hardware VHT/«Universal». At the PWI, experiments were carried out using EBW VHT. Aluminium sheets and titanium pipes were welded by 200–1200 W electron beam in 10^{-4} Torr vacuum. During the experiments the VHT positioning in the vacuum chamber was performed with the help of a special device. Experiments were carried out in two stages: first, the as-tested VHT with 200–350 W beam was used, and then the upgraded «Universal» of 700 W capacity. The separate experiments were made at up to 1200 W capacity.

During the first experiments with VHT, it was clear that its application for aluminium is impossible, due to insufficient electron beam power or the unit power density for melting.

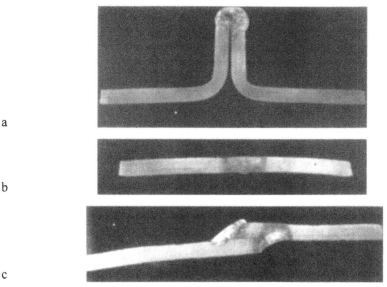

a

b

c

Figure 1 Welds made with the «Universal».

Only the upgraded VHT, called «Universal», made it possible because of its versatility to obtain good results at 700–1200 W beam power (Figure 1). (Maximum capacity of electron beam during conductance of experiment with VHT was 350 W, with hardware «Universal» — 750 W.) With a change in the beam angle of incidence to 30° from the vertical, the penetration was not changed. In manual welding these were inadmissible angular tolerances in the tool positioning.

Another situation is welding of titanium pipes. Acceptable results were obtained at 0.75 mm wall thickness of the tube and 40 m/h welding speed using VHT. At maximum 350 W beam power the tubes were continuously rotated and after 3 revolutions a visually sound welded joint was produced. However, metallographic examination showed coarse grain formation, due to high heat input, which reduced the strength by ≈ 17 %.

To evaluate the distribution of energy density, the penetration was investigated by keeping the electron beam in one point for several seconds. It has been found that the beam is not perfectly circular, but has an elliptical section.

To evaluate the depth of focus, comparative welding with different working distances was carried out. The minimum distance was 55 mm from the gun outlet. Figure 2 shows that in evaluation of weld width-depth ratio at different welding speeds, penetration conditions typical of the electron beam were not observed. In a zone of about 20 mm the beam preserves its properties as the penetration is provided mainly by a heat conductivity.

Evaluation of different weld shapes. Figures 1 and 3 show welds made by VHT of a radically new shape. The butt weld was of the best quality. In the weightlessness conditions the gap overlap should be better both with and without filler materials, than that under Earth conditions. Gap overlaping is especially improved in welding with a filler. It is difficult to evaluate the requirements for the welder's skill.

The problem of «corrosion» in orbital conditions remains open for the PWI specialists. This half-explained phenomenon led to cosmonaut-welders being recommended to make butt joints wherever possible. It is necessary to avoid any types of cleavage corrosion.

Welding experiments using a solid-body laser. Experimental CO_2-laser welding was also performed. The following properties detracted from its application:

• the energy density for welding requires 2–5 kW laser capacity;

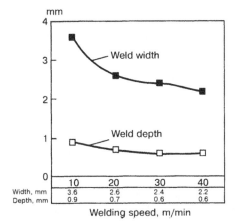

	10	20	30	40
Width, mm	3.6	2.6	2.4	2.2
Depth, mm	0.9	0.7	0.6	0.6

Welding speed, m/min

Figure 2 Weld shape at different welding parameters.

• the resultant required laser capacity will demand dimensions and mass unsuitable for space;

• the radiation of CO_2-laser in currently available now radiation sources possesses poor absorption properties on the aluminium surface, and uncontrollable reflection is hazardous for the astronaut;

• at present the laser radiation can be transmitted only with the help of rigid beam guides which are unsuitable for the manual welding.

From the above considerations the Nd:YAG laser was selected for laser welding. The investigations were performed using a pulsed laser of 500 W maximum capacity.

The purpose of investigations using a laser in vacuum was to determine whether they can provide other results of welding than those obtained in the atmosphere. Laser radiation

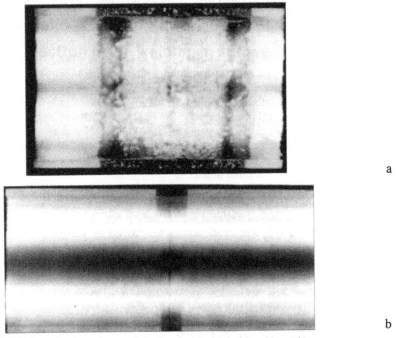

a

b

Figure 3 Welds on titanium pipes made by the electron beam (a) and laser (b) processes.

was introduced into a vacuum chamber through a quartz glass. The main problem was deposition on the glass disc and, due to this, increased absorption and failure. This deposition can be prevented only by using a shield against the vapours, but this means that the laser radiation should be introduced at 20–30° angle.

The results of welding titanium pipes showed that, owing to the better conditions of Nd:YAG laser focusing as compared with an electron beam, the heat input ensures a narrower weld with a smaller HAZ. Figure 3 gives a comparison of welds made using VHT and Nd:YAG laser. The loss in strength as compared with that of the parent metal is 15 or 17 %. Thus, the technological requirement arises to concentrate or minimize the heat input in EBW.

Metallurgical protection with vacuum was favourable for welds, and there were no oxidation.

Different pulsed conditions were investigated for aluminium welding. The maximum power of pulses was 1.2 kW at a pulse frequency of 35 Hz and 5 ms duration. Under these conditions the penetration depth was 0.8 mm, but the weld quality was not satisfactory. At higher pulse frequency and the same mean duration the pulse power and depth of penetration are similar.

The investigations showed that use of this type of lasers cannot provide sound joints of 1.6 mm thick aluminium sheets.

To obtain quality welding, a 2 kW laser capacity at continuous welding (CW) conditions, if possible, is required. Dimensions and mass of equipment do not meet the requirement to operate in space.

Practical realization of experiments. From the point of view of practical fulfilment of works manual welding (either of aluminium or titanium) cannot be widely applied. Remember that pipe must be welded at 360° in a spacesuit with a limited movement, appropriate habits of the astronaut are required. Manual welding of high-quality welds at definite temperature conditions can be used only in exceptional cases. More often it is repair welding of damaged places. It can be imagined, that it is necessary to weld-on prepared patches inside the object in case of leakage and then to strengthen them with sound outside welds for assurance. The simple longitudinal welds should be mastered first.

To create intricate structures manual welding can be used only as an auxiliary means in construction (welded spot or fastening mechanical junctions). Leak-tight welds with requirements for dynamic loads (for example, cyclic temperature loads) require mechanized or automated methods to guarantee proper energy input for the materials used.

Selection of welding method. The results of experiments state that at the present level of laser engineering preference should be given to EBW. Table 3 gives data that are competely in favour of EBW application. The efficiency factor of electrical energy conversion into beam energy (1.3:1.0, according to data of the PWI), and also beam energy (kinetic energy of electrons) into heat energy (~ 80 %) are an important advantage of the method. It is the depth of beam focus that is an important criterion for manual welding, as it presets the limit of penetration for a definite depth.

The high power for laser welding, given in Table 3, is a result of poor absorption of laser radiation on aluminium surface in particular. Another criterion follows from poor quality of welds produced with pulsed systems, where only proper formation of pulses can help to improve the quality [1]. The continuous-wave laser has proved itself for welding thin aluminium sheets [2]. In both methods similar reproducibility of parameters was observed. However the EBW method is more reliable on the basis of experience under on-land conditions, but with rapid progress of laser engineering this is only a matter of time.

The positive advantage of Nd:YAG laser welding is the narrower weld provided by the better focusing as compared with a weld produced by the electron beam using VHT. In this connection the mechanical properties of welded samples are improved, which should be defined as one of the requirements to EBW.

Table 3 EBW advantages.

Criterion of evaluation	Electron beam	Nd:YAG laser
Energy conversion	> 60 %	< 5 %
Electric power	2–3 kW	> 50 kW
Required beam power:		
for Al	1.5 kW	2 kW (CW)
for Ti	0.5 kW	1 kW (CW)
Mass of hardware	50 kg (max 100 kg)	> 500 kg
Potential hazard	X-ray radiation	Dissipated radiation
Noises		Deposition on optical elements
Beam deflection	Electromagnetic	Mechanical
Depth of focus	20 mm	1 mm
Focusing distance	Constant	Variable due to lens 50–200 mm

Further it is interesting to study laser on power diodes. Its efficiency factor is higher than that in Nd:YAG laser and absorption, in particular in aluminium processing is much better due to the shorter wave length. Characteristics of consumed power and weight are also improved.

Requirements for EBW equipment for space application. The welding equipment «Universal», developed at the PWI, meets space requirements by its main technical parameters. From the point of view of technology two factors will be of special attention in future:

• to avoid the melt overheating it is necessary to improve beam focusing, i.e. to decrease the diameter of a focal spot from 2 to 1 mm. This must be done without increase in accelerating voltage above 8 kV. Otherwise X-ray radiation, harmful to people, will occur. Consequently, the energy should be concentrated and removed quickly. This will not induce any problems in mechanized or automatic welding;

• in view of the first supposition in manual welding the device for fast beam deflection should be incorporated, which deflects a sharply focused beam so quickly, that only a partial melt overheating occurs. Coming from our experience of treatment of the external layer, 50–400 Hz frequency is required. The effect is reached at large tolerances in beam positioning simultaneously at lower overheating. Due to beam oscillation the heat treatment and brazing processes are simplified.

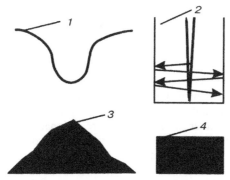

Figure 4 Change in energy input due to the beam focusing and deflection: *1* — defocused beam; *2* — beam is focused by a deflection device; *3* — amount of defocusing heat, 100 %; *4* — amount of deflection heat, 40–60 %.

Figure 4 demonstrates the effects obtained at the expense of beam deflection from the point of view of heat input, and, consequently, heating of the molten pool.

The gun design can be made in principle by a verified VHT model. Specific tasks and weld shapes can be preset for astronauts having no experience in welding, using updated software based on an appropriate set of optimum parameters as to the weld quality on a specific material.

Conclusions. After 8 months of project work, results on space welding were obtained which made it possible to give preference to EBW. The equipment requires modifications which should ensure reliable operation both in manual and automatic conditions. The beam deflection here is an important criterion.

Laser welding cannot now be used in space due to the dimensions and mass of the laser unit of the required capacity. There are problems with beam positioning, in particular, in manual welding, and also with metal vapours depositions on the optical elements. However, the new generation of lasers, for example, diode lasers, are promising.

1. Matsunawa, A. et al. (1993) Fusion and solidification characteristics in pulse-shaped YAG laser welding, *Tagungsunterlagen CISFFEL*, pp. 219–226.

2. Norris, I.M. et al. (1993) Supra kW Nd:YAG lasers. Their technical, economic and applications assessment, *Ibid.*, pp. 113–120.

Hybrid Laser-Microplasma Welding of Thin Sections of Metals*

B.E. Paton[1], V.S. Gvozdetsky[1], I.V. Krivtsun[1], A.A. Zagrebelny[1], V.F. Shulym[1] and V.L. Dzheppa[16]

Combined or hybrid processes implemented by the combined use of two different heat sources, e.g. laser beam and electric arc, are currently receiving increasingly wide acceptance. Initial investigations into laser-arc welding processes [1–5] showed that they possess a number of peculiarities which cannot be explained by simple superimposition of properties of the heat sources involved if taken separately. In particular, it was found that the combined effect exerted on metals by laser and arc heat sources was accompanied by a fundamental increase in the coefficient of utilisation of energy of both heat sources and stability of movement of the arc spot over the workpiece surface. All this allows a more than 1.5 times increase in the maximum depth of penetration, as compared with the corresponding laser process, which is especially important for the cases when low-power lasers are employed, and makes it possible to improve stability and almost double the productivity of the corresponding arc process.

Results obtained from utilisation of the constricted (plasma) arc instead of the freely burning one in the hybrid process turned out to be much more encouraging. Two diagrams of realisation of the laser-plasma welding are known in the art — one with a laser beam and plasma arc located at an angle to each other [6, 7], and the other with their coaxial arrangement [8–11]. The latter seems to be more rational, as it provides the required coaxiality of thermal and dynamic effects of the heat sources on the weld pool surface.

Unfortunately, employment of traditional plasmatrons for realisation of hybrid laser-plasma welding within the frames of the diagram under consideration is hardly possible. Coaxial arrangement of the laser beam and plasma arc requires the use of specialised devices,

* (2002) *The Paton Welding Journal*, No. 3, pp. 2–6.

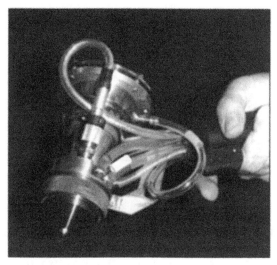

Figure 1 Integrated plasmatron for microplasma, laser and hybrid welding in automatic and manual modes.

i.e. integrated laser-arc plasmatrons [8–10, 12], the main peculiarity of which is design of the cathode unit (refractory tubular cathode or system of pin cathodes located on the circumference) to allow focused laser radiation to be introduced into the welding zone along the axis of the plasma-shaping channel.

This article considers results of tests of a prototype of the integrated laser-arc plasmatron intended for automatic and manual welding of different metals up to 3 mm thick, as well as results of studies of technological capabilities of hybrid welding of stainless steels, titanium and particularly aluminium alloy from the standpoint of its future application under space conditions.

Experimental equipment. The prototype of a versatile plasmatron for microplasma, laser and laser-microplasma welding in the automatic and manual modes was designed and manufactured for implementation of the process of hybrid welding of thin sections of metals (Figure 1). This device is a two-electrode plasmatron with a tungsten cathode 1.5 mm in diameter (for operation at straight polarity (SP)) and tungsten electrode 2.5 mm in diameter (for operation under conditions of alternating-polarity pulses of the electric current), installed inside a water-cooled casing. Electrodes are located diametrically at an angle of 21° to the plasmatron axis with a possibility of moving along their axes. The plasmatron has a replaceable plasma-shaping nozzle made from copper or molybdenum, with a diameter of the exit channel equal to 2–3 mm, and an external nozzle for shielding gas feeding.

The suggested plasmatron design allows a focused laser beam to be introduced into the welding zone along the axis of the plasma-shaping nozzle. For this purpose the casing is fitted in its upper part with an interfacing unit (Figure 2) to connect the plasmatron to the standard focusing system RSY-FM- D160Z HP/2 with an adjustable focal distance of 160±6 mm, which is connected to the light guide via the RSY-KM-B120 HPW/2 collimator. The plasmatron is designed for operation with up to 2 kW YAG laser (size of the focal spot is 0.6–1.0 mm) at straight-polarity currents of the microplasma arc equal to 50 A, and using alternating-polarity pulses of the electric current with an amplitude of 35 A.

Preliminary tests of the plasmatron in the arc mode, conducted by the PWI using an experimental power supply, included:

- identification of conditions for excitation and maintenance of the pilot and main arcs;
- checking functioning of the plasmatron at straight and reverse polarities;

Figure 2 Microplasma welding at straight polarity using the integrated plasmatron.

• preliminary evaluation of technological capabilities of the plasmatron in the micro-plasma welding mode.

With plasma-shaping nozzle channel diameters of 2–3 mm and plasma gas (argon) flow rates of 0.5–0.8 l/min the pilot arc was ignited in a stable manner both between each of the electrodes and the nozzle and between the two electrodes (pilot arc current 5–7 A, open-circuit voltage 60 V). The best shape of the pilot arc with the plasma flame going out of the nozzle exit section was observed with its burning between the plasmatron electrodes.

At a distance from the nozzle exit section to the sample surface equal to $h = 2$–3 mm, the SP main arc had a stable excitation, burning at an electric current of $I = 32$–53 A at an arc voltage of $U = 36$–29 V ($U_{o-c} = 80$ V). Increasing the open-circuit voltage to 100 V provided stable excitation and burning of the main arc at a distance to the sample surface increased to 6 mm.

Samples 1.0, 1.5, 2.0 and 3.0 mm thick of stainless steel 12Kh18N9T (304SS), titanium alloy OT-4 (3.5Al–1.5Mg) and aluminium alloy AMg3 (Al–3.5Mg) were used to evaluate the technological capabilities of the plasmatron in the microplasma welding mode.

The samples welded were mounted on a fixed welding table (using a copper backing), and the plasmatron was fixed on the bracket of a welding tractor and moved during welding operations (Figure 2). The technological experiments conducted included penetration of the samples and welding of butt joints. The plasma gas, shielding gas flow rate and the cooling water consumption were 0.6, 8.0, 1.5 l/min, respectively.

It was established that welded joints with full penetration of samples of stainless steel and titanium alloy, 1.0–1.5 mm thick, could be produced by microplasma welding with a SP arc using the plasmatron developed. In this case the welding parameters were as follows: $I =$ = 43–52 A, $U = 29$–25 V and welding speed $v_w = 0.24$–0.40 m/min. The width of the welds thus produced was no more than 3.5 mm. No marked undercuts or sags of the welds were seen, and the welds produced at a distance to the sample equal to 2 mm looked better.

The experiments conducted revealed the insufficiency of shielding of the weld pool surface, which resulted in a marked oxidation of the welds. As an increase in the shielding gas flow rate failed to provide an improvement of the shielding quality, certain modifications were made on the plasmatron: the diameter of the exit orifice of the shielding nozzle was increased and a fluoroplastic gas-distribution ring was made and installed into this nozzle. Although these measures provided some improvement of the quality of shielding of the welding zone, it was obviously insufficient (further optimisation of the weld shielding system is required).

Unfortunately, the attempts made at the first stage of the test to determine capabilities of the plasmatron in operation under conditions of the alternating-polarity current pulses failed because of unstable functioning of the available power unit.

Technological experiments. Technological experiments on utilisation of the integrated plasmatron for welding in the arc, laser and hybrid modes were conducted at BIAS in Germany. A standard power unit for arc and plasma welding of the MESSER TIG 450 AC/DC-P model with smooth regulation of the electric current from 5 to 450 A, including plasma and shielding gas feed and control system and closed loop of fluid cooling of the plasmatron with switching-on interlocking, was used as a power supply. The power supply provided stable ignition of the pilot arc, both between one of the electrodes and the plasmatron nozzle and between the two electrodes. In addition this power supply was equipped with an oscillator to ignite the main arc by breaking down the arc gap, which allowed the AC and DC main arc to be ignited further on without the use of the pilot arc.

A robot-manipulator of the KUKA Company was used to move the plasmatron during welding. The control system of the robot made it possible to programme all the parameters of the welding process, which, in addition to some advantages, also had certain drawbacks, as it did not allow adjustment of the process parameters during welding. In all the experiments the welding speed was set at a level of 0.25 m/min, which, firstly, is the mean value of the speed of manual welding operations, and, secondly, is the highest speed permitted by the limited power of the plasmatron. The distance from the exit section of the copper plasma-shaping nozzle to the sample was set to be equal to 2.0–2.5 mm, and the diameter of the plasmatron nozzle channel was selected to be equal to 2.5 mm. Samples of the same materials and thicknesses as those used for the tests conducted at the PWI were employed for the above technological experiments.

Microplasma welding. SP microplasma welding of samples of the above stainless steel and titanium alloy was carried out to determine the technological capabilities of the plasmatron in arc mode operation. Six butt joints of samples 1.0–1.5 mm thick were welded at different combinations of the arc current $I = 40$–51 A, and penetration of four samples 2–3 mm thick was performed for subsequent comparison of the depth and width of penetration in microplasma welding with the corresponding parameters of the welds produced by laser and hybrid welding.

Samples of aluminium alloy were welded by the AC microplasma arc using alternating-polarity pulses with a frequency of 125 Hz. The duration of the straight and reverse polarity pulses was 65 and 35 %, respectively. Welding was performed at arc currents of 25–35 A. It should be noted that at a microplasma arc current of 25 A the sample 3 mm thick was just cleaned without marked penetration. Welding of butt joints also provided a good cleaning of the surface, but no quality welds with full penetration were produced even at their thickness of 1 mm and an arc current of 35 A.

Laser welding. Upon completion of experiments in the arc mode, the plasmatron was connected to the focusing system RSY-FM-D160Z HP/2 (Figure 3) and then, via a flexible light guide, to the pulsed periodic-wave diode-pumped laser RS D4 044 (Rofin-Sinar Company) with a maximum power of 4.4 kW. The position of focus of the laser beam was adjusted (focal spot size 0.8–1.0 mm) relative to the sample surface at a distance from the plasmatron exit section equal to $h = 2$ mm.

The experiments started with welding and penetration of samples of stainless steel and titanium alloy. Prior to welding of aluminium samples, to prevent ingress of reflected radiation into the optical system of the laser, the plasmatron with the focusing system installed on it was turned to a angle of 8° to a vertical line (see Figure 3). This provided correction of position of focus of the laser beam with respect to the sample surface. Ten samples of the above materials 1.5–3.0 mm thick were treated at a laser power of 500–1500 W. As a result, full penetration of the stainless steel samples 2 mm thick was achieved

Figure 3 Integrated plasmatron with the laser focusing system connected to it.

at a laser power of 1 kW. For the titanium and aluminium alloy samples 3 mm thick the penetration depth was 1.2 mm at a laser power of 0.5 kW (OT-4) and 0.8 mm at a laser power of 1.5 kW (AMg3). In the case of welding butt joints on the aluminium alloy samples 1.5 mm thick, the attempts to achieve full penetration by increasing the laser power from 1.0 to 1.3 kW failed.

Hybrid welding. Prior to beginning of the welding experiments in the combined mode, the plasmatron with the connected focusing system was electrically insulated from the robot. The switching on trials revealed problems with the cooling system of the 4.4 kW RS D4 044 laser. Therefore, a continuous-wave laser RS CW 020 with a power of 2 kW, the beam of which was focused on the workpiece into a spot 0.6 mm in size, was connected to the plasmatron.

Penetration in the hybrid mode (laser + AC microplasma arc) was performed on aluminium alloy samples 3 mm thick. At an arc current amplitude of 25 A and laser beam power of 1.0–1.5 kW, the samples were penetrated to a depth of less than 1 mm, resulting in a strong blackening of the welds, although in the case of using only the AC microplasma arc this effect was not seen, and normal cleaning of the aluminium surface occurred. The plasmatron was disassembled (protective and plasma-shaping nozzles were removed) and leakage of cooling water was detected in a soldered joint in the lower part of the plasmatron casing. It is likely that this leakage was a cause of deterioration of shielding of the welding zone. Despite this fact, penetration of two aluminium alloy samples 3 mm thick was performed in the hybrid welding mode (the plasmatron was disconnected from the cooling system during those experiments). The hybrid welding conditions were as follows: laser power was 1.2 kW and amplitude of the microplasma arc current pulses was 25 and 35 A, respectively.

Experimental results. The developed integrated plasmatron for microplasma, laser and hybrid welding allows butt joints with full penetration of the stainless steel and titanium alloy samples up to 1.5 mm thick to be produced in operation in the DCSP arc mode at an arc current of 40–50 A and welding speed of 0.25 m/min.

The power of the plasmatron in welding aluminium alloy using alternating-polarity pulses (current amplitude 25 A) allows no more than cleaning of the surfaces of samples 3 mm thick, without visible traces of melting (Figure 4a). The attempts to produce butt joints in aluminium samples 1 mm thick showed that the above power was also insufficient for making the welds with full penetration. Therefore, the power of the plasmatron should be

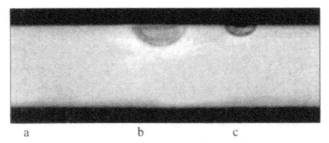

a   b   c

Figure 4 Macrosection of a sample of aluminium alloy 3.0 mm thick with penetration by microplasma (a), hybrid (b) and laser (c) welding (welding speed 0.25 m/min) using the alternating-polarity arc (current amplitude 25 A, frequency 125 Hz, plasma gas (Ar) flow rate 0.5 l/min) and YAG laser (radiation power 1.2 kW, focal spot size 0.6 mm).

increased (maximum value of the arc current should be increased at least to 50 A) for operation at alternating-polarity pulses.

The experiments with laser welding showed that at the selected welding speed values butt welded joints could be produced on stainless steel and titanium samples 1.5 mm thick at a laser power of 750–800 W. At this point we have to note some instability of the process, which results in formation of regions on the weld surface differing in appearance, although the welding process parameters were unchanged.

The attempts to provide full penetration in making butt joints on aluminium alloy samples 1.5 mm thick by increasing the laser power to 1.3 kW failed. Most probably, the laser power required to do it should be not less than 1.5 kW. In addition, in laser welding of aluminium the oxide film is not cleaned from the surface during welding, which results in an insufficiently high quality of the welds.

Penetration of samples of different materials, 2–3 mm thick, at the same speed (0.25 m/min) was performed for comparative evaluation of different welding methods (microplasma, laser and hybrid welding). In particular, in the mode of SP microplasma welding an arc current of 50 A the penetration depth was 0.6 mm for stainless steel and 2.0 mm for titanium alloy (the weld width was 2.8 and 5.5 mm, respectively). In the laser welding mode at a laser power of 1 kW and at the same welding speed, full penetration of the stainless steel sample 2 mm thick was achieved, while at a laser power of 0.5 kW the penetration depth was 0.8 mm for stainless steel and 1.2 mm for titanium alloy (at a weld width of 1.4 and 2.5 mm, respectively).

No visible melting was seen in penetration of aluminium alloy 3 mm thick by the AC microplasma arc (amplitude value of the current was 25 A) (see Figure 4a), whereas with an increase in amplitude of the current pulses to 35 A the penetration depth was 0.7 mm, weld width was 4 mm and penetration area was 2 mm^2 (Figure 5a). In laser penetration of the same sample at a speed of 0.25 m/min the penetration depth was 0.4 mm at a laser power of 1.2 kW (Figures 4c and 5b).

The combined use of the 1.2 kW laser beam and the AC microplasma arc (current amplitude 25 A) the depth of penetration of a sample 3 mm thick at a welding speed of 0.25 m/min is 0.8 mm, the weld width is 2 mm and penetration depth is 1.1 mm^2 (see Figure 4b), whereas in the case of laser welding (at the same beam power and welding speed) they were 0.4 mm, 1.2 mm and 0.35 mm^2, respectively (see Figure 4c), and in the case of microplasma welding using the 25 A arc there was no penetration at all (see Figure 4a).

Results of using the 35 A microplasma arc and laser beam of the same power (1.2 kW) in the hybrid process turned out to be even more interesting. In particular, full penetration was achieved on the AMg3 sample 3 mm thick with a weld width of 4.9 mm and cross sectional area of 10.6 mm^2 (see Figure 5c). While comparing the data obtained with the results of penetration of a sample in laser and microplasma welding performed separately

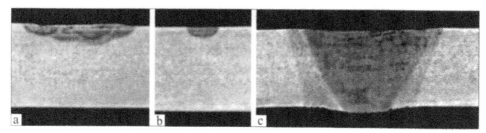

Figure 5 Macrosection of a sample of aluminium alloy 3.0 mm thick with penetration by microplasma
(a), laser (b) and hybrid (c) welding using the alternating-polarity arc (current 35 A)
and YAG laser (radiation power 1.2 kW).

(compare Figure 5a–c), it is clear that the penetration area in hybrid welding of aluminium
alloy is more than 4 times in excess of the sum of the corresponding areas in laser and
microplasma welding performed separately. This is indicative of a substantial increase in the
efficiency of utilisation of energy of the heat sources when they are combined in the hybrid
process.

Unfortunately, for the above reasons only one penetration sample was made in this mode.
Therefore, the results obtained require additional confirmation, for which it is necessary to
make a plasmatron which would allow welding in the mode of alternating-polarity pulses at
a current of 50 A and higher.

Conclusions. Preliminary studies of the technological capabilities of the integrated
plasmatron for microplasma, laser and laser-microplasma welding show the high potential
of hybrid welding of aluminium alloys using laser beam and microplasma arc burning in the
mode of alternating-polarity pulses of the electric current. On the one hand, this combination
enables cleaning of the aluminium surface during welding, which cannot be achieved in
purely laser welding and therefore improvement of quality of the produced welded joints.
On the other hand, the combined use of the two heat sources allows a substantial increase
in the efficiency of utilisation of energy of each of them, which makes hybrid welding very
attractive for such operations as repair under conditions of space flying vehicles.

We believe it would be very appropriate to continue the original investigations for
development of a higher-power integrated plasmatron, designed for current of up to 100 A
for operation in the mode of alternating-polarity pulses, and intended for use in combination
with a 1.5–2.0 kW diode laser, for experimental and theoretical investigations of the process
of hybrid welding of thin and medium sections of aluminium alloys, as well as in-depth
investigation of technological capabilities of this process.

*This study was performed in collaboration with the Bremen Institute for Applied Beam
Technologies (BIAS), Bremen, Germany.*

1. Steen, W.M. and Eboo, M. (1979) Arc augmented laser welding, *Metal Construction*,
   No. 7, pp. 332–335.

2. Steen, W.M. (1980) Arc augmented laser processing of materials, *J. Appl. Phys.*, No. 11,
   pp. 5636–5641.

3. Diebold, T.P. and Albright, C.E. (1984) «Laser–GTA» welding of aluminium alloy 5052,
   Welding J., No. 6, pp. 18–24.

4. Matsuda, J., Utsumi, A., Katsumura, M. et al. (1988) TIG or MIG arc augmented laser
   welding of thick mild steel plate, *Joining and Materials*, No. 1, pp. 31–34.

5. Gorny, S.G., Lopota, V.A., Redozubov, V.D. et al. (1989) Peculiarities of heating of metal
   in laser-arc welding, *Avtomatich. Svarka*, No. 1, pp. 73–74.

6. Walduck, R.P. and Biffin, J. (1994) Plasma arc augmented laser welding, *Welding & Metal Fabrication*, No. 4, pp. 172–176.

7. Walduck, R.P. *Enhanced laser beam welding*. Pat. 5.866.870, USA, Int. Cl. B 23 K 10/00, 26/00, publ. 02.02.99.

8. Paton, B.E. (1995) Upgrading welding methods — one of the ways of improving quality and cost effectiveness of welded structures, *Avtomatich. Svarka*, No. 11, pp. 3–11.

9. Dykhno, I.S., Krivtsun, I.V. and Ignatchenko, G.N. *Combined laser and plasma arc welding torch*. Pat. 5.700.989, USA, Int. Cl. B 23 K 26/00, 10/00, publ. 21.12.97.

10. Som, A.I. and Krivtsun, I.V. (2000) Laser + plasma: search for new possibilities in surfacing, *The Paton Welding J.*, No. 12, pp. 34–39.

11. Gvozdetsky, V.S., Krivtsun, I.V., Chizhenko, M.I. et al. (1995) Laser-arc discharge: Theory and application, *Welding and Surfacing Rev.*, Vol. 3, Part 3.3, Harwood A.P.

12. Krivtsun, I.V. and Chizhenko, M.I. (1997) Principles of design of laser-arc plasmatrons, *Avtomatich. Svarka*, No. 1, pp. 16–23.

Results of Experiments on Manual EBW in a Manned Space Simulation Test Chamber*

E.S. Mikhajlovskaya[1], V.F. Shulym[1] and A.A. Zagrebelny[1]

Experience of operation of long-term space vehicles, for instance, OC «Mir», revealed the possibility of various often unexpected situations arising, that require the use of specific technological processes, which include welding and allied technologies (cutting, brazing, etc.).

After conducting the first in the world experiment on welding in space, the efforts of the researchers of the PWI were focused on development of EBW as the most promising process for use under conditions of space and, in particular, development of the hardware and technology for manual EBW. The first manual electron beam gun was developed and tested in 1974 [1], followed by development of a VHT for manual electron beam welding, cutting, brazing and coating deposition, tested in open space in 1984 and 1986 [2]. Analysis of the samples obtained of the joints and coatings is given in [3–5]. Experimental results demonstrated the possibility of the cosmonauts performing welding operations in open space, but insufficient output power (up to 350 W) and accelerating voltage (5 kW) of VHT did not allow welding the aluminium alloys used in space vehicle structures. In the next modification of the hardware for manual EBW — the «Universal» — the above parameters were raised as much as possible: up to 750 W and 10 kV, respectively, by agreement with the customer.

In 1989–1990, in addition to thermal and dynamic testing of «Universal» hardware, also testing in the manned space simulation test chamber was performed to check the correctness of the used design solutions, providing the ability to perform the main manual welding operations. In this case attention was focused on evaluation of the techniques of handling the electron beam hand tools, organization of the workplace of an operator, wearing a spacesuit, as well as spacesuit protection. As regards technological issues, just the fundamental possibility of welding aluminium alloys and conducting manual EBW with filler wire feed was evaluated, whereas all the technological retrofitting was intended to be further on combined with welder-operator training sessions. The range of materials to be used during

* (2002) *The Paton Welding Journal*, No. 2, pp. 23–27.

Figure 1 Flexible sleeves with gloves (*1*) and fragment of a pressure helmet (*2*), mounted on the vacuum chamber flange (*3*).

the flight experiment was to be determined more precisely. Therefore, samples welded during testing, were not comprehensively studied, just visual examination was performed.

During preparation for the ISWE, an attempt was made to use flexible sleeves with gloves and a fragment of the pressure helmet, specially developed by ILC Dover, Inc., for preliminary retrofitting of manual welding operations, performed with «Universal» electron beam tools. They are shown in Figure 1. Testing demonstrated that a significant pressure gradient ($6.67 \cdot 10^{-3}$ Pa in the vacuum chamber) often resulted in a violation of the glove integrity. Moreover, their use could not allow a comprehensive assessment of the actions of an operator, wearing a spacesuit. Therefore, testing had to be restricted to performance by several operators (both Ukrainian and US) of welding of individual sections of a aluminum alloy 2219 cell (bodies of the modules of the ISS are honeycomb structures of alloy 2219 with cell sizes of $375 \times 375 \times 20$ mm in the US modules and of alloys AMg6 of $150 \times 150 \times \times 10$ mm in the Russian modules). The operators, not having sufficient skills of manual welding operation performance, noted the simplicity of the process and possibility of acquiring such skills.

Training sessions. Training of the primary and backup Russian crews to perform welding operations, using «Universal» system of electron beam hand tools, was carried out for preparation for the integrated «Flagman» experiment on board OC «Mir». Training was conducted[*] in a vacuum chamber of 50 m³ of TBK-50 facility in «Zvezda» Company (city of Tomilino, Moscow district, Russia), designed for retrofitting various cosmonaut EVA operations and fitted with «Orlan» spacesuit with a system for life support and body parameter monitoring during testing. The chamber is a horizontally located cylinder of 3.5 m diameter and ≈ 5.0 m length. A pumping system, using liquid-nitrogen filled freezing-out screen, maintained a pressure of $1.07 \cdot 10^{-1}$ Pa in the space simulation test chamber. Pressure was not measured directly in the zone where welding operations were performed, but it was assumed to be somewhat higher there, because of gas evolution and in-leakage from the spacesuit.

An all-purpose multifunctional platform «Flagman», mounted in the flight experiment configuration, was used to provide the «support» conditions and relative comfort during the

---

* Training of the Russian crews of OC «Mir», including G. Padalko, S. Avdeev, S. Zalyotin and A. Kaleri, was conducted in April–May, 1998.

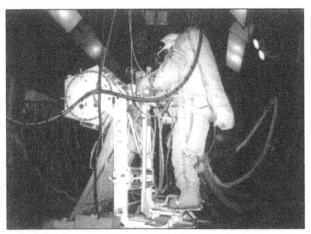

Figure 2 Work place of welder-operator in the space simulation test chamber.

operator's activity. It allows the operator to move on a mobile platform along the base guide rail (and the drum with samples, accordingly) for up to 2 m distance (Figure 2). Results of the first altitude simulation were used for retrofitting the tool weight-neutralising system (mounting it on an independent carriage), and an additional lamp was installed. This greatly facilitated the operator's task during subsequent training sessions, although imperfection of weight-neutralising devices and poor lighting could not be eliminated completely, which naturally affected the quality of the welded joints.

During training, samples of the same US materials were used, that were intended to be processed in the flight experiment and had already been delivered to OC «Mir». Their range is given in Table 1. Samples of the above materials were placed on replaceable cassettes, mounted on the faces of an all-purpose experimental drumlike sample holder of the work station, thus ensuring their convenient replacement, as well as welding operations performance.

In the course of training, the operators made butt and overlap joints of all the materials, performed cutting, as well as welding of covers to cells of bodies of the ISS US and Russian modules (simulation of repair operations in the case of a hole made in the body). Samples of aluminium alloy 5456 were processed by a tool with filler wire feed. In addition, holes in

Table 1 List of US materials, used during performance of processing experiments.

Material	Grade	Local analog	Sample thickness, mm
Stainless steel	304SS	12Kh18N9T	1.6; 1.0
(C — 0.12; Cr — 17.0–19.0; Ni — 8.0–10.0 wt.%)			
Titanium alloy	Ti–6Al–4V	VT6	1.6; 1.0
Aluminium alloy	2219Al	1201	1.6; 1.0
(Cu — 5.8–6.8; Mn — 0.2–0.4; Fe — 0.3 wt.%)			
Aluminuim alloy	5456Al	AMg5	1.6; 1.0
(Al–5.5Mg)			
Filler wire	5356Al	Same	1.0
(Al–5Mg)			
Braze alloy	Lithobraze720	PSr-72	1.0
(Ag — 72.0; Cu — 28.0 wt.%)			

titanium tubes were welded up and stainless steel tubes were brazed. 98 samples of welded and brazed joints were made, of which 21 samples were made available to the US party for studying, and 27 samples were selected for conducting research at the PWI by a program, co-ordinated by the Russian and US parties.

Investigation results. Visual inspection of the selected samples of welded and brazed joints was followed by X-ray inspection, the results of which were used to prepare samples of butt and overlap joints to conduct mechanical tests. In addition, the fracture surface was studied and metallographic examination and local X-ray microprobe analysis were performed.

Results of X-ray flaw detection, conducted in RAP-150/300 X-ray unit, revealed that welded joints of aluminium alloys, unlike those of stainless steel and titanium, are characterised by the presence of a great number of inner defects in the form of individual pores and cavities, as well as in clusters. Maximum size of individual pores was up to 2 mm. This is attributable to insufficient preparation of aluminium alloy samples for welding, as organisation of operations in the space simulation test chamber did not permit their mechanical cleaning (scraping) to be performed directly before welding, as well as a not quite favourable vacuum level in the zone of welding operations performance. In all probability, however, this was not the main cause, for, as follows from [6], higher porosity of weld metal was also observed in welding similar samples, using «Universal» hardware in the automatic mode, with their preliminary scraping and in the vacuum of $6.67 \cdot 10^{-3}$ Pa.

Based on flaw detection results it was possible to cut out just 8 samples of butt joints with a minimum number of inner defects for conducting mechanical testing. Such a number of samples were insufficient for statistical evaluation, and so we can only speak of the general level of mechanical properties of the produced joints. Results of tensile mechanical testing, conducted in R-50 tensile testing machine, are presented in Table 2.

Results of testing welded joints of stainless steel 304SS and titanium alloy Ti–6Al–4V confirmed the data, obtained earlier in welding local materials 12Kh18N9T and VT1, using VHT [3, 4]. A practically complete absence of inner defects and high strength of weld metal, close to that of the base metal for stainless steel, or even exceeding it for titanium alloy, are indicative of the applicability of the «Universal» electron beam hand tools for performance of the operations of repair welding of items from these materials up to 2 mm thick.

Unfortunately, investigation results do not permit the same statement with respect to aluminium alloys. Given below are some results of analysis of samples of welded joints on these materials, that to a certain extent confirm the above. As can be seen from Table 2, the level of strength properties of welded joints on both the aluminium alloys is low, strength lowering being less characteristic for samples of alloy 5456. Studying the fractography of the sample fracture

Table 2. Results of mechanical testing of butt joint samples.

Sample No.	Material	Ultimate tensile strength, σ_t, MPa
1	304SS	640.0
2	Ti–6Al–4V	1011.4
3	2219	267.5
4	2219	232.3
5	2219	294.0
6	2219	260.0
7	5456	247.0
8	5456	326.0

Note. Samples No.1 and 2 failed in the HAZ, the others — in the weld.

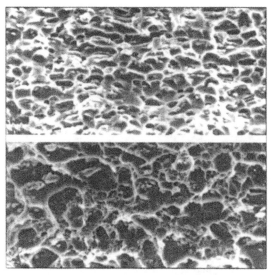

a

b

Figure 3 Microstructure of metal on the fracture surface of a welded butt joint on aluminium alloy 2219 (a) and base metal (b) after rupture testing (×800).

surface after performance of mechanical testing and simultaneous determination of the elemental composition were performed using a scanning electron microscope JSM-840, JEOL, with an X-ray energy spectrum analyser. The microstructure of the fracture surface metal in welded joints of alloy 2219 is that of a non-uniform tough fracture with a great number of pores and other inclusions, enriched in copper. As is seen from Figure 3a, microstructure of the metal of the surface of rupture running through the weld of a butt joint (sample No.3) is more finely dispersed, than that of the base metal (Figure 3b). Microstructure of the surface of rupture, running through the weld in joints of aluminium alloy 5456 is of a tough fracture type with a structure coarser, than that of the base metal. Coarse pores and numerous inclusions, enriched in iron, are found in the fracture zone.

A great number of pores and other inner defects in welded joints of aluminium alloys is largely accounted for in the results of their microstructural examinations in Neophot-2 optical microscope. Quantitative analysis of the secondary phases was performed using Omnimet image analyser, and the detectable phase composition and element distribution were determined in the scanning electron microscope JSM-840 in SEI mode, using a standard Link ZAF4/FLS program.

Microstructures of various zones of welded joints on aluminium alloy 2219 (Figures 4 and 5) demonstrate that the detected phases of the weld metal and HAZ have different dispersity. Considering the nature of distribution of copper, iron and manganese in the HAZ

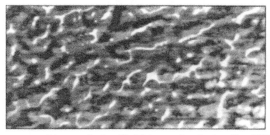

Figure 4 Microstructure of the metal of welded joint on aluminuim alloy 2219 of the large cell (SEI mode) (×900).

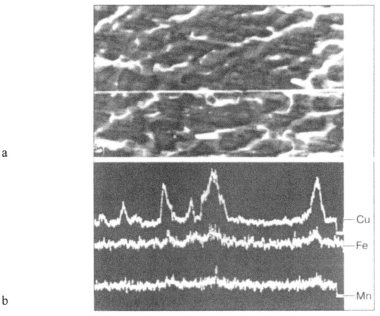

a

b

Figure 5 Microstructure of HAZ metal of welded joint on aluminium alloy 2219 of the large cell
(SEI mode) (a), and copper, iron and manganese distribution along the scanning line (b) (×900).

along the scanning line (Figure 5), it can be assumed that in addition to a homogeneous
structure of Al-based α-solid solution also a number of chemical compounds of $(FeMn)Al_6$
type form during welding. Examination of the microstructure of various zones of welded
joints on aluminium alloy 5456 revealed a somewhat greater amount of inclusions in the
HAZ, than in the weld metal. Proceeding from the nature of iron and manganese distribution
in the HAS metal along the scanning line (Figure 6), it can be assumed, that complex
compounds (for instance, AlFeSiMn), practically insoluble in aluminium, form during
welding. They are the cause for considerable porosity along the fusion line.

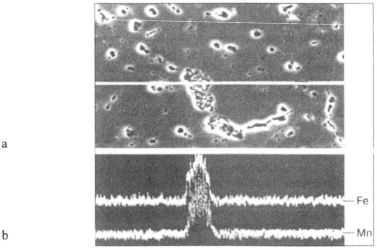

a

b

Figure 6 Microstructure of HAZ metal of welded joint on aluminium alloy 5456 of the small cell
(SEI mode) (a), and iron and manganese distribution along the scanning line (b) (×850).

Microstructural features of welded joints of aluminium alloys 2219 and 5456 indicate that weld metal develops a cellular-dendritic structure near the fusion line, and an equiaxed structure in its central part. The range of average values of welding speed was 10–19 m/h, when making butt joints, and 6–10 m/h in welding overlap joints and cells. However, even at such low welding speeds, complete penetration was not achieved to produce sound joints. Individual sections of welded butt joints are an exception, as well as of welds in welding a cell of 5456 alloy at minimal welding speeds (10.3 and 6.3 m/h, respectively).

Capabilities of «Universal» hardware provided a certain (up to 20 %) increase of the electron beam power, but because of the insufficient number of training sessions (three altitude simulations for the principal crew member and two for the backup), the operators were not able to fully master control of the weld pool behaviour and the techniques of welding operations performance with the electron beam tools at maximum levels of their power. Considering the imperfection of weight-neutralizing devices, it is rather difficult for the welder-operator, wearing a spacesuit, to maintain a constant speed of the hand tool displacement that allows avoiding the welded material burn through. Therefore, the maximum level of electron beam power (750 W) was only used in the final altitude simulations, when welding the US module cell of 2219 alloy. Their scope should be significantly increased during similar training sessions in the future (not less than five altitude simulations for each operator).

It should be noted that the level of strength properties of welded joints of aluminium alloys made with «Universal» hardware is somewhat higher than the strength of welded joints of alloy 1201 welded during experiments on board the flying laboratory at short-term zero gravity (237–245 MPa) [7]. On the other hand, according to [8], EBW of aluminium alloys of 1201 type provides welded joint strength on the level of 320–330 MPa. The results of the performed experiments lead to the conclusion that a certain lowering of strength of aluminium alloy welded joints made at zero gravity, must be related to a significant increase of their porosity [9, 10]. Therefore, when welding these alloys in space, a whole range of problems will have to be solved, concerning hardware support and technology of welding aluminium alloys that minimises pore formation in welded joints.

This, primarily, is surface preparation of the items to be welded that should be carried out immediately before welding. Magneto-abrasive cleaning, proposed in [11], can be used as one of the possible variants. The electron beam gun should provide the appropriate parameters of the welding process, considering that use of sharper-focused beams and increase of their specific power and welding speed may lead to lower porosity. Electron beam oscillation by a certain preset law, increasing the intensity of stirring of the weld pool molten metal, as well as application of vibration technologies, can have a positive effect.

Thus, the results of the training conducted, using «Universal» hardware, and subsequent studies of the produced samples, pointed to the need to improve the strength properties of aluminium alloy welded joints. To a certain extent, this is achievable, using «Universal» hardware at maximum power levels and appropriate welding speeds, which requires additional trials in the space simulation test chamber. A significant improvement of the above characteristics will probably make it necessary to change the electron beam gun parameters and welding modes, in particular to increase the electron beam specific power, thus providing greater penetration depth and higher welding speed.

To enable performance of installation and repair work in ISS under construction now, where the wall thickness is up to 3.2 mm, the electron beam power should be increased up to 1.5 to 2.0 kW. Performance of manual welding operations by a welder-operator wearing a spacesuit at electron beam power above 1.0 kW and high welding speeds seems problematic. Therefore, when developing the next generations of EBW hardware for space applications, it is expedient to envisage the possibility of its use both for manual and mechanised or automatic welding. Its parameters should be re-adjustable, respectively (for instance as described in [12]). The above hardware can be used in the manual mode for performance of auxiliary repair or installation operations, and in the automatic mode in welding critical welds, when certain requirements are made of dynamic loads, vacuum-tightness, etc.

Conclusions

1. Welder-operator training carried out in the manned space simulation test chamber confirmed the principal possibility of conducting the process of manual EBW in space, as well as the possibility of quickly acquiring the necessary skills compared to other welding processes. To provide the required level of operator training, the training sessions should include not less than five altitude simulations.

2. Results of investigation of samples of welded joints on stainless steel 304SS and titanium alloy Ti–6Al–4V, made during the training sessions, confirmed the possibility of producing welded joints with high properties, using «Universal» electron beam hand tools.

3. The results of the training and investigations conducted are indicative of the possibility of producing full penetration butt joints on aluminium alloys. Evaluation of the possibility of welding overlap joints and cells of the same materials requires further trials using «Universal» hardware at maximum power levels. The produced joints of aluminium alloys 2219 and 5456 are characterised by large amounts of inner defects and low strength properties.

4. To improve the quality of aluminium alloy welded joints it is necessary to provide a higher specific power of the electron beam in the next modifications of the hardware to allow increase of the penetration depth and welding speed.

5. In welding aluminium alloys in space, a range of problems will have to be solved related to preparation of the surface of the component immediately before the welding operations, minimising porosity as well as welded joint quality control.

1. Paton, B.E., Gavrish, S.S., Shulym, V.F. et al. (1999) *See pp. 190–212 in this Book.*

2. Nikitsky, V.P., Lapchinsky, V.F., Zagrebelny, A.A. et al. (1985) *See pp. 178–184 in this Book.*

3. Paton, B.E., Lapchinsky, V.F., Zagrebelny, A.A. et al. (1986) *See pp. 283–288 in this Book.*

4. Paton, B.E., Lapchinsky, V.F., Mikhajlovskaya, E.S. et al. (1998) *See pp. 289–296 in this Book.*

5. Paton, B.E., Lapchinsky, V.F., Mikhajlovskaya, E.S. et al. (1997) *See pp. 344–349 in this Book.*

6. Russell, C.C. and Anderson, R.G. (1999) Technological testing of tools for materials processing in space, *Avtomatich. Svarka*, No. 10, pp. 23–30.

7. Rabkin, D.M., Lapchinsky, V.F., Ternovoj, E.G. et al. (1985) *See pp. 229–236 in this Book.*

8. (1998) *Welding in aircraft construction.* Ed. by B.E. Paton, Kiev, MIIVTs, 695 pp.

9. Ternovoj, E.G., Bondarev, A.A., Lapchinsky, V.F. et al. (1976) Investigation of some aspects of weldability of aluminium alloys by the electron beam at zero gravity, *Space research in Ukraine*, Issue 9, pp. 5–11.

10. Ternovoj, E.G., Lapchinsky, V.F., Bondarev, A.A. et al. (1977) Investigation of the influence of zero gravity on weldability and quality of aluminium alloy joints, welded by the electron beam, In: *Space materials science and technology*, Moscow, Nauka, pp. 29–34.

11. Khomich, N.S. and Neznamova, L.O. (1999) Magneto-abrasive cleaning of the surface, *Avtomatich. Svarka*, No. 10, pp. 120–121.

12. Lapchinsky, V.F., Rieck, U., Sobisch, G. et al. (1993) *See pp. 297–304 in this Book.*

COATING DEPOSITION

Development of the Technique
of Thin-Film Coating Deposition in Space*

A.A. Zagrebenly[1], L.O. Neznamova[2], V.P. Nikitsky[2], A.L. Toptygin[6],
V.F. Shulym[1] and E.S. Lukash[1]

The practice of piloted flights in long-term orbital stations and the results of experimental studies of structural materials and coatings behaviour in space showed that the surface of space flying vehicles undergoes significant change during long-time operation. A surface layer down to 10 μm depth is the most prone to change, this affecting the optical, electro-physical and physico-mechanical properties of structural materials during their long-term service in open space [1–4].

Therefore one of the most important practical problems in space technology is deposition of multi-purpose coatings in the space environment [5]. Availability of hardware and technology for coating deposition under conditions of operation of various facilities in space would allow a significant increase in the fatigue life of modern and future space vehicles, as well as producing materials with unique properties under flight conditions.

The method of thermal evaporation and condensation of materials under vacuum was selected to apply thin-film coatings under the flight conditions, i.e. one of the most versatile, well-studied and widely accepted methods of producing high-quality coatings in ground-based technology, that allows fine adjustment of their composition, structure and physical properties [6].

This method was the basis for development and testing in 1979 on board OS «Saluyt-6» of «Isparitel» research unit, the specification of which is given below:

Deposited coating material	silver, copper, gold, alloys
Method of evaporation material heating	electron bombardment
Number of alternatively operating blocks, pcs	2
Power of each evaporation block, W	300–500
Weight of evaporation material in each block, g	not more than 5
Accelerating voltage, kV	0.8–2.0
Bombardment current, A	0.15–0.30
Duty cycle, s	not more than 450
Mains voltage, V	27^{+7}_{-4}
Consumed current, A	not more than 20
Unit weight, kg	25
Overall dimensions, mm:	
working block	$590 \times 252 \times 240$
control panel	$345 \times 290 \times 255$

Experiments conducted with this hardware demonstrated the applicability of thin film coatings in an orbital flight, and confirmed the correctness of selection of the structural and procedural solutions and, on the other hand, revealed some features of the evaporation — condensation processes running in space [7].

---

* (1989) *K.E. Tsiolkolvsky and space industrialisation*, Proceedings of 23rd Tsiolkolvsky Readings, Kaluga, 1988. Moscow, pp. 134–138.

In the majority of the cases the quality of coatings made in space differed only slightly from their ground analogues and satisfied the requirements of industry standards. Such parameters, however, as coating thickness, phase composition and substructure, had several differences. «Isparitel-M» hardware was developed for optimizing the technologies of coating deposition and revealing the characteristic features of the process. This made posible:

• to automatically set and maintain within the specified limits such process parameters, as anode current and cathode filament voltage, that provided constant speed of material evaporation;

• more flexible variation of the technological process parameters;

• to automatically maintain the exposure time;

• to perform deposition on a preheated substrate.

Specification of «Isparitel-M» hardware is given below [8]:

Deposited coating material	silver, copper, gold, alloys
Methods of heated material evaporation	electron bombardment
Number of alternatively working heaters, pcs	2
Power of each heater, W	40–400
Accelerating voltage, kV	4–6
Bombardment current, A	0.01–0.10
Mains voltage, V	27^{+7}_{-4}
Consumed current, A	not more than 20
Unit weight, kg	32
Overall dimensions of the unit, mm:	
working block	$590 \times 252 \times 240$
control panel	$380 \times 345 \times 305$

Experiments, conducted in «Isparitel-M» unit, allowed optimizing the technology of applying temperature-controlling coatings and coatings for brazing in an orbital flight, determining the technological process parameters more precisely, revealing the characteristic features of binary alloy evaporation at zero gravity. The unit evaporation block was used to test new crucible designs, allowing stabilization of the vapour flow.

The results, derived during testing «Isparitel» and «Isparitel-M» units, were used in development of a VHT [9, 10].

VHT allows the following technological processes to be performed: welding, severing, heating and brazing of metals, applying thin-film coatings by the method of thermal evaporation and condensation in open space, and has the following parameters:

Processed materials	steel, titanium and aluminium alloys
Thickness of processed materials, mm:	
welded	0.5–2.0
cut	0.1–1.0
heated and brazed	0.1–3.0
Coating materials	silver, gold, copper and their alloys
Methods of crucible preheating before evaporation	electron bombardment
Accelerating voltage, kV	4–5
Anode current, mA	10–70
Mains voltage, V	27^{+7}_{-4}

In addition to VHT, the earlier derived results on coating deposition in flight allowed development of the third generation hardware. In May 1987 a multi-purpose pilot production unit «Yantar» was tested in the «Mir». This hardware enabled test batches of metal foils to be produced with a low submicroporosity, coatings to be applied on a moving polymer substrate, multilayer condensates to be produced, and fundamental studies to be conducted in the field of evaporation of multicomponent systems at zero gravity and space vacuum.

During this series of investigations the following has been achieved:

• ability was confirmed to conduct directly in flight, the processes of applying thin-film coatings with specified characteristics and properties, not inferior to ground-based analogs;

• hardware was developed and produced for applying mono- and multilayer metal coatings by evaporation of metals and alloys;

• a VHT was developed and tested in space to perform processing operations on repair and restoration of orbital structures;

• multipurpose pilot production unit «Yantar» was developed, allowing production of test batches of film materials, as well as conducting fundamental studies of the processes of evaporation and condensation in an orbital flight.

1. (1978) Heat protection for the space shuttle, *Electron. Austral.*, **40**, No. 5, pp. 8–9.

2. Geis, T.V., Lei, O. and Sacles, T.M. (1978) *A review of solar all coverglass development*, pp. 91–96.

3. Stebfeus, H.D. and Höle, H.-M. (1979) Anwendung des thermischen Spritzens in den Luft- und Raumfahrt, *Schweißen und Schneiden*, **31**, No. 9, pp. 381–383.

4. Stadhaus, W. (1983) Oberflächenbeschichtungen in dem kosmischen Gerätetechnik, *Feingerätetechnik*, **32**, No. 2, pp. 531–536.

5. Paton, B.E. (1977) *See pp. 4–7 in this Book.*

6. (1977) *Thin film technology*. A handbook. Ed. by L. Mayssel and R. Glang, Moscow, Sov. Radio, 465 pp.

7. Dudko, D.A., Zagrebenly, A.A., Paton, V.E. et al. (1977) *See pp. 85–87 in this Book.*

8. Zhukov, G.V., Zagrebelny, A.A., Lapchinsky, V.F. et al. (1985) Hardware and technology for application of various-purpose thin-film coatings in space, In: *17th Gagarin Sci. Readings on Cosmonautics and Aviation*, Moscow, 1983–1984. Moscow, p. 300.

9. Paton, B.E., Dudko, D.A., Bernadsky, V.N. et al. (1977) *See pp. 174–178 in this Book.*

10. Paton, B.E., Dzhanibekov, V.A. and Savitskaya, S.E. (1986) *See pp. 11–18 in this Book.*

Equipment for Coating Deposition by Electron Beam Evaporation Process at Microgravity[*]

D.A. Dudko[1], A.A. Zagrebelny[1], V.F. Shulym[1], V.V. Stesin[1], P.P. Rusinov[1], E.S. Mikhajlovskaya[1], V.P. Nikitsky[2], L.O. Nezmanova[2] and B.N. Svechkin[2]

Construction and operation of long-term orbital stations and complexes in space posed a number of important tasks for space technology. A problem urgent even now is the development of the methods and means of repair of space facilities directly in orbital flight to extend their life and reduce the specific costs related to taking them to orbit. Not the least important of these tasks is finding the methods to restore the temperature-controlling, optical and protective coatings, which degrade during long-term operation in open space, to prevent violation of the normal operating mode of the space facilities.

Various coating technologies are very widely used in ground-based industry, and vacuum technologies have also been introduced, when evaporation and condensation of the

* (1999) *Avtomaticheskaya Svarka*, No. 10, pp. 44–50.

deposited materials proceeds in vacuum. These were exactly the technologies that formed the basis for development of the methods and equipment for coating application in space. Use of these technologies in space facilities, however, runs into certain difficulties.

The first is the limited power supply in space vehicles. For instance, just 500 W of power were allocated to conduct experiments on coating application on board OS «Salyut-6», this requiring development of hardware with a sufficiently high efficiency. The second is the need to contain the molten metal in the crucible at zero gravity, as vacuum coating technology envisages materials evaporation from the liquid phase. And, finally, there is the operating reliability of the hardware located in the immediate vicinity of the crucible with a temperature of up to 1500 °C in the absence of any active cooling means, and the need to eliminate heating of the surrounding elements of the station structure.

After study and comparison of various methods of materials evaporation, it was decided to select electron beam evaporation, which is characterised by an extremely high efficiency. Unfortunately, direct heating of the evaporation material by the electron beam had to be rejected. The point is that in the heated spot an additional force is applied to the molten metal that may lead to unfavourable consequences in the absence of gravity, namely molten material vapours penetrating into the cathode assembly zone, i.e. process disturbance and cathode damage. Therefore, a variant of indirect heating of the evaporation material was accepted, based on electron bombardment of the crucible bottom. With this variant the crucible is heated by bombarding the bottom by a defocused electron beam, thus preventing its burning-through and providing a more uniform heating. Rather simple measures allow complete elimination of penetration by the molten substance vapours into the cathode zone, without the electron beam bending. Implementation of such a variant required development of the following functionally important components (Figure 1):

• secondary power source, including a converter of DC voltage, supplied by the on-board power system, into AC voltage, as well as a system of stabilisation and control of the evaporation mode. The IED[12] developed a power converter with 90 % efficiency,

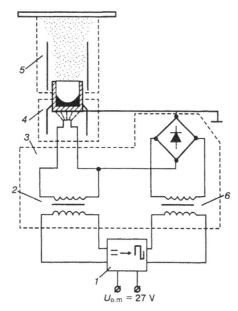

Figure 1 System of indirect heating of the evaporation material: *1* — secondary power source; *2* — filament transformer; *3* — high-voltage power source; *4* — electron beam gun; *5* — evaporator; *6* — anode transformer.

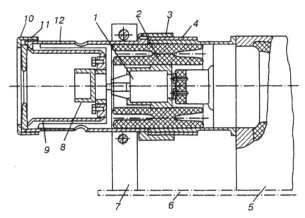

Figure 2 Fragment of a monoblock with electron beam gun and evaporator: *1* — cathode assembly; *2* — bushing; *3* — insulator; *4* — lower coupling nut; *5* — high-voltage block; *6* — platform; *7* — heat sink; *8* — crucible; *9* — sleeve; *10* — coupling nut; *11* — diaphragm; *12* — tubule.

resulting in about 70 % total efficiency of the entire system of electric energy conversion into thermal energy;

• high-voltage power source, converting the variable voltage of 20 kHz frequency into high (2 kV) rectified voltage (for powering the anode circuits) and low variable voltage (for powering the filament circuits of the electron beam guns);

• electron beam guns or electron injectors, forming a flow of electrons to preheat the crucible;

• evaporation devices, providing separation of the liquid and vapour phase of the evaporation material and creating vapour flows with specified parameters.

For maximum reliability and safety the high-voltage power source was made as a monoblock for evaporation and was combined with the electron beam gun and evaporator (Figure 2).

This design solution, like any other, alongside its indubitable advantages, also has a number of drawbacks. In this case, in view of the high heat transfer from the hot (up to 1500 °C) crucible to the high-voltage block elements, where the temperature should not exceed 65–70 °C, it was difficult to provide normal temperature conditions for the evaporation monoblock functioning. On the other hand, it was necessary to take into account the absence of liquid cooling contours on board the space facility in the zones provided by the customer for conducting the space experiments, possibility of direct solar radiation impact and vacuum environment, unfavourable for heat removal.

A wide range of laboratory studies has been conducted, and the methods to optimise the operating mode of the evaporation block were outlined:

• selection of materials with the required heat conductivity for gun elements and crucible fastening, providing the specified distribution of the heat flows;

• mounting a radiator in the form of a heat shunt removing part of the heat flow to the unit case;

• minimising the cross-section of the gun parts, the crucible fastening and creating additional barriers in the thermal flow paths;

• providing vacuum-shield insulation around the crucible in the form of concentric shields, only slightly contacting each other.

It should be noted that metallurgical units, available at the same time for space technology applications — «Versatile Electric Furnace» (USA) and «Splav» and «Kristall» (USSR) provided about 1100 °C temperature of the materials being melted at the same consumed power, and heated up to this temperature in them were not open crucibles, having high radiation and evaporation losses, but closed ampoules placed into special heat-absorbing cartridges.

A highly effective device developed by us allowed heating the evaporation materials up to a temperature of 1500 °C. However, its application in space required solving the problem of molten metal containment in the crucible.

At low gravity when the natural thermal convection is weakened and regular heat-mass exchange mechanisms are suppressed, the role of the surface and intermolecular forces becomes much more important, i.e. such phenomena as surface tension, adhesion and diffusion [1]. In addition, behaviour of the two-phase liquid–vapour system in metal evaporation at zero gravity differs from its behaviour on the ground. A stable vapour flow without the droplike phase should be provided to produce a high-quality coating. In the absence of gravity, separation of the liquid and gaseous phases is difficult and depends only on the surface forces (thermocapillary and concentration-capillary convection). In evaporation of pure metals just the thermocapillary convection proceeds, induced by thermocapillary stress that is determined by the surface tension gradient [2]. Under its impact, the liquid phase moves to the least heated zone, where this gradient is a maximum, while the gaseous phase moves to the most heated zone, this being related to energy conditions of this phase existence.

In other words, the molten metal containment within the crucible can be achieved by selecting such a shape and dimensions of the crucible, that the temperature gradient of the surface tension force, inevitably arising in electron beam heating of the crucible, initiated a stable movement of the liquid phase to the minimal temperature zone. The metal being evaporated should wet the crucible material well, because otherwise it forms a convex spherical surface not adhering to the crucible surface and may «float out» of it at the slightest disturbance.

Verification of these prerequisites during laboratory studies and testing at short-term zero gravity on board the flying laboratory allowed development of the crucible design, providing reliable containment of the molten metal and forming a stable vapour flow [3–5].

This resulted in development of a research unit «Isparitel», used in 1979–1981 on board OS «Salyut-6» to conduct several series of experiments on application of thin film coatings by the method of electron beam evaporation, producing about 200 sample of coatings from silver, copper, gold and their alloys. The unit appearance, its components, as well as installation on board the orbital station for experiment performance, are described in [1].

It should be noted, that a simplified variant of the unit was developed and manufactured, because of extremely limited time. This affected, primarily the control system that was not fitted with automatics or programming means, and therefore coating process control (switching on/off, and switching from one gun to another, sample replacement, holding the crucible preheating time and sample soaking, mode monitoring by the instrument) was performed manually by the operator. The unit could operate at two fixed levels of electron beam power (250 and 300 W), that could not be varied in time, and therefore the temperature of the crucible with the evaporation material was unstable and rose continuously. It is understandable that it was not easy for the operator, who has had limited training, to operate such a unit in a space fight, and there was a probability of him making mistakes and not being precise enough when doing the work. Therefore, the experiments conducted in this unit in space, could not provide any valid scientific data, and so the main results could be regarded to be verification of the basic ability to evaporate materials by the accepted sequence and correctness of the selected design solutions.

However, analysis of the produced coating samples and their comparison with the ground analogs revealed certain differences concerning mainly the binary alloy coatings [6–9]. To conduct a more thorough study it was necessary to carry out experiments on coating

Figure 3 High-voltage block with evaporators of «Isparitel-M» unit.

deposition in space at a new level of quality. In 1984 a new experimental coating unit «Isparitel-M» was made and delivered on board OS «Salyut-7», having wider processing capabilities [10].

The experimental procedure was the same: working block (WB) was installed in the docking chamber of the orbital station, and the control panel (CP) in the habitable module, so the unit design was unchanged, but with the same overall dimensions and configuration the WB and CP of «Isparitel-M» unit differed significantly from their predecessors. In particular, WB accommodated a new high-voltage source common for both the guns (Figure 3) with a higher accelerating voltage (5 kV), thus allowing the cathode assembly life to be extended at the same output power (300 W). Such a high-voltage block was later used in VHT [11].

Crucible design remained the same, but its thermal insulation was improved, evaporators were mounted on hinges, thus significantly lowering the high-voltage block heating and making replacement of the crucibles and cathode assemblies more convenient. Samples for coating could now be preheated up to the temperature of 150–250 °C, while the easily removable holder with samples could be replaced or taken off, when doing other processing programs. WB was fitted with a vacuum gauge and automatics module to monitor the processing parameters (for instance, temperature, pressure, etc.) by CP screen and telemetry system.

CP of «Isparitel-M» unit (see Figure 4 on p. 103) consisted of the following systems:

• secondary power source, i.e. power converter, inverting the on-board voltage into AC voltage of 20 kHz frequency;

• information-measurement complex, representing in the CP screen the condition and parameters of the unit functioning, as well as the main parameters of the technological process;

• programmer, allowing flexible embodiment of several variants of the technological process programs by setting the electron beam power after certain moments of time. Programs were entered, using a programming panel (PP), connected to the CP.

Cyclogram of one of the variants of the unit operation (coating on a preheated substrate) is shown in Figure 4. As is seen from the Figure, programming of heating power allowed considerably stabilising the crucible temperature (hatched line indicates its temperature at

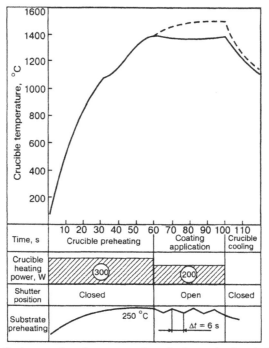

Figure 4 Cyclogram of coating onto a preheated substrate in «Isparitel-M» unit.

unchanged heating power of 300 W). Such cyclograms were verified in the laboratory for each evaporation material with mandatory monitoring of crucible and substrate temperatures with thermocouples (this is impossible in flight experiments). After that all the flight and ground-based experiments were conducted by verified programs.

During the whole year of 1984, the «Isparitel-M» unit was operated by several crews, who produced about one hundred samples of coatings of various materials (mostly copper-silver binary system). The results of analysis of these coating samples are described in [10]. In addition to that, «Overcooling» experiment on melting and solidification of silver-ger-manium eutectic alloy was conducted in the unit together with the Indian experts. Instead of the evaporators, the cosmonauts mounted special devices to produce a defocused electron beam, and instead of a six-position drum-manipulator — a replaceable block for melting, developed for this experiment. The block description and the experimental procedures are given in [12].

The operation of «Isparitel-M» unit confirmed the correctness of design solutions, embodied in it. Many of them were later used in VHT development and manufacturing. In particular it was fitted with the same high-voltage block and evaporation module with additional protection, preventing molten metal particles from flying out of the tool in case of careless handling by the operator.

In July 1984, cosmonauts V. Dzhanibekov and S. Savitskaya conducted successful testing of VHT in open space that involved welding, cutting, brazing and manual electron beam deposition of coatings [13]. The coating experiment was chiefly aimed at evaluation of the possibility of conducting such operations manually and testing the hardware perform-ance. To facilitate visual monitoring for the operator, silver was used as the evaporation material, and an aluminium plate blackened by anodising was the substrate. Experiment showed that manual deposition of coatings of an acceptable quality can be quite readily performed in open space even by an operator with a low level of training.

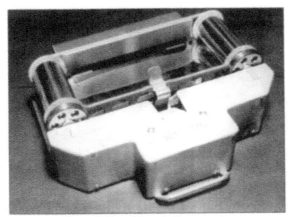

Figure 5 Tape drive of «Yantar» unit.

In order to continue investigation of the coating processes, «Yantar» unit, being a modification of «Isparitel-M», was delivered, this time, on board the «Mir» in 1987. While preserving the composition and control principles of its predecessor, it has acquired certain distinctive features. A six-position manipulator for coating flat samples was replaced by a tape drive, as the main working head (Figure 5), allowing coatings to be applied onto a moving polymer film (for instance, polyimide film PM-A 20 to 50 μm thick, moving at the speed of 20–40 m/h) for various purposes:

• studying the evaporation dynamics of binary alloys and multicomponent systems;

• providing the capability of alternative coating of several materials on one substrate without de-pressurizing the unit and studying the impact of atomic oxygen on the applied coatings;

• checking the ability to apply protective coatings (protectors) onto the earlier deposited temperature-controlling coatings;

• making test batches of metal foils with a low submicroporosity.

Instead of PP, the unit incorporated permanent storage for entering pre-checked experimental programs, thus preventing errors when entering a program.

Control system noise-immunity was improved to prevent malfunctions arising from operation of the high-frequency power converter that were observed in «Isparitel-M» unit.

These measures, however, turned out to be insufficient and hardware malfunctions continued to be observed. Therefore the program of experiments in «Yantar» unit had to be curtailed. Nonetheless, it was still possible to produce a large number of various coating samples.

After successful testing of VHT in space, the PWI developed and manufactured «Universal» hardware that is the prototype of a standard on-board system of electron beam hand tools for fulfilling specific technological tasks, arising in the orbital station operation [1].

Two of its four work tools (WT) are designed for coating application; for this purpose they incorporate special technological devices (TD) that can be readily dismantled later on and then the tools can be used for welding operations.

The first tool allows coatings deposition by evaporation from a crucible and is fitted with a rotating extension with four crucibles (Figure 6). This enables various material coatings to be applied or the stock of a certain material to be increased, as required, if the processed surfaces have considerable dimensions. After completion of the work with all the crucibles, or for other reasons, the operator can easily «shoot off» the extension, replacing it by another one, or use this tool further on as a welding tool.

Figure 6 Rotary extension with crucibles of the tool for coating («Universal»).

The second tool is designed for coating application by thermal evaporation of a material, fed in the form of wire into an open crucible, preheated by the electron beam. Its development is an attempt to eliminate the crucible method of material evaporation, as well as check the ability to evaporate aluminium by such a procedure. A replaceable device, mounted on the tool (Figure 7), allows feeding 0.8–1.0 mm wire into the heating zone at up to 5 mm/s rate. Laboratory testing, as well as testing in the space simulation test chamber, when manual manipulations were performed, demonstrated that such a tool enables application of silver coatings [14]. Evaporation of aluminium, having a lower vapour pressure, and of other metallic materials, requires a significant lowering of the wire feed rate or further fitting the tool with a device, providing its metered feed. This also requires finding some solution for the problem of material for the open crucible, since application of tungsten-cobalt alloys instead of molybdenum still did not permit a significant extension of the crucible service life. Therefore, application of aluminium coatings with such a tool during performance of repair-restoration work is still an urgent issue.

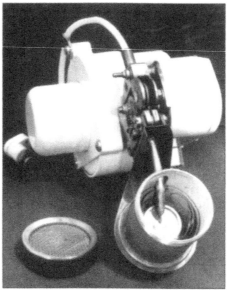

Figure 7 Evaporator with filler wire feed («Universal»).

Parameter	«Isparitel»	«Isparitel-M»	«Yantar»	VHT	«Universal»
Supply voltage, V	27^{+7}_{-4}	23–34	23–34	23–34	23–34
Maximum consumed current, A	20	20	20	20	20
Accelerating voltage, kV	2	5	5	5	8
Maximum current in electron beam, mA	150	65	65	65	Up to 120
Number of electron beam guns for evaporation, pcs	2	2	2	1	1
Maximum time of continuous operation of each gun, min	5	5	5	5	5
Evaporation materials	Silver, copper, gold, copper–silver	Copper, copper–silver, antimony-bismuth	Copper, copper–silver, copper–manganese, magnesium fluoride, silver–palladium	Silver	Silver, gold
Maximum weight of evaporation material, g	Not more than 5	5	5	5	5
Kind of substrate	Flat	Flat	Polyimide film	Plate	Plate
Number of substrates in a set, pcs	12	12	10 m long strip	2	Depending on modification
Substrate dimensions, mm	80×60	80×60	Width 70 mm	240×170	Same
Substrate heating, °C	Without heating	150–250	Without heating	Without heating	Without heating
Control	Manual	Programmed from PP	Programmed from ROM	Manual	Manual
Overall dimensions, mm:					
WB	235×250×530	240×260×550	240×260×530		
CP	290×350×400	295×355×405	295×355×405	500×450×300	
TD					470×420×375
WT					330×150×300
Weight, kg:					
WB	13.6	14.6	13.8	27.0	
CP	13.4	17.4	17.8		
TD					37.0
WT					8.0

The correctness and unambiguity of the obtained preliminary data can be confirmed by a series of check experiments, using hardware, made at a qualitatively new level. It will have to ensure reliable operation of the control system and high stability of process parameters, in particular, of the evaporated material temperature, measured and recorded during the coating process.

Specification of the coating hardware, developed and tested in space, is given in the Table.

Conclusions

1. A series of experiments on application of thin-film coatings, conducted in OS «Salyut-7» and OC «Mir», using «Isparitel», «Isparitel-M», «Yantar» and VHT hardware, confirmed the fundamental possibility of producing optical and temperature-controlling coatings in open space.

2. The quality of coatings produced in space is not inferior to their ground analogues and is on the level required by industry standards or higher in all probability, at the expense of a more favourable vacuum environment.

3. Analysis and investigation of samples revealed some special features of binary alloy evaporation at zero gravity.

4. Experiments involving VHT proved the applicability of such a method of coating deposition and hardware to perform high-quality repair and maintenance work.

5. The main design solutions embodied in the hardware have been verified and experimentally confirmed.

6. When the hardware was developed and previous experiments were prepared and conducted, it was not possible to create a stably operating device for evaporation of aluminium and some other non-metallic materials, in particular, silicon oxide. This should be the goal in development of a new generation of hardware and should allow performance of specific work on repair and restoration of protective temperature-controlling coatings in long-term orbital space facilities.

1. Paton, B.E. and Lapchinsky, V.F. (1998) *Welding and related technologies in space*, Kiev, Naukova Dumka, pp. 44–48, 92–100.

2. Shulym, V.F., Zagrebelny, A.A., Lapchinsky, V.F. et al. (1985) Hardware for materials evaporation in space, In: *Problems of space technology of metals*, Kiev, Naukova Dumka, pp. 57–63.

3. Zagrebelny, A.A., Lapchinsky, V.F., Stesin, V.V. et al. *Method and device for melting materials under zero gravity conditions*. USSR author's cert. 816070, Int. Cl. B 64 G 1/00, priority 10.01.80, publ. 21.11.80.

4. Kasich-Pilipenko, I.E., Pasichny, V.V., Malov, N.I. et al. *Evaporator*. USSR author's cert. 1110214, Int. Cl. B 64 G 1/00, priority 06.07.81, publ. 22.04.85.

5. Zagrebelny, A.A., Lapchinsky, V.F., Stesin, V.V. et al. *Evaporator*. USSR author's cert. 1485668, Int. Cl. B 64 G 1/00, priority 12.10.87, publ. 08.02.89.

6. Lukash, E.S., Lapchinsky, V.F., Shulym, V.F. et al. (1985) *See pp. 335–340 in this Book.*

7. Palatnik, L.S., Toptygin, A.L., Cheremskoj, P.G. et al. (1985) *See pp. 331–335 in this Book.*

8. Lukash, E.S., Shulym, V.F., Grigorenko, G.M. et al. (1986) *See pp. 340–344 in this Book.*

9. Toptygin, A.L., Arinkin, A.V., Savitsky, B.A. et al. (1986) Composition and structure of films of Ag–Cu alloys, condensed under conditions of orbital station flight, In: *Problems of space technology of metals*, Kiev, PWI, pp. 43–50.

10. Zagrebelny, A.A., Lukash, E.S., Nikitsky, V.P. et al. (1988) Development of the technique of thin-film coating application in flight conditions, In: *Proc. of 22nd Tsiolkovsky Readings*, Kaluga, 1987. Moscow, pp. 134–138.

11. Shelyagin, V.D., Mokhnach, V.K., Stesin, V.V. et al. *An apparatus for electron beam welding*. USSR author's cert. 1626542, Int. Cl. B 23 K 15/00, publ. 08.10.90.

12. Shulym, V.F., Lapchinsky, V.F., Lukash, E.S. et al. (1985) *See pp. 398–401 in this Book.*

13. Nikitsky, V.P., Lapchinsky, V.F., Zagrebelny, A.A. et al. (1985) *See pp. 178–184 in this Book.*

14. Paton, B.E., Lapchinsky, V.F., Mikhajlovskaya, E.S. et al. (1997) *See pp. 344–349 in this Book.*

Comparative Analysis of the Structure of Pure Metal Films, Condensed in Space and on the Ground[*]

L.S. Palatnik[6], A.L. Toptygin[6], P.G. Cheremskoj[6], B.A. Savitsky[6], A.V. Arinkin[6], V.P. Nikitsky[2], G.V. Zhukov[2], V.F. Lapchinsky[1] and V.F. Shulym[1]

This work is devoted to comparative study of the structure, substructure and submicroporosity of films of pure fcc-metals, condensed under the conditions of flight of OS «Salyut-6» and on the ground. The timeliness of conducting such studies is related to the need to solve the problem of development of a perfect technology of deposition and restoration in space of various-purpose coatings (temperature-controlling, protective, optical, etc.) [1].

The object of investigations were films of copper, silver and gold, condensed during several series of experiments in «Salyut-6» in «Isparitel» unit, developed at the PWI. Evaporation material was heated in a molybdenum crucible at the expense of bombardment of the latter by a beam of electrons formed by a special electron gun. Evaporator and crucible designs are given in work [2].

The evaporation block of the unit with the crucible and substrates is placed into the docking chamber, opening into the outside space at its depressurizing. Therefore, the composition of residual gases in the docking chamber is determined by the composition of the ambient atmosphere of the station. Substrates are fastened on a special drumlike holder, rotated to permit their successive processing during the experiment. Condensation was conducted onto unheated substrates of glass, titanium (polished and unpolished), titanium with a lacquer coating, and carbon-reinforced plastic. Film condensation time was varied between 1 and 200 s. This allowed making and studying films in the thickness range from units to thousands of nanometers. In some cases the total condensation time was up to 600–750 s due to repeated deposition of material onto its own sublayer.

Surface morphology, structure and substructure of the films were studied by metallographic analysis, scanning and transmission electron microscopy, X-ray diffractometric structural analysis, as well as the method of X-ray low-angle scattering (XLAS). Results of electron microscopy investigations of silver films, prepared in flight and of films, made on the ground, allow distinguishing four stages in the growth process:

- formation of nuclei and islet structure;
- islets growing together through coalescence;
- formation of channels;
- formation of a solid film.

[*] (1985) *Problems of space technology of metals*, Kiev, Naukova Dumka, pp. 88–93.

The following relationships are traceable for gold films. In this case, however, the recrystallisation processes proceed less intensively, therefore, gold films have a significantly more disperse structure, compared to that of silver films.

Results of electron microscopy examination of the microstructure of thin gold and silver films (up to 1000 Å) agree well with the results of X-ray diffraction analysis of macrostresses and substructure of thicker films, conducted by measurement of the crystalline lattice spacing and analysis of the X-ray line width.

Crystalline lattice spacing a of silver and gold coatings was determined by the position of maximum of $K_{\alpha 1}$-constituent lines (551), (333) and (420), recorded in «Dron-2» diffractometer in filtered radiation of copper and cobalt anodes. The inclined filming method ($\sin^2\psi$-method) [3] was used to analyze microstresses in unseparated films. Crystalline lattice spacing in unstressed condition was evaluated by computation, using a-$\sin^2\psi$ graphs, and by the data on self-standing films in a number of cases.

Results of inclined filming of flight and ground samples are given in the Table.

As can be seen from the Table, value a_0 for gold films in the unstressed state is close to the reference value a_r = 4.078 Å. Tensile stresses, observed in gold coatings according to [3], can be due to substrate heating at condensation (thermal stresses) and to condensate «shrinkage» as a result of partial annealing of crystalline lattice defects of solidification origin (structural stresses). Apparently, the fact that stresses in the film on a titanium substrate are significantly smaller than in the case of glass, is indicative (considering the higher heat conductivity of titanium) of a substantial contribution by the thermal component. The Table gives the results of an assessment of substrate heating at condensation by the magnitude of macrostresses on the assumption of the absence of the structural component.

In the case of silver, the crystalline lattice spacing is also close to the reference a_r = = 4.0863 Å. The tensile stress level in silver coatings is essentially lower, while ΔT value for space and ground samples is independent of the substrate material (glass or titanium) and is about 20–40 °C.

Crystalline lattice period and characteristics of silver and gold coating substructure.

Coating material	Substrate material	Condensation time, s	Thickness, μm	a_0, Å	Macrostress, kgf/mm²	ΔT, deg	Microstress, kgf/mm²	L, Å
Silver	Titanium	40 / 40	0.10 / 0.40	4.0859 / 4.0855	1.9 / 1.7	40 / 40	–	–
		100 / 100	0.60 / 2.00	4.0855 / 4.0855	0 / 0.3	0 / 10	–	–
		20×200 / 200	1.45 / 1.23	4.0859 / 4.0854	1.1 / 1.1	20 / 20	–	–
	Glass	20×200 / 20×200	1.27 / 0.56	4.0862 / 4.0862	1.5 / 3.0	20 / 40	–	≥ 1500 / 1000
Gold		20×200 / 20×200	0.12 / 0.65	4.078 (2θ) / 4.076 (2θ)	19.0 / 18.0	300 / 300	43 / 28	310 / 140
	Titanium	20×200 / 20×200	0.06 / 0.02	4.0791±0.005 / –	3 / –	60 / –	20 / 60	220 / –

Note. The numerator gives the data, obtained in space, and the denominator — the ground data.

Gold coatings are characterized by a significant widening of the lines. The width of the line (420) of thin gold samples was up to 5° (2θ), when filming was done in K_α-Co radiation. With greater thickness of the lines, the diffractograms become somewhat narrower, and a spectral doublet is observed. Silver films are characterized by a small widening of the lines, and splitting of K_α-doublet is practically the same as in the standard. This is indicative of a more perfect substructure of silver coatings.

The Table gives the data on the size of L blocks and microstresses, obtained on the basis of analysis of reflexes widening by the profile approximation method [3]. Two orders of reflections (111) and (200) were filmed to determine the role of the size of the blocks and microstresses (microdeformations). Filming was performed in K_α-Co radiation.

It follows from the Table, that gold coating substructure is characterized by a comparatively small size of blocks and high level of microstresses. In the case of silver, the observed effects are at the limit of sensitivity of the approximation method. Comparison of gold and silver films indicates that the latter have a much more perfect structure. Gold film substructure is indicative of a slowing down of the processes of recovery and recrystallisation.

A possible cause for the difference in the substructure of the gold and silver coatings are impurities entrapped from the residual atmosphere at condensation. Concentration of these impurities depends on the composition and pressure of residual gases, and to a great extent, also on the condensation rate. Estimation shows, that the rate of gold condensation is equal to (1–10) Å/s, whereas for silver it is (50–100) Å/s. Therefore at the same vacuum conditions, silver coatings should contain significantly fewer impurities than gold coatings. Such impurities may bind the vacancies, not letting them reach the surface, and, thus, lead to a noticeable reduction of the lattice spacing. It can be seen from the tabulated data, that in case of gold films a reduction of the lattice spacing is recorded in the ground coatings, compared to those made in space, and this difference is greater than for silver coatings. Also, the impurities prevent the movement of dislocations and grain boundaries, thus slowing down the recovery and recrystallisation processes. This results in silver films being structurally more perfect than gold films. At the same time, no significant differences are found in the substructure of flight and ground samples of gold and silver.

As was indicated above, after the channels have been filled and a continuous film has formed, it still has micro openings down to a considerable depth, thus leading to significant amount of porosity in the condensed films. Studying the porosity is of interest in itself, as the processes of pore formation and healing are the consequence of several mechanisms, namely diffusion-vacancy, shadowing effect, growing together or abutting of the structure elements being not close enough, their non-uniform growth, gas absorption, etc. The dominating role of each of these mechanisms is determined by the combination of physical and technological conditions of evaporation and condensation processes [4]. In this connection, comparison of porosity characteristics of the flight and ground samples is important for analysis of the influence of the orbital station flight conditions on the mechanism of pore formation.

For all the studied samples of silver, gold and copper films of various thicknesses it was found that their total porosity estimated by absorption of monochromatic radiation is up to several percent. The principal contribution to bulk concentration of pores in the studied flight and ground samples is made by micron-size macropores that are not revealed by the XLAS method. However, the number of submicropores is by several orders of magnitude higher than the number of macropores. Submicropores are mostly responsible for the observed low-angle scattering of X-rays in the studied samples.

Investigation of film submicroporosity was conducted by the XLAS method in a high-resolution low-angle diffractometer with U-shaped Kratki collimator. Figure 1 gives the indicatrices of XLAS beams for copper films, from which it follows that XLAS intensity

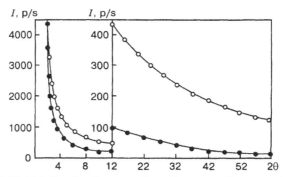

Figure 1 XLAS indicatrices for copper films: ● — space; ○ — ground samples.

and integral width of indicatrices have a significantly smaller role for space samples, compared to ground samples. This ratio of XLAS parameters in the flight and ground samples is also characteristic of silver and gold films.

Analysis of the asymptotic of XLAS indicatrices leads to the conclusion that silver and copper films mostly are found to contain submicropores, whose orientation and anisometry is predominantly determined by the orienting impact of the molecular flow. Indeed, in the ground samples of silver and copper they have an elongated shape and with their larger dimension are oriented in the direction close to the direction of the molecular flow incidence. In flight samples, however, the submicropores are practically equiaxial.

Unlike silver and copper films, gold films made on the ground, in addition to pores, oriented in the direction of the molecular flow incidence, also contain un-oriented sharply anisometric submicropores, chaotically distributed through the condensate volume. Their shape is unrelated to the direction of the molecular flow incidence. Gold films, prepared in space, also incorporate anisometric submicropores, however, the degree of their non-axiality is much smaller.

All the studied flight samples of silver demonstrate a smaller degree of polydispersity of the scattering inhomogeneities than the ground samples, which feature a pronounced polymodality of submicropore distribution. This is manifested in the form of several maxima on the curves of invariants of XLAS indicatrices (Figure 2a). Similar curves for the flight samples as a rule have one maximum (Figure 2b).

Bulk concentration of submicropores in the ground samples of the studied metals is 0.1 %, which is much higher than in the flight ones, where is does not exceed 0.04 %. The characteristic pore size in the flight samples is larger than in the ground samples (mean-root-square and characteristic size of pores in the films of gold is 220, copper — 650, and silver — $\sim$ 500–100 Å).

It is important to note that pores in the silver and copper films are much coarser than in the gold films, this being associated with the higher proneness of copper and silver to structure element coarsening at condensation. Indeed, as shown by the data of electron microscopy, recrystallisation in silver films proceeds more intensively than in gold films.

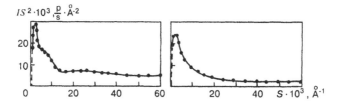

a   b

Figure 2 Invariants of XLAS indicatrices for silver films: a — space; b — ground samples.

Therefore films condensed under the conditions of flight of OS «Salyut-6», are not inferior to the ground samples by their structure features. The revealed differences in porosity characteristics are independent of condensation rate, kind of substrate or film material. These are attributable to inadequacy of the conditions of film preparation in the orbital station and in the ground-based laboratory, in particular the difference in the composition of the residual gases atmosphere at condensation. Further experiments must be staged to be able to unambiguously determine the causes for the differences found.

1. Paton, B.E. and Kubasov, V.N. (1979) *See pp. 8–11 in this Book.*

2. Shulym, V.F., Zagrebelny, A.A., Lapchinsky, V.F. et al. (1985) Hardware for materials evaporation in space, In: *Problems of space technology of metals*, Kiev, Naukova Dumka, pp. 57–63.

3. Palatnik, L.S., Fuks, M.Ya. and Kosevich, V.M. (1972) *Mechanism of formation and substructure of condensed films*, Moscow, Nauka.

4. Palatnik, L.S., Cheremskoj, P.G. and Fuks, M.Ya. (1982) *Pores in films*, Moscow, Energoizdat, 216 pp.

Some Results of Analysis of Coating Samples, Produced in Space*

E.S. Lukash[1], V.F. Lapchinsky[1], V.F. Shulym[1], A.A. Zagrebenly[1], L.O. Neznamova[2], V.P. Nikitsky[2] and G.V. Zhukov[2]

One of the most important practical problems for space technology is deposition of coatings in space. Development of commercial technologies for deposition of various coatings in space will allow an essential increase of fatigue life of the currently used and future space vehicles. Over recent years the most often expressed thought has also been that use of thin film technologies opens up broad perspectives in terms of creation of materials with unique properties in space [1–5].

The purpose of this work is to summarize the most preliminary results on the techniques of producing coating samples in space, as well as characterisation of the produced coating properties, conducted at the PWI.

Space experiments and their ground-based analogues were conducted in «Isparitel» unit (1979–1981) and «Isparitel-M» unit (spring of 1984).

The paper deals with the data, related to «Isparitel» unit. For obvious reasons this unit could not be used for fine experiments as its design features did not allow obtaining sufficiently valid information. A total of 186 coating samples of 60 × 80 mm size were produced in «Isparitel» unit. In the majority of the cases the quality of coatings produced in space differed only slightly from their ground analogues and met the requirements of industrial standards. However, some results of investigation of the properties of coatings, produced in this unit, both on the ground and in space, were unexpected. It is natural that these are exactly the results that are the most interesting. Let us consider two of them, concerning coating thickness and their phase composition.

Coating thickness was measured, using MII-4 microinterferometer. It turned out that coating thickness was non-uniformly distributed over the sample surface, this non-uniformity having a chaotic nature, and the difference in thickness often was up to 40 % for one and the same sample. Such a phenomenon was observed both in space and the ground samples.

* (1985) *Problems of space technology of metals*, Kiev, Naukova Dumka, pp. 81–87.

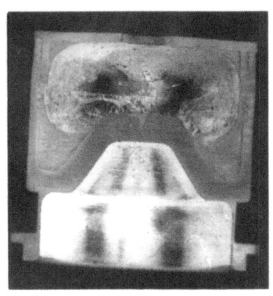

Figure 1 Cross-section of the crucible after alloy evaporation from it at zero gravity. Heating by the electron beam from below.

So it was natural to suppose that it was unrelated either to zero gravity or the space vacuum, but rather was determined by the design features of the crucible, used in «Isparitel» unit for evaporation at zero gravity (Figure 1). When this crucible was designed, it was assumed that the molten metal would be contained in the bottom recesses by surface tension forces, while evaporation would occur in a steady manner from the molten pool surface. As demonstrated by subsequent experiments, this mechanism operates in a reliable manner only at a certain level of crucible filling. In an overheated or almost empty crucible the molten metal can be non-uniformly distributed over the volume, thus leading to a non-uniform coating thickness. All this required a more careful approach to selection of the design of the crucible for evaporation at zero gravity, so that it on the one hand provided, a reliable containment of the molten metal, and on the other a uniform vapour flow that would be independent of the level of crucible filling.

An additional experiment was staged to select the optimal design. Coatings were produced on the ground in «Isparitel-M» unit. Standard glass photographic plates of 9 × × 12 cm size without photoemulsion were used as the substrate. Copper was the model material for evaporation.

Figure 2 gives the diagrams of coating thickness distribution, depending on the angle of the vapour flow incidence for several variants of the crucible design. The deposited coating was scratched through according to the schematic shown in Figure 3. Coating thickness was measured in points of intersection of the straight lines and the circumferences, using MII-4 interferometer.

Average thickness was determined by points *1–8* for the angle of vapour flow incidence, different from the normal. The magnitude of this deviation was found knowing the distance from the crucible to the substrate centre and the distance from the plate centre to each of points *1–8*. The same operation was conducted for a series of radial points at levels *A–F*. From Figure 2 it is seen that coatings produced during evaporation from the crucibles *C* (16 %) and *D* (10.5 %), have a thickness distribution, which is the closest to the cosinusoidal shape (dashed line within 20° angle). Coatings deposited from crucibles *A* and *B* have a much greater discrepancy in thickness within 20° angle, 32 and 26 %, respectively. These two

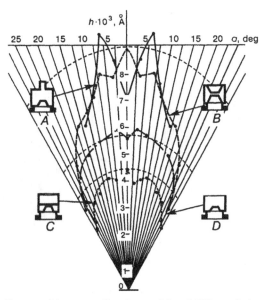

Figure 2 Directivity diagram of the vapour flow for crucibles of different design: *A, B* — crucible with «pipe on top» cover and a conical cover; *C* — standard crucible with a round hole in the cover; *D* — open crucible.

crucibles are characterized by a sharper directivity diagram of the vapour flow (within 7.5° angle), and deviations in coating thickness are not more than 13 % for crucible *A* and 6 % for crucible *B*. In view of the above crucible *C* was selected for experiments in «Isparitel-M» unit. Analysis of coating thickness in samples produced using this crucible in space fully confirmed the data obtained in the ground-based experiments. Therefore a crucible of such a design was recommended for performance of further work in «Isparitel-M» unit. No chaotic non-uniformity of the coating thickness was observed in the samples made in «Isparitel-M» unit. Crucibles *A* and *B* or their combination, apparently can be used to perform practical work, for instance, in a hand tool. Evaluation of the rate of evaporation and condensation was performed by determination of the amount of the evaporated metal

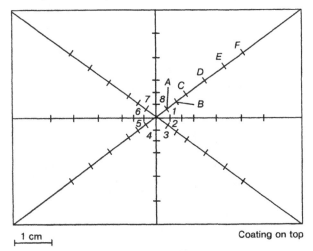

Figure 3 Schematic of coating thickness measurement.

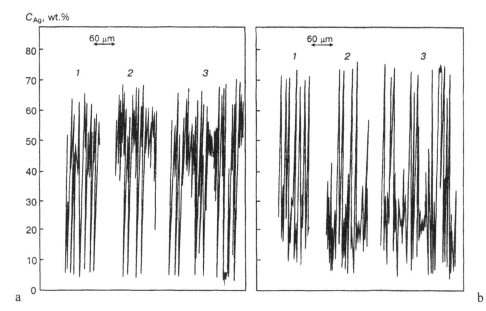

Figure 4 Distribution of the content of Ag (a) and Cu (b) across the crucible cross-section in the remaining Ag–50 vol.% Cu alloy after evaporation in space: *1* — left edge; *2* — middle; *3* — right edge.

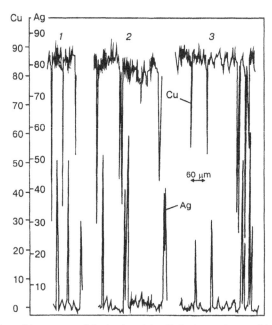

Figure 5 Distribution of the content of Cu (*top*) and Ag (*below*) over the crucible cross-section in the remaining Ag–50 vol.% Cu alloy after evaporation on the ground: *1* — left edge; *2* — middle; *3* — right edge.

by crucible weighing and measurement of coating thickness, respectively, taking into account the fact that the operator controlled the evaporation time. The obtained results demonstrated that in the majority of the cases the rates of evaporation and condensation in space are higher than their ground analogues in the corresponding modes.

Some differences in the space and ground samples were observed, during studies of the phase composition of the coatings and remaining material after evaporation of Ag–50 vol.% Cu alloy. The electron diffractometry method was used to determine that the coatings produced at the start of the process of evaporation of Ag–50 vol.% Cu alloy from the crucible, contain silver and silver oxides in the ground samples, and copper, silver and oxides of these metals in the space samples. Unfortunately the coatings produced in the middle and at the end of the process of evaporation from the crucible could not be studied because of their bad damage and contamination, but analysis of phase composition of the alloy remaining after evaporation in «Cameca» X-ray microanalyser demonstrated, that the remaining Ag–50 vol.% Cu alloy after evaporation in the conditions of space (Figure 4) preserved approximately the same composition, while after evaporation on the ground (Figure 5) it mostly contained copper. Photos of the microstructures of the remaining Ag–50 vol.% Cu alloy are shown in Figure 6, and further illustrate the results derived using the microprobe. Based on the derived data, the obtained structures can be defined as follows: photographs of the microstructure of the remaining alloy after evaporation on the ground (Figure 6a) show dendrites of solidified copper and traces of silver in the interdendritic space. The microstructure of the remaining alloy after evaporation in space (Figure 6b) is a typical eutectic structure, namely the dark-coloured acicular formations are copper, and the light field is silver. Sections for metallographic studies were prepared using the usual procedures [6]. Results of the conducted studies suggest that evaporation of the alloy of the given composition proceeds in a more uniform manner in space than on the ground.

These results studies were of indubitable interest, but need thorough verification and interpretation. Therefore the above experiments with Ag–Cu alloy were carried on in

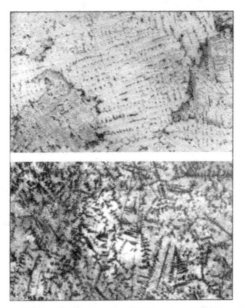

a

b

Figure 6 Microstructure of the remaining Ag–50 vol.% Cu alloy after evaporation on the ground (a) and in space (b) (×50). Chemical etching in the reagent (~ 10 ml HNO_3, 15 ml H_2SO_4, 14 ml chromium anhydride and 60 ml of distilled water) at etching time of 5 s.

«Isparitel-M» unit. Thin films were deposited at different levels of the crucible filling. Preliminary analysis of these films confirms the earlier derived results. We assume that it is expedient to continue these studies, possibly, with simpler binary alloys, as they may turn out to be very useful for understanding the features of mass exchange in liquids exposed to low-gravity fields.

1. (1980) *Space technology*. Ed. by L. Steg, Moscow, Mir, pp. 38–45, 129–143, 319–334.

2. Belyakov, I.T. and Borisov, Yu.D. (1974) *Technology in space*, Moscow, Mashinos-troyeniye, 290 pp.

3. (1974) Evaporation in space manufacturing, *AIAA Pap.*, No. 667, p. 14.

4. Dreis, A.L. and Paul, R.S. (1976) Evaporation, diffusion and convection in melts, being in a free state in a low-gravity field, *Raketn. Tekhnika i Kosmonavtika*, 4, No. 2, pp. 10–12.

5. Grishin, S.D., Leskov, L.V. and Savichev, V.V. (1978) *Space technology and fabrication*, Moscow, Znaniye.

6. Bogomolova, N.A. (1982) *Practical metallography*, Moscow, Vysshaya Shkola, 271 pp.

Investigation of the Influence of Gravity on the Composition of Coatings from Binary Ag–Cu Alloys[*]

E.S. Lukash[1], V.F. Shulym[1], G.M. Grigorenko[1], O.D. Smiyan[1] and L.O. Neznamova[2]

Experiments on deposition of thin films of braze metals based on Ag–Cu alloys on different substrates in space, conducted in the «Isparitel» unit in 1981 revealed, alongside important applied results, also some regularities which are of essential scientific interest, for instance, the anomalous nature of evaporation of binary alloys at microgravity [1]. During the following years these experiments were carried on in «Isparitel-M» unit. Evaporation of binary alloys with markedly different vapour pressure of the components on Earth is accompanied by a continuous change of the coating composition compared to the initial alloy [2–4]. The theoretical and experimental dependencies of the content of components in the coating on the time of evaporation of the following alloys are known: Ag–Cu, Sn–Cu, Sn–Pb, Sn–Ag, Ag–Al [5], Ni–Cr [6] and Al–Mg [7]. A number of methods are used in the technology of producing coatings from binary alloys, which allow making thin films of the specified relatively stable composition, namely, «flash» evaporation [8], evaporation from different sources [9, 10], and from solid state [11]. However, the disadvantages inherent in the above methods often lead to deterioration of the film quality. Therefore, the problem of production of coatings from binary alloys with reproduction of the initial material composition is still urgent.

The paper presents the results of investigation of the composition of coatings, produced in evaporation of Ag–Cu alloy in space and on Earth. The fact that diffusion is the main mechanism of mass transfer at zero gravity suggests that the process of binary alloy evaporation from the molten state in space is similar to evaporation from the «solid» state on Earth. If this assumption is confirmed, broad capabilities would be opened up for production of coatings from binary alloys with preservation of the initial material composition. This can turn out to be especially important for alloys in which one of the components

---

[*] (1986) *Problems of space technology of metals*, Kiev, PWI, pp. 22–28.

is present in an extremely small amount, whereas its vapour pressure is markedly different from the vapour pressure of the main component.

Space experiments and their ground analogs on evaporation of Cu–26 at.% Ag and Cu–10 at.% Ag alloys were conducted in «Isparitel-M» unit in the spring of 1984 by a procedure described in [12]. Glass (25 × 25 × 0.5 mm) and titanium (60 × 80 × 0.2 mm) substrates were used. The substrate surface was thoroughly degreased prior to coating deposition. It should be noted that the evaporation process was of a discrete nature. Coatings were successively deposited from one crucible onto 12 substrates, mounted on the rotating sample holder. The coating on each of the substrates was produced after preheating of the evaporation material (with the closed shutter). Exposure time was from 40 to 180 s at different periods of coating deposition. Each of the considered alloys was almost completely evaporated from the crucibles. The thickness of the produced coatings was measured in the interference microscope MII-4. Coatings on glass substrates were used for these purposes. Deposition rate during the entire process was evaluated by the results of thickness measurement, taking into account the time of exposure. Coating thickness was from 30 to 200 nm. The composition of the produced coatings was determined by Auger-spectrometry method. It is the only suitable method from the known compositional analysis methods, as it has an information depth of 2 to 5 nm (depending on the primary electron energy). LAS-2000 unit of Riber Company for three-dimensional surface analysis with Auger microprobe of cylindrical mirror type was used. Before the start of analysis, the sample surface was cleaned from adsorbates and surface contamination by a special low current ion gun CI-10 (ion energy of 1 keV, etching duration of 10 min) in the pre-chamber. During this time 0.5 to 1.0 nm layer was removed from the coating surface. Then the sample was moved into the analysis chamber with the atmospheric pressure of $1.3 \cdot 10^{-7}$ to $3.9 \cdot 10^{-8}$ Pa. The coating was etched off using a more powerful ion gun CI-40 (electron energy of 5 keV), which operated in the scanning mode. A 50 nm layer was etched off in 10 min. Information which was derived during investigations was plotted by an XY-recorder in the form of the energy spectrum. Measurement accuracy was 10 %. Metallographic examination of the remnants after evaporation of Cu–26 at.% Ag and Cu–10 at.% Ag alloys was conducted in «Neophot-2» microscope on microsections, prepared by the universally accepted procedures. VUP-4 unit was used to reveal the structure in the following mode of cathode etching: $I = 10$ mA; $U = 5$ kV; $P = 1.3 \cdot 10^{-2}$ Pa; $t = 20 \div 35$ min. The elemental composition of the remainder of the alloys after evaporation was determined in X-ray microanalyser «Cameca».

Change of silver content (and copper content, respectively) in the coatings during evaporation of Cu–26 at.% Ag alloy is illustrated by (see Figure 1 on p. 373). It is seen that in evaporation of Cu–26 at.% Ag alloy on Earth (see Figure 1, curve *1*, on p. 373) first silver intensively evaporates, whose vapour pressure is by an order of magnitude higher than that of the copper vapours at the same temperature. Under these conditions, temperature convection is the main cause for mass transfer. Therefore, continuous rather fast enrichment of the surface layer with the highly volatile component (silver) takes place. During evaporation, as follows from the Raoult law, a continuous change of the composition of the produced coatings occurs. In this case, all the silver which was in the alloy, had evaporated. X-ray analysis of the remainder of Cu–26 at.% Ag alloy after evaporation on Earth, showed it to contain pure copper. This is confirmed by the results of metallographic examination (Figure 1a).

At zero gravity, the process of material evaporation has its special features, as mass-transfer under these conditions is primarily determined by thermocapillary and chemical convection, as well as by diffusion. The first two processes have certain inertia in viscous liquids, as in intensity they are by several orders of magnitude lower than the normal

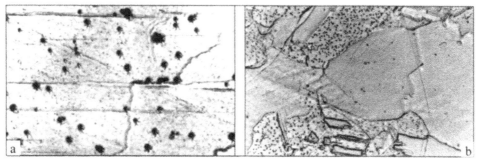

Figure 1 Microstructure of the remainder of Cu–26 at.% Ag alloy after evaporation on Earth (a) (×500) and in space (b) (×200).

temperature convection, and as they are related to surface tension, they are active only in the zone immediately adjacent to the interphases. Their role in mass transfer at short exposures is so small that it can be neglected. The main cause for mass transfer in this case is diffusion, although this process also has considerable inertia. In evaporation of binary alloys at zero gravity, such a state can be in place when the evaporation rate of the more volatile component and the rate of diffusion of this component to the surface, balance each other. A steady-state condition will set in, in which the composition of the produced coatings will remain unchanged. This will occur at some certain and constant temperature of the alloy being evaporated. It is natural that in this case a plateau should be present in the plot of the dependence of the alloy component content on evaporation time.

As was already mentioned, in order to solve the applied tasks, the duration of the substrates exposure to the vapour flow and the temperature of Ag–Cu alloy being evaporated, and, therefore, the rate of its evaporation, were varied during the experiments on coatings deposition in space and on Earth. The alloy temperature in the crucible was changed from 1250 up to 1550 °C. Therefore (see Figure 1, curve *2*, on p. 373) it has been so far impossible to get experimental confirmation of the theoretical postulates, except for one short-term section, when the alloy evaporation occurred without any change in the composition. The nature of the evaporation process could not affect the anticipated result, either.

However, analysis of the given results still allows tracing certain features of evaporation of the studied alloys at zero gravity. So, in determination of the layer-by-layer composition by the Auger-spectrometry method, the following regularity was found. If copper is present in the coating the surface is enriched in copper, and in coatings produced in space, the layer enriched with copper is several times thicker. This is confirmed by the run of curves *1* and *2* in Figure 1 on p. 373. It is seen that copper from Cu–26 at.% Ag alloy, starts evaporating in space earlier, than on Earth (curve *2* falls more steeply, than curve *1*). In each point of curves *1* and *2*, corresponding to a certain evaporation time, silver content in the space coatings is lower than in the ground ones, and that of copper is higher, respectively. In this connection, the results of evaluation of the evaporation rate of Cu–26 at.% Ag alloy at the same crucible temperature (≈ 1490 °C) in the space and ground experiments (Figure 2) are of interest. The form of curves *1* and *2* shows that the less volatile component of the alloy, i.e. copper, makes a more significant contribution to the change of the evaporation rate in the space experiment. In the ground variant the alloy evaporation rate is determined by pure silver evaporation rate for a rather long time (see Figure 2, initial section of the curve). By absolute value, the respective evaporation rates are higher in space than on Earth. Therefore, the same amount of the initial Cu–26 at.% Ag alloy evaporated faster in space (see Figure 1, curve *2*, on p. 373), than on Earth (see Figure 1, curve *1*, on p. 373).

The curves of the dependence of silver content (and copper content, respectively) in the coatings of Cu–10 at.% Ag alloy, produced in space and on Earth are, on the whole, similar

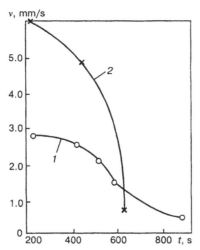

Figure 2 Change of evaporation rate of Cu–26 at.% Ag alloy during its evaporation under terrestrial (*1*) and space (*2*) conditions in the same modes.

to the respective curves for Cu–26 at.% Ag alloy. Because of a small content of silver in the initial alloy, however, its evaporation rate is determined primarily by the low-volatile component, that is copper. Therefore, in the remainder, after evaporation of Cu–10 at.% Ag alloy in space and on Earth (total evaporation time being the same as for Cu–26 at.% Ag alloy), the content of silver fractions by weight is 5–7 % (by the data of X-ray microprobe analysis). Also significant is the fact that in evaporation of Cu–10 at.% Ag alloy on Earth the time of pure silver evaporation from it, is longer than in a similar experiment for Cu–26 at.% Ag alloy. In evaporation of these alloys in space, the influence of the low-volatile component — copper, is manifested in almost the same way. This results from the fact that the cause of mass transfer on Earth is convection and in space it is diffusion. Note also the differences in the structure of the remainder of the above alloys after evaporation in space and on Earth. Comparison of copper microstructures, given in Figures 1a and b shows that the structure of the metal solidified in space is more perfect (etching pits are easily visible even with ×200 magnification). It is known [13] that in solidification of binary alloys, the grains in the structure of the space samples have somewhat larger dimensions compared to the structure in the ground samples. Comparison of microstructures of the remainder of Cu–10 at.% Ag alloy after evaporation on Earth (Figure 3a) and in space (Figure 3b) confirms this conclusion.

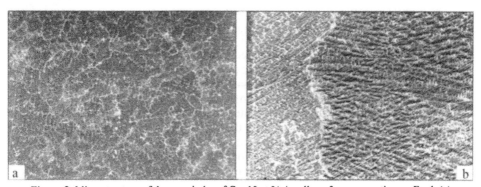

Figure 3 Microstructure of the remainder of Cu–10 at.% Ag alloy after evaporation on Earth (a) and in space (b) (×50).

In view of the above, it is possible to draw the conclusion that despite the preliminary nature of the performed investigations, they point to good prospects for further study of the influence of zero gravity on the process of binary alloy evaporation, both in terms of development of fundamental research and in terms of practical application.

1. Lukash, E.S., Lapchinsky, V.F., Shulym, V.F. et al. (1985) *See pp. 335–340 in this Book.*

2. Berry, R.W., Hall, P.M. and Harris, M.T. (1968) *Thin film technology*, New York, 221 pp.

3. (1977) *Thin film technology.* A handbook. Ed. by L. Mayssel and R. Glang, Moscow, Sov. Radio, 465 pp.

4. (1976) *Epitaxial growth.* Ed. by J.W. Matthews, New York, 542 pp.

5. Saur, E. and Unger, E. (1958) Spektralanalytische Untersuchung des Verdampfungsvorganges binärer Legierungen, *Zeitschrift Naturforschung*, **13a**, ss. 72–79.

6. Huijer, P., Langendam, W.T. and Lely, J.A. (1963) Vacuum deposition of resistors, *Philips Techn.*, **24**, pp. 144–149.

7. Böttcher, A. (1950) Über Herstellung und Aufbau aufgedampfter dünner Al–Ag- und Al–Mg-Schichten, *Z. Angew. Phys.*, No. 2, pp. 193–203.

8. Beckerman, M. and Bullard, R.L. (1962) In: *Proc. of Conf. on Electronic Compon.* Washington, pp. 53–56.

9. Mader, S. (1965) Metastable alloy films, *J. Vac. and Sci. Technol.*, No. 2, pp. 35–41.

10. Ostander, W.J. and Lewis, C.W. (1962) Electrical properties of metal-dielectric film, In: *Proc. of 8th Nat. Vac. Symp.*, Washington, 1961. New York, Pergamon Press, pp. 881–888.

11. Schneider, M. (1959) *Techn. Mitt. PTT*, **37**, No. 465, p. 470.

12. Shulym, V.F., Zagrebelny, A.A., Lapchinsky, V.F. et al. (1985) Hardware for materials evaporation in space, In: *Problems of space technology of metals*, Kiev, Naukova Dumka, pp. 57–63.

13. Ivanov, L.I., Kubasov, V.N., Pimenov, V.N. et al. (1977) Melting of copper-aluminium alloy and aluminium powder at zero gravity, *Fizika i Khimiya Obrab. Materialov*, No. 6, pp. 35–39.

Some Features of Formation of Silver Coatings at Different Gravity Levels*

B.E. Paton[1], V.F. Lapchinsky[1], E.S. Mikhajlovskaya[1], A.A. Gordonnaya[1], V.N. Sladkova[1], V.F. Shulym[1] and L.O. Neznamova[2]

When technological operations are performed in open space an extensive use of different kinds of coatings is required [1–4]. The need to apply reflective, optical and other coatings in space recently has already led to the emergence of a separate field, namely optical technology, within the space technology [5, 6].

A whole number of practical tasks, related to deposition of coatings in space, can be successfully solved using a VHT, designed by the employees of the PWI. In 1984 cosmonauts S.E. Savitskaya and V.A. Dzhanibekov tested the tool in open space [7, 8]. In addition to performance of cutting, brazing and welding of materials, samples with silver coatings were also produced. The main goal of the tests was demonstration of broad capabilities of the tool

* (1997) *Kosmichna Nauka i Tekhnologiya*, **3**, No. 3/4, pp. 71–75.

Stresses of the first kind and of lattice parameter.

No.	Production conditions	$\sigma \pm \Delta\sigma$, MPa	a, nm	$a_{table} - a$, nm
1	Space	-16 ± 20	0.40833	0.00027
2	Same	-25 ± 20	0.40805	0.00055
3	Earth	$+72 \pm 20$	0.40770	0.00090
4	Same	-52 ± 20	0.40800	0.00060

proper, and therefore the experiment on silver coatings deposition cannot be regarded as quite correct. However, under the conditions of an almost complete absence of data on the properties of coatings produced in open space, even the smallest information about it is of an indubitable interest. It should also be noted that the investigations, described below, were conducted right after the completion of the experiment. Their results, however, were described in classified reports in view of the specifics of those times.

Evaporation of silver of 99.99 purity was carried out from molybdenum crucibles with external electron beam heating of the crucible. In orbit the coating was applied at the pressure of 10^{-4}–10^{-5} Pa and microgravity level $g = 1 \cdot 10^{-5} g_0$ (g_0 is the gravity on the surface of Earth). On Earth the coating was produced in a vacuum chamber with the pressure of the residual atmosphere of $5 \cdot 10^{-2}$ Pa. Two substrates with a silver coating were produced, as well as their two ground analogs. Sheet rolled stock of D16 (Cu — 4.3; Mg — 1.0; Mn — 0.6 wt.%; Al — balance) alloy was used as substrates of $240 \times 180 \times 2$ mm size. Prior to coating deposition, each substrate had passed such stages of processing as degreasing and etching in an alkaline solution, washing in water, anodizing (180 g/l H_2SO_4 $T = 20$ °C, $I = 1.0$–1.5 A/dm^2, $U = 18$÷20 V, $t = 40$ min, washing in water, painting to TY 6-14-515–70), final washing in water and drying. As a result of such processing the substrates acquired a black mat colour. The initial direction of the rolled stock texture was quite clearly seen visually.

This present paper contains information on the influence of the conditions of production of silver coatings on stresses of the first kind, σ, lattice parameter a, and topography of their surface. Stresses of the first kind in silver coatings were determined by the change in the shift of the peak of line (420) Ag in the diffractograms, taken in Co-radiation in Dron-3 unit by the procedure from [9]. When measurements were taken, a considerable widening of the line was found, because of the finely dispersed structure of the coatings, which, naturally, somewhat lowered the measurement accuracy. It can be assumed that it was equal to 20 MPa in determination of stresses of the first kind. The error of the lattice parameter measurement was 0.00014 nm.

The Table shows values σ (average for 10 samples) and lattice parameter a in silver coatings, produced with VHT at different gravity levels. The data are indicative of the fact that the coatings, produced in space and on Earth, have both compressive and tensile stresses, most probably resulting from a pronounced relief of the initial substrate. However, by absolute value, stresses of the first kind in the coatings, produced on Earth, are equal to -30 – $+90$ MPa, and in the coatings, produced in space, they are within the range of measurement error. From the results given in the Table, it can be seen that the lattice parameter of silver coatings, produced in space, differs less from the tabulated value, than the one for the coatings, deposited on Earth. The table value of the lattice parameter of silver is 0.40860 nm.

Surface topography of silver coatings was studied in the scanning electron microscope SEM-515. Samples of 10×10 mm size were cut out, using guillotine type shears. In the

a

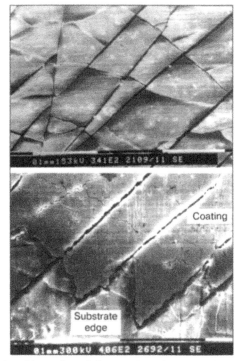

b

Figure 1 Topography of the surface of silver coatings, deposited using the VHT on Earth (a)
and in space (b) (edge of samples cut out for investigations).

photographs, given in Figure 1, it can be seen that the coatings fail in the locations adjacent
to the sample edges, as a result of the deformation impact in cutting. However, the nature of
fracture of the coatings, deposited in space and on Earth, is different. In the coatings,
deposited on Earth (Figure 1a), a dense net of cracks is found and there is much spallation
of the coating areas. The silver coatings, deposited in space, do not demonstrate any
spallation in cutting up to the very edge of the sample (Figure 1b), whereas the formed cracks
propagate predominantly along parallel lines. The surface of coatings, produced in space,
reproduces the substrate relief (Figure 2b). The initial texture of the substrate can be seen in
the photograph, selected from a large number of similar ones. This is not observed in
analogous coatings, produced on Earth (Figure 2a). Moreover, reproduction of the substrate
relief in silver coatings, produced in space, is found at thicknesses (600–800 nm) greater
than that of the corresponding coatings produced on the ground (350–500 nm). As was
already noted [10], with the same modes of evaporation the coatings deposited in space have
a thickness greater than that of their ground analogs, because of a higher rate of material
evaporation at zero gravity. The thickness of the coatings, studied in this paper, was assessed
by transverse macrosections.

Silver coatings produced on Earth are also characterized by the fact that the cracks
appearing in their fracture have the nature of cleavage (Figure 3a), and those in similar
coatings, produced in space, have the nature of tough fracture (Figure 3b). Illustration of the
nature of fracture of silver coatings, produced in space and on Earth, confirms the conclusion
that the coatings, produced at zero gravity, have lower stresses.

It is difficult to give a fully substantiated cause of the differences found in the properties
of silver coatings, produced at different gravity levels, because of the absence of control of
many parameters during the performance of the experiment. It can, however, be supposed

a

b

Figure 2 Topography of the surface of silver coatings, produced on Earth (coating thickness being 350–500 nm) (a) and in space (coating thickness being 600–800 nm) (b).

that, since in coating deposition the evaporation material initially stays in the liquid phase for a long time, then goes into the vapour phase and, finally, condenses on the substrate in the form of a hard film, the environment, as well as deposition rate, coating thickness, substrate temperature, energy of condensing particles [11] and their interaction essentially influence the properties of the produced coatings. It is shown in [12] that submicroporosity of silver coatings, produced in space, is smaller than of their appropriate ground analogs. Distribution of submicropores by dimensions is more uniform in the space samples than in the ground ones. Since stresses of the first kind, being macrostresses, are balanced through the entire volume of the studied coatings, the results, derived in this paper, correlate with the conclusions of [12], as it is known that (up to a certain limit) the more defects in the material, the higher its stresses. Figuratively speaking, the process of applying silver coatings at different gravity levels can be represented as follows: forces of electromagnetic interaction between the silver atoms act in such a way under the influence of different external factors, that on Earth the atoms have to «squeeze» into the space above the substrate, whereas in space they «are freely arranged». As a result of the conducted research the following conclusions can be made.

1. Stresses of the first kind that develop in silver coatings of 350–500 nm thickness, deposited using VHT on Earth on D16 alloy substrates, are equal to −30 – +90 MPa. In similar coatings of 600–800 nm thickness, produced in space, stresses of the first kind are practically absent.

2. Lattice parameter of silver coatings, produced in the space conditions, differs less from the lattice parameter of monolithic silver than in their ground analogs.

3. Unlike their ground analogs, silver coatings, produced in open space, reproduce the substrate relief also at thicknesses greater than in the corresponding ground samples.

a

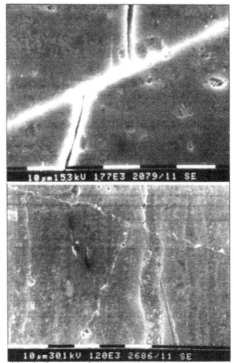

b

Figure 3 Nature of cracking in fracture of silver coatings, deposited on Earth (a) and in space (b).

4. In fracture of silver coatings, deposited on Earth, the cracks are of cleavage type, and in those made in open space they are of tough fracture type.

5. Analysis of the results of measurement of stresses of the first kind and lattice parameter of silver coatings and study of the topography of the coatings surface, allow the assumption to be made that silver coatings, produced under conditions of open space, will have a greater service life than similar coatings, produced on Earth and taken to space by a cargo spaceship.

6. In view of the above-said, there is every ground to believe that the space environment is more favourable for production of coatings than the terrestrial one. The authors support the opinion [5, 13] of the rationality of conducting the processing operations, related to coating deposition, directly in open space.

1. Paton, B.E. (1977) *See pp. 4–7 in this Book.*

2. Paton, B.E., Dudko, D.A. and Lapchinsky, V.F. (1984) Welding processes in space, In: *Welding and special electrometallurgy*, Kiev, Naukova Dumka, pp. 121–129.

3. Nikitsky, V.P. and Zhukov, G.V. (1984) Thin films in space systems and technology, In: *Ideas of K.E. Tsiolkovsky and problems of space manufacturing*, Proc. of 18th Tsiolkovsky Readings, Kaluga, 1983. Moscow, pp. 78–81.

4. Zagrebelny, A.A., Lukash, E.S., Nikitsky, V.P. et al. (1988) Development of the technique of thin-film coating application in flight conditions, In: *Proc. of 22nd Tsiolkovsky Readings*, Kaluga, 1987. Moscow, pp. 134–138.

5. Petrovsky, G.T. and Voronkov, G.L. (1984) *Optics technology in space*, Leningrad, Mashinostroyeniye, 158 pp.

6. (1980) Space optics, In: *Proc. of 9th Int. Congr. of Int. Com. on Optics*, Santa-Monica, 1972.

7. Nikitsky, V.P., Lapchinsky, V.F., Zagrebelny, A.A. et al. (1985) *See pp. 178–184 in this Book.*

8. Paton, B., Dzhanibekov, V., Savitskaya, S. et al. (1989) The test of the versatile hand electron beam tool in space, In: *Proc. of IIW Conf. on Welding under Extreme Conditions,* Helsinki, 1989. Helsinki, Pergamon Press, pp. 189–196.

9. Barret, Ch.S. and Massalsky, T.B. (1984) *Structure of metals,* Moscow, Metallurgiya, 255 pp.

10. Lukash, E.S., Lapchinsky, V.F., Shulym, V.F. et al. (1985) *See pp. 335–340 in this Book.*

11. Stetsenko, V.V., Movchan, B.A. and Michenko, V.A. (1983) Structure and reflective properties of silver coatings produced by direct electron beam evaporation and ion spraying, *Problems of Spec. Electrometallurgy,* Issue 19, pp. 47–49.

12. Cheremskoj, P.G., Toptygin, A.L., Panikarsky, A.S. et al. (1990) *See pp. 364–371 in this Book.*

13. Bruns, A.V., Grechko, G.M., Gubarev A.A. et al. (1977) Orbiting solar telescope on «Salyut-4» station, *Acta Astronautica,* No. 4, pp. 1121–1125.

Peculiarities of Formation of Structure and Composition of Binary Alloys in Evaporation under Microgravity Conditions[*]

L.S. Palatnik[6], A.L. Toptygin[6], A.V. Arinkin[6], A.A. Kozma[6], V.P. Nikitsky[2], G.V. Zhukov[2], V.F. Lapchinsky[1], E.S. Lukash[1], L.O. Neznamova[2] and I.S. Malashenko[1]

Investigation of evaporation and condensation processes under conditions of flight of orbital stations is imperative for development of a space technology for production of different-application vacuum complex-composition coatings. Brazing filler metals are based on alloys of the Ag–Cu system [1]. Sufficiently high solubility of copper (14.1 at.%) in silver at the eutectic temperature (779 °C) allows the effect of the concentration of components on the mechanism and kinetics of decomposition of solid solution, discontinuous (cellular) decomposition in particular, to be comprehensively studied [2]. During evaporation the Ag–Cu alloys obey Raoult's law [3], which substantially simplifies calculation of relationships used to determine the concentration of components in the condensing alloys.

Production of coatings. Films of Ag–Cu alloys were produced under conditions of flight of the OS «Salyut-6» in the course of two experiments, using the specialized equipment «Isparitel» developed by the PWI [4]. The evaporation unit of the installation with a crucible and substrates was placed in a lock chamber connected to extra-vehicular space. Evaporation of the Ag–Cu alloy with a copper concentration of 50 at.% (37 wt.%) in the initial alloy was performed from a molybdenum crucible of a special shape, heated up on the bottom side by electron beam. (This alloy is close to the eutectic composition of Ag–40 at.% (29 wt.%) Cu).

The same type of crucible, to which the initial alloy with a mass of 6.8 g was loaded, was used in both experiments. Six samples were produced in each experiment. The time of deposition of coatings was varied from 10 to 200 s. The coatings were deposited on unheated titanium substrates fixed in a special drum whose rotation allowed replacement of the substrates.

The temperature of the melt in the crucible during evaporation of the alloy varied, caused by peculiarities of operation of the heat source. The temperature conditions of evaporation

---

[*] (1988) Paper was presented at the 22nd Tsiolkovsky Readings, Kaluga, Sept. 15–18, 1987.

Table 1 Temperature conditions of evaporation of Ag–50 at.% Cu alloy.

Temperature of crucible, K	Time since beginning of operation of evaporator, s
1503	60
1573	90
1623	120
1673	180

are given in Table 1. A shutter over the crucible was opened 1 min after the beginning of the operation of the evaporator, when the temperature of the crucible reached 1503–1523 K.

Table 2 gives data on the time of exposure and current time of evaporation at the moment of completion of condensation of a given alloy, with no allowance for the time of heating up and cooling of the crucible. Taken together with the data of Table 1, they allow calculation of the temperature of evaporation and its variation during condensation of each sample.

Investigation of peculiarities of the structure and composition of coatings produced under space conditions was performed by comparing them with characteristics of their ground analogues. On Earth the coatings were produced using the «Isparitel» equipment in

Table 2 Conditions of formation, thickness and composition of Ag–Cu alloys.

Sample No.	Exposure (time of condensation), τ_{exp}, s	Current time of evaporation, τ_{ev}, s	Coating thickness, h, μm	Concentration of copper, C, %
1S1	10	10	0.070	1.5
1S2	20	30	0.005	–
1S3	30	60	–	–
1S4	40	100	0.160	0.5
1S5	100	200	0.200	6.5
1S6	200	400	0.800	13.0
2S1	10	410	0.190	1.0
2S2	20	430	–	–
2S3	30	460	–	–
2S4	40	500	0.070	2.0
2S5	100	600	0.300	10.0
2S6	200	800	0.750	14.0
1G1	10	10	0.300	1.5
1G2	20	30	0.140	1.5
1G3	30	60	0.150	1.0
1G4	40	100	0.310	1.7
1G5	100	200	1.100	3.0
1G6	200	400	0.600	2.5
2G1	10	410	0.100	0.5
2G2	20	430	0.130	0.2
2G3	30	460	0.145	1.2
2G4	40	500	0.420	2.0
2G5	100	600	0.370	2.5
2G6	200	800	0.650	5.0

Notes: Here and in Table 3 the first numeral in the sample designation means number of an experiment, the letter means the place of an experiment (S — space, G — ground) and the last numeral means number of a sample.

a vacuum chamber using the oil-vapour evacuation system. Thermal-physical conditions of deposition were identical in both cases.

Investigation methods and experimental results. *Thickness and composition of the Ag–Cu films.* Data on element composition and thickness of the coatings, non separated from the substrate, were obtained by fluorescent X-ray spectrometry. Measurements were conducted using the «Dron-0.5» diffractometer. After tube BSV-6 with a tungsten anode, the samples were placed into a holder of monochromator of goniometer GUR-4. The crystal analyser (flat graphite monochromator with $d_{(0002)} = 0.335$ nm) was installed in the goniometer holder for flat samples. The intensity corresponding to wave lengths of K_α–Ag and K_α–Cu was registered. The angle of incidence of primary radiation of the continuous spectrum of the tube and the outlet angle of secondary radiation θ were 6°, which allowed coatings of small thickness (to ≈ 50 nm) to be analysed.

Reference samples were massive plates of silver and copper with a purity of 99.9 %. Registration of the K_α-line of silver of thin coatings showed an increase in the background level caused by Bragg reflection of the continuous spectrum of the tube from a textured substrate. This circumstance was allowed for in estimation of the weight concentration of copper, C, and coating thickness h, which was determined in accordance with the following formulae:

$$I_{Ag} = I_{sub}\exp\left\{-2[\rho_{Ag}(1-C)+\mu_{Cu}]\beta h/\sin\theta\right\}$$
$$+ I_{Ag}^{em}\int_{\lambda_{Ag}}^{\lambda_0} I_{Ag}(\lambda, h, C)\frac{\lambda-\lambda_0}{\lambda^3}\,d\lambda / \int_{\lambda_{Ag}}^{\lambda_0} I_{Ag}(\lambda, \infty, 0)\frac{\lambda-\lambda_0}{\lambda^3}\,d\lambda;$$

$$I_{Cu} = I_{Cu}^{em}\int_{\lambda_{Cu}}^{\lambda_0} I_{Cu}(\lambda, h, C)\frac{\lambda-\lambda_0}{\lambda^3}\,d\lambda / \int_{\lambda_{Cu}}^{\lambda_0} (\lambda, \infty, 1)\frac{\lambda-\lambda_0}{\lambda^3}\,d\lambda,$$

where I_{Ag} and I_{Cu} are the measured values of the intensities; I_{sub} is the intensity of reflection from the substrate in the absence of the coating; μ_{Ag} and μ_{Cu} are the mass coefficients of attenuation of silver K_α-radiation for silver and copper, respectively; ρ is the density of the coating; λ_0 is the minimum wave length of the continuous spectrum; and λ_{Ag} and λ_{Cu} are the wave lengths of the ends of the absorption bands for silver and copper.

Relationships of I_{Ag} (λ, h, C) and I_{Cu} (λ, h, C) for monochromatic excitation radiation were derived by substituting the data of [5] to formula (8.7) of [6]. It was assumed that, without allowance for an absorption jump, the coefficients of attenuation and absorption could be described by dependence $\sim \lambda^3$. The above expressions do not account for a possible excitation of fluorescent radiation of copper by radiation of silver.

The above system of equations was numerically solved by a specially developed program. Testing the method on reference samples of the known thickness showed a satisfactory accuracy in determining thickness of the coatings, which was 20 % for the films with $h > 200$ nm. For such a thickness the absolute error in evaluation of the weight concentration of copper (at $C < 20$ %) was no more than 1 %.

Results of calculations of thickness of the investigated samples on the basis of the X-ray data are given in Figure 1 and Table 2. There is a substantial scatter of the values of thickness, which is associated with a variation of temperature in evaporation of the alloy and non-uniform distribution of thickness across the surface area of the samples.

As can be seen from Figure 1, thickness of the samples increases with an increase in the exposure time. In this case the range of values of thickness for the space samples overlaps the corresponding range for the earth samples, but lies at a lower level. This is indicative of

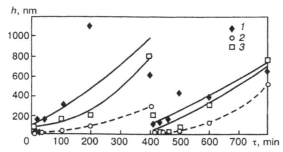

Figure 1 Dependence of thickness of the Ag–Cu alloy films upon the time of exposure: *1* — Earth;
2 — calculation; *3* — space.

the fact that thickness and, accordingly, rate of condensation of the films produced under space conditions are, as a rule, smaller than those of the earth analogues. This difference will be discussed below with allowance for the data on composition of the samples.

Consider dependence of the concentration of the Ag–Cu films upon the test time. As follows from Figure 2, for all experiments the concentration of copper in the samples increases with increase in the time of evaporation. As in evaluation of the concentration of copper the informative layer is 0.4 μm thick, and thickness of many samples for a condensation time of more than 100 s is in excess of the above value, the corresponding value relates not to the entire volume of the most massive samples but to their surface layer. Therefore, an increase in the time of evaporation leads to enrichment of the film surfaces facing the crucible with copper. However, the extent of this enrichment is much higher for the space samples, as compared with the earth ones. Thus, an increase in the time of evaporation from 10 to 400 s leads to an increase of the concentration of copper in the space samples from 1.5 to 13.0 wt.%, whereas in the earth samples the concentration of copper within the above time range changes from 1.5 to 3.0 wt.%. For space experiment No. 2, this increase is still higher (up to 14.0 wt.%). Therefore, as the alloy is evaporated from the crucible, the space samples become enriched with copper to a much higher extent than the earth ones.

A sudden change in the character of dependencies of the concentration of the film upon the time of evaporation of the alloy is seen within an evaporation time range of 400 – 410 s, despite the fact that the alloy evaporated in the crucible was not replaced. As shown by our calculations, this might be caused by the above-mentioned peculiarities of temperature conditions of operation of the evaporator. Figure 2 shows a calculated curve of dependence of the mean concentration of copper in the Ag–Cu films, plotted with allowance for variations of temperature during the process of condensation from one crucible. During calculations the concentration of the alloy was assumed to be uniform over the entire volume of the crucible. Comparing experimental and theoretical curves yields a satisfactory coincidence for the earth samples, especially at $\tau_{ev} \leq 400$ s, and shows that the space samples feature

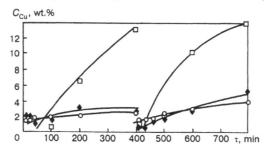

Figure 2 Dependence of the concentration of copper in the Ag–Cu alloy films upon the time of evaporation (see designations in Figure 1).

substantial differences in the value and character of dependence of the concentration upon the time of evaporation, as compared with the earth ones.

The noted differences in values of thickness of the films, their concentration of copper and character of variations in composition of the films with time can be attributed to the fact that during the evaporation process the surface of a melt under flight conditions is enriched with copper to a larger degree than under earth conditions. This results from the fact that under earth conditions the melt is characterized by more intensive convection processes occurring in it, which are induced by variations in the composition of the alloy during evaporation and those temperature gradients which are formed in the crucible of the design utilized. Under flight conditions (microgravity) these factors are mitigated to a fundamental degree, thus leading to an increase in the role of other mass transfer mechanisms, diffusion in particular. The rate of transfer of a high-volatility material (Ag) to the surface of the melt in space is much lower, which leads to its enrichment with copper. Because of the lower value of the vapour pressure of copper, as compared with silver (by one order of magnitude at a temperature of 1800 °C) [7], this enrichment leads to a decrease in the total rate of evaporation (under the same temperature conditions of heating of the crucible) and, there-fore, in thickness of the films, and causes an increase in the copper content of the condensate and change in the character of dependence of the concentration upon the time of evaporation for the space samples.

Structure and phase composition of the Ag–Cu films. Structure of the Ag–Cu films was analysed by the X-ray diffraction method. The samples of coatings were filmed using the «Dron-UM-1» diffractometer in filtered radiation of copper, cobalt and chromium anodes. Control of the diffractometer and processing of the data, followed by printing out by a digital printer, were done using the «Iskra-1256» computer facility. The concentration of solid solutions was estimated on the basis of values of the crystalline lattice parameter using the data of [7], according to which an increase in the lattice parameter of the Cu-based solid solution of silver is

$$\left(\frac{1}{C}\frac{\partial a}{\partial C}\right)_{Cu} \cong 0.56\cdot10^{-3} \text{ nm } (C, \text{ at.}\%),$$

and a decrease in the lattice parameter of silver with copper dissolved in it is

$$\left(\frac{1}{C}\frac{\partial a}{\partial C}\right)_{Ag} \cong 0.38\cdot10^{-3} \text{ nm.}$$

A plate of silver with a purity of 99.999 was used as a reference.

Analysis of diffraction patterns of flight and ground samples revealed substantial differences in their structure. Figure 3 shows typical diffraction patterns for the samples investigated within a range of angles equal to $2\theta = 42\div48°$, recorded in radiation of the cobalt anode. For all the samples investigated the reflection close to a position of maximum to reflection (111) of the reference silver sample ($2\theta = 44.56°$) is observed within the given range of the angles. The lines for the earth samples are comparatively narrow (Figure 3a). Unlike the earth samples, all the space samples are characterized by considerable widening of the diffraction lines. In addition, for the majority of the flight samples, the diffraction lines were of a doubled character (Figure 3b). This shape of the diffraction maximum seems to be associated with heterogeneity of the solid solution of copper in the lattice of silver. Also, this seems to account for a considerable widening of the lines, as compared with the earth samples. It should be noted that this «bifurcation» of the lines is more pronounced at large angle 2θ. However, because of a substantially lower intensity and considerable widening for lines with $h^2 + k^2 + l^2 > 4$, position of this component of the doublet cannot be analysed with

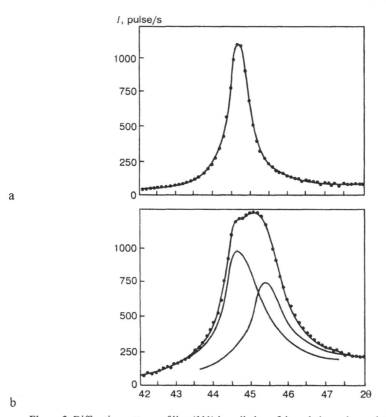

Figure 3 Diffraction patterns of line (111) in radiation of the cobalt anode: earth (a) and space (b) samples.

sufficient accuracy. A narrower component of the doublet on the side of smaller angles was registered for a number of the space samples to a sufficient degree of reliability for lines (311) and (222) in copper radiation (2θ ~ 80°). Therefore, the analysis was carried out on the basis of the intensity of line (111). The doublet was graphically split on the assumption of symmetry of the component on the side of small angles 2θ, based on the results of recording of lines (311) and (222). In addition, these records, as well as the shape of line (111) (presence of maximum or deflection of the profile near 2θ = 44.7° for λ — K_α-Co), served as the basis for selection of the initial position of maximum of the first component of the doublet. Later on, splitting was done graphically by the method of successive approximations. As a rule, 2–4 iterations were enough for this operation. Note that validity of such splitting was checked by the least squares method on an assumption of Gaussian shape of components of the doublet. Whereas for the flight samples the accuracy of measurements of the crystalline lattice spacing was ±0.0001 nm, for the components of the doublet on the side of small, a_{II}, and large, a_I, angles the accuracy of estimation of the crystalline lattice spacing was ±0.001 and 0.0005 nm, respectively.

Results of measurements of the lattice spacing are given in Table 3. In addition, the Table gives values of concentration C_I and C_{II} determined from the concentration dependencies of the lattice spacing of Ag-based alloys [8]. The accuracy of estimation of the concentration depends not only upon the accuracy of measurement of the lattice spacing, but also upon the concentration itself: in the case where $C > 10$ % the concentration was estimated by extrapolation of the data.

Table 3 Crystalline lattice spacing for the Ag–Cu films and concentration of copper in primary, C_I, and secondary, C_{II}, solid solutions.

Sample No.	Time of condensation, s	Side of crucible or substrate	a_I, nm	a_{II}, nm	C_I, at.%	C_{II}, at.%
1G5	100	Substrate	0.4081	–	1.4	–
1G6	200	Same	0.4083	–	0.9	–
1S5	100	»	0.4083	0.4044	14.0	0.9
1S6	200	»	0.4032	–	14.0	–
1S6	200	Crucible	0.4037	–	14.0	–
2G3	30	Substrate	0.4080	–	1.7	–
2G4	40	Same	0.4076	–	2.7	–
2G5	100	»	0.4074	–	3.2	–
2G6	200	»	0.4070	–	4.0	–
2S5	100	»	0.4083	0.4015	17.0	0.9
2S6	200	Crucible	0.4033	0.4025	16.0	0.9
2S6	200	Same	0.4083	0.4032	14.0	0.9

Note: Values of a_I and a_{II} are the crystalline lattice spacings of primary and secondary Ag-based solid solutions, respectively.

As established on the basis of the measurements, the space samples, unlike the earth ones, contain at least two phases, i.e. face-centred cubic Ag-based solid solution with a concentration of copper approaching equilibrium, and oversaturated solid solution with a concentration of copper equal to about 10 at.%. This phenomenon can be related to decomposition of the oversaturated solid solution. In fact, a very weak line corresponding to reflection (111) of copper was fixed for some samples while recording radiation of the cobalt anode.

Apparently, as early as in the process of condensation a discontinuous decomposition of the oversaturated solid solution occurs in the space samples to precipitate copper (or, to be more exact, Cu-based solid solution) and form a depleted equilibrium Ag-based solid solution. The fact that the earth samples do not have such a decomposition is attributable, in particular, to a lower concentration of copper in them. Such factors, non-controllable during the experiments, as composition of the atmosphere of residual gases (the lock chamber contains no vapours of diffusion oil) and temperature of the surface in condensation, may also have an effect. As susceptibility of the oversaturated solid solutions to decomposition increases with an increase in their concentration, it can be supposed that the fact that the oversaturated solid solution (silver, copper) of the space samples decomposes is caused primarily by an increased (compared with the ground samples) copper content of this solution.

Conclusion

Investigations conducted on the Ag–Cu films produced by deposition of the Ag–50 at.% Cu alloy show that evaporation of alloys under space and microgravity conditions is characterized by the fact that transfer of the high-volatility component to the surface of a melt is slowed down, which is caused by changes in the mechanisms of mass transfer in the liquid phase. As a result, the vapour phase of the Ag–Cu alloy and the condensed films are enriched with low-volatility copper component. In turn, this leads to an increase in concentration oversaturation of Ag–Cu solid solutions fixed in the films at low temperatures, which enhances the decomposition processes in the condensates of the coatings produced under the space conditions, whereas in the earth samples the oversaturated solid solution remains non-decomposed.

1. (1975) *A handbook of brazing*. Ed. by S.N. Lochmanov, I.E. Petrunin and V.P. Frolov, Moscow, Mashinostroyeniye, 407 pp.

2. Ageev, N., Hansen, M. and Sachs, G. (1930) Entmischung und Eigenschaftsänderungen übersättigter Silber-Kupferlegierungen, *Z. Angew. Phys.*, No. 66, p. 350.

3. (1977) *Thin film technology*. A handbook. Ed. by L. Mayssel and R. Glang, Moscow, Sov. Radio, 465 pp.

4. Paton, B.E. and Kubasov, V.N. (1979) *See pp. 8–11 in this Book*.

5. Blokhin, M.A. (1959) *X-ray spectrum investigation methods*, Moscow, Fizmatgiz, 386 pp.

6. Blokhin, M.A. and Shvejtser, I.G. (1982) *X-ray analysis handbook*, Moscow, Nauka, 376 pp.

7. (1995–1996) *A handbook of chemistry and physics*. Ed. by D.R. Lide, New York, London, Tokyo, CRC Press.

8. Pearson, W.P. (1958) *A handbook of lattice spacing and structures of metals*, New York, Pergamon Press, 718 pp.

Segregation Effects in Ag–Cu Coatings Deposited under Space and Earth Conditions*

A.L. Toptygin[6], E.S. Mikhajlovskaya[1], O.D. Smiyan[1], L.O. Neznamova[2], V.F. Shulym[1] and I.S. Malashenko[1]

Properties of thin coatings produced by evaporation and condensation of alloys, as well as their consistency in time depend to a large degree upon the uniformity of distribution of components in the bulk of coatings and upon the presence of non-metallic inclusions in them. In turn, these structural factors are determined by the purity of source materials, kinetics of evaporation of alloys, condition of the vacuum environment and processes of physical-chemical interaction that occur on the surface of a substrate in deposition of films during their initiation and growth. Allowance for all these phenomena is important for development of technologies for deposition of complex compositions of coatings under flight conditions of orbital stations.

This study is dedicated to investigation of element composition of Ag–Cu alloy coatings deposited under space and earth conditions to establish the main principles of distribution of basic components and impurities during the process of growth of condensates. Samples of coatings were produced in space on board the OC «Mir» and on Earth by thermal evaporation of alloys Ag–74 at.% Cu from cylindrical molybdenum crucibles, followed by their condensation on a continuously moving polyimide strip. The coatings were deposited using the «Yantar» unit. The coating procedure and calculation parameters are described in [1], they were identical for the space and earth samples.

As established in [2], in the case of melting of a material under zero-gravity conditions there is no temperature convection which determines mass transfer under the earth conditions. As a result, evaporation is accompanied by rapid depletion of the surface layer of the melt in the high-volatility component (silver in this case), the partial pressure of the vapours of which at the same temperatures is almost by an order of magnitude higher than that of copper [3].

---

* (1990) Paper was presented at the 5th Korolyov Readings of the 2nd Republ. Conf. on Fundamental and Applied Problems of Cosmonautics, Kiev.

Under zero-gravity conditions the mass transfer process is determined by diffusion, as well as thermocapillary and chemical convections. The latter in their intensity are by several orders of magnitude lower than normal temperature convection. They are associated with surface tension. This makes them effective only in the zone which directly adjoins the interfaces. At a short time of evaporation (seconds, minutes) their role is too small to take them into account. Therefore, the main mechanism of mass transfer under the decreased gravity conditions is diffusion [2]. In this connection, it is clear that during the coating process the rate of evaporation of the Ag–Cu alloy from the crucible continuously varies. Hence in space and on Earth, during the same period of time a different amount of the material of continuously varying composition was condensed on the strip substrate moving at a constant speed.

The samples produced on Earth and on board the OC «Mir» were selected to investigate coating compositions. Thickness of the Ag–Cu coatings investigated varied from 15 to 480 nm. That is why the coating compositions were studied by Auger spectrometry, where the informative layer is 1–2 nm deep, while the use of ion etching allowed profiles of distribution of different elements through thickness of the produced films to be plotted at any spacing.

Investigations were performed using the «Riber» instrument LAS-2000 intended for three-dimensional surface analysis. The instrument was equipped with the Auger micro-probe (5 μm diameter) of the type of a cylindrical mirror. Determined were the content of basic (Cu, Ag) and impurity (C, O, N, Cl) elements, as well as their distribution through thickness of a coating. Prior to analysis, the surface of a sample was cleaned from adsorbates and contaminants in the surface preparation chamber using the special low-current ion gun SI-10 ($E = 1$ keV) for 10 min. The layer removed from the coating surface during this time was 0.5–1.0 nm thick. Then the samples was placed into the analysis chamber with a pressure of $1.3 \cdot 10^{-7} - 3.9 \cdot 10^{-8}$ Pa. Layer-by-layer analysis (profiling) was done by etching with argon ions at $E = 3–5$ keV using an increased-power gun SI-40, which operated in the mode of scanning over the sample surface. After operation for a fixed period of time this gun was switched off and the element composition of the surface was analysed using the Auger spectrometer. The process of ion etching involving the SI-40 gun was continued for some more time, then the gun was switched off and the composition of the coating surface was again analysed, and so forth.

The samples were loaded into the analysis chamber in pairs, i.e. earth plus space samples. Etching of each pair of the samples was done in turns under constant conditions, and continued until the coating was etched off to expose the surface of the substrate, i.e. polyimide strip.

Concentration of elements under investigation was calculated with allowance for their relative sensitivity on the basis of the intensity of peaks in the energy spectrum fixed by a two-coordinate recorder.

Typical results of investigations into distribution of different elements through thickness of the coatings are shown in Figures 1 and 2. It follows from them that the earth and space coatings differed in content and distribution of elements through thickness of the film, as well as in thickness itself, despite the fact that the calculated conditions of their deposition on the substrate were identical. Coatings deposited on all the investigated samples on board the «Mir» differed in thickness from those deposited on the earth samples by more than a factor of two, independently of the deposition conditions (see Figure 1). The most probable cause of this difference lies in higher temperature of the crucible in evaporation of alloys on board the orbital complex, as compared with evaporation on Earth.

In addition to the basic elements of copper and silver, the resulting coatings were found to contain also a certain amount of impurity elements (C, O, Cl) and elements which belonged to the substrate material (N, C, O). Impurities of sulphur and chlorine, which were absent in

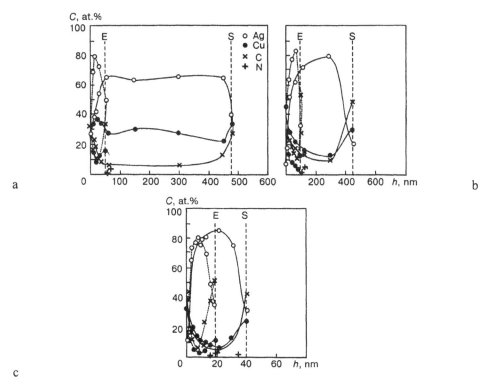

a

b

c

Figure 1 Profiles of variations in the concentration of copper, silver, carbon and nitrogen through thickness of the Ag–Cu coatings produced under space (solid lines) and Earth (dashed lines) conditions: a — sample No. 1; b — sample No. 2; c — sample No. 11 (vertical line indicates the coating–substrate interface).

the bulk of the above films, were detected on the external surface, and in some cases in the subsurface layer up to 5–8 nm thick.

Nitrogen was fixed only on the side of the substrate, whose constituent it was. Therefore, appearance of nitrogen in the spectrum, as well a dramatic increase in the carbon concentration could serve as the indicators of reaching the coating–substrate interface.

One of the key issues is distribution of such impurities as oxygen and carbon. As can be seen from Figures 1 and 2, the Ag–Cu coatings, and especially their surface layers, contain a large amount of carbon, i.e. up to 50 at.%, on the surface and 12–30 at.% in the bulk. The content of carbon dramatically decreases with etching, and it is much lower in the bulk of the films. Further increase in the time of etching is accompanied by an increase in the concentration of carbon, which is associated most probably with approaching the coating–substrate interface and capture of carbon from the surface of the substrate. Carbon fixed by Auger spectrometry is likely to be in a fixed condition. These could be molecules CO, CO_2, particles of hydrocarbons trapped in evaporation from the surrounding atmosphere, or other C-containing impurities entering the coating from the surface of the polyimide strip. In this case the concentration of carbon in the subsurface layers and in the bulk of the earth Ag–Cu coatings is much higher than in the flight analogues.

Enrichment of the surface layers of the Ag–Cu coatings with oxygen takes place as well, its concentration being much lower in the volume of the coatings. At the same time, the concentration of oxygen is higher in the flight samples (where it was fixed) than in the earth ones.

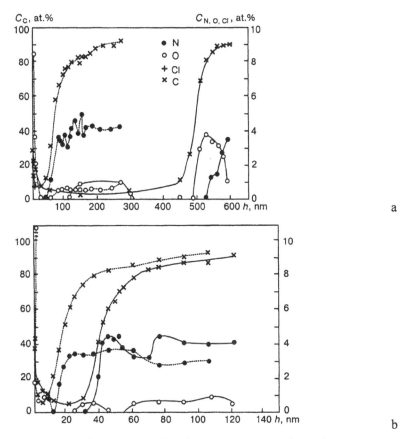

Figure 2 Profiles of variations in the concentration of impurity elements, such as carbon, oxygen, nitrogen and chlorine, through thickness of the Ag–Cu coatings produced under space (solid lines) and Earth (dashed lines) conditions: a — sample No. 1; b — sample No. 11.

Of special notice is the fact of a non-uniform distribution of the basic elements and impurities in the bulk of the coatings. As was indicated above, the carbon and oxygen content of the subsurface layers is much in excess of that in the bulk of the coatings. Besides, in some cases it was possible to fix variations in the concentration of these elements through thickness (see Figures 1 and 2). The same applies to other impurities, such as sulphur and chlorine. At the same time, the concentration of the basic elements is maximum in the bulk and decreases towards the surface, which is indicative of enrichment of the surface with foreign impurities.

This analysis of variations in the relative content of copper and silver through thickness of the coatings shows that copper and silver have different distribution in the volume. Figure 3 shows profiles of variations in the relative concentrations of copper and silver through thickness of the coatings in terms of the two-component system with no allowance for impurities. These curves enable details to be revealed of the mutual effect of components in the melt and coating during evaporation and deposition of a binary alloy. It can be seen from Figure 3 that the relative content of copper increases in the subsurface layer of the coating, i.e. the surface is enriched with copper, as compared with the volume. This applies to both interfaces: coating–vacuum and coating–substrate. The Table gives values of averaged volume, C_v, and surface, C_s, concentrations of copper for the Ag–Cu coatings with

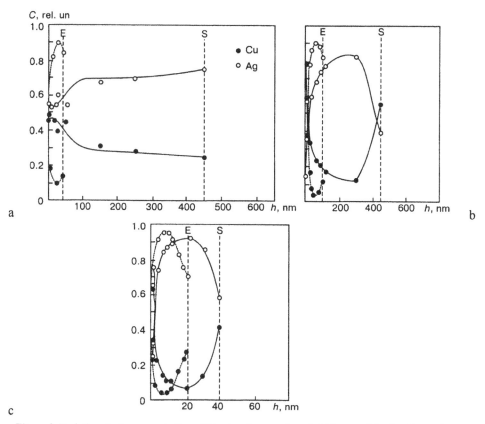

Figure 3 Variations in the concentration of Cu–Ag elements though thickness of the Ag–Cu coating on the polyimide substrate (see the plots). Designations are the same as in Figure 1.

a differing silver content. The Table also gives values of the surface enrichment (segregation) coefficient for copper, calculated as a ratio of the surface to volume concentration: $K_s = = C_s/C_v$. It follows from the Table that the segregation coefficient depends upon the content of components in a coating and increases with decrease in the concentration of copper in the volume.

Analysis of the data given in Figure 3 and the Table shows that in the subsurface layers of the Ag–Cu coating there occurs a redistribution of its components, which is not associated with distillation of the alloy during evaporation. Otherwise the layers adjoining the substrate would have been enriched with silver, and the volume concentration of copper in the samples deposited later would have been much higher. Moreover, thermodynamic considerations and calculations based on the electron theory in an approximation of a strong bond [4] show that silver should have segregated on the external surface of the Ag–Cu coatings. But we have the opposite phenomenon. It is likely that the effect detected has no simple explanation, but is a result of superposition of a number of diffusion processes occurring directly during condensation of a two-component vapour flow in the course of formation of the final structure and phase composition of the Ag–Cu coatings. Such processes include, in addition to the primary ones, i.e. initiation and growth, also recrystallization, decomposition of the oversaturated solid solutions (OSS) and precipitation of the second phase by the continuous and intermittent mechanism, as shown by our X-ray diffraction and electron microscopy studies. Besides, under actual vacuum

Concentration of copper in volume and surface layers of the Ag–Cu coatings (in terms of binary system).

Sample No.	Coating thickness, δ, nm	Concentration of copper, at.%		Surface segregation coefficient $K_s = C_s/C_v$
		Volume, $\overline{C_v}$	Surface, $\overline{C_s}$	
1	480	30.10	46.810	1.550
	60	10.30	50.000	4.850
2	30	8.53	77.750	9.110
	15	5.17	61.650	11.920
3	60	48.41	82.320	1.700
	80	10.23	65.225	6.380
4	50	73.11	95.100	1.300
	15	9.30	53.930	5.800
5	120	98.35	99.030	1.006
	72	65.70	80.910	1.230
6	66	100.00	100.000	1.000
	100	46.80	84.700	1.810
7	154	100.00	100.000	1.000
	22	74.55	79.100	1.060
8	100	99.00	99.000	1.000
	130	93.20	96.500	1.035
9	30	9.74	62.000	6.370
	15	5.50	49.620	9.020
10	40	8.24	75.660	9.180
	15	4.06	35.060	8.640
11	36	54.82	95.960	1.750
	18	4.42	48.760	11.030

Note: Data obtained in space are given in the numerator, and those obtained on Earth are given in the denominator.

conditions the possibility exists of the occurrence of chemisorption, oxidation and other effects of interaction of residual gases with the atmosphere.

Despite the complexity of the general picture of what is happening in this case, we can distinguish the dominating factors which determine the final results at each stage of formation of the coatings. One of such factors is condensation-enhanced diffusion detected in [5, 6] dedicated to investigation of two-layer film systems Au–Ag. This effect shows up in acceleration of the processes of volume and grain boundary diffusion as a result of generation of vacancies formed on the surface of growth and in the subsurface layers directly during condensation. The presence of such a generation source leads to formation of a flow of vacancies to the drains, which can be the free surface of growth, the opposite surface of a film, grain boundaries, dislocations and other defects. Intensive saturation of the surface with vacancies leads to a substantial increase in diffusion activity of the films. The directed flows of the vacancies cause the opposite flows of atoms. In this case, inequality of partial coefficients of diffusion of components may result in an enhancement of the flow of some atoms and weakening of the flow of others, which leads to segregation phenomena [7]. The above patterns of relationships are applicable both to volume and grain boundary diffusion. Moreover, in the last case the effect of the condensation-enhanced diffusion is stronger [6].

In our experiments deposition was carried out without preheating of the substrate at temperatures of up to 40–50 °C. Under these conditions the volume diffusion is frozen, and the segregation effect we observed can develop only because of grain boundary diffusion, which is favoured, as indicated by electron microscopy, by a high degree of dispersion of the grain structure of the Ag–Cu coatings. In this case increase in the copper content is accompanied by a decrease in the size of grains, which is minimum in coatings of the hypereutectic and non-eutectic composition. The grain boundary diffusion processes are substantially facilitated, and the diffusion paths are small, because of small values of thickness of the coatings, under conditions of a dispersed structure (the number of boundaries is very large). That is why even for a short time of evaporation (~ 3 s) it is possible that redistribution of components can take place in them under the effect of the vacancy flow.

Often the grain boundaries of the produced coatings are located normal to their surface, which favours distribution of the vacancy flow in a direction from the surface to substrate. It is a known fact that the diffusion homogenization occurring by the vacancy mechanism, because of the differencies in partial coefficients of diffusion of components, results in a flow of vacancies responsible for the Kirkindall effect, which compensates for differencies in the flows of atoms from different parts of a sample. This flow of vacancies is directed towards a more mobile component. If an extra vacancy flow caused by external factors, stresses and other types of non-equilibrium, i.e. the so-called «vacancy wind» (according to Manning [8]), emerges in the system, it can stimulate the so-called «ascending diffusion» [9] and activate the segregation phenomena by increasing the partial coefficient of diffusion of a high-mobility component. In our case this component is silver. Therefore, the vacancy flow directed from the growth surface to the interface with the substrate promotes removal of silver from this interface and its enrichment with copper.

The fact that the coating–substrate interface is also enriched with copper is not at variance with the above, as occurrence of the condensation-enhanced grain boundary diffusion is inseparably linked with migration of the grain boundaries, i.e. recrystallization [10]. It is reported [11] that vacancies are generated in ultra-dispersed systems during recrystallization. That is why the «vacancy wind» near the growth surface may change its direction, whereas near the interface with the substrate it will become stronger. Reversal of the vacancy flow near the surface of the coatings results in their enrichment with copper. Occurrence of recrystallization processes in the Cu–Ag films, accompanied by decomposition of OSS, is supported by our electron microscopy and X-ray diffraction analysis data, which show in particular that the films with a content of Cu < 10–15 at.% exhibit growth of grains of Ag-based solid solution with their plane (111) oriented in parallel to the substrate. As follows from the Table, K_s reaches high values in such films. Also of note is the increased porosity of these samples, revealed by analysis of electron microphotographs of the structure. This porosity was not detected in coatings with a content of Cu ≥ 25–30 at.%, which is indicative of the existence of the intensive vacancy flows directed to the silver-rich regions.

The probability of occurrence of the condensation-enhanced diffusion in the system under consideration is evidenced by cleaning (full or partial) of the volume of a coating from foreign impurities, carbon in particular, which was present on the surface of earth samples, because the evaporation process was performed in «oil» vacuum, and the residual gases contained large amounts of hydrocarbons. The «Mir» environment also contains a large amount of impurities and contaminants which are evolved from vapours of materials used at the orbital complex, including from the substrate material (N, O, C, Cl and S). The above effect of the condensation-enhanced diffusion and generation of vacancies by the surface during condensation promotes redistribution of impurity elements towards the vacuum–coating interface. Recrystallization initiated by the vacancy flows also favours displacement of the impurities from the bulk of the coatings and their escape to the free surface, owing to a substantial decrease in the area of the grain boundaries, where most of the non-metallic inclusions are located.

It should be noted in conclusion that the effects of a segregation separation of components in the Ag–Cu coatings, revealed in this study, take place at low temperatures and a short time of condensation, which is possible only because of their kinetic nature. Analysis of Auger spectrometry and structural data supports the vacancy mechanism of formation of segregations of copper near both external boundaries of the coating. According to this mechanism, formation of segregations is caused by the different mobility of silver and copper atoms during the condensation-enhanced diffusion initiated by non-equilibrium vacancies. The flows of such vacancies are generated on the surface and in the bulk of the film in the course of formation of a coating and structural transformations (recrystallization and decomposition of OSS) during condensation. These processes also promote cleaning of the films from foreign inclusions.

It should also be noted that the results of this study, along with the results of [12], can serve as the source data for evaluation of parameters of deposition of coatings under earth and space conditions with a preset profile of distribution of components through thickness and with the minimum possible content of foreign non-metallic impurities in the actual process of their formation.

1. Zagrebelny, A.A., Neznamova, L.O., Nikitsky, V.P. et al. (1989) *See pp. 319–321 in this Book.*

2. Lukash, E.S., Shulym, V.F., Grigorenko, G.M. et al. (1986) *See pp. 340–344 in this Book.*

3. Larikov, L.N. and Yurchenko, Yu.F. (1985) Diffusion in metals and alloys. Structure and properties of metals and alloys, In: *Thermal properties of metal and alloys. A handbook,* Kiev, Naukova Dumka, 367 pp.

4. Murhersee, S. and Moran-Lopez, I.I. (1987) Surface segregation in transition metal alloys and in bimetallic alloy dusters, *Surface Sci.*, pp. 1135–1142.

5. Kosevich, V.M., Kosmachev, S.M., Karpovsky, M.V. et al. (1986) Condensation-enhanced grain boundary diffusion in two-layer gold-silver films, *Poverkhnost*, No. 8, pp. 151–152.

6. Kosevich, V.M., Kosmachev, S.M., Karpovsky, M.V. et al. (1987) Diffusion in two-component gold-silver films developing during condensation, *Ibid.*, No. 1, pp. 111–116.

7. Bokshtein, S.Z. (1973) *Diffusion and structure of metals*, Moscow, Metallurgiya, 208 pp.

8. Manning, J. (1971) *Kinetics of diffusion of atoms in crystals*, Moscow, Mir, 280 pp.

9. Geguzin, Ya.Ye. (1979) *Diffusion zone*, Moscow, Nauka, 343 pp.

10. Kosmachev, S.M., Karpovsky, M.V. and Klimenko, V.N. (1987) Investigation of diffusion-induced migration of grain boundaries in two-layer films, In: *Abstr. of pap. of 2nd All-Union Conf. on Structure and Electron Properties of Grain Boundaries in Metals and Semiconductors*, Russia, Voronezh, 1987. Voronezh, 117 pp.

11. Trusov, L.I., Novikov, V.I., Lopukhov, Yu.A. et al. (1984) Recrystallization in ultra dispersed systems, *Poroshkovaya Metallurgiya*, No. 5, pp. 28–34.

12. Toptygin, A.L., Neznamova, L.O. and Malashenko, I.S. (1990) *See pp. 56–64 in this Book.*

Volume-Structural and Phase Micro-Heterogeneities of Electron Density in Metal Films Condensed under Flight and Earth Conditions*

P.G. Cheremskoj[6], A.L. Toptygin[6], A.S. Panikarsky[6], A.V. Arinkin[6], L.O. Neznamova[2], V.F. Shulym[1] and E.S. Mikhajlovskaya[1]

Investigations were conducted to study films of copper and Cu–Ag alloy < 1 μm thick deposited on fixed and moving thin polyimide substrates under earth and flight (on board the OC «Mir») conditions, using the «Yantar» unit equipped with an evaporator developed by the PWI.

The basic investigation methods included X-ray low-angle scattering (XLAS) and X-ray diffractometry. XLAS indicatrices were recorded using high-vacuum low-angle X-ray diffractometer with an increased resolution in a molybdenum anode radiation. The procedure used for low-angle recording, methods of the XLAS indicatrix processing and evaluation of characteristics of revealed micro-heterogeneities of electron density are described in [1].

Effect of flight conditions on the character of XLAS in copper films. Copper films were investigated in the state not separated from the polyimide substrate. XLAS, having an intensity of $5 \cdot 10^{-4} - 1 \cdot 10^{-3}$ of that of the primary beam, was detected even at an insignificant thickness of the films. XLAS indicatrices shown in Figure 1 were normalized to the same scattering volume, as the films were somewhat different in thickness (Table 1). It is seen that in condensation under earth conditions the XLAS intensity of the films condensed onto the fixed substrate is in excess of that of the films condensed onto the moving substrate over the entire angle scattering range. (Fixed position of a substrate was caused by a stop of the film-pulling mechanism at reversing for deposition of the next layer through thickness during condensation of copper, or by a stop prior to deposition of the next layer along the length during condensation of Cu–Ag alloys.) In addition, both types of the films are also different

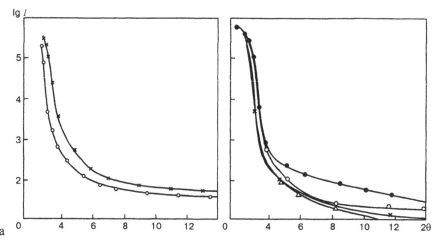

Figure 1 XLAS indicatrices of copper films condensed under earth (a) and flight (b) conditions:
●, ✕ — moving and fixed substrates, respectively, 0.25 μm thick (a); ● — $h = 0.175$; ○ — $h = 0.08$; ✕ — $h = 0.05$; △ — $h = 0.03$ (b) μm.

---

* (1990) Paper was presented at the 5th Korolyov Readings of the 2nd Republ. Conf. on Fundamental and Applied Problems of Cosmonautics, Kiev.

in the angular distribution of the XLAS intensity. In condensation onto the moving substrate, in a range of super low angles, the XLAS intensity falls more steeply, which is indicative of a different character of size distribution of electron density scattering heterogeneities in the films investigated. According to the data of previous investigations, submicro- and micropores of a condensation origin are the dominating type of scattering heterogeneities in the condensed films of pure metals. Processing of the XLAS indicatrices by the tangent method (according to Guinier) was done with allowance for this circumstance. Results of processing of the profile of the XLAS indicatrices are given in Table 1.

Despite considerable polydispersity of submicro-pores, attributed to the small thickness of the films, the tangent method made it possible to detect most reliably only those submicro-pores the size of which was not in excess of 200 Å, and the set of which was conditionally subdivided into two size fractions: $R_1 < 25$ Å and 25Å $< R_2 < 200$ Å. The total volume and amount of submicro-pores in the films condensed onto the fixed substrate are markedly higher than in films condensed onto the moving substrate. The main difference lies in the relative contribution made by coarser submicro-pores. In the films deposited on the fixed substrate, their volume concentration is 4 times as high as that in the films condensed on the moving substrate; this difference being less pronounced for finer submicro-pores (see Table 1). This is indicative of the fact that the mechanism of pore formation processes is very sensitive to the condensation conditions. It is well known that the main mechanisms of formation of pores of a condensation origin include the following: diffusion-vacancy effect of «shading», looseness of alignment and coalescence of structural elements, gas evolution and absorption, as well as non-uniform shrinkage. The non-uniform contribution by the «shading» effect is the most probable cause of the observed differences in submicro-porosity of the films investigated.

In condensation onto a substrate which performs multiple reverse-translation movements with respect to the evaporator, the role of the «shading» effect during the pore formation process decreases because of an equally probable distribution of the condensate coming onto the substrate from a molecular flow at different condensation angles. This is evidenced also by the fact that the degree of anisotropy of XLAS, caused by anisometry and preferred orientation of submicro-pores along the molecular flow, is much less pronounced in the films condensed onto the moving substrate than in the films deposited onto the fixed substrate. In the case of the fixed substrate the conditions created are more favourable for the diffusion-vacancy growth of fine submicro-pores, because of the continuous replenishment of the growing films with excess vacancies from the molecular flow at a given temperature of the substrate. It is this fact that causes the large volume contribution by coarse submicro-pores into deposition on the fixed substrate.

For the films condensed under flight conditions, we managed to establish the character of variations in the intensity of XLAS, dispersion and volume concentration of submicro-pores with growth of thickness of the condensate. No substantial differences were revealed in the level of the XLAS intensities for films of an equivalent thickness condensed under flight and earth conditions on the moving substrate. However, size distribution of the submicro-pores detected was of a different character. The level of the XLAS intensity in the flight films grows with increase in thickness of the condensate (Figure 1).

This dependence, as was established by previous investigations, is characteristic also of the films condensed under earth conditions. However, XLAS in the flight films, especially at comparatively high angles of scattering, is more isotropic than in the earth films. This is in agreement with the earlier conclusions that condensation under the flight conditions leads to formation of more equiaxial submicro-pores. Over the investigated range of thickness the total volume concentrations of submicro-pores were commensurable in level for both earth and flight films (see Table 1). However, of note is the difference between the volume and

Table 1 Calculation data of processing (according to Guinier) of XLAS indicatrices for copper films condensed on earth and flight samples.

Sample No.	Sample characteristic	Coating thickness, μm	$I_1(0)$, p/s	$I_2(0)$, p/s	R_1, Å	R_2, Å	$N_1 \cdot 10^7$	$N_2 \cdot 10^5$	$\Sigma N_i \cdot 10^7$
1	Earth, I/1, MS	0.250	0.800	19.00	25	195	18.0	0.35	18.0
2	Earth, I/1, FS	0.250	1.000	13.00	25	105	29.0	9.90	29.0
1	Flight, I/1, MS	0.175	1.350	160.00	40	170	3.6	930.00	13.0
2	Flight, I/1, MS	0.008	0.275	21.50	27	164	7.8	1.30	7.8
3	Flight, I/1, MS	0.030	0.200	7.00	28	114	4.5	3.50	4.5
4	Flight, I/1, MS	0.050	0.180	3.35	27	96	5.1	4.60	5.1

Table 1 (cont.)

Sample No.	Sample characteristic	$V_1 \cdot 10^{12}$, cm³	$V_2 \cdot 10^{12}$, cm³	$\Sigma V_i \cdot 10^{12}$	$C_1 \cdot 10^{-2}$, %	$C_2 \cdot 10^{-3}$, %	$\Sigma C_i \cdot 10^{-3}$	$\widetilde{R}_{av}$
1	Earth, I/1, MS	17.0	1.1	18.0	1.3	0.88	1.4	35
2	Earth, I/1, FS	24.0	4.9	29.0	1.9	3.90	2.3	40
1	Flight, I/1, MS	9.6	17.0	27.0	1.1	1.90	3.0	120
2	Flight, I/1, MS	6.4	2.3	8.7	1.6	5.80	2.2	65
3	Flight, I/1, MS	4.1	2.2	6.3	2.7	1.50	1.2	60
4	Flight, I/1, MS	4.2	1.7	5.9	1.7	1.40	1.8	50

Notes. Here and below FS stands for the fixed substrate and MS — for the moving substrate; $I_1(0)$, $I_2(0)$ is the intensity of XLAS at a zero angle of scattering for submicro-pores with inertia radii, respectively; $R_1 \leq 40$ Å and $40 < R_2 < 200$ Å ; N_1 and N_2 are the quantitative and C_1 and C_2 — volume concentrations of submicro-pores, respectively, with inertia radii $R_1 \leq 40$ Å and $40 < R_2 < 200$ Å ; V_1 and V_2 are the volumes of pores of fractions I and II, respectively; ΣV_i is the total volume of submicro-pores; ΣN_i and ΣC_i are the total quantitative and volume concentrations of submicro-pores; R_{av} is the mean-square average weighted size of submicro-pores.

quantitative contributions made by coarser submicro-pores. Their content in the flight films is much higher than in the earth films, the quantitative contribution by fine submicro-pores being lower (compare N_1 and N_2 in Table 1). This result is also in agreement with that obtained earlier for silver films. The mean-square average weighted size of submicro-pores in the flight films is higher than in the earth ones and continuously grows with an increase in thickness of the films.

Therefore, condensation under flight conditions leads to formation in the films of more equiaxial and coarser submicro-pores, as compared with earth conditions. However, no substantial difference was revealed in the level of total micro-porosity for the above types of films.

XLAS in Cu–Ag films condensed under earth and flight conditions. Evaporation of Cu–Ag alloy from one crucible onto a moving polyimide strip leads to the fact that alloys with a different concentration are condensed on different regions of the strip. Because silver and copper have different vapour pressures at the initial stages of deposition, the condensate formed is a solid solution rich in silver (Table 2). The films deposited at later stages of the evaporation process are substantially rich in copper. Because of a considerably lower evaporation rate of the charge depleted in silver, their condensation rate and thickness are

Table 2 Results of estimation of the concentration of silver in samples and individual phases on the basis of the X-ray diffractometry data.

Experiment	Sample	τ, s	$a_{c.g}$, Å	$\bar{a}$, Å	a_1, Å	a_2, Å
Flight, II/1	1 FS	30	4.0620	4.657	4.0640	–
	2 FS	30	4.0720	4.064	4.0720	–
	3 MS	32	4.0790	–	4.0858	–
	4 MS	50	4.0800	4.077	4.0840	–
	5 MS	67	4.0530	4.046	4.0770	4.013
	6 MS	100	3.63230	–	3.6210	–
Earth, II/1	1 FS	30	4.0770	–	4.0860	–
	2 MS	32	4.0840	4.082	4.0860	–
	3 MS	66	4.0210	–	4.0858	–
	4 MS	97	4.0500	4.035	4.0810	4.039
	5 MS	127	4.0470	4.039	4.0690	–
	6 MS	130	4.0620	4.057	–	–
Earth, II/2	1 MS	55	4.0810	–	4.0858	–
	2 FS	85	4.0570	4.050	4.0770	4.043
	3 FS	85	4.0438	4.036	–	4.032
	4 MS	88	4.0230	4.011	–	4.026
	5 FS	130	3.6240	3.626	3.628	–

Table 2 (cont.)

Experiment	Sample	$C_{c.g}$, at.%	C, at.%	C_1, at.%	C_2, at.%
Flight, II/1	1 FS	93.0	92.0	93.5	–
	2 FS	–	–	–	–
	3 MS	97.5	–	99.0	–
	4 MS	96.0	97.5	98.5	–
	5 MS	91.0	89.0	89.5	81.0
	6 MS	3.0	–	1.0	–
Earth, II/1	1 FS	97.0	–	100.0	–
	2 MS	99.0	98.0	100.0	–
	3 MS	86.0	–	99.0	–
	4 MS	90.0	86.5	–	–
	5 MS	89.5	87.5	98.0	87.5
	6 MS	93.0	92.0	95.0	–
Earth, II/2	1 MS	98.0	–	99.0	–
	2 FS	92.0	90.5	97.5	88.5
	3 FS	89.0	87.0	–	87.5
	4 MS	83.5	80.0	–	84.0
	5 FS	2.0	2.0	25.0	–

Notes. Here and below τ is the time of evaporation allowing for the time of preheating of the crucible and reaching the working conditions (30 s); $a_{c.g}$ is the lattice spacing determined from the centre of gravity of the intensity curve; $\bar{a}$ is the mean spacing; a_1 and a_2 are the lattice spacings of the primary and secondary solutions, respectively; $C_{c.g}$ is the concentration of the primary, C_1, and secondary, C_2, solutions, respectively.

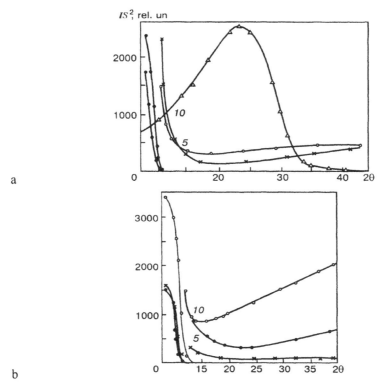

Figure 2 Invariants of XLAS indicatrices for Cu–Ag films condensed under earth (a) and flight (b) conditions: Δ, X — on the side of copper and silver, respectively (a); O — intermediate region; Δ, O — fixed substrate; X — moving substrate; X, O, ● — on the side of copper and silver, respectively; ●, X — fixed substrate; O — moving substrate (b).

much lower than those of the films rich in silver. That is why, they are more prone to oxidation and gas saturation. Small thickness and presence of amorphous oxides lead to a substantial decrease in their reflectivity and hamper recording of the Wolf–Bragg reflections. This makes evaluation of phase composition of such films much more difficult. Table 2 shows that the Cu-based solid solution is fixed in these films.

As well as for the earth copper films, the intensity of XLAS of the thickest of the silver-rich films deposited on the fixed substrate over the entire angle range is in excess of the intensity of scattering of the films, similar in composition, condensed onto the moving substrate (Figure 2). Unlike the films of pure metals, XLAS of condensates of alloys may be caused not only by their submicro-porosity, but also by concentration of submicro-heterogeneities localized in micro-volumes, which are larger in size than the radiation wave length.

According to the data of X-ray diffractometry, no marked indications of decomposition of solid solution was revealed in the case of condensation onto the moving substrate under the earth conditions, unlike the case of the films of a similar composition deposited onto the fixed substrate, where an incomplete decomposition of initial solid solution was detected. Such films are close in concentration composition but differ in thickness.

As to the level of intensity and degree of anisotropy of XLAS, the earth films deposited on the moving substrate, rich in silver, hardly differ from the earth copper films condensed in an analogous manner (see Figure 2). According to the results of [2], this is characteristic of the condensed solid solutions which contain no localized concentration heterogeneities.

As well as in the earth films of pure metals, XLAS of such solutions is caused primarily by the presence of submicro- and micro-pores of a condensation origin.

Submicro-porosity of the earth Cu–Ag films increases in the case of deposition on the fixed substrate, which is one of the causes of a rise in their scattering ability. An incomplete decomposition of solid solutions in the above films, detected by X-ray diffractometry, leads in turn to an increase in XLAS. This is likely to be attributable to a larger increment in the intensity of XLAS in a change of the moving and fixed substrates, as compared with similar copper films.

It might be supposed that an increment in the level of intensity of XLAS (Figure 2a), observed in the analogue Cu–Ag films, is caused to a large degree by the concentration heterogeneities formed as a result of an incomplete decomposition of solid solution.

Scattering ability of thin, but copper-rich films condensed onto the moving substrate is by two orders of magnitude higher, and their mean size of heterogeneities is smaller than those of the films produced in an identical manner, but silver rich and much thicker (Table 3). Most probably this is caused not only by the high degree of heterogeneities of such condensates, resulting from oxidation and gas absorption, but also by their branched submicro-discontinuity (island-type structure), which is characteristic of thin metal coatings on dielectrics. This is also proved by electron microscopy. In this case, the curves of invariant of XLAS indicatrices clearly show a characteristic bimodality of size distribution of the volume of scattering heterogeneities. It may be indicative of electron density heterogeneities of a different nature present in the films.

Condensates, more or less enriched with silver, were studied on the films produced under flight conditions. Condensation onto the fixed substrate results in the formation of comparatively stable Ag-based solid solutions, which exhibit no diffractometry detectable indication of decomposition. However, the level of their XLAS intensity turned out to be very sensitive to the concentration of silver (Figure 2b).

If we assume that the processes of pore formation in condensates of pure metals and non-decomposed solid solutions are identical in character, in the case of enrichment of solid solution with silver the detected increase in scattering ability of the films will not be a surprise, because it is caused by a natural increase in a gradient of electron density at the pore–matrix interfaces. In this case, it is likely that an increase in the contribution by coarse heterogeneities does reflect some peculiarities of the processes of pore formation in films of differing compositions. We think it is impossible to explain these peculiarities within the framework of the investigations performed.

The largest difference in character and level of the XLAS intensities is observed for flight films of close compositions condensed onto fixed and moving substrates (see Figure 2b), especially in a range of high scattering angles, where a dramatic increase in scattering ability of the films takes place. This is accompanied by a decrease in mean size of micro-heterogeneities, resulting from an increase in the specific contribution made by the smallest of them (see Table 3).

Because of a small thickness of the films, we, unfortunately, failed to register XLAS over the entire available range of scattering angles, as the level of the intensity at «tails» of XLAS indicatrices is commensurate with the background level of recording devices. Therefore, an ascending «tail» of invariant cannot have an unambiguous interpretation. This could be the presence of either a second maximum, which is indicative of bimodality of heterogeneities, or disturbances characteristic of the asymptotic form of XLAS indicatrices, which are typical, in particular, for strongly non-equiaxial and mutually disoriented scattering centres.

Phase heterogeneities of Cu–Ag alloy. X-ray diffraction analysis of the samples made under the space conditions and in analogous earth experiments was conducted using the «Dron-UM-1» device in cobalt anode radiation. As shown by phase analysis, coatings contain fcc solid solutions based on silver and copper. No other phases were detected by the

Table 3 Calculation data of processing of XLAS indicatrices and diffraction patterns for the Cu–Ag films condensed under earth and flight conditions.

Sample characteristic	Coating thickness, μm	$I_1(0)$, p/s	$I_2(0)$, p/s	R_1, Å	R_2, Å	$\bar{a}$, Å	a_1, Å	a_2, Å	C, at.%	C_1, at.%	C_2, at.%
Earth I/1, FS	0.250	0.96	12.5	20	110	4.039	4.081	4.039	87.5	98.0	87.5
Earth I/1, MS	0.250	0.60	40.0	25	140	4.011	–	4.026	80.0	–	84.0
Earth I/1, MS	0.040	0.33	1900.0	40	70	–	–	–	–	–	–
Flight II/2, FS	0.250	0.26	12.0	30	115	–	–	–	–	–	–
Flight II/2, FS	0.158	0.46	19.5	21	110	–	–	–	–	–	–
Flight II/1, MS	0.250	2.80	5.6	24	86	4.057	4.064	–	92.0	–	93.5

X-ray method. The Cu–Al alloy coatings, like in the earlier experiments carried out under the «Isparitel» program, are characterized by the presence of axial texture with axis (111) oriented along a normal to the surface. This circumstance allowed us to limit our studies to detailed analysis of the diffraction pattern within a range of angles 2θ from 40 to 55°, covering the entire range of probable positions of diffraction maxima (111) for solid solutions of Cu–Ag and pure silver and copper. An increase in the time of operation of evaporators is accompanied by a shift of the position of maximum (111) in diffraction patterns of the samples towards higher angles 2θ, which is indicative of a decrease in the silver content of a coating. Unfortunately, the small thickness of the condensate in the majority of cases hampers X-ray analysis of non-separated samples, as the diffraction pattern of a coating is superimposed as a rule on a non-linear background caused by scattering on the polymeric substrate. Note that near $2\theta \approx 52°$ a number of the samples of substantial thickness exhibit a low maximum (200) of the Ag-based solid solution. Further analysis distinguished this maximum from the general curve of distribution of the intensity.

At marked concentrations of copper (more than 10 at.%) one can easily see formation of two solid solutions, which are attributable to ageing (decomposition) of a primary oversaturated solid solution. At lower concentrations of copper, decomposition of solid solution is accompanied as a rule by a clearly defined asymmetry of the central part of a line or extended one-sided «tails». The latter may be caused both by heterogeneity of solid solution (copper segregation) and small size 50–100 Å of the fcc phase particles rich in copper. As evidenced by data from electron microscopy and Auger analysis, both may take place in the Cu–Ag films.

Estimation of the concentration of silver (and copper) in the samples was done by determining the «centre of gravity» of a general profile, from which a mean spacing of the crystalline lattice, $a_{c.g.}$, was calculated. The $\bar{a}$ value, corresponding to the position of the «centre of gravity» of the intensity profile, was determined in the case of a complex shape of the profile and/or large length of the diffraction maximum, allowing for varying angular factors and structural amplitude (atomic factor).

Data on the $a_{c.g}$ and $\bar{a}$ values, as well as the corresponding estimates of the concentration of silver, $C_{c.g}$ and C, are given in Table 2. According to the character of distribution of the intensity, in analysis of the concentration characteristics of a sample the preference should be given to a value which is shown in the Table without brackets. In addition, this Table also gives values of the crystalline lattice spacing, a_1 and a_2, as well as the corresponding concentration estimates, C_1 and C_2, of the fcc solid solutions, obtained on the basis of data on positions of the observed maxima of distribution of the intensity. Within the framework of one experiment the samples were arranged in an ascending order of time of operation of the evaporator. As can be seen from Table 2, an increase in the time of operation of the evaporator is accompanied by a general increase in the concentration of copper. Comparison with the data of Auger element analysis shows a qualitative agreement. All quantitative differences can be attributed to the fact that data of analysis of the X-ray diffractometry curves, given in Table 2, apply to the concentration of silver (copper) only in the crystalline lattice of the fcc phases. In the case where condensation results in formation of amorphous oxides or other compounds of more reactive copper, or if segregation of copper occurs along the boundaries of fine grains, these data should be increased (in silver), as compared with the Auger spectrometry data. At a small thickness and substantial background of the polymeric substrate, such an increase might be expected for a structure with fine precipitates enriched with copper, as reflection from the latter can hardly be distinguished from the background. Note that the mean concentration of silver in a coating ($C_{c.g}$ or C) is as a rule lower than C_1, corresponding to a maximum of the intensity. This effect seems to be associated with heterogeneity of a solid solution and/or presence of precipitates of the copper-rich phase. The general character of the structure, judging from the diffractometry data, corresponds to the data obtained earlier for other substrates, whereas the ability to decompose is characteristic of alloys with a high copper content (provided that coatings are sufficiently thick). In thin coatings the stabilizing effect of the surface and impurities occurs, thus leading to a delay of decomposition.

1. Cheremskoj, P.G. (1985) *Methods for investigation of porosity of solids*, Moscow, Energoatomizdat, 102 pp.

2. Palatnik, L.S., Cheremskoj, P.G. and Fuks, M.Ya. (1982) *Pores in films*, Moscow, Energoizdat, 216 pp.

Coatings Produced in Near-Earth Orbit and Prospects of Their Application in Microelectronics*

D.A. Dudko[1], E.S. Mikhajlovskaya[1], V.F. Shulym[1] and L.O. Neznamova[2]

For a long time cosmonautics have extensively used various coatings [1–3] for protection of metallic materials [4] and polymers [5] employed in fabrication of flying vehicles and large-sized structures. This solves the problems of providing optical reflecting surfaces [6] and protecting surfaces from the effect of atomic oxygen, ultraviolet radiation, flows of charged particles, dust, etc. It is known [7, 8] that the space environment has a favourable effect on formation of sound coatings with a better reflecting ability, as compared with earth, owing to removing oxide contaminants from metal; higher electrical conductivity, more uniformly distributed over the coating surface, provided by a decreased porosity when condensation is performed in a high

* (1999) *Avtomaticheskaya Svarka*, No. 10, pp. 51–55.

vacuum; improved ductility and fatigue properties; and lower internal stresses, which is attributable to general purity and structural perfection of the coatings.

This article analyses the general feasibility of development of a commercial space electron beam technology for deposition of thin coatings for the needs of modern microelectronics, based on the results of investigation of different properties of the coatings produced in space over a period of many years.

The space experiments were designed and performed by associates of the PWI in collaboration with the S.P. Korolyov RSC «Energiya»[2].

Initial experiments on electron beam evaporation of pure metals [9] were conducted using the «Isparitel» unit under flight conditions on board the OS «Salyut-6». Films 0.02–1.00 μm, including silver, gold and copper, were deposited on the substrates of glass and titanium. Surface morphology, structure and substructure of these coatings were investigated [10] by metallography, scanning and transmission electron microscopy, X-ray diffraction analysis and X-ray low-angle scattering.

No fundamental differences were detected in substructures of flight and earth coatings of gold and silver. The silver coatings were noted to have a more perfect substructure, as compared with the gold ones. The character of substructure of the latter is indicative of hindrance of their recovery and recrystallization processes. The concentration of submicropores in the investigated earth coatings is about 0.1 vol.%, which is much higher than in the similar coatings produced in space, where this concentration is no more than 0.04 vol.%. Besides, the flight samples have pores of a larger size. For example, their mean square and characteristic size in gold, copper and silver films is approximately 22, 65 and 50–100 nm, respectively. Pores in the silver and copper films are coarser than in the gold ones, which is associated with a higher susceptibility of copper and silver to coarsening of structural elements during condensation.

The «Isparitel-M» unit, having wider technological capabilities, was applied in spring 1984 for production in space of coatings of alloys, in addition to coatings of pure metals. As shown by the investigation results [11, 12], the rates of evaporation of a material under space conditions are several times higher than in similar earth experiments, and distribution of thickness of a coating over the sample surfaces is more uniform. The process of evaporation of alloy Cu–26 at.% Ag in that experiment was of a discrete character, as the material contained in a crucible was successively deposited on glass substrates located on a rotating drum. Thickness of the coatings was 30–200 nm. Composition of the coatings was determined by Auger spectrometry. The character of variations in the silver content of the coatings during the entire process of evaporation from one crucible is shown in Figure 1 [12]. Curve *2* has a steeper slope than curve *1*, i.e. being a lower-volatility component of alloy Cu–26 at.% Ag, copper starts evaporating in space earlier than on earth. Curve *2* has a small inflection approximately at the level of composition of the initial alloy, which suggests that there is a short-time period during which the composition of the initial alloy persists. Metallography of remains of alloys in the crucibles after evaporation in space and on earth showed the presence of pure copper in both cases. Therefore, the fact that in space all silver of the alloy evaporated earlier than on earth suggests that the rate of evaporation in this case was higher.

Silver coatings produced using VHT at different levels of gravity [13] are characterized by the following differences:

• stresses of the first kind, which are formed in earth coatings 350–500 nm thick, are –30 –+90 MPa. Similar coatings 600–800 nm thick produced in space have almost no stresses of the first kind;

• lattice parameter of silver coatings produced in space is less different from that of monolithic silver than with coatings produced on earth;

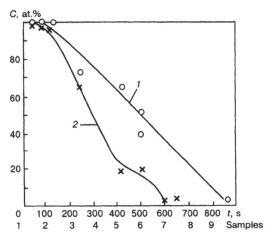

C, at.%

Figure 1 Dependence of silver content in the coating on the time of evaporation of Cu–26 at.% Ag alloy: *1* — on Earth; *2* — in space.

• silver coatings produced in space, unlike the earth analogues, reproduce relief of the substrate even at larger thicknesses than of the corresponding earth samples;

• in fracture of silver coatings deposited under earth conditions, cracks have the character of cleavage, whereas cracks in similar coatings deposited under the open space conditions have the character of tough fracture.

In 1989 the experiments on deposition of coatings of alloys Ag–50 wt.% Cu and Ag–20 wt.% Pd on a continuously moving polyimide substrate in space and on earth were conducted using the «Yantar» unit. Figure 2 shows schematic of cutting out of specimens for investigation of element composition of the coatings of alloy Ag–50 wt.% Cu by Auger spectrometry. Unfortunately, while taking a cassette out from the unit, the initial regions of the coated strip were heavily damaged. So, they were discarded. Figure 3 shows the time dependence of element composition of coatings produced in space and on earth in evaporation of alloy Ag–50 wt.% Cu. It is likely that in evaporation of this alloy its components are differently distributed under different gravity conditions. Coatings produced on earth are characterized by a composition which continuously varies during formation, whereas for coatings produced in space the character of the curves allows their extrapolation with great reserve, so that at a level of about 50 wt.% the curves are characterized by inflection. This is very apparent, even despite the fact that the evaporation process was too short because of heavy overheating of the crucible (up to 1600 °C), and composition of the produced coatings should insignificantly change with time because of susceptibility of alloy Ag–Cu to ageing. Also, this proves the fact that, like in a discrete process, the rate of evaporation of the alloy in space is higher than on earth, i.e. silver was evaporated from alloy Ag–Cu in space earlier (see Figures 1 and 3).

In coatings produced of binary alloys on earth, their constant composition during the evaporation process is achieved by artificial suppression of convection flows in molten

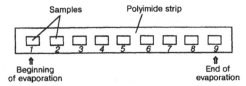

Figure 2 Schematic of cutting of a polyimide strip with the Ag–50 wt.% Cu coating deposited in space and on earth using the «Yantar» unit.

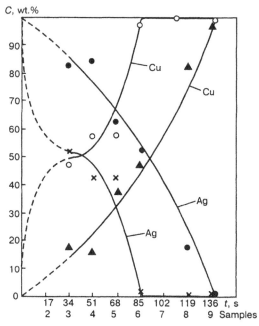

Figure 3 Dependence of the concentration of elements in the Ag–50 wt.% Cu coatings deposited in space (X — Ag, ○ — Cu) and on earth (● — Ag, ▲ — Cu) upon the time of evaporation.

metal, which makes the existing methods power-consuming and of low cost effectiveness. Melting of a material in space is accompanied by natural suppression of thermal convection, i.e. the process which has the highest effect on mass transfer. Theoretically, in evaporation of a binary alloy from the liquid state in space there could be a moment when the rate of evaporation of a material and the rate of diffusion of an alloying component of the alloy to the surface become equal, and, hence, composition of the coating thus produced corresponds to that of the initial alloy. Figuratively, it can be assumed that the process of evaporation from the liquid state in space is similar to evaporation from the solid state on earth.

The choice of alloy Ag–Pd for the evaporation experiments was based on the fact that this system has a simple constitutional diagram and is not subjected to ageing with time. Also, it was of interest to evaporate in space an alloy whose components have substantially different vapour pressure values. In the Ag–Cu system these values differ by two orders of magnitude, whereas in the Ag–Pd system this difference amounts to three orders of magnitude. It is likely that it was this factor that played a decisive role in making the experiment on evaporation of the Ag–20 wt.% Pd alloy a failure. Moreover, in that case parameters of heating of the crucible were deliberately set too low to avoid its overheating.

Thickness of the formed coatings estimated from the Auger profiles is given in the Table (numbers of the samples correspond to those given in Figure 2).

It is likely that the rate of evaporation of silver in space was several times higher than that on earth. The palladium content of the coating was determined by secondary ion mass spectrometry. The data obtained are just at the breaking point of sensitivity of the method. Therefore, it can be concluded that all palladium remained in the crucibles. However, by comparing microstructures shown in Figure 4, it can be assumed that evaporation of silver from the alloy on earth was dominated by the convection processes (structure of the residue in Figure 4b is cast, containing lots of closed labyrinths), whereas evaporation in space was dominated by the diffusion processes (no closed labyrinths are seen in Figure 4c).

Thickness of Ag–20 wt.% Pd coating and palladium content.

Sample No.	Conditions of production	Coating thickness, nm	Pd content, at.%
3	In space	70.0	0.47
4		60.0	0.65
5		4.5	0.28
6		1.8	0.13
3	On earth	17.0	0.16
5		10.0	0.22
7		3.0	0.22
9		2.0	0.18

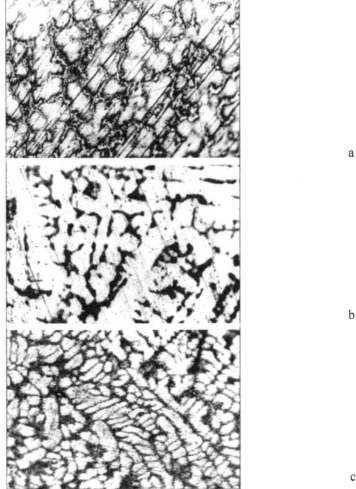

a

b

c

Figure 4 Microstructures of residues of alloy Ag–20 wt.% Pd after evaporation: a — initial; b — after evaporation on earth; c — after evaporation in space (×250).

As shown by analysis of results of the investigations conducted, as compared with their earth analogues, thin coatings produced in space by the electron beam technology are characterized by a lower submicroporosity and lower level of internal stresses. Besides, evaporation of binary alloys from the liquid phase in space may occur without separation of their components and at a higher rate.

The results proved an existing opinion that the space environment is favourable for formation of various coatings. This circumstance, as well as high requirements to structural purity of materials used in a new generation of modern micro- and nanoelectronic devices account for the current attempts to use the space environment for deposition of thin films on semiconductor materials.

Experiments on the technology of molecular epitaxy conducted in orbit by American specialists [14] showed that the epitaxial GaAlAs and GaAl films have a more perfect structure than their earth analogues. The technology of molecular epitaxy will make it possible to use liquid- and gas-phase source materials for production of epitaxial structures [15], such as compounds of the type of A_3B_5 and A_2B_6, synthesized on semiconductor silicon substrates.

However, the earth problems of modern microelectronics are not limited to those associated with producing epitaxial layers. High requirements for structural purity of materials persist also in manufacture of film integrated circuits, where thin ($< 1\ \mu m$) coatings of metals and alloys are used to manufacture conductors, capacitors, induction elements, resistors and contact pads. In particular, at present for the manufacture of microminiature resistors and working storage elements the preference in modern microelectronics is given to coatings of binary alloys, which, owing to their high specific electrical resistance and low natural noise, can be employed at an operating voltage of up to several hundreds of volts and at frequencies of up to several hundreds of MHz. Even now the future manufacture of micro- and nanoelectronic devices on earth is seen as a general integrated system based on three principles [16]:

1. Super high cleanness of the processes in a super clean environment (super high vacuum).

2. Formation of the atomic-flat and super clean surface of a plate free from natural oxides, particles, organic and metallic contaminants.

3. Precision control (monitoring) of parameters of the processes and, therefore, their full automation [17].

Development of fully integrated production lines with simultaneous progress of space technology will allow the use in open space of such satellite-plants where the initial vacuum achieved can amount to 10^{-13}–10^{-14} Torr. We are sure that the electron beam technology used for deposition of various coatings will play the priority role among a wide variety of technologies, and that a new generation of integrated circuits will be made by using the coatings, the properties of which are considered in this article.

1. Nikitsky, V.P. and Zhukov, G.V. (1983) Thin films in space equipment and technology, In: *K.E. Tsiolkovsky and problems of space manufacturing*, Proc. of 17th Tsiolkovsky Readings, Kaluga, 1982. Moscow, pp. 78–81.

2. Zagrebelny, A.A., Neznamova, L.O., Nikitsky, V.P. et al. (1989) *See pp. 319–321 in this Book.*

3. Smarsly, W. and Strobel, C. (1996) Schichten und Schichtsysteme für die zukünftigen Werkstoffe in der Luft- und Raumfahrt, *Galvanotechik*, **87**, No. 67, pp. 1828–1847.

4. Nguye, D.C. (1994) Innovation in coating technology for space applications, *Ceram. Qaurt.*, **63**, No. 1, pp. 25–26.

5. Klembergsapieha, J.E., Wertheimer, M.R. and Zimcik, D.E. (1989) Plasma-deposited multipurpose protective coatings for space applications, *ESA Bull.*, **13**, No. 2, pp. 117–126.

6. Osantowski, J.S. and Fleetwood, C.M. (1991) Optical coating technology for extreme UV-range, Coll. pap. of 28th COSPAR Plen. Meet. of Astrophys., Hague, 1990. *Adv. Space Res.*, **11**, No. 117, p. 201.

7. Palatnik, L.S., Nikitsky, V.P., Toptygin, A.L. et al. (1993) Effect of the space environment on stability of structure and properties of films Bi–Cr–Ca–Cu–O, *Fisika Tv. Tela*, **35**, No. 11, pp. 3025–3034.

8. Lippman, M.E. (1972) In space fabrication of TaIn-film structure, In: *NASA Construct. Rep.*, 1969. Washington, pp. 12–18.

9. Shulym, V.F., Zagrebelny, A.A., Lapchinsky, V.F. et al. (1985) Hardware for materials evaporation in space, In: *Problems of space technology of metals*, Kiev, Naukova Dumka, pp. 57–63.

10. Palatnik, L.S., Toptygin, A.L., Cheremskoj, P.G. et al. (1985) *See pp. 331–335 in this Book.*

11. Lukash, E.S., Lapchinsky, V.F., Shulym, V.F. et al. (1985) *See pp. 335–340 in this Book.*

12. Lukash, E.S., Shulym, V.F., Grigorenko, G.M. et al. (1986) *See pp. 340–344 in this Book.*

13. Paton, B.E., Lapchinsky, V.F., Mikhajlovskaya, E.S. et al. (1997) *See pp. 344–349 in this Book.*

14. (1995) Troublesome TSC endeavour is landing, *Flight Int.*, **148**, No. 4491, p. 24.

15. Markov, E.V., Antropov, V.Yu., Biryukov, V.M. et al. (1997) Space materials for microelectronics, In: *Proc. of Joint 10th Eur. and 6th Russian Symp. of Phys. Sci. in Microgravity*, St.-Petersburg, 1997. St.-Petersburg, pp. 11–20.

16. Valiev, K.A. and Orlikovsky, A.A. (1996) SSIL technology: main tendencies in development, *Elektronika: Nauka, Tekhnologiya, Biznes*, No. 5/6, p. 3.

17. Ohmi, T. (1993) Materials for space, *IEEE Trans. Reliab.*, **81**, No. 5, pp. 616–689.

Ability to Restore Coatings under Actual Conditions of Space*

B.E. Paton[1], E.S. Mikhajlovskaya[1], V.F. Shulym[1], A.A. Zagrebelny[1], V.P. Nikitsky[2], I.V. Churilo[2] and L.O. Neznamova[2]

The problem of extending service life of coatings used in structures of flying vehicles (FV) is still urgent, despite the fact that the current earth technologies are being continuously improved. The aerospace industry uses coatings of the most diverse functional purposes, such as: temperature-control (reflective and absorbing), corrosion-resistant, protective, etc. Among them we can distinguish protective coatings which prevent degradation of FV structural materials under different operational conditions, in particular, in low earth orbits (LEO), where such factors of the outside influence as atomic oxygen [1–3], micrometeorites and dust [4–6], moisture [7], as well as thermal-chemical and mechanical processes occurring on the materials surfaces acquire the decisive role, and lead to significant oxidation and erosion. The density of atomic oxygen in LEO is small ($\sim 10^9$ at/cm^3), corresponding to the gas density at a residual pressure of 10^{-5} Pa. However, the intensity of its flows bombarding the FV surface is significant (10^6 at/cm^2), in view of the high orbital speed (≥ 8 km/s) [8].

* (2000) *The Paton Welding Journal*, No. 1, pp. 32–37.

Under these conditions the impact of the environment on the FV surface is comparable with its bombardment by high-energy (~5 eV) atoms of oxygen. As a result, metallic materials are characterised by strong oxidation, and the non-metallic ones — by erosion and weight losses. According to calculations [1], the degree of degradation of graphite-epoxy composite material for FV with a long functioning period will be about 0.3 mm during the 11 years cycle of the activity of the sun, and about 0.9 mm during 30 years. Considering that the design thickness of structural elements of this material is ≈ 1.0 to 2.0 mm, in three decades it can be reduced by more than a half, this being inadmissible.

Numerous life tests of different materials were conducted by Russian and Ukrainian experts both on earth and on board the orbital stations «Salyut-6», «Salyut-7» and OC «Mir».

The objects of flight studies were temperature-control coatings, polymer and structural materials, solar battery (SB) elements, optical antireflection coatings. Performance of the materials tested was varied both directly under flight conditions and in the earth test laboratories. Physical parameters determining performance of a material as a whole were taken as the criterion of the degree of degradation of materials in each concrete case. For instance, more than 25 types of ceramic and more than 20 lacquer-paint temperature-control coatings were tested. Exposure to open space environment was from 3 months to 3 to 6 years. As a rule, during long-term exposure the coatings and materials lost their initial characteristics, changed their properties, and in some cases a significant degradation of the working surfaces was found.

Analysis of the tested polymer materials (fluoroplastic and polyimide films) also points to degradation of their properties during long-term service in space environment.

During the flight period of three orbital stations, more than 200 different structural materials were tested, and mostly by non-destructive testing methods directly under flight conditions. Their performance during a service life of up to 10 years was confirmed (for instance, stability of thermal radiation properties of the coatings of radiators of temperature control systems was preserved). However, deterioration of some properties should also be noted. This concerns, in particular, degradation of the surface of carbon fibre reinforced plastics, as well as deterioration of some physical properties of non-metallic structural materials. A graph of variations of dielectric properties of model polymer matrices based on ED-16 epoxy is given as an example (Figure 1). As can be seen from the Figure, all the samples demonstrated a smooth decrease in capacity during the entire period of exposure (9 years), which is indicative of a decrease in their strength properties.

As chemical reactions, rather than the mechanical effect of microparticles [9], prevail in the mechanism of polymer erosion in LEO, thin protective coatings can be applied to prevent destructive surface processes [10, 11]. These coatings, as was mentioned above, are now widely used in practice. For instance, flexible SB substrates and outer temperature-control surfaces of FV of kapton are protected from erosion by Al_2O_3, SiC_2 and other coatings [12], and coatings of magnesium fluoride, silicon nitride, indium oxide alloyed with tin, etc. are applied onto the metallic (silver, aluminium) reflective surfaces of mirrors of the solar-energy power systems for their protection from oxidation [13–15]. Considering that in the current FV designs the service life should be about 30 years [16], obvious is the need for periodic restoration of functional properties of the coatings, as well as preventive maintenance of their structural elements. These operations can be performed on Earth, taking FV to orbit by the modern transportation means. However, in view of the high cost of carrier rocket launches, it is more rational to perform restoration and repair work directly under the conditions of their service.

This paper analyses the possibility of using electron beam technology of materials evaporation for deposition of coatings in orbit, also with the aim of their protection from degradation, as well as performance of repair-restoration work under these conditions.

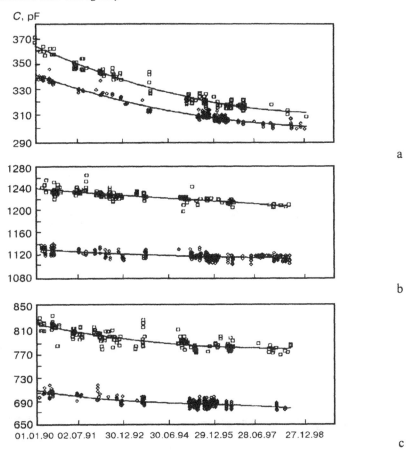

Figure 1 Dependence of capacitance upon the time for three types of samples: a — sample of PCM with ED-DADFM binder and glass cloth; b — sample of ED-DADFM binder; c — sample of ED-2MDA binder.

Development of electron beam technology for production of coatings under microgravity conditions was based on the results of numerous earth and flight experiments, which had been conducted by staff of the PWI and the S.P. Korolyov RSC «Energiya»[2] together with other organisations for many years. During those experiments such process tasks as selection of the evaporation method, containment of molten material in the crucible and others, were solved, and properties of the produced coatings were studied. So, according to [17], the rate of formation of coatings of pure silver (model material) at $g = 1 \cdot 10^{-5} g_0$ (g_0 is the gravity on the Earth surface), and at a residual atmospheric pressure of 10^{-4}–10^{-5} Pa is several times higher than under earth conditions. Stresses of the 1st kind formed in silver coatings of 350 to 500 nm thickness, deposited on a substrate of alloy D16 under earth conditions, are –30 – +90 MPa; in similar coatings of 600 to 800 nm thickness produced in LEO there are almost no such stresses. The difference in the level of macrostresses in the above coatings is illustrated in Figure 2. It can be seen that cracks initiating from a cut made with a «guillotine» type cutter have the form of cleavage in the coatings produced on Earth (Figure 2a and b) and the form of ductile fracture in the coatings produced in orbit (Figure 2c and d).

Submicroporosity of silver coatings produced under microgravity conditions is smaller than that of their earth analogues [18]. Distribution of submicropores by size in the samples

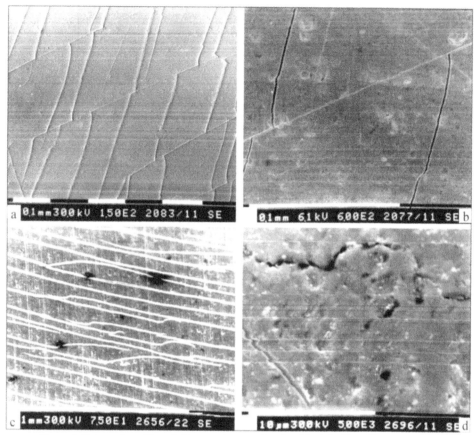

Figure 2 Nature of cracking of silver coatings produced on Earth (a, b) and in space (c, d).

produced under microgravity is more uniform than in the samples produced on Earth. Similar results were also derived for gold and copper coatings [19].

Influence of microgravity conditions on the process of evaporation of binary Ag–Cu alloys [20] shows up in that the mass transfer mechanism in the molten material is in this case determined by diffusion, and not by convection, as it is on Earth. Therefore, production of coatings of binary alloys without separation of their components is really possible, as parameters of the process can be selected so that the evaporation rate of the less volatile component of the alloy and the rate of its diffusion into the evaporation surface become equal. Therefore, the coatings produced directly in orbit can have a long service life before loss of their service properties, and before their restoration becomes necessary. The process of coating production under microgravity conditions will be more effective owing to a decrease in porosity and, hence, electrical conductivity which is higher and more uniformly distributed over the area, better total purity and structural perfection, and, therefore, better ductility properties and lower internal stresses. One should also take into account the fact that under LEO conditions the rate of material evaporation increases several times, and when the evaporation process proceeds from the molten state it is possible to produce coatings from binary alloys with preservation of their stoichiometry.

Analysis of the results of the performed studies confirmed the effectiveness of using electron beam technology for production of coatings directly in orbit and allowed the development and manufacture on its basis of the appropriate equipment designed with allowance for the problems related to performance of the work in orbit, including containment of molten metal in the crucible, energy deficit (~500 W), absence of additional cooling, ensuring safety of the operator, etc. An optimal combination of all the above factors was achieved in a design whose block-diagram is shown in Figure 3. The hardware operation is based on indirect heating of the evaporation material by the electron beam. Extensive laboratory investigations led to creation of a highly effective device which allows heating of the materials being evaporated to a temperature of up to 1500 °C. It was used as a basis for the design and manufacture of stationary units for production of coatings under micro-gravity, namely «Isparitel» and «Isparitel-M» [21, 22]. A modification of one of them is shown in Figure 2 on p. 186. During 1979 to 1984 the units were tested to advantage on board the «Salyut-6» and «Salyut-7» orbital stations. Testing of the hardware for coating application was carried on in 1987 and 1989 on board the OC «Mir» using the «Yantar» unit, which permitted production of coatings on a polymer strip continuously moving with a speed of 20 to 40 m/h, which, naturally, does not exclude the possibility, in principle, of movement of the evaporation unit proper at the same speed. Successful flight tests showed that this hardware could be used as stationary equipment as part of the standard set of FV for periodical restoration of service properties of reflective and other surfaces. In July 1984 the electron beam VHT was tested in open space. It was used for performance of different process operations [23], including coating. As stated by cosmonauts S.E. Savitskaya and V.A. Dzhanibekov, the tool is very simple and convenient in operation.

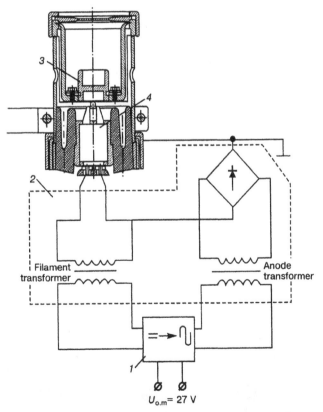

Figure 3 Schematic of indirect heating of an evaporation material and block-diagram of the device:
1 — secondary power supply; *2* — high-voltage power supply; *3* — evaporation device;
4 — electron beam gun.

In order to perform specific process tasks arising during FV service, especially in emergency situations or in difficult-of-access places, the PWI developed and manufactured «Universal» hardware based on the electron beam hand tool. Two of the four tools, which are part of the hardware, are designed for coating application. The first tool is equipped with a removable rotary attachment with four crucibles, thus allowing coatings of different materials to be deposited alternately or, if required, the stock of one material to be increased, if the surfaces being treated have considerable dimensions. The second tool allows evaporation of a material fed in the form of wire into the open crucible heated by the electron beam. Operational diagram and appearance of the tool are shown in Figure 4. Evaporation material in the form of 0.8 to 1.0 mm diameter wire is fed into the tungsten-cobalt crucible with the

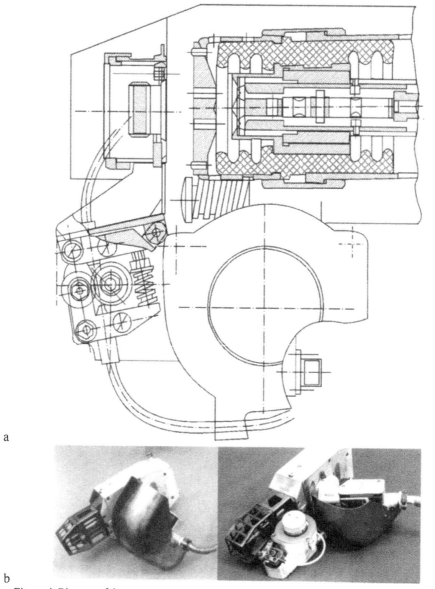

a

b

Figure 4 Diagram of the evaporator (a) and appearance of the tool (b) with evaporation material feed.

speed of up to 5 mm/s. Laboratory tests, as well as performance of manual operations by the operator in the space simulation test chamber, proved the possibility of using such tools for deposition of silver coatings under vacuum. For evaporation of aluminium, whose vapour pressure is by an order of magnitude lower than that of silver, it is necessary to significantly decrease the wire feed speed or complete the tool with a device providing dosed (pulsed) feed. In the case of evaporation of aluminium, also because of its aggressiveness with respect to other materials, it is necessary to carefully select a material for the manufacture of the crucible. Solving these problems will allow the above tool to be used for aluminium coating deposition, this being a very urgent but yet unsolved issue in conducting the package of repair-welding operations.

Process characteristics of the coating hardware, developed and tested in orbit, are given in the Table (see p. 329).

The versatility of the created coating hardware enables the following processes to be implemented in orbit:

• scheduled maintenance to maintain performance of the reflective and other surfaces of FV using hardware which is part of the standard set of FV and which functions in the automatic mode;

• deposition of reflective coatings on large areas of solar-energy power systems using manipulators and remotely controlled robots;

• cleaning of the surface prior to coating deposition, not excluding the possibility of the use [24] of other cleaning methods, such as ion-chemical or magnetic-abrasive one;

• performance of such processing operations as welding, brazing, cutting, heat treatment, etc., in addition to coating application;

• repair-restoration work during the operator's EVA, especially in difficult-of-access places.

Conclusions

1. Electron beam technology for evaporation of materials under microgravity conditions has been developed, based on numerous laboratory and flight experiments for production and investigation of properties of the coatings.

2. Adequacy of the main design solutions of the hardware developed on the base of this technology was confirmed by flight tests of the coating hardware on board the «Salyut-6», «Salyut-7» orbital stations and OC «Mir».

3. Results of analysis of the coatings produced in the flight experiments allow a conclusion that they can be applied during FV service in microgravity environment and have quality not inferior to that achieved under the earth conditions.

4. It is shown that the created versatile equipment holds promise for use as stationary units in a set of flight equipment of FV for performance of operations associated with restoration of coatings that lost their service properties.

5. Manual modification of the device is envisaged in the versatile equipment for performance of repair-restoration work in orbit in difficult-of-access places during the operator's EVA.

6. Different modifications of the developed versatile hardware can be used in the future for production of reflective surfaces in the construction of mirrors of solar-energy power systems and in developments of various settlements on the Moon or other planets.

1. Lege, L.Zh. and Veinsentein, J.T. (1987) Protection of space flying vehicles from the impact of atomic oxygen, *Aerokosm. Tekhnika*, No. 2, pp. 7–11.

2. (1990) Atomic oxygen erosion on LDEF, *Dep. Daily*, **166**, No. 30, p. 28.

3. Moag, C.R., van Easbeer, M., Deshapande, S.P. et al. (1997) The contamination environment at the Mir Space Station as measured during the EUROMIR'95 mission, In: *Proc. of 7th Int. Symp. on Space Environment*, Toulouse, 1997. Noordwijk, pp. 301–308.

4. McKay, D.S., Barrett, R.A. and Bernhard, R.P. (1987) Impact damage to solar maximum satellite caused by micrometeorites and microparticle orbital debris, *Meteoritics*, **22**, No. 4, pp. 453–456.

5. Nahra, H.K. (1988) Assessment of the effects of space debris and meteoroids environment on the space station solar array assembly, In: *Proc. of 20th IEEE Photovolt. Spec. Conf.*, Las Vegas, 1988. New York, pp. 868–873.

6. Coombs, C.K., Atkinson, D.R., Wagner, J.D. et al. (1992) Damage areals on LDEF aluminium panels: preliminary results, *Lunar and Planet Sci.*, No. 3, pp. 245–246.

7. Morganti, F., Marchetti, M. and Reibaldi, G. (1984) Effects of moisture and thermal ageing on structural stability of sandwich panels, *Acta Astronautica*, No. 7, p. 11.

8. Zimcik, D.G. and Maar, K.P. (1989) Consequences of reactions between atomic oxygen and materials being on board the space Shuttle in STS-41G mission, *Aerokosm. Tekhnika*, No. 5, pp. 111–119.

9. McLean, P.D., Wiebe, W., Garton, A. et al. (1986) Exposure of epoxy-amine plastics to the space environment, *Astronavtika i Raketodinamika*, No. 39, pp. 45–48.

10. Murad, E. (1989) Spacecraft interactions as influenced by thermochemical considerations, *J. Spacecraft and Rockets*, **26**, No. 3, pp. 145–150.

11. Sykes, G.F., Funk, J.G. and Clemp, W.S. (1986) Assessment of space environment induced microdamage in toughened composite materials. In: *Proc. of 18th Int. SAMPE Techn. Conf.*, Seattle, 1986. Covina, Vol. 18, pp. 520–534.

12. Banks, B.A., Mirtich, M.J., Rutledge, S.K. et al. (1985) Protection of solar array blankets from attack by low earth orbital atomic oxygen, In: *Proc. of 18th IEEE Photovolt. Spec. Conf.*, Las Vegas, 1985. New York, pp. 381–386.

13. De Rooy, A. (1985) The degradation of metal surfaces by atomic oxygen, In: *Proc. of 3rd Eur. Symp. on Space Environment*, Noordwijk, 1985. Paris, pp. 99–108.

14. Gulino, D.A., Egger, R.A. and Banholzer, W.F. (1987) Oxidation — resistant reflective surfaces for solar dynamic power generation in near earth orbit, *J. Vac. Sci. and Technol.*, **5**, No. 4, pp. 2737–2741.

15. Gulino, D.A. (1988) Space station solar concentrator materials research, *J. Adv. Materials and Processes*, **3**, No. 2, pp. 261–277.

16. Zimcik, D.G. and Kavanagh, J. (1990) Material requirements for the space station mobile servicing system, *Can. Aeronaut. and Space*, **36**, No. 1, pp. 11–17.

17. Paton, B.E., Lapchinsky, V.F., Mikhajlovskaya, E.S. et al. (1997) *See pp. 344–349 in this Book.*

18. Cheremskoj, P.G., Toptygin, A.L., Panikarsky, A.S. et al. (1990) *See pp. 364–371 in this Book.*

19. Palatnik, L.S., Toptygin, A.L., Cheremskoj, P.G. et al. (1985) *See pp. 331–335 in this Book.*

20. Lukash, E.S., Shulym, V.F., Grigorenko, G.M. et al. (1986) *See pp. 340–344 in this Book.*

21. Zagrebelny, A.A., Lapchinsky, V.F., Stesin, V.V. et al. *Evaporator.* USSR author's cert. 1485668, Int. Cl. B 64 G 1/00, priority 12.10.87, publ. 08.02.89.

22. Zagrebelny, A.A., Lukash, E.S., Nikitsky, V.P. et al. (1988) Development of the technique of thin-film coating application in flight conditions, In: *Proc. of 22nd Tsiolkovsky Readings*, Kaluga, 1987. Moscow, pp. 134–138.

23. Nikitsky, V.P., Lapchinsky, V.F., Zagrebelny, A.A. et al. (1985) *See pp. 178–184 in this Book.*

24. Grinkevich, A.D., Babenko, P.L. and Prisnyakov, V.F. (1986) Ion-chemical cleaning of the portholes of a space vehicle, In: *Problems of space technology of metals*, Kiev, PWI, pp. 28–33.

Aspects of Implementation of an Experiment on Synthesis of Semiconductor Films in Space*

A.I. Nikiforov[8], A.O. Pchelyakov[8], O.P. Pchelyakov[8], L.V. Sokolov[8], V.G. Khoroshevsky[8], V.I. Berzhaty[2], L.L. Zvorykin[2], A.V. Markov[2], I.V. Churilo[2] and A.A. Zagrebelny[1]

Prospects for synthesis of multilayer compound semiconductors from molecular beams under the conditions of an orbital flight of space vehicles are considered. Advantages are demonstrated of using deep vacuum, formed as a result of manifestation of the molecular shield effect, to produce new thin-film materials with unique properties. A possible variant of the architecture of the system of automatic control of the experiment and processing of its results is proposed.

Introduction. The ability to create high vacuum in open space near orbital stations using the «molecular shield» effect was theoretically predicted and experimentally confirmed [1–3]. Development of high technologies involving the use of vacuum currently is one of the most intensively advancing areas of space semiconductor materials science. Such technologies primarily include molecular-beam epitaxy (MBE) with crucible and gas sources of molecular beams.

Development and commercial application of MBE process have clearly demonstrated that it is the best method of producing multilayer structures with boundary smoothness at an atomic level, and precisely assigned layer thickness, composition and alloying profile. Application of a highly-sensitive electron-probe and an optical means of monitoring the synthesis of these structures ensures high reproducibility of the specified parameters.

There exist a great variety of applications of such structures in the new generation of semiconductor devices, where the principle of operation (unlike traditional microelectronics) is based on the wave nature of the electron. In implementation of the MBE process in ground-based vacuum units, however, the limiting factors for producing high-quality structures are vacuum level and purity and pumping system efficiency, as well as the presence of the vacuum chamber walls, accumulating and evolving the components of molecular beams and residual gas atmosphere. These defficiencies can be eliminated when a processing unit, without a chamber, is taken out into open space to the wake area of the molecular shield [3–7].

Main prerequisites and current status of the problem. As shown by analysis of studes, available in scientific literature and the Internet, the problem of semiconductor materials science involving high-vacuum technologies in open space, is of great fundamental and applied importance. Its solution will determine the progress not only in studying the processes of producing thin-film crystalline coatings and multilayer-heterostructures, when they are grown from molecular beams in ultrahigh vacuum, but also in the development of processing facilities and integrated manufacturing of semiconductor electronics of the XXI century [8]. This work is currently being performed only in the USA and Russia. US studies were initiated in 1989 by the Space Vacuum Epitaxy Center (Houston University). In Russia

---

* (2001) Paper was presented at the 5th Russian Conf. on Semiconductor Physics, Russia, Nizhnij Novgorod.

they were started in 1996 in the Department of Molecular Epitaxy of the IPSC SD RAS[8] in co-operation with the S.P. Korolyov RSC «Energiya»[2] and Research Institute «Research Centre» under «Epitaxy» Program and «Shield» Project.

The idea of using a molecular shield to create an ultrahigh vacuum at low orbits was conceived by US scientists, who with NASA support performed theoretical analysis of the condition of the gaseous medium around a semi-spherical shield, flying in space [1, 2], and formulated the concept of an orbital laboratory with a super rarefied atmosphere [3]. It should also be noted, that development of eight reusable units to produce advanced materials in space was planned as far back, as the eighties under NASA SPACELAB programme, and one of them was a «molecular shield». However, a lack of interest in this device in industry led to the elimination of this programme from NASA's five-year plan for 1980–1984. This idea was proposed again by A. Ignatiev and C.W. Chu in 1985 to conduct experiments on MBE.

In 1988 the Space Vacuum Epitaxy Center was opened at the University of Houston, and implementation of the programme of these investigations began in 1989. Following a four-year period of ground-based studies, in 1994–1996 the first space experiments on synthesis of epitaxial structures based on gallium arsenide were carried out under the «Wake Shield Facility» (WSF) Project [4, 5]. So far the US government subsidies for each of the stages of WSF Project have been about 15 mln USD. From 1999 a commercial SPACEHAB Company became involved in funding this project, which had invested 275 mln USD into space projects that looked potentially profitable. In 2000 more than 5 mln USD were invested into expansion of the Space Vacuum Epitaxy Center (Houston) [6]. The next flight of the reusable module, carrying WSF unit on board is scheduled for the end of the year 2001. This time the intent is to use a new generation unit with a high efficiency to actually apply the epitaxial structures in devices on the ground. The final goal of this project is setting up by the year 2005 a mini-factory in orbit to manufacture semiconductor films with record properties for opto-, micro- and nanoelectronic devices. The intent is to raise the capacity of such a unit to 3500 wafers per year, thus allowing manufacturing about 10 mln device structures annually.

Project commercialisation is based on the high efficiency of the unit, quality and reproducibility of structure parameters in individual processing of the wafers. Calculations, performed by A. Ignatiev demonstrate, that manufacturing the device structures in space will cover the cost of material transportation and generate profit. In addition, great importance is attached to the ability to perform unique materials science studies, using an automated MBE unit under conditions of pure ultrahigh vacuum in space to develop advanced superpure materials.

The results of the first experiments of WSF Project and funding of further work on commercial manufacturing of epitaxial materials in space confirm the good prospects for development of this new space technology.

Features of the Russian project. In 1996 we started working on the «Epitaxy of heterogeneous structures on silicon in space» Project. The main goal of the Project is scientific and technical substantiation and ground-based retrofitting of the programmes and procedures of a flight experiment to produce heretostructures for alternative substrates, containing thin layers of expensive compound semiconductors on the surface of large diameter silicon wafers. The S.P. Korolyov RSC «Energiya»[2] staff conducted methodological design studies of the level of rarefaction behind the protective shield in an orbital flight of piloted objects [5]. Free-molecular incident cross-flow of upper atmospheric gases was considered at 250 to 400 km altitudes of the protective shield of a shape similar to a flat disc. It is assumed that the levels of gas evolution from the «shadow» surface of the shield into the aerodynamic wake area do not exceed 10^{-12} cm^{3}cm/cm^{2}s atm. Such a level of gas evolution is characteristic of pre-degassed metal walls of vacuum chambers (for instance, of polished stainless steel), or for thin-walled structures in the flight along the shadow side of the orbit at temperatures of about 100 K. The corresponding partial pressure (by atomic oxygen) does not exceed 10^{-12} mm Hg.

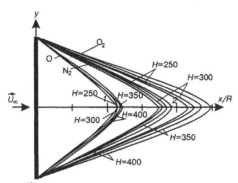

Figure 1 Isobars $p(\tilde{x}) = 10^{-12}$ mm Hg in the field of vision behind the shield for three main gaseous atmospheric components at different flight altitudes H, km ($\alpha = 0°$).

The protective shield, oriented normal to the flight direction, cuts out a conical zone of aerodynamic wake, where the flow is formed by gas particles penetrating into this zone from the incident flow, and particles of its own gas evolution from the disc inner surface. For the flow conditions at large Knudsen numbers, these flows are statistically independent, and, therefore, the general flow parameters in the wake are determined by their superposition.

Flow isobars behind the shield exposed to cross-flow have been analysed at flight altitudes of 250 to 400 km, taking into account gas evolution from its surface and the effect of screening of the incident flow.

It is demonstrated that the above levels of gas evolution into the wake area behind the shield only slightly influence the configuration of the high vacuum area boundaries (isobar $p = 10^{-12}$ mm Hg (Figure 1). In this area a zone stable as to impinging angles ($\alpha \leq 15°$) (see Figure 3 on p.81) can be outlined. At shield transverse dimensions of ~ 3 m this zone conditionally is a cylindrical area of about 0.7 m diameter and length of ~ 1.5 m from the shield bottom, i.e. it can readily accommodate the processing and scientific hardware, as well as instrumentation.

Calculation results also demonstrate, that only «fast» molecules of the light components of the upper atmosphere (He and H) penetrate into the rarefaction zone behind the shield from the environment. The velocities of their thermal motion are essentially higher than the orbital flight velocity (7800 m/s). Their partial pressures at flight altitudes of the orbital vehicles and stations ($H = 300$ km) are 5 to 6 orders of magnitude lower compared to the above partial pressure of gas evolution molecules. Theoretically the total pressure in this area, determined by hydrogen and helium, can be less than 10^{-14} mm Hg (at partial pressure of hydrogen $< 10^{-14}$ and of helium $< 10^{-18}$ mm Hg and almost infinite pumping rate for all the gaseous medium components including inert gases) [9]. For comparison, it should be noted that in ground-based super high-vacuum processing units with cryopumps the limit of rarefaction of not higher, than 10^{-12} mm Hg is achieved. The ultimate achievable vacuum at the inlet in one of world's best helium pumps (V.A. Grazhulis, M.P. Larin) was at the level of 10^{-13} mm Hg [10].

However, when work on organising the «space» vacuum behind the protective shield of piloted space vehicles is performed, it is necessary to take into account their ambient outer atmosphere (AOA). AOA is a complex dynamic formation, including the gaseous, aerosol and finely-dispersed phases and having a negative influence on the results of astrophysical, geophysical, materials science and engineering studies and experiments. Aerosol and disperse particles have characteristic dimensions between 0.1 mm and several millimetres and are usually detected at distances of up to 15 m from the vehicle surface [7, 10].

In AOA stationary state, when no dynamic operations have been conducted for a long time (for instance, at gravity stabilisation) AOA gaseous phase pressure at the vehicle surface is estimated to be ~ 10^{-6}–10^{-5} mm Hg, which is confirmed by the data of flight measurements in OC «Mir». At about 10 m distance from «Mir» surface the pressure in AOA does not exceed values of the order of 10^{-7} mm Hg. During dynamic operations, when the propulsion systems are functioning, the pressure in AOA rises abruptly by 2 to 4 orders of magnitude, compared to the background conditions, and then relaxes to the initial condition.

These circumstances point to the need to apply special extension devices that allow experimental and measuring instruments to function beyond the AOA during performance of materials science, processing, astrophysical and geophysical, investigations in the RM ISS. The PWI and S.P. Korolyov RSC «Energiya»[2] have co-developed a specialised «Tyulpan» unit, fitted with an extension device to provide undisturbed incident flow around the protective shield at orbiting altitudes. The main components of the «Tyulpan» experimental set up (Figure 2) include:

- reusable extension and folding device (REFD);
- protective shield (PS) with a rotary platform (RP);
- payload platform (PP);
- anchoring platform (AP);
- instrumentation and research instruments, also mounted on payload platform;
- power and information cables.

Performance of technological investigations on MBE in the «Tyulpan» unit at orbiting altitudes allows using the following advantageous orbital flight factors:

- deep vacuum and practically unlimited efficiency of pumping of the working molecular beam components, providing a unique possibility for a super fast change of the gaseous phase composition in the growth zone on the substrate surface. These factors enable the production of heterojunctions with ideally sharp profiles;

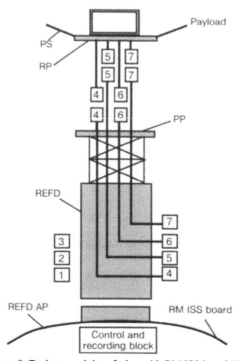

Figure 2 Block-diagram of «Tyulpan» unit interfacing with RM ISS board (1–7 are the connectors).

• complete absence of working chamber walls, accumulating and releasing the components of the molecular beams and residual gas atmosphere, will allow the neutralising of the influence of the so-called «memory» effect, as well as produce multilayer structures containing a large number of layers of dissimilar composition with precisely specified alloying profiles;

• ability to remove the processing fixture elements and analytical means from the epitaxial growth zone and increase the distance from the substrate to the molecular beam sources. These factors are the determinant ones for increasing the number of independent molecular beam sources and the homogeneity of the layers across the area with increase of substrate diameter;

• ability to use toxic volatile liquids and gases (hydrides, metal-organic compounds) as initial materials for film synthesis without environmental contamination. These compounds quickly disperse to safe concentrations and easily dissociate into safe components under the impact of ionising solar radiation;

• microgravity is a factor that is not used directly in this project, unlike most other projects.

In keeping with the work plan for «Epitaxy» Program, in 1996 we conducted experimental and theoretical studies of the processes of producing semiconductor heteroepitaxial films of germanium and gallium arsenide on silicon substrates by the method of MBE on the ground. Variations of the produced heterosystem structure, depending on the parameters of the process of their synthesis, were studied. The goal of the work in the first stage was determination of physical factors, limiting the ultimately achievable parameters of heterosystems, synthesized in the actual ground-based ultrahigh-vacuum MBE units. Among such factors the two most important ones were found to be limitation of the pumping rate and presence of the vacuum chamber walls, accumulating the deposited materials and releasing them to the substrate in an uncontrollable manner during deposition of the layers. The associated impossibility to combine in one vacuum volume the processes of producing A_3B_5, A_2B_6 compounds and elements of the 4th group necessitates moving the substrates from one growth module to another.

Conducting such processes requires sophisticated ultrahigh-vacuum units of the cluster type. On the other hand, substrate moving from chamber to chamber may lead to contamination of atomically pure growth surface and prevent production of structurally-perfect buffer layers and films of gallium arsenide on silicon. Experiments were conducted to determine the dynamics of variation of the multilayer structure properties with improvement of the vacuum conditions in the processing volume of commercial unit of «Katun» type. It is found that with improvement of vacuum conditions (total pressure of residual gases of 10^{-12} mm Hg) the concentration of the electrically-active background impurity in silicon films reaches saturation with the increase of the number of experiments. The level of minimal concentration of the impurity is $5 \cdot 10^{-13}$ cm^{-3}. Quantitative characteristics of the process of transfer of non-volatile dope components (boron in silicon) over the surfaces of the vacuum chamber walls were determined.

Further work by the performers of this project and the IPSC SD RAS[8] is aimed at development and testing of ground-based prototypes of all the technological systems of MBE unit, designed for taking out into space. This implies further development of the results of many years of work on creation of basic technologies of MBE of elemental and compound semiconductors and three generations of commercial locally-made equipment for MBE. The decisive factor in the project implementation is making use of unique experience in space materials research and manufacturing the units for growing bulk crystals in space (Research Institute «Research Centre»). Significant achievements in development of processing fixtures, instrumentation, power systems and all the on-board means of automatics and telemetry (the PWI and S.P. Korolyov RSC «Energiya»[2]) will be also used. Unlike the US project, oriented chiefly to application of gallium arsenide plates as substrates, the Russian programme is focused on using a less expensive (15 times) and more light-weight (2.3 times) substrate material of large area (silicon wafers of up to 200 mm diameter). The intent is to produce a heteroepitaxial layer of gallium arsenide directly before synthesis of the device structures.

As a result of the project implementation it is intended to set up an experimental orbital mini-factory to manufacture highly perfect alternative substrate material and multilayer heterojunctions, based on compound semiconductors of A_4B_4, A_3B_5, and A_2B_6 type, on large-diameter silicon wafers for intergral opto-, micro- and nanoelectronics applications.

Procedure of processing the information, received during films synthesis by a video channel from a recording fast electron diffractometer. The procedure was developed and tried out at the IPSC SD RAS[8] during synthesis of Ge–Si nanostructures that was conducted in MBE unit «Katun-S», fitted with two electron beam evaporators for Si and Ge (Figure 3). The dope (Sb and B_2O_3) evaporates from the effusion cells. The analytical part of the chamber consists of a quadrupole mass-spectrometer, quartz thickness gauge and fast electron diffractometer of up to 20 kV energy. During the growth process, the diffraction pattern is recorded by the CCD camera and the image is entered into a personal computer. Specialised software developed by us, allows both the image and the selected areas of the diffraction

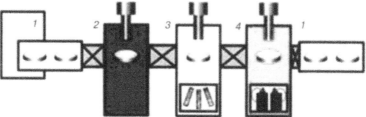

Figure 3 General view and structure of a three-chamber unit «Katun» for MBE: *1* — load chambers for substrate loading-unloading; *2* — substrate cleaning chamber; *3* — chamber for growing multilayer semiconductor structures A_3B_5 (GaAlAs, GaInAs, etc.); *4* — chamber for MBE of germanium on silicon.

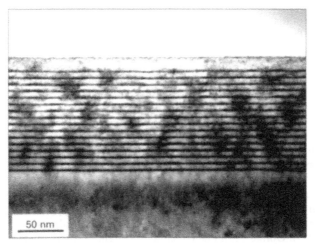

Figure 4 16-layer structure with germanium quantum points in silicon matrix.

pattern to be followed with a speed of 10 frames per second. Ge growth rate was 10 mol. layer/min, temperature was varied from 200 up to 700 °C. Used as substrates were silicon wafers with (100) ± 0.5° surface. Germanium film growth was preceded by high-temperature annealing of the substrate and growing a silicon buffer layer.

Changes in the diffraction pattern qualitatively reflect the changes in the growing film morphology. Quantitative information can be obtained by recording the diffraction pattern intensity. Variation of the diffraction pattern intensity along a line crossing the streak and three-dimensional reflexes was also recorded to determine the moment of plastic relaxation (Figure 4).

It is shown that recording the variation of the diffraction pattern intensity provides valuable quantitative information on the change of growing film morphology. Data on stresses in the growing layer are of specific interest, as they are the main driving force for the observed morphological changes. These stresses could be evaluated by a change of the lattice constant of the growing germanium film during transition from the layered to islet growth, as well as by the change of their shape. It is seen from Figure 4 that the distance between the reflexes, corresponding to reflections 01 and $0\bar{1}$, varies in the process of growing. This may mean that interatomic distance $a_{\parallel}$ of germanium film in (100) plane changes its magnitude. Reflexes from the superstructure (2×1) undergo a similar change of angular position. Comparing the derived dependence with the phase diagram and kind of diffraction pattern, it is possible to single out areas, corresponding to different phases of germanium film growth on silicon (100) surface.

Data acquisition and video information processing were conducted, using a specially developed software package, which allowed the analysis in the model experiment of the required capacity of the communication channel between the experimental unit in the space station and the ground-based terminal with the control and monitoring systems.

Robust integrated systems for control of the process of semiconductor film synthesis. A control system is one of the major components in a space mini-factory for manufacturing epitaxial semiconductor films. A modern system of control of technological processes is made up by means of computing hardware and software. Therefore, the architectural flexibility of the latter determines the effectiveness of the mini-factory as a whole. Specific conditions of the space station require creation of integrated distributed control systems and application of computer systems with sufficiently complete architectural properties.

The concept of distributed computer systems (CS) with a programmable structure [11, 12], designed by the Siberian Division of the Russian Academy of Sciences has a potential for development. This concept allows, in particular, the avoidance of the obsolescence of the control system due to the capability of evolutionary improvement of CS structure and inclusion of the most advanced microprocessor means (instead of the existing ones or in addition to them) and does not require high initial and total expenses.

CS with a programmable structure is understood to be a collective of elementary machines, with their functional interaction provided through a program-adjustable communications network. The simplest configuration of an elementary machine (EM) is a processor and a local commutator (for connection to the neighbouring EM).

The following major architectural features of computer systems with a programmable structure should be noted.

CS architecture type are multiple instruction–multiple data (MIMD) streams; MIMD architecture can be transformed into multiple instruction–single data (MISD) and single instruction–single data (SISD) architectures by CS program readjustment.

CS productivity. Unlike a computer, computer systems have no fundamental limitations in increasing their productivity. Productivity increase in them is achieved chiefly by increasing the number of EM.

Scalability of computer systems is their ability to increase and reduce resources, as well as vary their productivity. The complexity (labour consumption) of the problems (of simulation, control, monitoring, decision taking, etc.) will become greater during the mini-factory operation. The computer system should have the architectural property of scalability so as to remain an adequate means for solving complex problems with time. This implies, in particular, that the productivity, achieved by the CS with a certain number of machines, can be increased, by adding one or several more EM.

The property of increasing the CS productivity offers a potential possibility for unlimited widening of the range of mini-factory management problems and solving problems of any a priori assigned complexity. However, practical realization of such a capability requires the algorithm of a complex problem solution to satisfy the condition of locality: machine-machine data passing should only slightly influence the time of problem solution. This can be achieved through large-block paralleling of complex problems and (or) hardware, that allow the combining of machine-machine data passing with computations.

Reconfigurability CS. CS structural and functional flexibility is due to the extensive capabilities of the systems of static and dynamic reconfiguring.

CS static reconfiguring is provided by varying CM number, their structure and components, selection of link numbers for the local commutator for connection to other EM, capability of constructing structures in the form of graphs, belonging to different classes, admissibility of applying for communication the channels of various types, different physical nature and different length, etc. CS ability of static reconfiguring enables adaptation of the control system for the range of problems to be solved both at the stage of mini-factory commissioning and during its upgrading.

CS dynamic reconfiguring is achieved by the capability of forming in the systems the subsystems in which the structures and functional organization are adequate to the current multiprogram situation and structures of the problems being solved. Therefore CS capability of dynamic reconfiguring leads to its high versatility, when the preset level of productivity is achieved while solving a broad range of problems; modes (sharing, batch processing, etc.), methods of controlling the computation process (centralized, decentralized, etc.), block-diagrams (isolated computing machines, systems of several processors and one computer, systems of one computer and several storage devices, etc.) and methods of data processing (conveyor, matrix, distributed, etc.), known in computer science, are implemented.

CS capability of dynamic reconfiguring is due to the complete embodiment of the principle of structure programmability. Such a capability of CS enables it to perform automatic restructuring during operation to carry out data exchange between EM, «adjust» the condition of functional units and components in elementary machines to achieve the adequacy of CS to the totality of processes simultaneously proceeding in it.

BS reliability and robustness. These two notions are semantically close, both used to characterize CS architectural features, required to perform its functions. Each of them, however, reflects CS specific features of using the sound resources during data processing.

CS reliability is understood to be the capability of automatic (program) adjustment of such block-diagrams and organizing their functioning, which under conditions of failures and restoration of machines, provide a preset productivity level in the case of realizing parallel programs for solving complex problems or. In other words, CS reliability is the ability to use a fixed number of sound EM. This notion characterizes the CS capability to process information in the presence of a fixed structural redundancy (represented by part of the machines) and when using parallel programs with a fixed number of branches.

When CS reliability is studied, a failure is understood to be an event, such that the system is no longer capable of fulfilling its functions related to realization of a parallel program with a specified number of branches. If CS is in the failure mode, the number of faulty EM exceeds the number of EM, constituting the structural redundancy. The notion of CS reliability falls into the universally-accepted notion of system reliability, and the block-diagrams created within the CS for a reliable realization of parallel programs with a fixed number of branches act as virtual systems and are sufficiently close to the systems with a (loaded) reserve.

CS robustness means the property of program adjustment and organizing the functioning of such block-diagrams that under conditions of EM failure and restoration guarantee, in the case of running of a parallel program, the productivity remains within specified limits and all sound machines can be used. The notion of CS robustness characterizes the system capability to organize fault-tolerant computations, or, in other words, to realize parallel programs, allowing variation of the number of branches within known limits.

When CS robustness is considered, complete or partial failures are differentiated. CS complete failure is an event resulting in the system being no longer capable of fulfilling a parallel program with a variable number of branches. A partial failure is an event when machine failures occur, but the CS preserves the ability to implement a parallel program with a variable number of branches. Complete failure makes the system productivity zero, while partial failure leads to just a certain lowering of productivity, i.e. longer time required for running a parallel program with a variable number of branches.

Robust CS allow using hardware redundancy on the level of separate functional units and components of elementary machines, this redundancy, however, having just an auxiliary role. *It should be noted, that in a robust CS the total productivity of all the sound machines is being used at any moment of system functioning.* The latter means that problem solution programs should have the property of adaptability (for the number of sound EM) and data redundancy.

When developing a viable integrated system for controlling the mini-production of semiconductor films in the orbital flight of a space station, the intent is to use the multi-year experience of the Siberian Division of the Russian Academy of Sciences and local industry in development of computer systems with a programmable structure [11–14]:

- first «Minsk-222» CS with uni-dimensional topology (1965):
- family of space-division systems ASTRA (1970–1973);
- multi-minicomputer systems MINIMAX (with two-dimensional topology, 1975) and
SUMMA (where each EM was connected to the three nearby ones, 1977);

- families of multi-microprocessor systems: MICROS (1986), MICROS-2 (1992) and MICROS (1996);
  - supercomputers of MCS family: MVS-100 (1995), MVS-1000 (1999).

1. Hueser, J.E. and Brock, F.J. (1976) Theoretical analysis of the density within an orbiting molecular shield, *J. Vac. Sci. and Technol.*, **13**, No. 3, pp. 702–710.

2. Melfi, L.T., Outlaw, R.A., Hueser, J.E. et al. (1976) Molecular shield: an orbiting low-density materials laboratory, *Ibid.*, pp. 698–701.

3. Ignatiev, A. (1995) The wake shield facility and space-based film science and technology, *Earth Space Rev.*, **2**, No. 2, pp. 10–17.

4. (1997) The wake shield facility flies again, *Compound Semiconductors*, No. 1, pp. 11–12.

5. Neu, G., Teisserire, M., Freundlich, A. et al. (1999) Donor-acceptor-pair spectroscopy of GaAs grown in space ultravacuum, *Ibid.*, No. 2, pp. 33–34.

6. Darvin, J. (2000) Spinning off in space, *Houston Business J. Exclusive reports*, No.5.

7. Antropov, V.Yu., Biryukov, V.M., Berzhaty, V.I. et al. (1996) Epitaxy of A_3B_5 and A_2B_6 compounds from molecular beams in ultrahigh vacuum behind the molecular shield, In: *Proc. of 2nd Russian-American Symp. according to «Mir»–NASA Program*, Korolyov, 1996. Moscow, pp. 12–14.

8. Valiev, K.A. and Orlikovsky, A.A. (1996) SSIL technology: main tendencies in development, *Elektronika: Nauka, Tekhnologiya, Biznes*, No. 5/6, p. 3.

9. Naumann, R.J. (1987) On the path to absolute vacuum, *Aerokosm. Tekhnika*, No. 10, pp. 129–132.

10. Grazhulis, V.A. (1999) Tender of RF SCST on development of UHVM, In: *Promising Technologies*, Vol. 3, Issue 21, pp. 1–12.

11. Khoroshevsky, V.G. (1987) *Engineering analysis of functioning of computer machines and systems*, Moscow, Radio i Svyaz, 255 pp.

12. Khoroshevsky, V.G. (1997) Computing systems with a programmable structure, *Inform. Tekhnologii i Vych. Sistemy*, No. 2, pp. 13–23.

13. Khoshevsky, V.G. (1998) MICROS: a family of large-scale distributed programmable structure computer systems, In: *Proc. of 6th Int. Workshop on Distributed Data Processing*, Russia, Novosibirsk, 1987. Novosibirsk, SD RAS, pp. 65–76.

14. Levin, V.K. Locally produced supercomputers of MCS family, http://parallel.ru/mvs/levin.html

MELTING AND SOLIDIFICATION

Study of the Features of Melt Solidification at Zero Gravity Using Optically Transparent Models*

A.A. Zagrebelny[1], V.F. Lapchinsky[1], V.V. Stesin[1] and Yu.M. Chernitsky[1]

One of the key areas of space technology is the study of the features of melting and solidification of metals in low-gravity fields. Without a thorough investigation of these features, further effective development of space technology will be extremely difficult. On the other hand, the range of investigation methods currently available for the technologists, is limited. As a rule, the majority of the experiments are reduced to melting and solidification of metals or alloys in fields of various gravity levels and subsequent study of the micro- and macrostructure of the produced ingots. Despite the extreme interest and usefulness of such experiments, they are certainly insufficient for revealing the full picture of the phenomena at zero gravity. The main disadvantage of such a procedure, in our opinion, is the fact that only the final result of the experiment is accessible to the researcher for study, i.e. the outcome of all those numerous and complicated processes which proceeded in the metal during melting and solidification.

When studying these processes in a space environment, which is new for technologists, and characterized by the unique property of zero gravity whose influence on the melt behaviour is little known to us to date, the wish of the researchers to observe the kinetics of the phenomena occurring is natural. Only in this case it becomes possible to trace the entire sequence of transformations experienced by the material being studied. Such an ability is especially important if a fast process of metal welding is being investigated or methods of active control of solidification are being studied.

At present the most effective method of investigation of any process in its kinetics, is filming with subsequent interpretation of the cinegrams. However, filming of the behaviour of the surface of molten and solidified metal does not provide enough information. It is, as a rule, only possible to follow the displacement of the solidification front and the behaviour of the liquid and gas interphase, accessible for observation. The rest of the system of complex phenomena occurring in the metal bulk and being critical for understanding the process as a whole, such as capillary phenomena, convective flows of different nature, phases formation, crystal growth, etc. remains hidden from the observer's eyes. Therefore, of great interest is the study of these processes on optically transparent models, simulating the solidifying metals.

From two possible variants of simulating media, namely, oversaturated salt solutions and salt melts, the latter were chosen, as the salt melts much more accurately simulate such important thermo-physical properties of the metals as toughness, surface tension, entropy of phase transformations, high temperature gradient over the solidification front, etc. Also important was the fact that simulation of metal solidification in welding on models from salt melts had been quite well verified on the ground [1, 2].

We used sodium and potassium saltpetre ($NaNO_3$, KNO_3) in their pure form and in the form of two-component alloys with different percentage of the components, as the model of the molten metal. The rather high temperature of these salts melting (581 and 614 K, respectively) provides a more or less satisfactory simulation of the cooling conditions and temperature gradient in solidification. These salt melts have a density close to that of the

* (1984) *Space research in Ukraine*, Issue 18, pp. 14–17.

light metals ($2.1 \cdot 10^3$ and $2.0 \cdot 10^3$ kg/m^3), and rather high surface tension ($119 \cdot 10^{-3}$ and $110 \cdot 10^{-3}$ N/m).

Alloys of these two salts have an equilibrium diagram, characteristic of metals, when the components are fully soluble in the molten state, have limited solubility in the solid state and form a eutectic. This allows simulation of the conditions of solidification of pure metals and eutectic alloys.

Disadvantages of these salts are their low heat conductivity and considerable sensitivity to overheating above the melting temperature, thus requiring a thorough verification of the experimental procedure and frequent replacement of the samples. Otherwise, the salts start slowly evolving oxygen already at the melting temperature, forming $NaNO_2$ and KNO_2, while greater overheating leads to formation of Na_2O and K_2O oxides with nitrogen evolution.

A 16 mm camera with filming speed of 8-48 frames/s was used to study the solidification process in its dynamics. Filming was performed through one of the eye-pieces of the binocular stereoscopic microscope. The second eye-piece of the microscope was used for visual observation of the process (visual observation could also be conducted through the viewfinder of the camera).

The optical system of microscope — camera in principle provided a magnification in a range of ×9–×250. However, the 12 to 50 times magnifications were the ones the most often used, the field of vision of the optical system being 100–10 mm^2, respectively.

Two schematics of model illumination, namely by transmitted and reflected light, were used depending on the investigation goals (Figure 1). Each of these schematics has its advantages and disadvantages. Therefore, the most complete picture of the studied process can be derived, as a rule, with an integrated use of both the schematics. During the experiments contactless heating of the samples being melted with a Ni–Cr-alloy heater of 60 W power was used, the heater being located at 2–4 mm distance from the model surface. When the process of solidification of a stationary molten pool was studied, the heater was switched off after production of a pool of the required dimensions, and then was quickly removed by an electric drive from the microscope field of vision. In the case of a moving pool (welding simulation), displacement of a hot heater above the model surface was performed with an assigned speed.

Optically transparent models were made in the form of salt plates of 0.5–1.5 mm thickness, which were placed on a substrate from mica or high-temperature resistant glass (Figure 2).

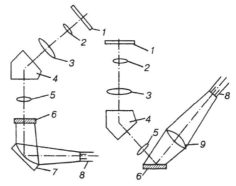

a   b

Figure 1 Optical schematics of filming in the transmitted (a) and reflected (b) light: *1* — film; *2* — camera lens; *3* — microscope eye-piece; *4* — prism; *5* — microscope objective lens; *6* — transparent model; *7* — mirror concentrator; *8* — lamp; *9* — lens concentrator.

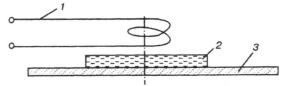

Figure 2 Schematic of optical transparent model: *1* — heater; *2* — model; *3* — substrate.

The thickness of the model plate and of the substrate is very important for provision of the optimal conditions of observation, especially in the case of filming in transmitted light. In this case the optimal results are, as a rule, achieved with the minimal thickness of the model and the substrate. Minimal thermal inertia of the solidification process is also provided in this case.

In order to conduct zero gravity experiments, the entire set of the research tools (salt models, microscope, camera, heater with a device for its displacement, control panel, lamp and optical lighting systems) was mounted as one set-up, which was placed on board the flying laboratory, providing for 20–30 s the overloads from 10^{-2} up to $5 \cdot 10^{-3} g_0$ [3] ($g_0 = 9.8$ m/s^2). Such a set up permits studying the following features of melt solidification under different conditions of heat removal (Figure 3): free solidification of the melt on the substrate (Figure 3a); melt solidification on the substrate with forced heat removal along one (Figure 3b) or two (Figure 3c) axes; melt solidification on the substrate under the conditions of a forced radial heat removal (Figure 3d); melt solidification on the substrate in the skull with the heat source moving at different speeds (Figure 3e) (simulation of fusion welding process).

Study of the features of solidification of a melt containing solid phase particles in suspension, is also highly interesting. When working with optically transparent models, these particles can be Na_2O and K_2O inclusions formed during the salt decomposition. The dimensions of such inclusions usually are not more than $5 \cdot 10^{-2}$ mm.

Despite the short time of zero-gravity condition and complexity of synchronizing the moment of its impact with the model melting, the performed experiments demonstrated the indubitable potential of such an experimental procedure for future application.

A number of interesting phenomena could be observed during the experiments. It was found that the rate of free solidification of the melt at $g = 10^{-2} g_0$ is reduced several times compared to $g = g_0$. One of the causes for this phenomenon, apparently, is an abrupt reduction of convective heat removal from the molten melt surface at zero gravity, as forced blowing

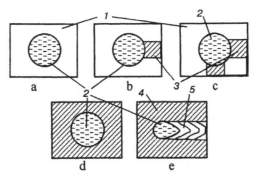

Figure 3 Diagram of experiments: a — free solidification; b, c — solidification with heat removal along one axis and along two axes, respectively; d — solidification with radial heat removal; e — solidification in the skull with a moving heat source: *1* — substrate; *2* — molten model; *3* — heat sink; *4* — unmolten model; *5* — remelted model.

of the surface with an air flow velocity of 1.0 m/s again resulted in an increase of the solidification rate, although not up to the initial value.

Preliminary analysis of the behaviour of oxide particles dispersed in the melt, leads to the conclusion about a certain difference of the process of convective stirring of the melt near the solidification boundary at zero gravity and at different levels of acceleration acting on the melt. In a number of cases, differences in the direction of growth and the shape of the crystals were also observed, especially at free solidification at a low rate. All the observations and conclusions made are only of a preliminary nature so far. It does not appear possible to make any final conclusions proceeding from the conducted experiments, because of a fundamental imperfection of the procedure of working with the melts under the conditions of the flying laboratory. Nonetheless, preliminary analysis of the derived experimental data permitted the outlining of ways of further improvement of this procedure. It can be anticipated that staging such experiments under conditions of long-term zero gravity in space vehicles at the overload level of $(10^{-4}-10^{-5}\ g_0)$ will allow detection and more precise determination of not only the above-mentioned, but also of finer phenomena related, in particular, to the influence of gravity on impurities diffusion along the phase boundaries and, consequently, on the solidification rate, perfection of ingot structure, etc.

1. Wittke, K. (1968) Modellirung der Primärkristallisation beim Schmelzschweißen, *Schweißtechnik*, No. 7, p. 18.

2. Boldyrev, A.M. (1976) On the mechanism of weld metal structure formation by introduction of low frequency oscillations into the weld pool, *Svarochn. Proizvodstvo*, No. 2, pp. 52–54.

3. Belyakov, I.T. and Borisov, Yu.D. (1974) *Technology in space*, Moscow, Mashinostroyeniye, 290 pp.

Experience of the Use of an Electron Beam Unit for Remelting Materials in Space[*]

V.F. Shulym[1], V.F. Lapchinsky[1], E.S. Lukash[1], V.V. Stesin[1] and G.V. Zhukov[2]

Development of space technology of metals requires the use of various research units for conducting metallurgical experiments. Of high interest in this respect are power-effective electron beam units, which among other things have another important advantage, i.e. they offer the possibility of treating materials in a vacuum avoiding sealed ampoules.

The device under consideration presents a pioneering attempt to use an electron beam as a heat source for melting of metals in space. The device is one of the removable modules of the multipurpose research unit «Isparitel-M». This module was designed and manufactured specifically for the joint Soviet-Indian experiment «Overcooling» (Figure 1).

Spherical samples of material *1*, which are to be remelted or heat treated, are placed between the loops of thermocouples (in this case, a chromel-alumel thermocouple) installed in a quartz glass cylindrical cell *2*. The samples are located in such a way that they are in the focus of a molybdenum radiator *3* made as a semi-cylinder with a radius of 17 mm. All these elements are housed in a cylindrical multilayer container, i.e. vacuum insulator *5* made from titanium and molybdenum foil. The container has flap cover *7* used for replacement of samples of the material treated, which at the same time serves as a counter-reflector, and hole *6* for introduction of the electron beam.

* (1985) *Problems of space technology of metals*, Kiev, Naukova Dumka, pp. 22–25.

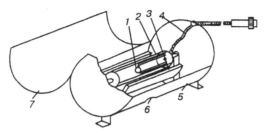

Figure 1 Schematic of isothermal container designed for metallurgical experiments
(see designations in the text).

The defocused electron beam, through hole 6 in the container, heats up the molybdenum radiator, which in turn heats the treated material samples by infrared radiation. Heat removal from the material heated due to thermal conductivity of the thermocouples can be minimized by using small-diameter thermocouples placed in the thin-layer flexible glass insulator *4*. The entire container is mounted on an aluminium plate, i.e. radiator, using titanium brackets made from a thin sheet. This limits leakage of heat to the surrounding elements of the structure, thus reducing it to a value which is not in excess of 8 % of heat input. Therefore, at time of the heating cycle equal to 30 min and temperature in focus of the radiators equal to 800 °C, the temperature of the plate is not in excess of 80 °C.

In the Soviet-Indian experiment «Overcooling» the device described was used for melting Ag–Cu (T_{melt} = 650 °C). Choice of the melting conditions was based on the data on a temperature field of a given design of the heater and optimal cyclogram of heating and cooling, depending upon the power of the heater (electron beam gun), which were determined as a result of optimization of the experiment under earth conditions. It was shown that the heater was almost isothermal along the longitudinal axis (Figure 2). Deviations of temperature from the mean level along the longitudinal axis are normally not in excess of ±20 °C. A characteristic feature is that, in heating, the maximum temperature is observed at the centre of the heater, immediately above the hole for introduction of the beam. In cooling the temperature above the hole, on the contrary, is a minimum. This is attributable to the fact that the hole weakens the effect of the screen-vacuum insulation.

The heater investigated has very large temperature gradients in a radial direction. This is associated with the fact that deviations from the axis lead to displacement of the samples out of focus of the radiator. For example, displacement of the centre of the samples to 4–5 mm from the radiator axis led to a decrease of 100 °C or more in their temperature at the same power of the electron beam bombarding the radiator. Temperature of the samples during heating and cooling in the given device varies following the exponential law, $T = f(e)^{t/\tau}$, where T is the temperature of the samples, t is the time and τ is the time constant of the heating system.

Time constant τ depends upon many factors, bur firstly it upon the power of the electron beam, mass and heat capacity of the radiator, and mass and heat capacity of the samples treated. At a preset mass of the samples and a certain chemical composition, the cyclogram

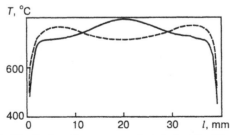

Figure 2 Distribution of temperature along the length of the radiator.

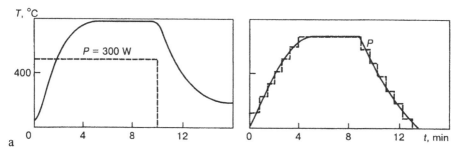

a   b

Figure 3 Cyclogram of non-controlled (a) and controlled (b) conditions of heating and cooling.

of heating and cooling can be regulated by varying power of the electron beam and mass of the radiator. Here it is possible to use either controlled or non-controlled conditions of heating and cooling. The non-controlled conditions provide immediate setting of the electron beam power required for achieving the desirable temperature, and immediate switching off of the beam upon completion of heat treatment. The cyclogram of these conditions is shown in Figure 3a. In this case the mean rate of heating and cooling of the samples is determined only by the radiator mass. Maximum values of the heating and cooling rates can be achieved at a minimum mass of the radiator.

Under the controlled conditions, in the heating and cooling regions the power of the electron beam that heats the radiator varies following the preliminarily set program. The «Isparitel-M» unit, to which the device under consideration is connected, allows discrete settings of 30 levels of the beam power for a time of 30 min. The cyclogram of heating and cooling of the samples, obtained in this case, is shown in Figure 3b. The described heating and cooling cycles can be in various combinations. In particular, the cyclogram (Figure 4) with two power levels in heating and passive cooling was selected for the «Overcooling» experiment. Shoulders in the heating and cooling regions are caused by phase transitions. The highest mean rates of variations in temperature, which can be achieved by means of the described device using the «Isparitel-M» unit, are as follows: 5 °C/s in heating and 3 °C/s in cooling. The lowest values depend upon the temperature to which a sample is heated (e.g. for $T = 800$ °C it is 0.3 °C/s in heating and 0.2 °C/s in cooling).

After simple readjustment, this device allows also a gradient heating of a cylindrical metal sample. For this the sample should be heated from one end and heat should be removed from the other end (at which the sample is fixed to the base platform). Using a low-power levitation device to keep the material under zero gravity, it is possible in principle to perform crucibleless melting.

Therefore, this heating device offers wide possibilities and can be applied in the cases where a short duration of treatment of materials and comparatively high rates of heating and cooling of samples are a requirement. The total duration of the heating and cooling cycles is limited by the life of a cathode assembly of the electron beam gun and can be increased

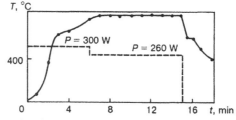

Figure 4 Cyclogram of heating and cooling of the samples in the «Overcooling» experiment.

to 90 min. The device can be used with the «Isparitel-M» and «Yantar» (UN-100, UN-105) as a removable module.

Experiments on Solidification of Aluminium Melts in «Isparitel-M» Unit*

V.N. Pimenov[17], V.F. Shulym[1], S.A. Maslyaev[17] and S.Ya. Betsofen[17]

Studies [1, 2] reported different solidification textures in the direction of the temperature gradient grad T, found in samples after ground-based and space experiments. Grain orientation of α_{Al}-solution in Al–0.4 wt.% Cu [1] and Al–5 wt.% Cu [2] melts in flight samples satisfied the condition of $<110>\alpha_{Al} \parallel$ grad T, and the condition of $<100>\alpha_{Al} \parallel$ grad T in the ground samples. Such a change of solidification texture was associated with the influence of the type of mass transfer in the liquid on the nature of temperature fluctuations on the crystal-melt interphase and with the different conditions of thermodynamic equilibrium on this interphase on Earth and at zero gravity.

On the other hand, in [3, 4] it was reported that the orientation of grad T vector relative to the vector of free fall acceleration $\mathbf{g}_0$ in the experiments on the ground, influences some structural features of the material produced. Moreover, the distribution of the field of temperatures and concentrations in the melt essentially depends on the orientation of heat flow vector relative to $\mathbf{g}_0$ vector [5–7].

In view of the above-said, the following question is of great interest: whether the solidification texture in the alloys depends on the relative orientation of the vectors of heat flow and gravitation, as well as how sensitive is it to the mechanism of mass transfer in the liquid and the value of free fall acceleration $\mathbf{g}_0$. The reply to this question can only be provided as a result of a series of ground-based and space model experiments.

This work presents the results of the first stage of the planned investigations, namely during the ground-based optimization of the experimental modes in «Isparitel-M» unit, the solidification texture was studied in the alloys in the direction of the temperature gradient along the sample in two variants of the experiments, namely the horizontal and the vertical one.

Experimental procedure. Ground-based experiments were conducted in a replaceable module of «Isparitel-M» unit, which was used in the Soviet-Indian experiment «Overcooling» [8, 9]. The diagram of the heating device is shown in Figure 1. A defocused electron beam through a hole in the container wall, heated a molybdenum radiator of a cylindrical

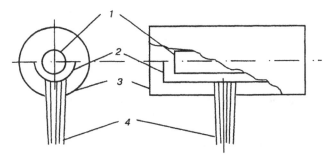

Figure 1 Schematics of the heating device of the replaceable module of «Isparitel-M» unit: *1* — ampoule; *2* — radiator; *3* — counter-reflector; *4* — electron beam.

* (1986) *Fizika i Khimiya Obrabotki Materialov*, No. 6, pp. 41–46.

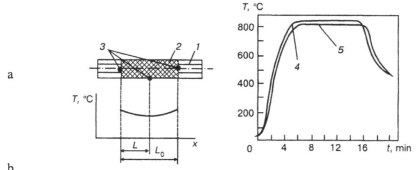

a

b   c

Figure 2 Diagram of the quartz ampoule (a), temperature distribution along its axis after switching off the electron beam (b) and experiment cyclogram (c): *1* — quartz tube; *2* — crucible with sample; *3* — position of thermocouples during the experiment; *4* — temperature in the heating zone center; *5* — temperature at the end face walls of the crucible.

form which by infrared radiation heated the ampoule with the sample of the material being treated, which was in the radiator focal point. The required temperature gradient along the heater length of ~10 °C was created at the stage of cooling by means of the predominant heat sink in the central part of the heating zone through a hole in the vacuum shield insulation. In this case the minimum temperature was in the centre of the longitudinal axis of the heater.

Cylindrical samples of Al–0.2 wt.% Cu alloy and of pure aluminium of the following dimensions: 0.6 cm diameter and length L_0 of 2.0 cm, were used as the objects of study. The samples were placed into graphite crucibles, with inner dimensions: 0.65 cm diameter and length L of 2.5 cm. The crucibles were closed with a cover and inserted into the quartz ampoules. In order to perform temperature measurement during the experiments, quartz tubes were soldered into the ampoule end faces, the tubes being the inlets for thermocouples and channels for pumping down the internal volume of the ampoules. Temperature measurements were taken in three points, namely at the crucible side walls and in the centre. The diagram of the ampoule, temperature distribution along its axis, and the cyclogram of the experiments, are shown in Figure 2.

After the experiments, during which melting of the initial samples and melt solidification from the centre towards the edges occurred, transverse macrosections of each sample in three sections were prepared, namely the central one and the two end face ones at 2 mm distance from the edge. This was followed by X-ray analysis of each of the macrosections, namely the method of reverse pole figures was used to determine the solidification texture in the direction of the temperature gradient. Moreover, copper distribution along the longitudinal axis of the samples, was determined by the method of X-ray microprobe analysis for Al–0.2 wt.% Cu alloy.

Results and discussion. X-ray analysis of the samples of aluminium and Al–0.2 wt.% Cu alloy after the horizontal experiment, showed that none of the studied macrosections had the solidification texture in grad T direction. On the other hand, solidification texture in grad T direction was found in the samples after the vertical experiment, both in pure aluminium and in the alloy. In this case the condition of <100> ∥ grad T was satisfied in each of the studied sections. In the end face sections, however, the predominant orientation of aluminium grains had a more pronounced nature, than in the central section. So, the fraction of aluminium grains oriented in <100> ∥ grad T direction in the central section, was ≈ 25 %, whereas in the end face sections it was ~ 35–50 % of the total number of grains in the section plane. Thus, for the vertical experiment, the results of this work agree with the results of [1, 2]. However, if in [1] a similar solidification texture in grad T direction was

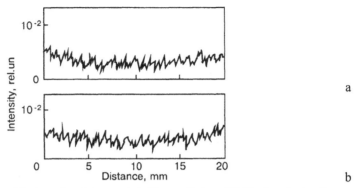

Figure 3 Distribution of the intensity of characteristic X-ray radiation of CuK$_\alpha$ along the length of Al–0.2 wt.% Cu sample for vertical (a) and horizontal (b) experiments.

also found for the horizontal sample of Al–0.4 wt.% Cu alloy, no such relationship was found in this case.

As regards copper distribution in Al–0.2 wt.% Cu alloy along the samples length, the respective results are given in Figure 3. Analysis showed this distribution to be approximately the same for the conditions being compared.

Discussion of results. In order to interpret the derived results, let us consider the features of mass transfer in the melt for the vertical and horizontal experiments. In the first case (grad T ∥ g_0), only gravity type convection was found in solidification, namely thermal convection in aluminium melt, and in Al–0.2 wt.% Cu melt also the concentration convection, associated with impurity ousting (coefficient of copper distribution in aluminium $k <$ < 1) and its accumulation ahead of the solidification front. Now for the case of grad T ⊥ g_0, alongside the above kinds of convective flow, non-gravity type convection was also found, namely the thermocapillary and concentration capillary (Marangoni effect) convection, associated with the dependence of surface tension on the temperature and concentration, respectively [5, 10]. The time for establishment of the convective flow in the melt volume [11] is $\tau_\upsilon \sim L^2/\upsilon$ (L — characteristic linear dimension, υ — kinematic viscosity) and is equal to $\cong 120$ s. This time is several times longer than the melt solidification time τ_c during which the concentration and concentration-capillary convection develop, and is comparable with time τ_m of melt stirring by thermal and thermocapillary convection at the cooling stage ($\tau_m \cong$ $\cong 150$ s). Therefore, only the last two kinds of convective flow are considered below, and the «concentration effects» are not taken into account.

The Raleigh number which characterizes the intensity of thermal convection, is equal to

$$Ra = \frac{g_0 \beta L^3 \Delta T}{\upsilon a}, \tag{1}$$

where β is the coefficient of bulk thermal expansion; a is the thermal diffusivity; ΔT is the characteristic temperatures difference.

For our experiments, Ra ≈ 300 ($\beta = 1 \cdot 10^{-4}$ deg^{-1}, $\upsilon = 3 \cdot 10^{-3}$ cm^2/s, $a = 2.4 \cdot 10^{-1}$ cm^2/s, $L = 0.6$ cm, $\Delta T = 10$ K were assumed). The characteristic flow velocity in this case was [12]

$$u_0(Ra) \approx (Ra/Pr)^{\frac{1}{2}} uL^{-1} 1/2 \, uL^{-1}, \tag{2}$$

where Pr $= \upsilon/a$ — Prandtl number. For this case $u_0(Ra) \approx 0.8$ cm/s.

The intensity of thermocapillary convection is determined by Marangoni number

$$\mathrm{Ma} = \frac{\partial \sigma}{\partial T} \frac{L_s \Delta T}{\rho \upsilon a}, \tag{3}$$

where σ is the surface tension; ρ is the density; L_s is the characteristic length of free surface. The value of the maximum flow velocity on the free surface is equal to [11]

$$u_s(\mathrm{Ma}) \approx \upsilon(\mathrm{Ma/Pr})^{\frac{2}{3}} L_s^{-1}. \tag{4}$$

Whence for the horizontal experiment, assuming $(\partial\sigma/\partial T) = -0.16$ dyn/(cm·deg) [13], $\rho_{Al} = 2.3$ g/cm^3, $L_s = 1$ cm, the other parameters being the same as in assessment of Raleigh number, we have Ma ≈ 950 and $u_s(\mathrm{Ma}) \approx 5.5$ cm/s. The mean mass rate of stirring of the melt, entrained into motion by the flow of liquid on the surface, is approximately equal to [11]

$$u_0(\mathrm{Ma}) \approx 2\upsilon(\mathrm{Ma/Pr})^{\frac{1}{3}} R^{-1}, \tag{5}$$

where R is the crucible radius; $u_0(\mathrm{Ma})$ is the flow velocity in the volume outside the subsurface boundary layer $\delta(\mathrm{Ma})$ at thermocapillary convection. The thickness of the boundary layer is equal to

$$\delta(\mathrm{Ma}) \cong (\mathrm{Ma/}Pr)^{-\frac{1}{3}} L_s. \tag{6}$$

For this case we have $u_0(\mathrm{Ma}) \cong 0.85$ cm/s and $\delta(\mathrm{Ma}) \cong 2 \cdot 10^{-2}$ cm.

Features of flow and mass transfer. The performed estimates allow qualitative analysis of the features of liquid flow and mass transfer in the vertical and horizontal experiments.

In the first case, aluminium melt solidification which proceeded in the direction from the sample centre towards its end faces, was accompanied by rather intensive convective stirring in the lower part of the sample, the flow being of a laminar nature (Reynolds number $\mathrm{Re} = (u_0 L/\upsilon) < 10^3$) (Figure 4a). In the upper half of the melt, the material was transferred predominantly by diffusion, as the heavier component was below. In both the cases solid phase growth proceeded in quasi-equilibrium conditions under which the growing crystal assumes such a form that its surface energy was minimum:

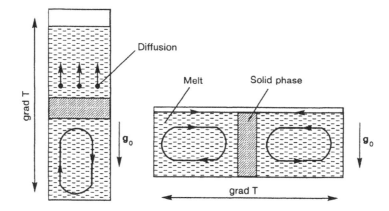

a   b

Figure 4 Schematics of convective flows and mass transfer in the melt at directional solidification for vertical experiment with Al–0.2 wt.% Cu alloy (a) and horizontal experiment with pure aluminium and Al–0.2 wt.% Cu alloy (b).

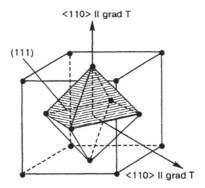

Figure 5 Schematics of cutting of a crystal with fcc-lattice in its directional growth under the conditions of the gravity pull in <100> ∥ grad T direction (<110> ∥ grad T case corresponds to zero gravity experiments [1, 2]).

$$\sum_{i=1}^{n} \sigma_i S_i,$$

where σ_i and S_i are the surface tension and the area of any face, respectively.

For the metals with fcc-lattice (as in our case), this condition is satisfied when the surface of a freely growing crystal has the form of an octahedron with (111) faces. This is connected to σ being smaller in planes of type (111), than in planes of another type [14]. With directional solidification this situation is implemented, if crystal growth in grad T direction, coincides with <100> crystallographic direction (Figure 5).

This is exactly the situation which was observed by us when studying the solidification texture of samples of aluminium and Al–0.2 wt.% Cu alloy after the vertical experiment. In this case, a somewhat more pronounced disorientation of grains in the central section of samples compared to end face sections, could result from distortion of isotherm lines in this region, because of the impact of the lateral heat sink through a hole in the vacuum shield insulation.

The picture was different in the case of the horizontal position of the sample. The intensity of melt stirring in the volume was markedly enhanced at the expense of Marangoni surface convection (see Figure 4b), namely the total average velocity of the bulk convective flow increased by approximately twice. This increased the probability of the laminar flow losing its stability and going into the turbulent mode. The presence of turbulent vortices in which the velocity, pressure and temperature are chaotically pulsating, promotes spontaneous formation of solid phase nuclei in the overcooled melt [10, 15]. The conditions of these nuclei growth can essentially differ from the conditions of the directional growth of the crystal in the solidification front, whereas their random orientation is in no way connected with grad T direction. Moreover, more intensive stirring of the melt promoted, on the one hand, temperature equalizing along the sample, and on the other hand, increase of temperature of the upper (lighter) layers of liquid and creation of a transverse temperature gradient. This resulted in the change of the nature of heat sink in solidification (compared to the conditions in the vertical experiment), and the axial direction of the solid phase growth was disturbed.

Thus, the absence of solidification texture in the samples after the horizontal experiments was the consequence of the influence of thermocapillary convection.

Now, as regards longitudinal distribution of copper in the samples being compared, its nature confirms the presence of convective stirring of the melt in the vertical and horizontal experiments. A more detailed study of the features of impurities segregation in different stages of the experiments under consideration, can be conducted later by the method of numerical simulation of the process.

Conclusions

1. The solidification texture and impurities distribution were studied in samples of aluminium and Al–0.2 wt.% Cu alloy after the vertical and horizontal experiments in «Isparitel-M» unit; it was found that copper distribution is approximately similar under the conditions being compared, while the solidification texture in <100> || grad T direction is only found with the vertical position of the samples.

2. It is shown that the cause for the absence of the predominant orientation of grains in grad T direction under the conditions of the horizontal experiment, is thermocapillary Marangoni convection, leading to the change of the temperature field, because of the loss of flow stability and transition to the turbulent mode.

1. Favier, I.I. (1982) Instabilite de l'interface solide/liquide: limites et aspects morphologiques en conditions de microgravite, In: *Abstr. of pap. of 24ᵗʰ COSPAR Symp. on Fundamental Aspects of Materials in Space*, Ottawa, 1982. Ottawa, p. 25.

2. Pimenov, V.N., Maslyaev, S.A., Sasinovskaya, I.P. et al. (1985) Solidification of Al-base alloys at zero gravity, *Fisika i Khimiya Obrab. Materialov*, No. 1, pp. 65–68.

3. Zemskov, V.S., Barmin, I.V., Raukhman, M.R. et al. (1983) Experiments on growing alloyed indium antimonide under conditions of orbital flight of space complex «Salyut-6»–«Soyuz», In: *Technological experiments at zero gravity*, Sverdlovsk, UNTs AN SSSR, pp. 30–46.

4. Khryapov, V.T., Tatarinov, V.A., Kulchitskaya, T.V. et al. (1983) Growing of bulk single-crystals of germanium by the method of directional solidification at zero gravity, *Ibid.*, pp. 59–71.

5. Avduevsky, V.S., Barmin, I.V., Grishin, S.D. et al. (1980) *Problems of space manufacture*, Moscow, Mashinostroyeniye, 224 pp.

6. Ostrakh, S. (1980) Role of convection in technological processes conducted under microgravity conditions, In: *Space technology*. Ed. by L. Steg, Moscow, Mir, pp. 9–37.

7. Haynes, I. (1979) Fundamental and applied aspect of fluid physics under microgravity, In: *Proc. of 3ʳᵈ Eur. Symp. on Material Sciences in Space*, Grenoble, 1972. Grenoble, pp. 275–279.

8. Konovalov, B. (1984) Orbital metallurgy, *Izvestiya*, April 11.

9. Tarasov, A. (1984) The stars are shining for the «Beacons», *Komsomol. Pravda*, March 22.

10. Ivanov, L.I., Zemskov, V.S., Kubasov, V.N. et al. (1979) *Melting, solidification and phase formation at zero gravity*, Moscow, Nauka, 255 pp.

11. Anisimov, N.Yu. and Leskov, L.V. (1983) Physical features of directional solidification technological process at zero gravity, In: *Technological experiments at zero gravity*, Sverdlovsk, UNTs AN SSSR, pp. 124–139.

12. Leskov, L.V. and Savichev, V.V. (1982) Investigation of the features of technological processes in space vehicles, In: *Hydrodynamics and heat mass exchange at zero gravity*, Moscow, Nauka, pp. 173–186.

13. Nizhenko, V.I. and Floka, L.I. (1981) *Surface tension of liquid metals*, Moscow, Metallurgiya, 208 pp.

14. Lifshits, B.G. (1963) *Metallography*, Moscow, Metallurgizdat, 422 pp.

15. Chalmers, B. (1968) *Theory of solidification*, Moscow, Metallurgiya, 287 pp.

Electron Beam Crucibleless Zone Melting
of Silicon Single Crystals[*]

B.E. Paton[1], E.A. Asnis[1], S.P. Zabolotin[1], P.I. Baransky[18],
V.M. Babich[18] and M.Ya. Skorokhod[18]

The advantages of crucibleless zone melting using a disk-shaped electron beam in space, under conditions of microgravity, as compared with other methods of zone melting involving induction heaters and resistance furnaces (see, e.g., review papers [1–3]), include:

• high power efficiency (efficiency of the process amounts to 80 %) and, hence, relatively low power consumption. This is very important, especially for space objects designed for long-time functioning under microgravity conditions;

• homogeneous nucleation of single crystals under conditions of crucibleless zone melting (CZM), which provides high purity and homogeneity as compared with single crystals grown under conditions of mostly heterogeneous nucleation, involving crucibles and ampoules;

• ease of control and maintenance of preset sizes of the molten zone, which is very important for conditions of practical implementation of CZM, etc.

The principle of operation of a unit designed for growing semiconductor and other single crystals under earth and microgravity conditions by the CZM method using a disk-shaped electron beam is described in [4–6].

Exceptionally high sensitivity of electrophysical, optical and other properties of semiconductor crystals to very low concentrations of doping (and residual) impurities is determined not only by the ratio of concentration of impurity atoms to atoms of the crystal matrix $[1:(10^{10} - 10^{12})]$, but also by peculiarities of their spatial distribution in the crystalline lattice. All this puts the process of solidification of semiconductor single crystals made by the CZM method in the category of a topical object of research and, moreover, allows it to be used as a tool that possesses wide capabilities in getting a better understanding of physics of phase transitions under specific microgravity conditions [7], which can hardly be realised in full under earth conditions.

In view of the paramount importance of the shape of the solidification front, as only a solidification front close to a flat one can ensure (both under typical conditions of 1 g_0 and under microgravity conditions) formation of perfect crystals, the task posed included identification and practical implementation of the CZM parameters (and, accordingly, topology of the thermal field), which would provide realisation of exactly such a shape of the front.

The X-ray diffraction pattern of the longitudinal section of a single crystal grown under such conditions, i.e. crystallographic surface orientation (001) (Figure 1), obtained in MoK_α-radiation, proves that parameters of growth of the crystal used in these experiments provide solidification front close to the flat one. Also, this X-ray pattern shows that region *II* of the ingot (after electron beam remelting) has a higher level of perfection than region *I*, which was not subjected to remelting.

Detailed study of structural perfection of the crystal (using metallography, in addition to radiography) is worthy of special attention and will be considered separately with regard to investigation of the effect of temperature gradients on the stressed state of the resulting crystals.

As proved by investigation of the longitudinal distribution of specific electrical resistance $\rho(x)$ in the samples produced, it is essential that, depending upon the melting

* (1999) *Transactions of the National Academy of Sciences of Ukraine*, No. 7, pp. 108–112. (Published in a broadened version.)

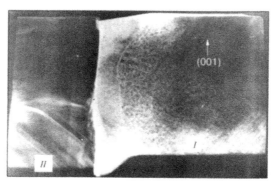

Figure 1 X-ray pattern of a silicon sample: longitudinal section; crystallographic surface orientation
(001); $t = 820$ μm; MoK_α×8.8 (the zone moved in a direction of $I \rightarrow II$).

parameters, doping level and thermal annealing that preceded CZM, the donor-type impurity,
i.e. phosphorus, which is typical for n-Si:P, is displaced differently but efficiently enough,
by the solidification front as a result of single-pass remelting of the crystal (Figure 2).

To check the stability of the solidification process within the framework of the selected
parameters, a p-Si:B crystal (containing a boron impurity) was subjected to crucibleless zone
remelting. This was preliminarily grown by the Czochralski method and therefore contained
a residual impurity of electrically neutral oxygen atoms ($\approx 5 \cdot 10^7$ cm^{-3}) and a certain amount
of thermal donors of an oxygen origin.

Electron beam crucibleless remelting of this crystal was to lead to a decrease in the level
of compensation of the crystal and, accordingly, to a decrease in its specific electrical
resistance ρ, which was proved by the experiment conducted (Figure 3). Very small
fluctuations of specific electrical resistance ρ(x) within the material remelted are caused by
some instability of solidification conditions. This fact should be taken into account in

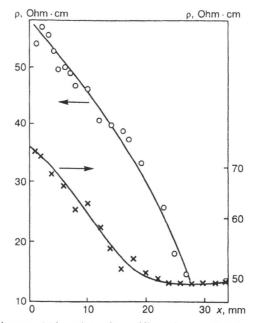

Figure 2 Distribution ρ = ρ(x) in regions of two different ingots n-Si:P doped with a phosphorus
impurity (displacement of the zone coincides with a direction of axis 0x in this Figure).

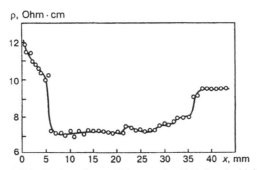

Figure 3 Distribution of $\rho = \rho(x)$ in crystal p-Si:B doped with boron.

development of the melting technology and ensuring the required automation of regulation of the melting parameters, using computers for programming of the solidification process with feedback to control it.

Another single crystal was grown (under the same conditions) from the other billet cut from the same ingot p-Si:B. This single crystal was intended for a detailed study of distribution of $\rho = \rho(x)$ in a transverse section of the sample on three disks (each 10 mm in diameter and about 1 mm thick) cut from three different locations along the crystal. The following results were obtained on each of these disks (by 12 measurements made along the diameter): 8.56 ± 0.12, 8.56 ± 0.12 and 8.41 ± 0.10 Ohm·cm. This shows that spread of the specific electrical resistance in a transverse section of the sample does not exceed 1.4 %. This result is an additional indication of the fact that the solidification front in the samples investigated, produced by electron beam crucibleless zone remelting is really close to the flat one.

Therefore, the investigations conducted on CZM of silicon single crystals, using a disk-shaped electron beam, show that the solidification front formed by keeping to the appropriate melting technology is rather close in shape to the flat one, which is one of the most important and desired conditions for production of perfect and homogeneous crystals. In addition, per pass of the zone along the ingot, the process described is characterised by a sufficiently high degree of purity of the crystals, containing doping or residual impurities, with segregation coefficients other than unity.

It might be expected that under microgravity conditions, where the convective mass transfer, which in solidification under earth conditions is controlled by the processes of thermal and concentration convection, will be almost fully suppressed, and the stability of parameters which determine the shape of the solidification front would be increased, which in turn would promote an increase in perfection of the material solidifying under such conditions.

Microgravity conditions will allow handling of a number of fundamental problems associated with production of silicon single crystals, such as: investigation of the physics of the process of solidification and phase transitions with manifestations of thermocapillary convection characteristic of these conditions, and specific features of heat and mass transfer, and, in this connection, study of the specific issues related to the technology of production of semiconductor materials in space.

Investigation of the solidification process under microgravity conditions will most probably allow an adjustment of solidification technology for earth conditions, as well as improvement of the process equipment.

The above method for purification of crystals from doping and residual (or background) impurities through pushing them from the solid into the liquid phase by the solidification front (under conditions of ingot recrystallisation) is far from being versatile. It is efficient only in cases with the impurity segregation coefficient $K = N_{sol} / N_{liq} < 1$ (N_{sol} and N_{liq} are the equilibrium concentrations of a corresponding impurity in the solid and liquid phases). If K is close to one (as is the case in reality, e.g. when silicon single crystals contain a boron impurity, where $K \approx 0.8$), purification of the crystal from such an impurity is of low efficiency. Given this circumstance, as well as taking into account the specific behaviour of the oxygen impurity in the bulk of silicon single crystals under conditions of their heat treatment [8, 9], which is an integral part of processes of manufacture of devices based on them, as well as a point of reliable control of the method of doping these crystals, it is necessary to subject them to deep purification to remove, first of all, the background oxygen impurity.

However, there are often cases when it is necessary to subject silicon crystals to deep purification not only from the background impurities, but also from the doping impurities, for example, for preparation of a source material for making high-efficiency nuclear radiation counters [10] or for transmutation doping [11].

Consider the dependence of the efficiency of removal of the background (oxygen) and doping (phosphorus) impurities from the bulk of the silicon crystals upon the level of vacuum under conditions of electron beam crucibless zone recrystallisation (CZR).

Experiments in series I used single crystal grown by the Czochralski method from the melt, which in an initial state had the following concentration of oxygen impurity: $N_0 \approx$ $\approx (3 \div 6) \cdot 10^{17}$ cm^{-3}, and was doped with a boron impurity (p-Si:B). This choice of an investigation object is based, first of all, on the peculiarities of the boron doping impurity, i.e. its low vapour pressure and, hence, low volatility in vacuum [12], as well as on the proximity of its segregation coefficient to unity. On the other hand, this choice is based on a requirement for keeping electrophysical characteristics (such as specific resistance, etc.) after recrystallisation close to the initial ones and providing varied oxygen concentrations controlled by the IR absorption in the «purest» form.

Concentrations of internodal optically active oxygen in silicon crystals were determined from the IR absorption spectra using the following formula:

$$[O_i]_{opt} = K \cdot \alpha_{max} \, (1106 \text{ cm}^{-1}),$$

where $K = 2.45 \cdot 10^{17}$ cm^{-2}.

Plane-parallel washers (*1–4*) 3 mm thick were cut from the electron beam CZR ingot. A schematic of their location is shown in Figure 4. The IR absorption within a range of wave lengths corresponding to frequencies of 1050–1200 cm^{-1} was studied at $T \approx 293$ K on each of them after grinding and mechanical and chemical polishing. Results of these measurements are shown by curves *1–4* in Figure 4. Based on the values of the $N_0^{(i)}$ concentrations (where $i = 1$–4) at locations of the washers (see the captions of Figure 4) after electron beam recrystallisation, as well as estimations made, we come to the conclusion that at a location of washer *2* the $N_0^{(2)}$ concentration decreased by a factor of $\approx 4.5 \cdot 10^{17}/6 \cdot 10^{16} \approx 7.5$ within a period of time required for the zone to make one pass (i.e. from the lower to upper end of a sample). At the location of washer *3*, the degassing coefficient of oxygen atoms is even higher. Therefore, it provided, as seen from the experimental data, a decrease of $\approx 6 \cdot 10^{17}/5 \cdot 10^{15} \approx 120$ times in $N_0^{(3)}$, as compared with its initial value. As the most significant parameters which determine recrystallisation conditions, except for pressure in the vacuum chamber, remained almost unchanged during the entire recrystallisation process, an increase by a factor of $\approx 120/7.5 \approx 6$ in the efficiency of removal of the oxygen impurity during the recrystallisation process, i.e. transition from $N_0^{(2)}$ to $N_0^{(3)}$, can be related only to $a \approx 1.5$ times decrease in pressure in the vacuum chamber from $3.4 \cdot 10^{-5}$ to $2.3 \cdot 10^{-5}$ mm Hg during a corresponding period of time. Note that a substantial acceleration of desorption in vacuum with time, caused by a decrease in pressure inside a vacuum unit, is well known for

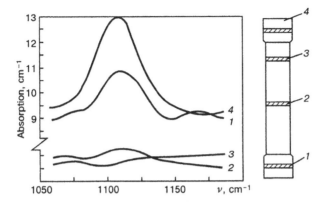

Figure 4 Concentrations of oxygen impurity in individual washers (*1–4*), calculated from the IR-absorption data, corresponded to the following values, at/cm^3: *1* — $3 \cdot 10^{17}$; *2* — $6 \cdot 10^{16}$; *3* — $5 \cdot 10^{15}$; *4* — $6 \cdot 10^{17}$. Numbers of the curves correspond to numbers of the washers shown in the cutting diagram. Washers *1* and *4* were cut from the non-recrystallised region of a material and washers *2* and *3* were cut from the recrystallised region.

metallurgists as well. For example, the author of review [13] states that «duration of holding to remove 50 % As from steel (initial content is 0.06–0.2 % As) reduces from 7 h at 10^{-3} mm Hg to 2 h at 10^{-5} mm Hg». The only difference is that in our case the indicator of the efficiency of removal of the oxygen impurity in vacuum from the bulk of silicon is the concentration of the impurity atoms remaining in the crystal, rather than time.

Crystal *n*-Si:P (doped with a phosphorus impurity) with a specific resistance of $\rho_{300} K \approx$ ≈ 7 Ohm·cm (in the initial state) over the entire length of the ingot (straight line *1* in Figure 5) was used to conduct series II of the experiments.

After electron beam remelting of the ingot (curve *2* in Figure 5) its specific resistance measured at ≈ 293 K grew dramatically by a factor of ≈ 13.6. And if this growth of ρ is associated with evaporation of the phosphorus impurity in vacuum, the concentration of charge carriers in the bulk of the crystal should be changed accordingly. Measurements of

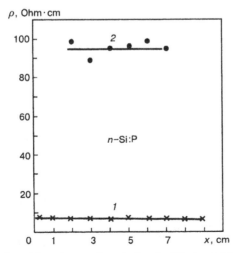

Figure 5 Distribution of specific resistance along the length of the ingot: $\times$ — in the initial state (*1*); ● — after electron beam remelting (recrystallisation) (*2*). Regions 0–1 and 8–9 of the ingot were not subjected to recrystallisation.

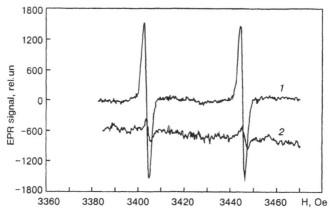

Figure 6 EPR spectra characterising region of the ingot not subjected to zone recrystallisation (*1*)
and central part of the ingot after electron beam remelting (recrystallisation) in vacuum (*2*).

the Hall effect made at ≈ 293 K, as with measurements of ρ, also showed that the concentration of the charge carriers after recrystallisation decreased in the bulk of the crystal by a factor of $\approx 9.4 \cdot 10^{14}$ cm^{-3}/$7.6 \cdot 10^{13}$ cm$^{-3} \approx 12.4$, which coincides with variations in ρ with an accuracy to a change of 13 % in mobility of the carriers. For the sake of reliability, it was necessary just to make sure that the donor impurity in the n-Si:P crystal investigated was really the phosphorus impurity.

Both types of electron paramagnetic resonance (EPR) in the form of a doublet split over the scale of magnetic field H into 42 Oe (curve *1*, Figure 6) and its variation in amplitude values after electron beam remelting of the ingot finally convinced us that the crystal was really doped with the phosphorus impurity in that experiment. Moreover, standard processing of curves *1* and *2* in Figure 6, allowing, naturally, for the value of the EPR signal from a reference sample with known content of the EPR active centres, led to the following relationship:

$$N_{EPR}^{(1)}/N_{EPR}^{(2)} \equiv N_P^{(1)}/N_P^{(2)} \approx 10.1.$$

Its value differs only by 23 % from the above relationship of the charge carrier concentration for the Hall data measured at 293 K on samples made from the corresponding washers used to measure the EPM spectra. These washers were cut from both recrystallised and non-recrystallised regions of the ingot. It is most likely that the above difference in the data is associated with the possible effect of recrystallisation on the numerical value of the Hall factor, which could not be taken into account in processing of the results (therefore, for calculations in both cases the value of the Hall factor was assumed to be equal to unity). These results unambiguously prove that variations in electro-physical (in this case also in EPR) properties of the n-Si:P crystal in its electron beam remelting by the CZR method, are almost completely determined by desorption of doping atoms of phosphorus from the bulk of the crystal into vacuum.

Summarising the additional results, the following conclusions were drawn:

• In order to produce silicon crystals freer from oxygen impurity, as well as other gas impurities, by the CZR method, the recrystallisation process should be conducted in vacuum, rather than in an inert gas atmosphere.

In the case of EZR the recrystallisation process should be carried out in vacuum. As proved by experience, in this case the level of purity of the crystal regarding the oxygen impurity dramatically increases with a decrease in pressure of residual gases inside the work space of the unit. Taking into account this circumstance, as well as the fact that at the values of pressure equal to $\approx 3 \cdot 10^{-5}$ mm Hg, at which the experiments were conducted, the concentration of air molecules at room temperature is $\approx 10^{12}$ cm^{-3}, which is almost two orders of magnitude in excess of the concentration of an electrically active impurity in the purest crystals, i.e. Si $\approx 10^{10}$ cm^{-3}, it can be reasonably concluded that semiconductor materials science still has much potential in terms of improvement of vacuum and, therefore, deeper purification of silicon crystals from gas impurities of the oxygen type, among other things.

- The above results make it possible to consider even now the ultra-high vacuum ($\approx 10^{-12} \div 10^{-10}$ mm Hg), which is formed in the aerodynamic trace of a molecular screen under orbital flight conditions [14–16], as one of the attractive possibilities in terms of practical realisation to be combined with electron beam CZR in order to produce high-quality silicon single crystals, the level of purity of which in the concentration of the background oxygen impurity is almost unachievable under earth conditions.

There is no doubt that deep purification of semiconductor materials, i.e. both elementary semiconductors of the type of silicon and germanium and individual components used for synthesis of sophisticated compound semiconductors (heterostructures and superlattices) will become more topical with time, in view of the transition of applied electronics to utilisation of nanostructures and functional elements of microelectronics of submicron sizes. This places this problem in a category of the highest priority not only for semiconductor materials science, but also for solid-state electronics as a whole.

1. Regel, L.L. (1987) Results of science and technology, In: *Space explorations*, Moscow, VINITI, Vol. 29, Part 2, 296 pp.

2. Regel, L.L. (1991) Results of science and technology, *Ibid.*, Vol. 36, Part 4, 364 pp.

3. Milvidsky, M.G., Verezub, N.A., Kartavykh, A.V. et al. (1997) Growing of semiconductors in space: results, problems, prospects, *Krislallografiya*, **42**, No. 5, pp. 913–923.

4. Paton, B.E. and Lapchinsky, V.F. (1998) *Welding and related technologies in space*, Kiev, Naukova Dumka, 184 pp.

5. Lapchinsky, V.F. (1994) Some aspects of material processing in space, In: *Transact. of Int. Center for Gravity Materials Science and Applications on Some Potential CIS Microgravity Collaborators*, Potsdam, New York, Clarkson Univ.

6. Lapchinsky, V.F., Asnis, E.A. and Zabolotin, S.P. (1996) Processing of materials in space by electron beam heating, In: *Proc. of 4th Ukraine–Russia–China Symp. on Space Science and Technology*, Kiev, 1996. Kiev, pp. 548–550.

7. Avduevsky, V.S. (1985) Main tasks of investigation of hydrodynamics and heat transfer under zero-gravity conditions, *Izv. AN SSSR*, Series Physics, **49**, No. 4, pp. 627–634.

8. Baransky, P.I., Babich, V.M., Baran, N.P. et al. (1983) Investigation of formation conditions on thermal donors-I and -II in oxygen-containing n-type silicon within the temperature range 400 to 800 °C, *Phys. Stat. Sol.* (A), **78**, pp. 733–739.

9. Babich, V.M., Bletskan, N.I. and Venger, E.F. (1997) *Oxygen in silicon single crystals*, Kiev, Interpress, 240 pp.

10. Akimov, Yu.K., Ignatiev, O.V., Kalinin, A.I. et al. (1980) *Semiconductor detectors in experimental physics*, Moscow, Energoatomizdat, 344 pp.

11. Baransky, P.I., Bugaj, A.A., Girij, V.A. et al. (1984) *Transmutation doping of silicon: production, physical properties, application*, Kiev, IF AN USSR, Issue 28, pp. 1–60.

12. (1962) *Vacuum metallurgy*. Ed. by A.M. Samarin, Moscow, GNTI, 516 pp.

13. Polyakov, A.Yu. (1962) Thermodynamic principles of application of vacuum, In: *Vacuum metallurgy*. Ed. by A.M. Samarin, Moscow, GNTI, pp. 59–63.

14. Melfi, L.T., Ontlaw, R.A, Hueser, J.E. et al. (1976) Molecular shield: an orbiting low-density materials laboratory, *J. Vac. Sci. and Technol.*, **13**, No. 3, pp. 698–701.

15. Nauman, R.J. (1987) On the path to absolute vacuum, *Aerokosm. Tekhnika*, No. 10, pp. 129–132.

16. Berzhaty, V.I., Zvorykin, L.L., Ivanov, A.I. et al. (1999) Prospects for realisation of vacuum technologies under orbital flight conditions, *Avtomatich. Svarka*, No. 10, pp. 108–116.

Chapter 4

STUDY OF THE MAGNETOSPHERE OF EARTH

Chapter 4. Study of the Magnetosphere of Earth

This Chapter includes studies which describe one of the main lines of exploration of near-earth space, i.e. investigation of the iono- and magnetosphere of Earth. Active experiments with injection of a high-power electron beam into the magnetosphere were conducted using up-to-date electron beam equipment. They resulted in obtaining data on plasma physics and provided explanations of a number of geophysical phenomena.

Active Experiments in the Near-Earth Space[*]

**B.E. Paton[1], D.A. Dudko[1], V.K. Lebedev[1], Yu.N. Lankin[1], O.K. Nazarenko[1],
V.D. Shelyagin[1], V.V. Stesin[1], V.I. Pekker[1], E.N. Bajshtruk[1], V.K. Mokhnach[1],
Yu.V. Neporozhny[1], N.N. Yurchenko[12] and G.F. Pazeev[12]**

Modern electron beam equipment provides unique possibilities for space exploration. It is a very powerful and flexible tool which allows not only probing, but also perturbation of limited ranges of the near-earth plasma. The accuracy of measurement and regulation of the targeted effects brings active methods of space exploration involving electron beam guns closer to the methods used under earth conditions.

In active experiments involving artificial injection of electrons, an electron accelerator (electron gun) is mounted aboard a rocket or satellite to inject electrons in short pulses to distinguish the injection effects from the natural phenomena. The experiments on injection of electrons make it possible to derive data from the area of physics of plasma of the magnetosphere, and in particular to determine configuration of force lines of the magnetic and electric fields, study the effects that occur when the electron beam passes through a space plasma, and investigate the process of ejection of the electron beam into the dense atmospheric layers.

In the former Soviet Union the organisers and scientific leaders of active experiments in the ionosphere and magnetosphere were the SRI[5] and the ITMIRAS[21]. The PWI, having experience in making industrial electron beam welding equipment since 1958, as well as the flight one for the world-first technological space experiment «Vulkan», was also involved in the program of active experiments to be conducted in the ionosphere. Electron beam units with output parameters close to those of commercial welding units were developed for these experiments. However, they were much superior in their technical characteristics to the earth analogues. The units developed were characterised by minimum dimensions, weight and power consumption, stable output parameters, a fully automatic work cycle, possibility of measuring and transmitting the main operating parameters to the earth through the telemetry channels, and a high reliability.

«Zarnitsa» hardware. Experiment «Zarnitsa-1» was the first Soviet active experiment involving injection of a high-power electron beam into the earth magnetosphere [1, 2]. Block diagram of the hardware is shown in Figure 1, and its general view — in Figure 2. As compared with the «Vulkan» unit, «Zarnitsa» was characterised by a substantial increase in power of the electron beam, i.e. up to 3.3 kW. Design of the guns and the accelerating voltage remained unchanged.

Three electron beam guns, operating in parallel, were used to generate an injection current of up to 0.5 A. The accelerating voltage source, anode voltage converter and storage battery were designed for the correspondingly increased power. Unlike the «Vulkan» unit, in «Zarnitsa» all the guns operated simultaneously and were adjusted approximately to the same current. The marginal emission current allowed an injection current of 450 mA to be generated even by one gun in the case of failure of cathodes of the rest of the guns.

The power source of the electron beam guns, consisting of high-potential filament transformer *4*, high-voltage full-wave semiconductor rectifier *2*, anode transformer *3* and ballast resistors *10*, was made integral with the electron beam guns.

The voltage converter developed at the IED[12] served as a square voltage generator to power the anode transformer. It is a transistor-based unit working at a frequency of 1 kHz.

The filament transformer is powered from converter *5*, and the injection current is regulated by varying the filament power of cathodes of the electron beam guns using the

---

[*] (1999) *Avtomaticheskaya Svarka*, No. 10, pp. 74–80.

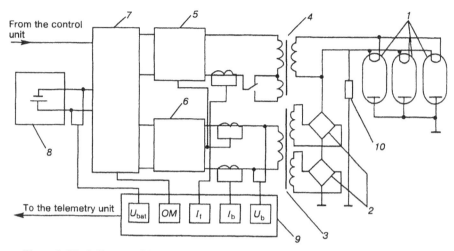

From the control unit

To the telemetry unit

| U_{bat} | OM | I_f | I_b | U_b |

Figure 1 Block diagram of the electron beam hardware «Zarnitsa»: *1* — electron beam guns; *2* — high-voltage rectifier; *3* — anode transformer; *4* — filament transformer; *5* — filament current converter; *6* — anode inverter; *7* — control unit; *8* — storage battery; *9* — unit of converters of telemetry parameters (battery voltage U_{bat}, hardware operation mode programmer *OM*, filament current I_f, beam current I_b, beam accelerating voltage U_b); *10* — ballast resistors.

supply voltage pulse-width modulator. Negative feedback by the load current of anode inverter *6* is used to stabilise the injection current at a preset level.

Program of the experiment provided for two modes of operation of the unit: at an injection current of 280 and 450 mA. In this case the current was modulated by amplitude at a frequency of 0.5 Hz and modulation depth of 100 %.

Switching of the power and control circuits, modulation and stabilisation of the injection current, as well as protection of the inter-electrode gap of the electron beam guns from high-voltage breakdown with an automatic restart after removal of a cause of the breakdown, are provided by control unit *7*. Commands to switch on the hardware and change to the other operational mode arrive at unit *7* from the command-apparatus of the rocket.

Reliability of the control circuit is ensured by using high-quality components, selection of a load factor equal to no more than 0.5, and redundancy of each element. The circuit is made so that it allows pre-flight testing of each of the duplicated elements, for which a special instrumentation module was developed. All wires are loop backed to ensure protection from failures of the type of breaks.

Storage battery *8* is used as a direct current source.

Operation of the hardware is monitored by measurement and transmission of its main parameters to earth via the telemetry channels. The appropriate electric values are converted in unit *9* into unified electric signals of the direct current, which are galvanically isolated from the power circuits of the hardware. Then these voltages are fed to the operation telemetry unit.

The hardware was installed on board the meteorological solid-fuel rocket MR-12, which was launched on May 29, 1973, from the Kapustin Yar space launch site along a ballistic trajectory with an apogee of 180 km. According to the program of the experiment, the hardware was switched on upon leaving the dense atmosphere layers at a height of about 100 km. Pulsed injection of electrons was done along the lines of the magnetic field of Earth. Owing to rotation of the rocket about its axis, the injection angle was varied per revolution from 30 to 85°. In the first operation mode with a duration of 60 s, the injection current in a pulse was 300 mA at an electron energy of 8.9 keV. After switching to the second mode, the

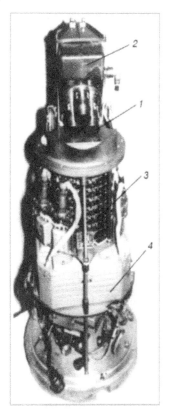

Figure 2 General view of the electron beam hardware «Zarnitsa»: *1* — electron beam guns;
2 — high-voltage transformer, high-voltage rectifier and filament transformer; *3* — power inverter;
4 — storage battery.

injection current increased to 440 mA, and the electron energy fell to 6.4 keV because of a decrease in voltage of the storage battery, caused by an increase in the load current. Operation in the second mode lasted 103 s, after which the injector was switched over to a continuous mode of generation at the same parameters of the electron beam. The injector was working in this mode for another 70 s until the rocket entered into the comparatively dense atmosphere layers, when the decreased vacuum made further operation of the electron beam guns impossible.

Figure 3 shows samples of records of the telemetry data transmitted from the rocket. Several stations with optical (including TV) and radio physical observation means [3] were located in the launch area to examine effects of artificial ejection of electrons into the atmosphere. The ground-based observations allowed fixation and tracing of the evolution of aurora (from the shape of the beam along the magnetic force lines of the geomagnetic field) formed by exciting ionospheric atoms and molecules by the electron beam. As a result, about three hundred photos were obtained, one of which is shown in Figure 4. Maximum brightness of aurorae amounted to a 5 stellar magnitude. The data obtained indicate that the electron beam passes at least several tens of kilometres (from the rocket to a deceleration region) by retaining the same electron bunching. Aurora of the type of the corona discharge, which was detected along the rocket flight trajectory and formed only during injection of electrons, was also of a high interest.

The radar units used in the experiment provided data on radio waves scattered from ionisation heterogeneities, generated in the ionosphere at altitudes of about 110 km by the

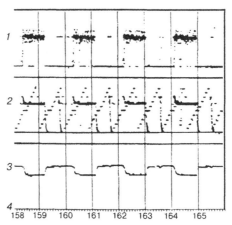

Figure 3 Record of telemetry parameters of the experiment «Zarnitsa-1»: *1* — beam current; *2* — accelerating voltage; *3* — storage battery voltage; *4* — time, s.

beam of the injected electrons (similar to natural scattering of radio waves in polar aurorae). Of high interest is a detected radio emission at a frequency of 44.5 mHz, which was formed during injection of electrons and modulated at the rocket rotation frequency.

The second launch of the hardware under the «Zarnitsa-2» experiment program was performed on September 11, 1975. The program of the injector operation was identical to that of the first experiment. «Zarnitsa-2» was another substantial step forward in the area of the active experiments. The data obtained on potential of the rocket were extremely important for understanding the function of the discharge processes in a near-rocket space, the information on radio emission being of an exceptional interest.

ARAKS hardware. The experience accumulated in the development of electron beam facilities installed in rockets and satellites made it possible to build an electron injector [3–5], characterised by record-braking properties for that time, for the joint Soviet–French experiment ARAKS. ARAKS is an English abbreviation of Artificial Radiation and Aurora between Kerguelen and Sogra, defining a purpose and contents of the experiment: artificial aurora and radiation on the Kerguelen island and in the Sorga settlement. At the same time,

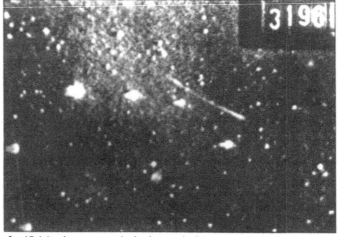

Figure 4 Ray of artificial polar aurora at the background of the starry sky in the «Zarnitsa-1» experiment. The ray is 30 km long and 2 km in diameter.

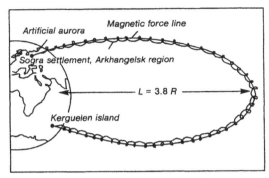

Figure 5 Magnetically conjugate points of the Kerguelen island and Sogra settlement of the Arkhangelsk region (*L* — distance to the magnetic force line in the plane of equator, expressed in the earth radii *R*).

Araks is the name of a river in Armenia, where an agreement on performing the experiment was signed. The French side was represented in the experiment by the Centre of Space Emissions and the National Centre of Space Exploration (CNES).

The choice of the Kerguelen island (France) and the Sogra settlement (Arkhangelsk region, Russia) was based on the fact that they represent a unique pair of magnetically conjugate points, both being located on land and in high latitudes of different hemispheres (Figure 5).

Two French rockets «Eridan» were used in the experiment. The research equipment of each included about 50 different devices with a total weight of more than 450 kg, the key device among them being the electron injector developed by the PWI. As to its main technological parameters, the injector is analogous to the industrial device U-250 with the U-530 gun, which was the best for that time, i.e. it provided a beam power of 13.5 kW, beam deflection to angles of ±90°, and allowed a pulse-mode operation. However, as to its dimension-weight and reliability characteristics, it was much superior to the ground analogue. So, the weight of the device without a storage battery was 120 kg, its height was 1.02 m and diameter — 0.5 m. To compare, the U-250 device had the following parameters: weight of the power cabinet — 180 kg, dimensions — 1.7 × 1.6 m; weight of the control cabinet — 250 kg, dimensions — 1.6 × 0.5 m; weight of the electron beam gun U-530 — 14.3 kg, dimensions — 0.33 × 0.16 m.

The electron accelerator is an independent device designed for injection of an electron beam at a current of 0.5 A and power of 15 and 27 keV, following a preset program. Physically, the accelerator is made of three interconnected units (Figure 6): power storage battery, instrumentation module and high-voltage unit with the electron beam gun. The high-voltage unit operates in vacuum and has the form of a monolithic module sealed with a compound based on epoxy resins. The instrumentation module houses a low-voltage part of the hardware. Block diagram of the hardware is shown in Figure 7. The power part of the accelerator is actuated from standard silver-zinc storage cells *11* with a capacity of 50 A·h. The storage battery loaded by a current of 400 A provides voltage of 36–38 V at an open-circuit voltage of 65–67 V. Control circuits and electron units are powered from control battery *15*. The program-time device (PTD) also has an independent power source, i.e. storage battery *16*.

The hardware uses a specially developed high-reliability vibration resistant triode electron-beam gun *1* with a disk-shaped lanthanum-boride cathode heated by electron bombardment. The gun is designed so that it can operate consistently under conditions of a comparatively low vacuum and features stability of three-dimensional characteristics of the beam within a wide range of accelerating voltage variations. The electromagnetic deflection

Figure 6 General view of the electron beam hardware ARAKS: *1* — electron beam gun; *2* — electromagnetic deflection system; *3* — high-voltage power source; *4* — instrument container; *5* — storage battery container.

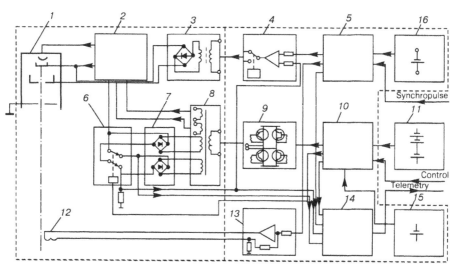

Figure 7 Block diagram of the electron beam hardware ARAKS: *1* — electron beam gun; *2* — filament source and cathode bombardment unit; *3* — beam current regulator output unit; *4* — beam current regulator; *5* — PTD; *6* — high-voltage switch; *7* — high-voltage rectifier; *8* — high-voltage transformer; *9* — power inverter; *10* — control unit; *11* — power storage battery; *12* — deflection coil; *13* — beam electromagnetic deflection current regulator; *14* — unit of converters of telemetry parameters; *15* — control unit storage battery; *16* — PTD storage battery.

system with a clearly defined pole pieces is used to inject the beam in three directions, i.e. +70, 0 and −70°.

The accelerating voltage source of the gun uses wide-band step-up transformer *8* and semiconductor rectifier *7*. Multiple sectionalisation of the source, ingenious design and technology of making of the high-voltage transformer windings and circuits for its connection to the high-voltage rectifier are employed for improving operational reliability. These measures, as well as addition of a special adjustment link provided a high-voltage power source which is non-resonant within a wide range of load current variations. The high-voltage rectifier circuit remains serviceable even if up to 15 % of its elements fail. The choice of electromagnetic and thermal loads of elements of the accelerating voltage source was based on a requirement of continuous consumption of power of 15 kW for 10 min.

Special high-voltage switch of the wafer-type with oil isolation *6* was developed to ensure two values of the accelerating voltage.

The unit for heating the electron bombardment cathode is located in the high-voltage source. It is a bombardment voltage source and bombardment current stabiliser. Corresponding windings are included into the magnetic circuit of the power transformer. The bombardment current stabiliser based on the magnetic amplifier provides accuracy of maintaining the current at a level of 13 % at supply voltage variations of 50 %. The beam current of the electron gun is regulated by supplying a control voltage gradually regulated from 0 to 8 kV to the cathode — cathode electrode gap. This is done using current regulator output unit *3* built-in into the high-voltage source.

Conversion of the direct voltage of the power storage battery into the accelerating voltage with a frequency of 1000 Hz to power the step-up transformer of the accelerating voltage source is done using power inverter *9* developed by the IED[12]. The inverter unit comprises a transistorised converter and protection units. The transistorised inverter was designed using a three-stage circuit consisting of a master oscillator, power amplifier and inverter. Regulated saturation of switching transistors via feedback circuits is used to optimise operation of power transistors over a wide range of load currents and temperature.

Beam current regulator *4* provides stabilisation and regulation of the electron beam gun current following a preset program, independently of fluctuations in the accelerating and supply voltages, variations in the ambient temperature, etc. [6]. The range of the beam current regulation is 0.05 to 0.5 A, and accuracy of stabilisation of the current is not worse than ±1 % at its variations without the regulator equal to ±30 %.

Deflection current regulator *13* is used to stabilise the angle of deflection of the electron beam in the case of fluctuations of the storage battery voltage and variations in resistance of the deflection coils caused by their heating.

Programmed variations in the electron beam gun current and deflection of the electron beam synchronised by the external quartz clock are provided by PTD developed by the PWI and manufactured by CNES. Some of the measurement instruments were located in a container separated from the main instrumentation module. Prior to launching, PTD of the injector and PTD of the separated container were synchronised. Quartz stabilisation allowed readings of the separated container instruments to be reliably related to operation of the electron injector.

Detailed earth tests and analysis of the telemetry data proved the designs used to be correct and operation of the injector to be highly reliable.

The electron injector was placed in the nose part of the French solid-fuel ballistic rocket «Eridan» which was launched from the Kerguelen island on January 26 and February 15, 1975. The set of flight instruments aboard the rocket performed detection of electrons and wave emissions, as well as control of the rocket potential. Magnetic and electric wave components were measured over a wide range. The active method, i.e. ejection of caesium plasma from the plasma generator developed by the I.V. Kurchatov AEI[22], was used to

Figure 8 Separated nose cone with research equipment and electron injector prior to mating.

compensate for a positive charge of the rocket body excited in departure of electrons. The separated nose cone with the research equipment and electron injector is shown in Figure 8. Trajectory variations were carried out using the Soviet radar unit installed on the Kerguelen island.

The first rocket was launched in a plane of the geomegnetic meridian to the north from Kerguelen. The focus was on investigation of interaction of waves and particles, as the waves are more effectively amplified in injection along the magnetic field, and on the processes relating to physics of polar aurorae.

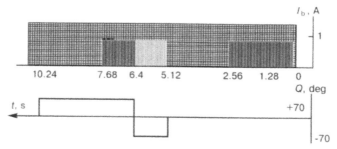

Figure 9 Cycle of operation of the ARAKS hardware: I_b — electron beam current; Q — angle of deflection of the electron beam with respect to the gun axis; t — time.

In compliance with the program of the experiment, the accelerator worked according to a cycle shown in Figure 9. The accelerating voltage was 27 kV during the first two thirds of the flight, and then was switched to 15 kV.

The second rocket was launched to the geomagnetic east to study the eastward azimuthal drift of electrons («electron echo») and the effect of the electric field on movement of these electrons. Like in the first launch, the focus was on investigation of distribution of particles by energies and angles after scattering in the atmosphere under the rocket.

Analysis of the available telemetry data showed that injectors provided calculated parameters of the electron beam and performed in full their program. The set of the instruments installed aboard the rocket provided detection of electrons and wave emissions, as well as control of the rocket potential. Special instruments were installed on board the Soviet research ship «Borovichi» to receive data from the Soviet flight instrument «Spektr» in the region of the Kerguelen island.

Each launch of «Eridan» was immediately preceded by launches of the «Arcas» rocket to a height of 80 km over the Kerguelen island. «Arcas» included parachuted detectors of X-rays caused by operation of the electron gun. These experiments were performed by Houston University, USA.

Twelve TV-units of ultra high sensitivity were installed in the Arkhangelsk region to fix the artificial polar aurorae. In addition, optical observations during the second launch were carried out from the flying lab «Yak-40». The ITMIRAS[21] radar units and radio spectrograph worked in the Kostroma and Vologda regions.

Poor weather conditions did not allow obtaining optical data, such as those which had been received during the «Zarnitsa» experiments. At the same time, the radar observations were a success. The points of arrival of the beam in the USSR territory allowed different calculation models of the magnetic field of Earth to be experimentally verified. The time difference between injection of electrons in the ionosphere over the Kerguelen island and registration of their arrival by the radar units at the magnetically conjugate point substantially differed from the calculated one. Duration of the radar signals was longer than the respective time of the beam injection. These results require further theoretical substantiation.

Interesting data were obtained concerning initiation and development of instability of plasma, i.e. the processes which are of a paramount importance for solving the problem of controllable fusion reactions. Scientific results of the ARAKS experiment are described in detail in [5]. It should be noted that ARAKS made a substantial contribution not only into plasma physics, but also into geophysics. Many new, unexpected and hard-to-explain effects were detected. They should be analysed in detail before we can state that we are in a position of performing really controllable (active) experiments in the iono- and magnetosphere of Earth.

It is opportune to note that the experience gained from creation of the flight electron beam equipment had a great influence on development of commercial equipment for welding and special electrometallurgy. Facilities which are developed now use an increasingly large number of components and assemblies which were initially tested in space hardware.

1. Paton, B.E., Nazarenko, O.K., Patsiora, S.K. et al. (1975) *See pp. 47–52 in this Book.*

2. Paton, B.E., Nazarenko, O.K., Chalov, V.I. et al. (1970) *See pp. 160–168 in this Book.*

3. Zhulin, I.A., Cambou, F. and Sagdeev, R.Z. (1977) Active experiments in the ionosphere and magnetopshere, In: *Science and Mankind*, Moscow, Znaniye, pp. 216–233.

4. Paton, B.E., Dudko, D.A., Bernadsky, V.N. et al. (1978) *See pp. 434–438 in this Book.*

5. Cambou, F., Lavergnat, J., Migulin, V.V. et al. (1978) *See pp. 468–473 in this Book.*

6. Lankin, Yu.N. and Bajshtruk, E.N. (1974) Electron beam gun current stabiliser, *Avtomatich. Svarka*, No. 2, pp. 72–73.

ARAKS — Controlled or Puzzling Experiment?*

F. Cambou[19], J. Lavergnat[20], V.V. Migulin[21], A.I. Morozov[22], B.E. Paton[1],
R. Pellat[23], A.Kh. Pyatsi[24], H. Reme[19], R.Z. Sagdeev[5],
W.R. Sheldon[25] and I.A. Zhulin[21]

The injection of an electron beam into the magnetosphere has been considered a straightforward technique for studying the large-scale structure of the Earth environment. The large current released in the magnetosphere by the ARAKS experiment has produced many results which are not yet well understood.

ARAKS (Artificial Radiation and Aurora between Kerguelen and Soviet Union), the Franco-Soviet project, was designed to study the injection of an electron beam into the ionosphere and magnetosphere. The first phase was the launch of two «Eridan» rockets from the Kerguelen island (70° 22′ W, 49° 35′ S) on 26 January and 15 February, 1975. The final stage of each rocket included two complementary experiment systems: an electron gun, indirect potential measuring devices, particle flux detectors, and a cone ejected at a speed of 40 m·s^{-1} from the main payload. This cone carried antennas to detect radio waves generated by the electron beam and its subsequent interaction with the ionosphere. Great efforts were made to have a truly controlled experiment: it was possible to change the energy and intensity of the electron beam, to vary the pitch angle during injection and accurately determine the relative trajectories of the nose cone and the main payload. This could be accomplished as previous experiments had already yielded extensive results [1–3]. Many ground-based measurement facilities were set up in conjunction with this experiment, with emphasis on optical and radar measurements in the Northern hemisphere, at the magnetically conjugate point of the Kerguelen island, as well as on the very low frequncy (VLF) and very high frequncy (VHF) measurements at both points. In addition, the «Arcas» rocket placed parachute deployed X-ray detectors at ~ 80 km above Kerguelen just before each «Eridan» launch, to verify that there was no large geomagnetic disturbance under way, and again while the electron gun was operating. The X-ray experiment was performed by the University of Houston[25].

Objectives. The study of the injection of energetic electrons in the upper atmosphere can be divided into three, each subject requiring a different technique of study, although they are complementary from a physical point of view: first, ionisation and visible phenomena (aurora) at the magnetically conjugate point of the launch site, second, dynamics of injected particles, and third, radio waves generated by the electron beam and the effects of wave-particle interactions on the beam itself. A satisfactory compromise was reached between these objectives and the constraints imposed by a space experiment. Two energies were chosen (27 keV and 15 keV) to determine the influence of magnetospheric electrostatic fields on particle trajectories.

Three pitch angles (0, 70, 140°) were used to study the following:

• the production of artificial aurorae and the atmospheric backscattering of the particles at the conjugate point;

• the magnetic reflection at the mirror point;

• atmospheric backscattering of the injected electrons in the southern hemisphere.

A variable pulse duration (20 ms, 1.28 and 2.56 s) was adapted in order to have either an accurate definition of the injection angle (20 ms) or a large amount of energy deposited in the atmosphere (2.56 s and a current of 0.5 A).

* (1978) *Nature*, **271**, No. 5647, pp. 723–726.

One of the rockets was fired towards magnetic east in order to compensate for the curvature and gradient drift of reflected electrons in the magnetic field. The other rocket was fired towards the north (26 January), first, in order to reduce the size of the impact area of the beam at the conjugate point and thus to facilitate observation of luminous phenomena, and second to obtain more information from the observation of the beam emission, due to a better trajectory of the cone with respect to the main payload.

We shall emphasise here those results which remain problematic, either in comparison with previous experiments or in relation to theoretical ideas prevailing in this field.

Phenomena observed at the conjugate point. Climatic conditions did not permit definitive observations of the optical effects in the atmosphere: the first launch could not take place during astronomical twilight, and during the second launch there was some cloud cover. It is, therefore, difficult to draw conclusions and we can only say that the brightness of the artificial aurorae in any case was not brighter than that of a *mag 7* star (the TV device sensitivity permitted the observation of stars down to *mag 9*).

Artificial aurorae were, however, clearly detected in the magnetically conjugate area by means of pulsed and continuous wave radars operating at 23 and 44 MHz, in the Kostroma and Vologda areas, Russia (during the whole first flight and during the last third part of the second flight). The measurement precision is 2 km in the north-south direction and 10 km in the east-west direction. For the first launch the results agree well with predictions made from the POGO (8/71) and GSFC (12/66) geomagnetic field models. For the second launch, the agreement between the measured and theoretical conjugate point is less satisfactory; however, we note that the magnetic activity index was higher during the second launch ($K_1 =$ $= 4$ *VS*, where $K_1 = 1$ for the first launch).

The variation in distance perpendicular to the line of sight of radar echoes is caused by the lateral movement of the rocket during electron injection. The detailed study of radar echoes coming from the impact of electrons in the neutral atmosphere is more important. We detected the presence of two types of radar echoes, which could be distinguished by a variation in the Doppler effect corresponding to a speed variation from 30 to 200 m·s^{-1}. The spectra obtained for the low-velocity component are narrower than the equivalent spectra observed in natural radio aurorae.

The time difference between the onsets of the injection of electrons into the ionosphere at Kerguelen and the onsets of the radar echo at the conjugate is surprising. Figure 1 shows the scattering of radar echo delays for the first flight for different durations of the gun pulse sequences with specific electron pitch angle of injection. The calculated time for the transit between conjugate points is 0.65 s for 27 keV electrons, and 0.85 s for 15 keV electrons. Apart from the abnormally high delay times shown in Figure 1, we also observed that radar echo durations were longer than the corresponding gun pulse durations (up to more than 10 s). Both unexpected delays and durations of radar echoes could be due to beam behaviour close to the rocket or beam-plasma interactions along the beam path or ionospheric processes at the conjugate points. The ejection of electrons without a return current would necessarily carry the main payload, including the gun, to a very high positive potential which would thus eliminate electron emission. When such an injection occurs in a rarefied environment, ionised or not, there exist processes capable of generating these return currents (collection of thermal electrons, ionisation of neutral gas and so on). To facilitate collection of the return currents, two techniques may be used: either to increase the collection surface [1], or to increase the conductivity of the environment by injecting a plasma (argon plasma was used in the «Electron Echo» experiments [2] and a caesium plasma in ARAKS). This experiment had clearly shown that plasma source causes a large scale disturbance in the payload environment, consequently exerting an influence on gun neutralisation.

The AGC (Automatic Gain Control) of the telemetry signals of the «Eridan» and «Arcas» payloads shows similar disturbances during the first minute of the active phase of

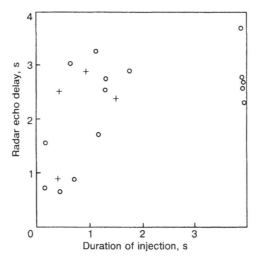

Figure 1 Radar echo delay (≈ ±0.01 s) for different duration of the gun pulse sequences with specific electron pitch angle of injection: ○ — 0°; + — 70°.

the 26 January ARAKS experiment, although the telemetry frequencies are quite different (250 and 1680 MHz) and the two payloads are 80 km apart on the same geomagnetic field line. Figure 2 shows part of these data. A large disturbance occurred in the «Eridan» signal at about 2 h 39 min 10 s UT when the plasma source was turned on and intensified 1.5 s later when the first long duration gun pulse occurred.

Among the electrons constituting the return current we can distinguish a population on all altitudes having a higher temperature than that of ionospheric plasma (0–300 eV) and another population of energetic electrons (1–3 keV) at higher altitudes (> 120–130 km). Available data are not sufficient to draw certain conclusions on the rocket potential: ~ 125 V or 1.5–2.0 kV above 120–130 km; on lower altitudes it is possible to deduce from particle measurements that the rocket potential value is less than ≈ 100 V.

The increase of the electron density near the rocket is confirmed by the existence of high frequency radio waves (10–75 MHz) with a wide bandwidth. Figure 3 shows the radio noise

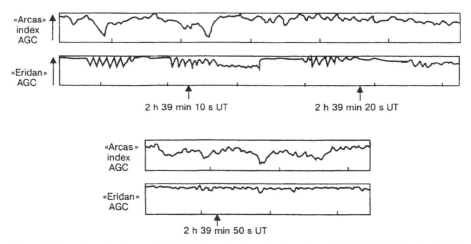

Figure 2 Telemetry signal strength (AGC) from the X-ray («Arcas») and ARAKS («Eridan») payloads. In each case the highest signal is best and the lowest is poor, about half way between these two extremes the signal is unsuitable for the recovery of useful data.

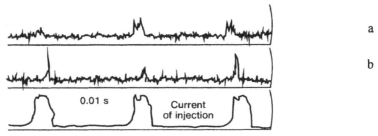

Figure 3 Radio noise at Kerguelen: a — 50 MHz; b — 75 MHz. The lowest curve represents the current of the injected electron beam.

data at two frequencies (50 and 75 MHz) detected at Kerguelen when the electron gun was operating. This noise had already been observed in the «Zarnitsa-1» experiment [3]; with ARAKS, it was generated in a direction very different from that perpendicular to the magnetic field lines. The measurements should coincide with the presence of a halo around the rocket, visible from the ground when it is at a low altitude. The present results show that the mechanisms of neutralisation, which exist whatever other conditions are involved, are so complex that these observations do not permit a full understanding of them.

Atmospheric electron scattering. Intense electron fluxes ($E > 8$ keV) were observed on the rocket by the wide angle detectors during the gun firings. For downward injection, the measured electron fluxes depend on the pitch angle of injection and on the rocket altitude. The measured fluxes at the same altitude were different for the two flights, perhaps because of large differences in the atmospheric density (Figure 4). There is a good agreement between the intensities of measured fluxes and the values calculated by a Monte–Carlo method for downward injection. For upward injection, the measured intensities of electron fluxes are several orders of magnitude higher than those calculated by the Monte–Carlo method. Thus it seems that electron scattering by the atmosphere is an important process during downward injection; at present, there is no satisfactory interpretation of the results obtained during upward injections.

Wave particle interactions; wave emission. As expected, the electron beam generates radio waves when penetrating the plasma. Among the various devices used for the study of waves generated by the beam, we used for the first time a wide bandwidth telemetry system transmitting a signal waveform up to 5 MHz. Thus we could measure the time evolution of wave spectra with high resolution in the frequency/time domain. It was found in the 0.1–1.0 MHz range that the time structure of the radio impulse repeats exactly that of the gun impulse, while this is not the case for higher frequency waves.

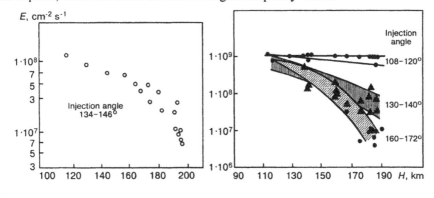

Figure 4 Electron fluxes ($E > 8$ keV) observed by the wide angle detectors for downward injections on 26 January 1975 (a) and 15 February 1975 (b).

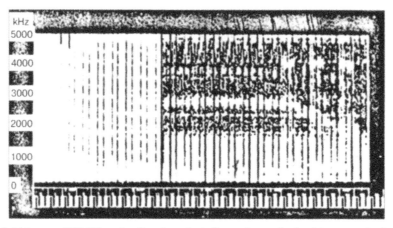

Figure 5 26 January 1975 HF results. Grey intensity indicates the amplitude of the waves the frequency of which is along the ordinate. The abscissa is the time of the flight and at the bottom the gun sequence is represented.

Figure 5 gives a general representation — analogous to a «sonagram» of the HF part of the received waves. It must be noted that during the north flight the lateral separation of the nose cone from the beam trajectory varies from 200 m up to 1.5 km. The Table summarises the observations comparing them to previous ones [2]. For the frequencies located between the plasma frequency and the upper hybrid frequency the emission due to the gun shows mainly a wave front (Figure 6) probably due to the coherent part of the spontaneous radiation [4, 5]. In the whistler mode, that is, for frequencies lower than the electron gyrofrequency, ARAKS gives an important result: even at a large distance perpendicular to the beam a continuous radiation is observed. This suggests the existence of some mechanisms by which part of the unstable modes trapped in the beam are converted into radiated modes.

For very low frequencies, a very large difference appears during ARAKS which until now has been unexplained: the caesium plasma source generates a quasimonochromatic electrostatic wave (no magnetic component can be detected). The frequency of which changes from 4.5 kHz down to 3.5 kHz. The caesium may have reached the ejectable cone where these measurements were being made.

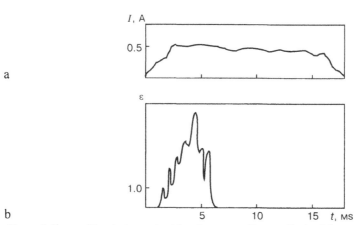

Figure 6 Shape of the electron pulse (a) and corresponding amplitude of the waves generated close to the plasma frequency in arbitrary units (b).

Comparison of «Electron Echo» experiments with ARAKS experiment.

Frequency characteristics	«Electron Echo 1» ($I = 70$ mA)	«Electron Echo 2» ($I = 70$ mA)	ARAKS ($I = 0.5$ A)
$f_p < f < f_h$	Wave front $V < 3$ μV	Continuous emission	Wave front $V < 200$ μV
$f \sim n f_b$	Experiment conducted $V < 100$ μV	Experiment conducted	$\omega \sim 4\omega_b$ mainly for the first flight
$f < f_b$	Wave front	Continuous emission $V \sim 300$ μV	Continuous emission even at $V \sim 500$ μV
LF	Wide-band noise around the LHR (argon source)		Wide-band noise with E and V component (electron gun). Monochromatic electrostatic wave (caesium source) $V \sim 100$ μV

In conclusion, ARAKS has made a remarkable contribution to plasma physics (discharge of a plasma beam generation of instabilities in an unlimited environment) as well as in geophysics (topography of the geomagnetic fields, analogy with VLF hiss). Some of these points are difficult to explain and we must examine them in detail before claiming that we are really able to perform «controlled» experiments.

1. Hess, W. N. (1969) *Science*, No. 164, pp. 1512–1513.
2. Winckler, J. R. (1974) *Space Sci. Rev.*, No. 15, pp. 751–780.
3. Cambou, F. et al. (1975) *Space Rev.*, No. 15.
4. Alekhin, J. U. et al. (1973) *Cos. Electrodyn.*, No. 3, pp. 406–415.
5. Pellat, R. et al. (1973) *C.R. Acad. Sci.*, No. 276B, pp. 685-687.

A Device for Measuring Gas Pressure During Vacuum Welding[*]

E.N. Bajshtruk[1], Yu.N. Lankin[1] and Yu.A. Masalov[1]

A special vacuum meter for measuring vacuum in the weld zone has been developed for the equipment of «Vulkan» type. It represents a thermocouple pressure gauge tube, a thermo-emf amplifier and a stabilizer of the filament current of the pressure gauge tube. Figure 1 shows the dependence of the value of the device output voltage on the pressure in the form of a calibration curve (for dry air at +20 °C).

The circuit of the vacuum meter (Figure 2) includes: a filament current stabilizer operating on transistors *T10, T11, T12*; voltage converter operating on transistors *T1, T2*; voltage converter operating on transistors *T13, T14* for supplying current to the filament of the thermocouple tube; a thermo-emf modulator operating on transistors *T3, T4*; an amplifier operating on transistors *T5, T6, T7*; a demodulator, operating on transistors *T8, T9*; a resistance temperature-sensitive element *R20* fitted to the bulb of the thermocouple pressure gauge tube *L1*.

* (1971) *Avtomaticheskaya Svarka*, No. 12, pp. 56–57.

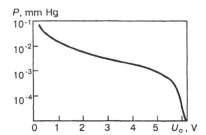

Figure 1 Calibration curve of instrument.

The diagram operates as follows. A voltage proportional to the filament current is passed from the resistance $R23$ and $R24$ and compared with a reference voltage equal to the difference between the voltage of stabilizer diodes $D5$ and $D6$. The mismatch obtained is amplified in voltage by the transistor $T10$ and in power by the two-channel emitter follower $T11$ and $T12$. The output voltage from $T12$ is converted by a modulator $T13$, $T14$ into square shape AC voltage of 10 kHz frequency. This voltage is transformed and rectified by the diodes $D7$ and $D8$ and is fed to the filament of the tube $L1$. The modulator $T13$, $T14$ energizes from the voltage converter, operating on transistors $T1$, $T2$. Thus, there is a closed system of the automatic adjustment of the filament current of $L1$ at a very high accuracy of current stabilizing in changing the filament resistance, fluctuations in the supply voltage and the ambient temperature. The divider, operating on resistors $R25$, $R26$ compensates for the supply voltage, thus increasing the accuracy of the filament stabilizing during fluctuations of the supply voltage. The filament current of the pressure gauge tube is set by a potentiometer $R23$.

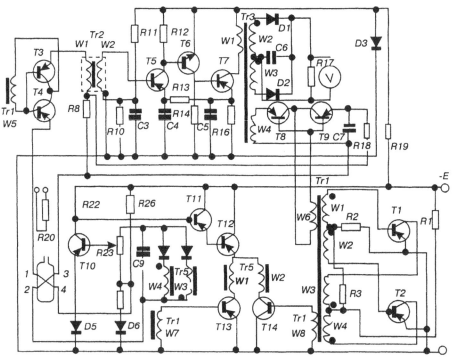

Figure 2 Schematic diagram of the instrument.

The thermocouple emf from the pressure gauge tube *L1* is amplified by an amplifier made according to a modulator–AC amplifier–demodulated circuit. The modulator *T3, T4* is supplied from the voltage converter *T1, T2*. The AC amplifier *T5, T6, T7* has direct connections and a large overall DC negative feedback to stabilize the transistor working conditions. The output voltage of the amplifier, proportional to the pressure being measured, is rectified by the diodes *D1, D2* and is supplied to an indicator.

The transistors *T8, T9* make up the modulator of the overall negative feedback as to the output voltage; this improves the linearity and stability of the amplifier.

If the ambient temperature during the device service differs greatly from +20 °C (the temperature for which the calibration curve in Figure 1 was recorded), an adjustment should be made to vacuum meter readings. The dependence of the temperature of the thermocouple pressure gauge tube filament and, consequently, the output voltage of the device on the temperature of the tube bulb can be found using the equation of the tube thermal balance.

The thermal balance equation for thermal pressure gauges without an allowance for the gas convection or losses for the filament heat conductivity can be presented in the form [1]

$$a_1(T_f - T_b) + a_2(T_f^4 - T_b^4) = 1. \tag{1}$$

We shall find the numerical values of the coefficients a_1 and a_2 by approximating with equation (1) to the experimental calibration curve for the tube (Figure 1). In the range $1 \cdot 10^{-1} \div \div 1 \cdot 10^{-4}$ mm Hg this curve is approximated by the following equation:

$$0.294p\,(T_f - T_b) + 0.374 \cdot 10^{-10}\,(T_f^4 - T_b^4) = 1, \tag{2}$$

hence,

$$p = \frac{1 - 0.374 \cdot 10^{-10}\,(T_f^4 - T_b^4)}{0.294\,(T_f - T_b)}. \tag{3}$$

We shall find temperature T_f from the device readings V and the calibration curve for the chromel-coppel thermocouple used in the LT-4M tube. The ambient temperature T_b is determined from the readings of a resistance thermometer *R20*. Substituting these data into the formula (3) we shall find the pressure corrected for the temperature (in mm Hg).

Technical data of vacuum meter:

• range of pressures measured by vacuum meter — $2 \cdot 10^{-1} \div 10^{-3}$ mm Hg;

• change in filament current of thermocouple tube heater in changing the supply voltage by ±20 % and also during 1 h operation is not more than ±0.1 %;

• range of filament current adjustment in thermocouple tube — $110 \div 120$ mA;

• measuring scheme in the device converts emf of the pressure gauge tube thermocouple into a DC signal at $0 \div 6$.V;

• the given error in measurement of emf of of the therocouple by the device measuring circuit is not more than ±2 %;

• the time constant of the device in the range $1 \cdot 10^{-2} - 1 \cdot 10^{-4}$ mm Hg is determined from expression $T = 2.5 \div 9.5 \lg p$ (where T — time constant, s; p — pressure, mm Hg);

• vacuum meter is supplied from DC mains of voltage 27 V ±20 %;

• range of ambient temperature — from 0 up to 40 °C.

1. (1967) Vostrov, G.A. and Rozanov, L.N. *Vacuum meters*, Leningrad, Mashinostroyeniye.

Powerful Electron Accelerator
for Active Space Experiments*

B.E. Paton[1], D.A. Dudko[1], V.N. Bernadsky[1], G.B. Asoyants[1], Yu.N. Lankin[1],
O.K. Nazarenko[1], V.D. Shelyagin[1], V.V. Pekker[1], V.V. Stesin[1], V.I. Kirienko[1],
E.N. Bajshtruk[1], V.K. Mokhnach[1], Yu.V. Neporozhny[1],
Yu.I. Drabovich[12] and G.F. Pazeev[12]

Introduction. Considerable experience at the PWI in the development of electron beam equipment for welding and electrometallurgy enabled the design of a series of electron beam accelerators of various amounts of power and energy which met the requirements for satellite and rocket equipment [1]. Using the electron beam device in the «Vulkan» system [2], the cosmonaut V. Kubasov accomplished, for the first time, an experiment on welding in space on board the «Soyuz-6» spaceship. In 1973 and 1975 a 3.5 kW electron accelerator, developed at the Institute, allowed the creation of an artificial polar aurora, «Zarnitza». The latest development of this kind is the electron accelerator for the joint Soviet-French experiment ARAKS.

    General description. The ARAKS electron accelerator represents a self-contained apparatus designed for injection of an 0.5 A electron beam. Energies of 15 and 27 keV can be selected using the onboard program. The accelerator is designed in the form of three interconnected units (see Figure 6 on p. 422): power storage battery, instrument module and a high-voltage unit with the electron beam gun. The high-voltage unit operates in vacuum and is designed as a monolithic module sealed with an epoxy resin based compound [3]. In the instrument module the low-voltage part of the apparatus is mounted. The height of the accelerator is 1020 mm, its diameter is 500 mm and its weight is 152 kg including storage cells filled with electrolyte. A block diagram of the apparatus is given in Figure 1. Standard silver-zinc batteries of 50 A·h capacity were used for supplying the power to the accelerator. The storage battery of 400 A current provides a voltage of 36–38 V at an emf of 65–67 V. Control circuits and electronic units of the accelerator are supplied from a control battery of 3 A·h capacity with a rated voltage of 27 V.

    Design and development of electron gun. For the injection of electrons a triode electron gun is used in the apparatus. The main difficulties encountered in the development of the electron gun resulted from the requirements that the maximum beam current remain the same (500 mA) and that the three-dimensional characteristics display minimum variations when reducing the accelerating voltage from 30 to 15 kV, taking into account an impulse modulation of the beam with the peak current value. A comparatively low and unstable vacuum (10^{-3} to 10^{-5} Torr) in the region of beam formation could result in the development of anomalous non-stationary processes (breakdowns and discharges) and in changing the three-dimensional characteristics of the beam.

    Taking into account all these requirements a lanthanum-boride disc cathode, heated by electron bombardment, was used in the electron gun. Such a solution ensures high reliability and longevity of the cathode unit, as well as stability of the three-dimensional time characteristics of the beam, provided that the probability of development of anomalous non-stationary processes can be minimised. This condition was met by enabling the beam current to pass through the anode aperture even with wide variations of the potential on the near-cathode electrode and of the accelerating voltage.

    To enable beam injection into three different directions (+ 70, 0 and –70°) with minimum variations of the three-dimensional characteristic of the beam, we developed a special

---

* (1978) *Space Science Instrumentation*, No. 4, pp. 131–137.

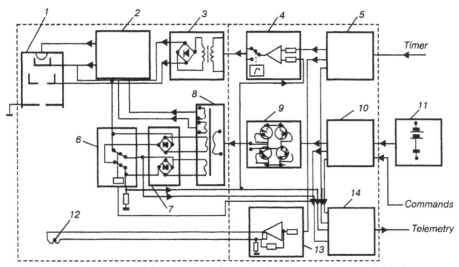

Figure 1 Block diagram of the accelerator: *1* — electron beam gun; *2* — cathode heating unit;
3 — output unit of the current regulator; *4* — current regulator; *5* — programmer; *6* — high-voltage
rectifier; *7* — high-voltage transformer; *8* — inverter unit; *9* — control unit; *10* — storage battery;
11 — deflection coil; *12* — current regulator of the deflection coil; *13* — deflection coil current regulator
unit; *14* — telemetry unit.

electromagnetic deflection system with distinct pole pieces. The deflection system is characterised by small weight (400 g) and small stray fields required by the proximity of certain devices.

A wide-band step-up transformer and semiconductor rectifier are used in the accelerating voltage power source of the gun. To increase reliability of operation, we introduced multiple sectionalisation of the power source and used a unique design and fabrication technique for the high-voltage transformer windings and its interconnection with the high-voltage rectifier. These measures, as well as the introduction of a special correcting unit, enabled us to produce a non-resonant high-voltage power source, whose output voltage profile is close to the rectified rectangular profile desired, within a wide range of load current variations. The high-voltage rectifier remains operational even when up to 15 % of its elements are defective.

Electromagnetic and thermal loads for the elements of the accelerating voltage power source are determined from the condition of continuous power consumption of 15 kW during 10 min. The allowable power of the load as a function of time of operation is given by the formula:

$$P_{max} = 47\sqrt{(1/t)} \ \text{kW},$$

where t is the operation time in minutes. Maximum load in this case should not exceed 25 kW.

A special high-voltage switch with oil insulation has been developed to ensure two values of the accelerating voltage. During switching the rectifier sections are switched from a series to a parallel connection. Application of this high-voltage switch allowed us to considerably decrease weight and size of the accelerator.

The cathode heating unit is placed in the high-voltage power source. It provides the electron bombardment voltage and stabilises the bombardment current. The traditional high-voltage transformers for the heater and for the bombardment voltage are not used in the ARAKS accelerator. The proper windings are mounted on the magnetic core of the power transformer. The bombardment current stabiliser, located on the base of the magnetic amplifier, stabilises the current to a value of 13 % when the supply voltage varies by 50 %.

It should be noted that due to internal positive feedback of the cathode circuits and heater, a system without stabilisation of the bombardment current is found to be unworkable. Regulation of the beam current of the electron gun is performed by supplying into the region between cathode and the near-cathode electrode a control potential which is smoothly regulated from 0 up to 8 kV. For this purpose an output unit of the current regulator is provided, which is mounted into the high-voltage source. This unit provides a galvanic separation between high- and low-voltage circuits and increases the output voltage of the current regulator to the level required.

Development of several versions of the design of the high-voltage unit indicated that the most acceptable insulation material is epoxy compound, whose thermal conductivity and heat resistance are high enough. Besides, the epoxy compound provides for sealing of the unit and mechanically strengthens it as well. The high voltage design permits a considerable decrease in heterogeneity of electric fields and reduction of heat and mechanical loads for the most important units.

Power inverter. The transformation of a constant voltage of the power storage battery into 1000 Hz alternating voltage for supply of the step-up transformer of the accelerating voltage source is performed by a power inverter developed by the IED[12]. The inverter unit includes a transistor converter and protection and control circuitry. The transistor converter is designed using a three-stage scheme consisting of a master oscillator, a power amplifier and an inverter [4]. The system used for control of the switching transistor saturation depth and the feedback system enabled us to optimise the operation of the inverter power transistors over a wide range of load currents and temperatures [5]. Power loss for open-circuit operation does not exceed 0.3 % of the maximum output power, while the inverter efficiency is not less than 0.94 for the specified temperature variation and load currents ranging from almost zero up to the maximum current of $I_{k\,max} = 630$ A.

The reliability required has been attained by means of providing the optimum operating conditions for the primary inverter components, by applying subsystem redundancy, and also by sectionalisation of the power components, thus introducing reserve redundancy.

Electron beam current stabilisation. Since the gun current is proportional to the accelerating voltage to the 3/2 power, in order to provide a constant current over a wide range of accelerating voltages without having to readjust the optics of the gun, automatic stabilisation is required. A current beam regulator provides stabilisation and regulation of the electron beam gun current according to a preset programme, irrespective of fluctuations of the accelerating and supply voltages, changes in the electron beam gun parameters, variation of the temperature of the surrounding medium, etc. The regulator has been designed according to a scheme using negative feedback from the load current of the accelerating voltage source. Since current passage of the gun approaches 100 %, the current of the high-voltage source is practically equal to the beam current. The regulation range of the beam current is 0.05–0.5A. The residual current at complete cut-off of the gun, with the accelerating voltage at 30 kV, is less than 10^{-3} A. Accuracy of current stabilisation is better than ±1 % at a current variation without the regulator of ±30 %.

Figure 2 shows an oscillation of a current impulse of $20 \cdot 10^{-3}$ s duration, programmed and obtained by the regulator.

For stabilisation of the electron beam angle against fluctuations of storage battery voltage and resistance changes in the deflection coils due to heating, a current deflection regulator is used. This regulator also corrects the deflection coil current against variations of accelerating voltage in such a manner as to leave the angle of deflection unchanged. The stability of the deflection angle with variation of the accelerating voltage of ±10 % is better than ±2 %.

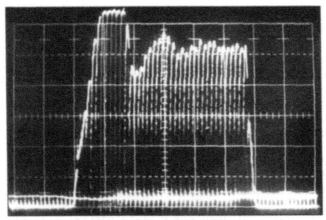

Figure 2 Oscillogram of the accelerator current pulse. Scale: *vertical* — 0.1 A/div, *horizontal* — 3.3 ms/div.

Electron beam programming. Programmed variation of the gun current and deflection of the electron beam are accomplished by a programmer developed by the PWI and designed and fabricated at CNES (France). 50 Hz positive synchronisation pulses are used by the programmer. Frequency dividers transform this sequence into impulses with intervals of 40 ms, 100 ms, 1.28 s and 12.8 s. Decoders provide the periodic waveforms required at the outputs of various control inputs. One operational cycle of the accelerator programme is shown in Figure 3.

Telemetry. The telemetry unit transforms the instrument parameter values into analog voltages between 0 and 5 V. The following accelerator parameters are telemetered: electron beam current, accelerating voltage, output voltage of the programmer, voltage on the gun control electrode, voltages and currents of power and control storage batteries, pressure in the instrument module and execution of the control commands by the accelerator. Sampling of the first three parameters occurs at 1 kHz, and the remaining signals are sampled at 32 Hz.

High voltage breakdown protection. A serious problem arising in the development and testing of the accelerator is protection from high voltage breakdowns of inter-electrode gaps in the gun because of vacuum deterioration, contamination, etc. Breakdowns in the gun

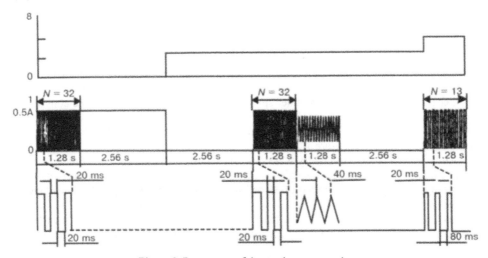

Figure 3 Programme of the accelerator operation.

are equivalent to a short circuit at a high voltage source output, accompanied by high currents, overvoltages and very high levels of interference. To prevent breakdowns in the gun the inverter is provided with a special unit switching off the accelerating voltage within $2 \cdot 10^{-3}$ s. The device restarts the inverter operation after being switched off for 0.5 s. The cathode bombardment unit, the high-voltage source and the output unit of the current regulator are also provided with special protection circuits. Interference and overvoltages arising at breakdowns not only disturb operation of the programmer, but often lead to failure of semiconductor components. Conventional methods reducing interference effects–shielding, twisting of wires, application of filters, voltage limiters, etc., are inefficient under these conditions. A radical solution of the problem is complete galvanic separation of high-voltage and low-voltage circuits.

Ground testing as well as successful operation in flight confirmed the validity of this design approach for the accelerator.

1. Paton, B.E., Nazarenko, O.K., Patsiora, S.K. et al. (1975) *See pp. 47–52 in this Book.*

2. Paton, B.E., Nazarenko, O.K., Chalov, V.I. et al. (1970) *See pp. 160–168 in this Book.*

3. Shelyagin, V.D., Lebedev, V.K., Zaruba, I.I. et al. *Devise for electron beam welding.* USSR author's cert. 206991, Int. Cl. B 23 K, priority 22.02.66, publ. 08.12.67.

4. Pazeev, G.F., Drabovich, Yu.I. and Mikhalskaya, V.F. *Transistorised inverter.* USSR author's cert. 413589, Int. Cl. B 64 G 9/00, publ. 03.02.74.

5. Drabovich, Yu.I., Ponomaryov, I.G. and Pazeev, G.F. *Compensation direct voltage stabiliser of a series type.* USSR author's cert. 408289, Int. Cl. B 23 K 15/00, publ. 07.08.73.

High-Power Small-Size Electron Beam Device for Technological Operations and Physical Experiments in Space*

B.E. Paton[1], V.D. Shelyagin[1], O.K. Nazarenko[1], Yu.N. Lankin[1], V.K. Mokhnach[1], Yu.V. Neporozhny[1], E.N. Bajshtruk[1], V.I. Kirienko[1], V.I. Pekker[1], Yu.I. Drabovich[12] and G.F. Pazeev[12]

The PWI in collaboration with the IED[12] developed a high-power small-size electron beam device and tested it under earth conditions and near-earth space. Such devices were utilised for injection of electrons into space during the Soviet–French geophysical experiment under the ARAKS project. Such equipment can be used for electron beam welding or performance of other technological operations under space conditions. Physically the device consists of three parts of approximately equal weight: primary power supply *I*; the instrumentation module with inverter unit *II*, and the high-voltage electron accelerator *III*. A block diagram of the device is shown in Figure 1.

An independent storage battery was employed in the ARAKS experiment as a *primary power supply.* The battery was of single application and was charged under earth conditions. The storage battery was assembled of 36 cell jars of the silver-zinc type. Rated voltage at the discharge current of 400–20 A was 40 V at a cell open-circuit voltage of 65–67 V. Further increase in voltage was limited by the parameters of the transistors used in the inverter. It should be noted that, according to the experimental conditions, the electron beam should be

* (1984) *Space research in Ukraine*, Issue 18, pp. 3–9.

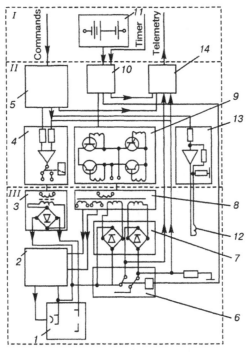

Figure 1 Block-diagram of electron beam device: *1* — the electron beam gun; *2* — cathode heating unit; *3* — high-potential part of the beam current regulator; *4* — beam current regulator; *5* — control unit; *6* — high-voltage switch; *7* — high-voltage rectifier; *8* — high-voltage transformer; *9* — inverter unit; *10* — program-time device; *11* — power storage battery; *12* — deflection coil; *13* — deflection coil current regulator; *14* — telemetry unit (see designations *I–III* in the text).

injected following a complex program, an operation cycle of which is shown in Figure 9 (see p. 424). The power of the electron beam, energy of electrons and direction of injection had to be periodically varied, the time of variations being varied as well. Voltage of the battery was varied from 67 to 38 V with variations in power consumption from 0.01 to 1.00. Such variations in voltage of the primary power supply created large difficulties for development of many components of the device.

An auxiliary battery consisting of 18 cells of the silver-zinc type was added to the device to ensure a more stable power supply to the electron units, the control circuits of the inverter and the switches. The rated voltage of this battery was 27 V, the total energy capacity of the power battery being about 10400 kJ. This energy could be used in the course of the experiment which lasted up to 380–400 s. The weight of the primary power supply in a sealed container was 40 kg, the diameter of the container being 500 mm and height — 235 mm.

The instrumentation module housed the inverter unit, electronic units of the beam and deflection current regulators, program-time device, telemetry unit, and control unit with its switching equipment and storage battery.

The instrumentation module is sealed, and prior to flight is filled with dry nitrogen at a pressure of 1.2 atm. The weight of the module is 55 kg, diameter is 500 mm and height is 490 mm.

The inverter unit provides conversion of the direct voltage of the primary power supply into an alternating square-wave one, delays supply of the alternating voltage to a step-up transformer by the duration of heating up of a cathode of the electron beam gun at switching on of the electron beam equipment, stops supply of the alternating voltage to the step-up transformer at electrical breakdown of the inter-electrode gaps of the electron beam gun, and resets the alternating voltage after a preset time interval. In accordance with the functions it performs, the inverter unit can be conventionally subdivided into two parts: transistorised direct to alternating voltage converter, and switch-on delay and short-circuit protection device. The weight of the inverter unit is 40 kg.

The transistorised direct to alternating voltage converter designed for operation as part of the flight electron beam equipment should feature stable operation over a wide-range of variations in the supply voltage, load current and temperature. In the ARAKS experiment, the voltage of the power supply was varied by a factor of 1.8. The load current was periodically varied from zero to a maximum value, and in some modes of operations — to a short-circuit current. The possibility of removal of heat from the converter to structural elements of the instrumentation module was extremely limited. Therefore, the converter had to operate during the entire experiment almost completely with the intrinsic heat capacity of its own structure. The weight of the converter in accordance with general requirements to the flight electron beam equipment should be a minimum. It follows from the above-said that efficiency of the converter should be a maximum in all operational modes.

At optimal saturation coefficient g_{opt} of power transistors, their power factors are minimum [1]. For high-power silicon transistors used to build converters, g_{opt} ranges from 1.4 to 2.5, depending upon operational conditions. In transistorised converters assembled on the basis of a push-pull circuit, saturation coefficient g_{opt} of the transistors varies over very wide ranges, depending upon the operational conditions, and may amount to 100 and more. This leads to a substantial decrease in efficiency, increase in dimensions and weight [2]. In the direct to alternating voltage converter under consideration, the saturation coefficient of power transistors during operation is automatically maintained at a level close to the optimal one. This is done by the method of maintaining g_{opt} based on utilisation of the dependence of voltage U_{c-e} between the collector and emitter electrodes of the transistors upon the depth of their saturation. The main point of the method is the fact that voltage U_{c-e} is constantly kept at a preset level when the transistors are open [3]. The level of stabilisation of U_{c-e} is selected so that the saturation coefficient of the transistors is optimal at a maximum load current $I_{l\,max}$. Al lower load currents I_l the saturation coefficient of the transistors decreases and approaches unity at $I_l \rightarrow 0$. This is caused by the fact that voltage U_{c-e} depends not only upon g, but also, to a considerable degree, upon I_l.

The converter switches on the master oscillator (MO), power amplifier (PA), power amplifier supply voltage regulator, transistor saturation data reading circuit (SDR), reference element and preset voltage source. Power is supplied from two voltage sources. The step-up transformer serves as a load. Maintaining the saturation coefficient at a level close to the optimal one is done as follows. Voltage between the collector and emitters, U_{c-e}, in the open state of the transistors is measured during operation of the converter using the SDR circuit. Output voltage of this circuit is fed to the reference element, where it is compared with the preset one, and the error voltage is fed to the input of the power amplifier supply voltage regulator to change its output voltage. This causes a change in the PA output voltage and current flowing through basic circuits of power transistors of the converter. In turn, this leads to a change in voltage U_{c-e} of such transistors. The process continues until U_{c-e} becomes equal to the preset value. If the preset voltage is selected so that it fits g_{opt} for $I_{l\,max}$, efficiency of the transistors at any load current is within a range of $I - g_{opt}$. Figure 2 shows an experimental dependence of efficiency of the converter used in the ARAKS experiment upon the current. It is seen from the Figure that within a range of currents from a maximum value of 630 A to 0.1 $I_{l\,max}$ the efficiency of the converter was not lower than 0.94, while at a rated current of about 430 A the efficiency amounted to 0.96. The total power consumed by the inverter unit at open-circuit voltage and maximum values of supply voltage of power and control circuits was not in excess of 180 W, which was about 1 % of the rated output power.

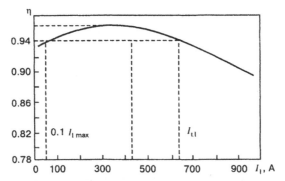

Figure 2 Experimental dependence of the efficiency of the transistorised converter upon the current in the primary winding of a high-voltage transformer.

The switch-on delay and short-circuit protection device includes current pickup, time-setting network and relay assembly. The current pickup is connected to the circuit of primary winding of the step-up transformer, and contact set of the relay assembly is switched on by the master oscillator and the converter power amplifier.

Switching on the supply voltage initiates charging of the time-setting network capacitor. When voltage at the capacitor reaches a certain value, a switching on signal is fed from the output of the time-setting circuit to the relay assembly. Its contact set connects MO to input of PA to feed voltage to the primary winding of the step-up transformer. At the same time, the time-setting network comes back to the initial state.

When the anode current of an electron beam gun and, hence, the current in the primary winding of the step-up transformer achieve a maximum value, a signal is fed from the current pickup to the relay assembly and time-setting network to switch off the relay assembly and switch on the time-setting network. The contact set switches off PA and MO. Voltage at the primary winding of the step-up transformer and current flowing through it become equal to zero. In a certain period of time the time-setting network gives a signal to switch on the relay assembly, then the alternating voltage will again be fed to the primary winding of the step-up transformer.

Parallel connection of transistors is used to switch currents with a frequency of 1000 Hz in the inverter unit. In this case reliable operation was ensured by stand-by redundancy and special sorting of transistors used at the stage of incoming control [4]. Quick-break fuses were connected to the emitting and base circuits of each of the transistors to ensure automatic switching off of the faulty ones [5].

The high-voltage electron accelerator is in vacuum during operation. Physically, it is made in the form of a monolithic module in which the following units are made integral and sealed using an epoxy compound: electron beam gun, broad-band step-up transformer and semiconductor rectifier, high-voltage switch, power electron heating stabilisation units of the gun, high-potential part of the beam current regulation unit, accelerating voltage dividers and beam current pickups, high-voltage rectifier and beam current regulator protection devices. The weight of the high-voltage electron accelerator is 42 kg.

Electron beam gun. A three-electrode electron beam gun was used, which allowed a pulsed modulation of the beam current and stabilisation of its value at a high accuracy (up to ±1 %). The injection gun is made in such a way that, at zero potential at the control electrode and an accelerating voltage of 15 kV, the gun provides a beam current equal to 0.55–0.65 A, and at 27 kV the beam current is 1.3 A. To ensure a complete blanking of the gun at an accelerating voltage of 30 kV, a voltage of 6 kV should be supplied to the control electrode.

In the ARAKS hardware, voltage of the primary power supply increased to a substantial degree with decrease in the consumed power. Accordingly the accelerating voltage increased to 50–56 kV. Therefore the voltage at the control electrode of the gun had to be increased to 9 kV. The cathode of the gun was made from lanthanum boride and was electron heated, power of heating being 40 W. The gun forms a beam with a diameter of about 1.5 mm in the region of a crossover with a half divergence angle of about 1×10^{-2} rad. The gun can be equipped with a coil for additional electromagnetic focusing. The deflection system allows the beam to be deflected along one coordinate within the limits of $\pm120°$. The weight of the gun with the deflection system is 2.6 kg.

The accelerating voltage source consists of a broad-band transformer, semiconductor rectifier, high-voltage switch, voltage dividers and current pickups.

Transformers constitute a considerable part of the high-voltage source. Their weight can be decreased by increasing electromagnetic loads and frequency.

In the ARAKS hardware, the rated frequency of output voltage of the inverter was 1000 Hz, and voltage had a rectangular shape. Choice of this frequency was based on a comprehensive study of its effect on all components of the device. Further increase in frequency was limited by the absence of high-voltage high-frequency rectifying diodes, which were not available at that time. The progress of semiconductor engineering will allow the frequency to be increased to 10–50 kHz.

Increase in electromagnetic loads leads to a decrease in weight and dimensions of the transformer, although the losses in it increase. Because of the short time of the experiment, the choice of electromagnetic loads based on the condition of permissible overheating was not approved as a criterion for calculations. It is advisable to choose electromagnetic loads based on the condition of a minimum weight of the storage cell–transformer system, followed by monitoring overheating of the device.

Calculations show that within the accepted frequency range the induction should be selected so that it is as close as possible to that of saturation, being not in excess of this value at permissible variations in voltage. Dependence of the system weight upon the current density has a quiet minimum. Under the accepted conditions, in the case of short duration of operation, the optimal current density can be considered with a sufficient accuracy to have an inversely proportional dependence.

The rectangular shape of the curve of primary voltage has a substantial effect on output parameters of the high-voltage source. If a transformer is designed with no allowance for free oscillations of the windings, the amplitude of the secondary voltage can be 2.5 times as high as the calculated amplitude by the voltage transformation coefficient, depending upon the load of the high-voltage source. Excessive voltages can cause failure of valves, break-down of the transformer windings or discharge gap in the gun, etc.

The existing calculation methods and methods of adjustment of a frequency characteristic of radio frequency and pulse transformers fail to be sufficiently helpful in our case. The weight of transformers as calculated by the above methods is usually larger. Decrease in overvoltages by matching resistances of the loads and source due to artificial increase in resistance of the source, as well as addition of padding resistors on the high-voltage side, are inexpedient, as they lead to fundamental power losses.

The special calculation method [6] and design of a high-voltage accelerator were developed. A desirable increase in the attenuation coefficient and frequency of natural oscillations of the resonant circuit of the accelerator is done through correction of distribution stray capacitances [7]. In this design the high-voltage insulation worked not at the alternating voltage but at the direct one, which is uniformly distributed over all the elements. This allowed the weight of the high-voltage insulation to be decreased, reliability of operation to be improved, and a high-voltage source to be built of minimum weight with a stable secondary voltage over the entire range of variations in the beam current.

Multiple sectionalisation of the source was done, and a special method and technology of manufacture of the high-voltage winding were employed to improve reliability of its operation.

According to the experimental conditions, the accelerator should inject electrons at two accelerating voltages of 13.5 and 27 kV and corresponding currents of 1.0 and 0.5 A.

Switching of the modes is done using a special small-size high-voltage switch of a wafer type. It switches over two groups of the high-voltage rectifiers of bridges from a parallel to series connection.

The high-voltage source houses a unit of electron heating of the gun cathode and a unit of stabilization of the electron heating current. Cathode filament supply and electron bombardment windings are located on cores of the power transformer, and the electron bombardment current is stabilised by a magnetic amplifier with a parametric support element connected to its control winding [8]. The electron heating unit is protected from discharge of stray capacitances of the source, should breakdowns occur in the gun. The high-potential modulator unit is built into the source to regulate the beam current of the electron beam gun. This unit includes a high-frequency isolation transformer, rectifier and modulator elements which prevent failure in the case of breakdowns in the gun. The frequency of the supply voltage of the modulator is 20–25 kHz, and the amplitude of the signal fed to the control electrode of the gun is 10 kV. The device comprises different voltage dividers and current pickups which give indications of operation of the injector.

Automatic stabilisation is required to ensure constancy of the beam current over a wide range of variations in accelerating voltage without readjustment of optics of the gun. The regulator of the beam current provides stabilisation and regulation following a preset program of the electron beam gun current independently of fluctuations of the accelerating and supply voltages, variations in parameters of the electron beam gun, temperature, etc. A high-voltage source current almost equal to the beam current was used as a regulated parameter, because, as shown by the experiments, deposition of the beam at the anode is less than 1 % of the total current. A signal proportional to the beam current is compared with the preset one, their difference is amplified and converted into an alternating voltage with a frequency of 30 kHz, and fed to an isolation high-potential transformer. Rectified voltage of this transformer is fed to the control electrode of the electron beam gun. The regulator provides a current regulation range of 0.05 to 0.5 A, the residual current at a complete blanking of the gun and accelerating voltage of 30 kV is no more than $1 \cdot 10^{-3}$ A. Accuracy of stabilisation of the current amounts to $\pm 1\%$ at a variation in the current without a regulator equal to ± 30 %. Dynamic characteristics of the regulator are shown in Figure 3, which gives an oscillogram of reproducing a pulse of the current by the regulator. High-frequency (2 kHz) pulsations of the current are caused by pulsations of the accelerating voltage. A deflection current regulator is used to stabilise the angle of deviation of the electron beam at fluctuations of voltage of the storage batteries and variation in resistance of the deflection coils because of heating. This regulator is made following the normal diagram of the current source with a deep negative feedback by a regulated parameter. The regulator performs a correction of the deflection coils current at a variation in the accelerating voltage. As a result, stability of the deflection angle is ± 1 %, variations in the accelerating voltage being ± 10 %. The programmed variation in the gun current and deflection of the electron beam synchronised by external quartz clock are provided by the program-time device (PTD). Synchronising positive pulses with a frequency of 50 Hz are fed to the PTD cathode. Frequency dividers convert this voltage into pulses with an interval of 0.04, 0.10, 1.28 and 12.80 s. Periodic voltages are formed by means of decoders at outputs of the corresponding control channels. One cycle of the program of operation of the accelerator set by PTD is shown in Figure 9 (see p. 424).

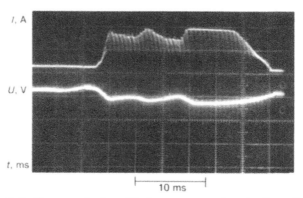

Figure 3 Oscillograms of pulses of the beam current and accelerating voltage.

The unit of conversion of parameters checked by telemetry provides their conversion into a unified analog voltage of 0–5 V. The following parameters of the accelerator are checked by telemetry: electron beam current, accelerating voltage, PTD output voltage, voltage at the control electrode of the gun, voltages and currents of the power and control storage batteries, pressure in the instrumentation module and passage of external commands of the accelerator.

A serious problem that arose in development and testing of the accelerator was its protection from high-voltage breakdowns of the inter-electrode gaps in the gun, which are caused by deterioration of vacuum, contamination, etc. A breakdown in the gun is equivalent to a short-circuit at the output of the high-voltage source, which is accompanied by high currents, overvoltages and a very high level of noise. At breakdowns in the gun the inverter switches off the accelerating voltage during a time of $3 \cdot 10^{-3}$ s. The automatic resetting device resumes operation of the inverter 0.5 s after switching off. The cathode bombardment unit, high-voltage source and output unit of the current regulator are equipped with special protection circuits. Noises and overvoltages arising in breakdowns, in the case where no special precautions are made, make operation of the digital control equipment impossible and lead even to failure of the semiconductor elements. Traditional methods for decreasing the effect of noise, such as screening, twisting of wires, using filters, voltage limiters, etc., turned out to be inefficient. The problem was solved by complete galvanic isolation of high- and low-voltage circuits. Galvanic isolation of analog signals was done using the modulator–transformer–demodulator circuit, and logic signals were transferred via optrons.

Earth and successful flight tests confirmed the reliability and expediency of the accepted design of the device. Relationships we selected proved to be correct, the system was reliably reproduced in accordance with the program.

1. Drabovich, Yu.I. (1967) Comparison of switching on circuits of transistors in the state of saturation, In: *Electromagnetic processes in conversion devices*, Kiev, Naukova Dumka, pp. 186–192.

2. Pazeev, G.F. (1975) Dependence of efficiency of the transistorised supply direct to alternating voltage converter upon the load current value, In: *Up-to-date tasks of converting equipment*, Kiev, IED, Part 6, pp. 335–367.

3. Pazeev, G.F., Drabovich, Yu.I. and Mikhalskaya, V.F. *Transistorised inverter*. USSR author's cert. 413589, Int. Cl. B 64 G 9/00, publ. 03.02.74.

4. Maslobojshchikov, V.S. and Drabovich, Yu.I. (1986) Distribution of currents in parallel-connected transistors in the state of saturation, In: *Issues of theory and calculation of conversion engineering devices*, Kiev, Naukova Dumka, pp. 78–82.

5. Drabovich, Yu.I., Ponomaryov, I.G. and Pazeev, G.F. *Compensation direct voltage stabiliser of a series type.* USSR author's cert. 408289, Int. Cl. B 23 K 15/00, publ. 07.08.73.

6. Shelyagin, V.D. (1972) *Investigation and development of some types of power sources for electron beam welding,* Synopsis of Cand. Techn. Sci. Thesis, Kiev, 36 pp.

7. Shelyagin, V.D. *Power supply device for electron beam unit.* USSR author's cert. 280720, Int. Cl. B 64 G 9/00, priority 15.07.69, publ. 23.06.70.

8. Shelyagin, V.D., Mokhnach, V.K., Popov, V.A. et al. *Cathode unit of electron-optical welding system.* USSR author's cert. 493094, Int. Cl. B 23 K 15/00, priority 09.08.73, publ. 28.07.75.

Chapter 5

LARGE-SIZED STRUCTURES
IN SPACE

Chapter 5. Large-Sized Structures in Space

Exploration of space is now at the stage of long-time operation of orbital manned stations, expansion of their functional capabilities and carrying out scientific experiments, which require fabrication, erection and other operations of large-sized structures in space.

Technologies, equipment and tools, involving welding and other methods for making permanent joints, are available which allow construction of such structures in open space. The use of welding or brazing for joining structural members enabled fabrication of rigid truss structures of a stable shape to solve various problems of space technology.

Technologies were developed for the fabrication of welded transformable (shell) structures which found wide application, along with truss structures. The use of transformable structures makes it possible to overcome various difficulties associated with transportation of space vehicles to orbit and their subsequent assembly.

The problem of power supply to orbital stations has been practically solved by the application of large-sized reusable solar batteries, which are assembled and deployed directly in orbit.

Large-sized structures of different shapes are the basis of large orbital complexes, which are especially needed now in a period of rapid scientific-and-technical progress.

On Application of Transformable Welded Structures for Space Systems and Constructions*

B.E. Paton[1], V.M. Balitsky[1], V.N. Bernadsky[1] and V.N. Samilov[1]

Progress of engineering is often accompanied by a contradiction between the logical need of increasing the weight and dimensions of different objects and limited capabilities of their movement and transportation over short and long distances. This contradiction is especially strong in aerospace engineering, where the weight and dimensions of the objects essentially influence the practicability and harmony of the engineering solution.

Dimensions of the systems are primarily determined by the process tasks and functional features of the space object. Selection and specifying of the dimensions in many cases is influenced by the awareness of certain limitations in the technical capabilities of the task implementation. For instance, the recommended dimensions of areas and volumes of premises for space expeditions differ drastically from each other. Despite the psychological aspects of selection of the initial parameters, solution of the future tasks can hardly be achieved without an essential increase in the dimensions of the space systems.

Large-sized objects in space and on other planets can be created using three basic geometries, characterised by simplicity of their determinant features. These geometries can exist independently or in all the possible combinations. Firstly, space objects can be assembled from individual finished sections or modules. The dimensions of the modules are determined by the weight and dimensional characteristics of the cargo ship. The complete readiness of the modules allows their use immediately after docking and pressurising of the docking unit.

Secondly, space objects can be delivered by the cargo ship in a compact folded state. As a result of conversion or transformation of their shape and dimensions, the system is brought into the working condition and is ready for use.

Thirdly, space objects can be assembled from individual, as a rule, enlarged elements. Unlike the first two, this method is the most difficult to implement, as it involves considerable consumption of skilled labour and time.

The diversity of the outstanding tasks in the conquest of space predetermines the use of all these geometries, as is the case on Earth. The rationality of applying a particular one of them is determined in each case by the specifics and totality of the initial conditions, and guarantees of the reliability and cost effectiveness of the engineering solution.

Incrementing the object by finished modules is made possible, naturally, by a simple and regular operation of combining the individual viable units. This solution is quite perfect. This is how the orbital laboratories «Soyuz–Soyuz», and then «Soyuz–Salyut» and «Soyuz–Apollo» were created. Wide use of modular systems in different project proposals is logical in the investigation and conquest of space, for instance, in lunar bases [1] and large orbital stations [2].

Transformable structures are becoming more and more important alongside modular systems. Convertible or transformable structures are manufactured on Earth in a compact folded state. In such a convenient form for transportation they are delivered to their destination and are transformed there into the working condition, in terms of dimensions and shape. The degree of fitness for purpose of the transformed structure depends on the perfection of its engineering solution.

Looking for optimal solutions, the developers of space systems are more and more often turning to the principle of transformation. A number of transformable structures are in

---

* (1978) *Production and behaviour of materials in space*, Moscow, Nauka, pp. 29–36.

operation currently and are successfully fulfilling their purpose. Transformable antenna systems are the most widely represented in the practical performance of space flights. Space vehicles of «Venera» type are fitted with a deploying parabolic aerial. Rod [3, 4] and parabolic aerials from elastically coiled tubes [5] are becoming accepted.

The spherical surface of a passive communications satellite is formed [6] by straightening a metal net folded into a tight package, which occurs under thermal impact, due to the initial shape memory of the material of the net deformed rods. Shape memory materials are highly promising for development of transformable structures. While preserving the continuity of the item body in the folded state, they simultaneously possess a latent ability of thermal energy conversion, thus providing the sought change of shape.

Metallic three-dimensional latticed systems for antennas of a specified configuration and curvature are produced as a result of deployment of pre-folded structures [7], which occupy quite a small volume. Rigid three-dimensional shapes are transformed into foldable shapes by incorporating special hinges into all the components and temporarily into the middles of some part of the panels. The most accessible and usual is compact packaging of structures made of easily bent soft materials. Therefore, quite numerous are the developments of expanding structures from cloths and films [8, 9]. Soft shells are formed by pressure, created inside closed cavities. Being incapable of taking up the compressive force, the structures of soft materials preserve their shape mainly at the expense of excess pressure.

The possibility of reduction of the deformation of inflatable structures and simultaneous improvement of the protective qualities of the walls is implemented in development of multilayer shells with a filler. Foam materials are introduced into the interwall spaces and specially envisaged cavities, which later on solidify, turning flexible structures into rigid ones.

The basic carrying and protecting elements of space vehicles and apparatus are all kinds of metallic shells. The ability of temporarily changing the shape and, therefore, also the dimensions of the welded metallic shells, would allow expansion of the field of transformable structures application and would yield significant benefits in terms of delivery of large-sized objects and systems.

The geometrical laws of surfaces bending can be used for analytical studies of the change of the shape of thin-walled metallic shells [10, 11]. Surface bending is such a continuous deformation, when its metrics characterised by a quadratic form, remains unchanged.

If surface Q assigned in orthogonal curvilinear coordinates α, β is converted into a family of surfaces Q', then the first quadratic form expressed as

$$ds^2 = A^2 d\alpha^2 + B^2 d\beta^2, \tag{1}$$

where A and B are the coefficients of the first quadratic form, which are α and β functions of the surface, remains unchanged for each of points α, β of the surface. In this case, there exists a mutual correspondence between these surfaces, at which any two corresponding curves on them have the same lengths. The isometry of transformation achieved here satisfies the condition of equality of zero membrane deformations, so critical for changing the shape of thin-walled shells.

The first quadratic form determines the internal geometry of the surface, characterising it with an accuracy up to bending. In order to choose a certain hardened surface, it is necessary to know the second quadratic form that characterises the main curvatures on the surface and, therefore, its geometrical connection with a three-dimensional space.

The interconnection between the first and second quadratic forms is established in the most complete manner by Codazzi–Gauss differential equations:

$$\frac{\partial}{\partial\alpha}\left(\frac{1}{A}\frac{\partial B}{\partial\alpha}\right)+\frac{\partial}{\partial\beta}\left(\frac{1}{B}\frac{\partial A}{\partial\beta}\right)=-\frac{1}{R_1 R_2}AB,$$

$$\frac{\partial}{\partial\alpha}\left(\frac{1}{R_2}B\right)=\frac{1}{R_1}\frac{\partial B}{\partial\alpha}, \qquad \frac{\partial}{\partial\beta}\left(\frac{1}{R_1}A\right)=\frac{1}{R_2}\frac{\partial A}{\partial\beta}. \tag{2}$$

The geometrical meaning of the equations consists in that the values A, B, R_1, R_2 as functions of coefficients α, β of surface points, cannot be arbitrarily assigned. For every surface the main curvature radii R_1 and R_2 are connected with coefficients A and B of first quadratic form of relationships (2), while the complete curvature is fully determined only when the first quadratic form is assigned (1).

During the process of change of shape at geometrical bending of the surface, the condition should be satisfied that in the areas of continuity of initial surface Q, the surface of family $Q'(t)$ also preserved its continuity. Ruptures, however, can develop for lines tangential to the surfaces being bent. In deformation, folds and ridges can form along certain lines, that will be either fixed on the surface, or will shift over the surface with change of parameter t.

Thus, two classes of isometric transformations can be singled out. One of them belongs to closed surfaces, to surfaces fixed on the contour or having connections in their area. Such surfaces can be converted only by local bends, accompanied by disturbances of regularity in the places of formation of the ridges, they are bendable only in the class of piecewise-regular surfaces.

Piecewise-regular isometric conversion is produced at dissection of the convex surface by a plane, and reflection of part of the surface relative to this plane, at which the surface moves from one side of the plane to the other one. Figure 1 shows a photograph of a welded spherical shell in the form of circumferential waves corresponding to its multiple mirror reflection.

Broader possibilities of isometric change of shape are inherent to an open surface not having in its area or on its boundary, any bonds limiting the displacements. Transformation of such surfaces can be performed not only by local bending, but also by global deformations, leading to a change of curvature of the surface in the entire area.

The laws of geometrical bending of the surfaces for the welded metallic shells are implemented in practice in the developments of the PWI. The method of coiling flat welded panels and pipes became widely accepted especially in tank construction. Change of the

Figure 1 Welded metal spherical shell in the folded state.

Figure 2 Two-cone shell in the folded (a) and transformed (b) states.

shape of the flat panels is the simplest, particular case of the general law of shell bending, when one of its main curvatures is always zero and the surface is deployable into a plane. Portable tubular aerials are deformed by the same principle due to the high elasticity of the material.

Multi-cone shells can be named as an example of an unusual solution for transformable structures, produced by bending flat panels. Figure 2 shows the photographs of two flat metal disks, whose shape is the development of a conical surface. The discs are connected by a weld around a circular contour. If the edges of the cut-out are drawn together, by the impact of either external forces or the forces of elasticity, or shape memory of the material, the flat elements turn into a shell made up of two cones. A flat system of several disks can be turned into a multi-cone shell in a similar manner. In order to achieve the leak-proofness of the formed volumes, it is necessary to weld the closing butt along the cone generatrices.

The laws of bending of surfaces of a non-zero Gaussian curvature allow folding and deployment of the most diverse kinds of shells. Figure 3 shows a welded metal shell of a

Figure 3 Welded parabolic shell in the transformed (a) and folded (b) states.

a

b

Figure 4 Toroidal shell in the folded (a) and transformed (b) states.

dual curvature in the folded state. After transformation from the folded state, the shell resumes the parabolic form of its surface, so badly needed for the antennas and solar concentrators.

Changing the curvature by certain laws allows a thin-walled metal toroidal shell to be folded into a comparatively compact coil without fracture. The shell folded in the form of a spiral (Figure 4) is transformed for instance by internal pressure into a toroidal shell of a regular shape, having temporarily sealed the section end faces for this purpose.

Investigation of the processes of changing of the structure shape yielded solutions, incorporating features inherent to the shell and truss systems. Two similar welded metal panels (Figure 5), placed one onto the other, are joined by a tight butt weld around the entire contour. Lines of a lower rigidity, having a lower bending resistance, i.e. linear hinges, are singled out on the panels, for instance, by increasing the rigidity of certain areas of the surface.

A purposeful change of the welded structure shape can be achieved by arranging linear hinges on the panel surface in a certain pattern and by excess pressure, induced in a closed cavity between the panels. A three-dimensional structure with a complex-shaped surface is formed of flat panels. Surface areas, remaining flat, are its truss elements, while those which became curvilinear, are the shell elements. Absence of any significant resistance in transformation of a truss-shell structure is indicative of the preserved isometry of its median surface during the change of its geometry. The degree of transformation is usually evaluated by the ratio of the two extreme values of the most characteristic changing dimension of the structure. Coefficient of transformation of the truss-shell systems by their height varies between 15 and 25.

If a unit system, formed of two flat panels, turns out to be insufficient for solving a particular task, it is possible to combine such unit elements into a multi-section systems. The volumes of multi-section systems corresponding to one or several sections, can be easily separated from each other by air-tight membrane-partitions.

Transformable welded structures have a common feature of organisation and regularity of the transformation processes. These qualities enable elements of controllability in the

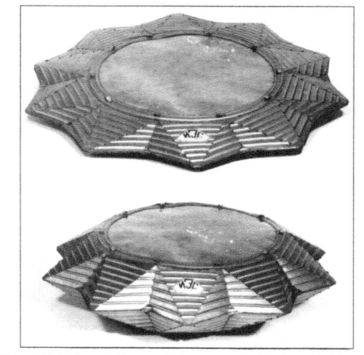

a

b

Figure 5 Welded truss-shell structure in the transportation (a) and working (b) states.

transformation processes to be found. The possibilities for automation and control in erection of large space systems are essentially inherent in the transformable structures.

Substantiation and regularities of transformation of welded metal shell structures in the case, when they have not yet found practical acceptance, have been verified and confirmed by the experience of using the working models of welded structures, which rather completely preserve the scale of dimensions, process parameters and qualitative characteristics of the used materials in relation to the anticipated full-scale prototypes.

1. Stambler, I. (1963) *Space, Aeronautics*, **40**, No. 7, pp. 60–67.

2. Parker, P.J. (1973) *Space Flight*, **15**, No. 3, pp. 82–86.

3. (1967) Extensible STEMS in space, *Space Flight*, **9**, No. 10, pp. 346–347.

4. (1970) The Satellites ATS-F and ATS-G, *Electronics*, **43**, No. 9, pp.7–9.

5. Pimrott, F.P.I. (1965) *Machine Design*, **37**, No. 28, p. 156.

6. Johnson, I.F., Reiser, D. and Overevik, G.S. Pat. 3.391.882, USA, publ. 09.07.68.

7. Vaughan, D.H. (1969) *Abstr. of pap. of 15th Ann. AAS Int. Conf.*, Denever, 1969. Denever, Paper NVC-4.

8. Forbes, F.W. (1964) *Space, Aeronautics*, **42**, No. 7, pp. 62–68.

9. Yarimovych, M.I. and Forbes, F.W. (1969) Expandable structures for space applications, In: *Proc. of 8th Int. Symp. on Space Technol. and Sci.*, Tokyo, 1969. Tokyo, pp. 59–63.

10. Rashevsky, P.K. (1956) *Course of differential geometry*, Moscow, Gostekhizdat, 230 pp.

11. Kagan, V.F. (1948) *Principles of the theory of surfaces*, Moscow, Leningrad, Gostekhizdat, 153 pp.

Development of a Transformable Structure of the Docking Module and Technology for Its Fabrication*

V.N. Samilov[1], O.Yu. Gonchar[1] and E.Yu. Burmenko[1]

Engineers often come across contradicting requirements for increasing sizes and weight of objects and limited transportation capabilities. This contradiction is especially pronounced in space engineering, where dimensions and weight of objects have a fundamental effect on the feasibility of their implementation. Here we consider two variants of shell structures capable of changing their dimensions and volume.

Such structures occupy a minimum room during transportation. At the same time, they can create extra volumes on board the space objects. Application of such a structure is considered by an example of a transformable module.

The first variant is based on the principle of isometric transformation of surface of a flat corrugated disk into that of a circular cone. The corrugated disk (Figure 1) is a complex surface with concentric ring corrugations. The conical surface with a certain half-opening angle can be produced depending upon the ratio of the pitch to depth of the corrugations. Pressure is generated inside a volume bound the disk and end faces to transform the corrugated disk into a conical shell (Figure 2). The structure of a required dimension and volume is made by successively joining several disks on the ring edges of small and large bases (Figure 3). Frames are welded into the circular joints between the disks to ensure the required stiffness and strength. If necessary, the structure can be divided into sections using partitions. During transformation of the structure of the corrugated disks the edges are displaced along the axis parallel to themselves or with a small rocking.

Development of a transformable module required a thorough consideration to be given to selection of a material for its fabrication. The material had to have a number of properties to ensure a defect-free formation of the corrugated disks and their subsequent

Figure 1 Flat corrugated disk.

* (1986) *Problems of space technology of metals*, Kiev, PWI, pp. 67–71.

Figure 2 Conical shell transformed from a corrugated disk.

transformation into conical shells, low specific weight, high ductility and strength and good weldability. Experiments were conducted with materials having different physical-mechanical properties, first of all with stainless steel and commercial titanium alloys.

As proved by the experiments, other conditions being equal, materials with elongation of 20 to 30 % turned out to be most appropriate. In this case, no defects resulted from bending around a mandrel equal to 8–10 thicknesses of a blank, followed by unbending. A valuable property of these materials is their ability to preserve high ductile properties at low temperatures. The increased sensitivity of titanium alloys to gas saturation in heating using

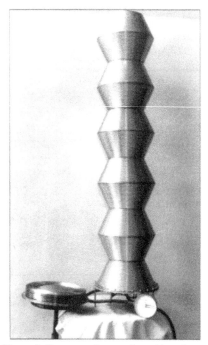

Figure 3 Structure of a transformable volume.

Figure 4 Hyperboloid fold.

the welding arc requires that special consideration be given to shielding of the welding zone and weld regions that cool down.

The second variant of making a transformable module casing consists in isometric transformation of hyperboloid folds into a cylindrical shell. The hyperboloid fold is a multi-drop surface bound by two types of stiffeners (Figure 4). Straight-line stiffeners of the fold (inside and outside) can be represented as generating lines of two hyperboloids of revolution of two nappes, located one inside the other.

Owing to the isometric surfaces of the hyperboloid fold and cylinder, a folded structure is transformed into a cylindrical surface under the effect of pressure created inside its cavity, the ring bases being turned about the axis of rotation of the shell relative to each other. During the process of deployment of the structure the folds are straightened and the straight-line stiffeners are bent over the cylindrical surface. Thus, the multi-section transformable structure is produced by joining several folds on the ring end edges. Circular stiffeners or partitions are installed between the sections and welded.

Folds with different directions of inclination of the stiffeners relative to their bases can provide multi-section systems with an almost absolute elimination of the relative turn of ends of the structure.

Development of the technology and equipment for formation of a complex folded surface is an independent problem of fabrication of these types of the structures.

There are two methods that were developed for formation of the folds:

• the first methods consists in successive bending of triangular panels on a flat sheet, followed by welding and assembly of the closing joint;

• the second method provides for formation of triangular panels directly on the closed cylindrical surface.

The preference should be given to the second method, as it allows production of the fully finished transformable member. Its point is that the cylindrical shell is subjected to successive uniform deformation in a radial direction along the lines of all long stiffeners of a future fold using a spinner. Gradually increasing regular dents are formed on the surface

Figure 5 Structure consisting of the hyperboloid folds during the formation process.

Figure 6 Module casing in the design position.

of the cylinder, transferring to the triangular panels bound by rectangular inside and outside stiffeners of the forming hyperboloid fold.

The work was done on structures with a diameter of 2000 mm, section height of 1240 mm and wall thickness of 1 mm. In this case the required angle of inclination of stiffeners was 53°. Twenty short and twenty long stiffeners delineated forty triangular panels. The module casings consisting of two hyperboloid folds were made and tested. The height of the module in a transportation position was no more than 330 mm.

Figures 5 and 6 show a model of the module casing on a 1:5 scale, transformed of the hyperboloid folds in the intermediate and design positions, respectively.

Conclusions

1. Structures capable of decreasing many times their dimensions and volume during transportation can be fabricated using the principle of isometric transformation of surfaces.

2. The structures can be fabricated from readily available and common materials.

3. The methods developed for formation of corrugated disks and hyperboloid folds provide the high quality of the products.

4. Transformable structures can be utilised for a wide range of objects.

All-Welded Transformable Metal Structures[*]

B.E. Paton[1], V.N. Samilov[1], I.S. Pilishenko[1], O.Yu. Gonchar[1] and E.Yu. Burmenko[1]

Progress in space science and technology, as well as practical exploration of space which started in the 1960s, posed difficult and exciting tasks for scientists and engineers concerning investigation and exploration of near-earth space and, in the future, the nearest planets.

Flights of the manned spacecraft and long-service orbital stations «Salyut», «Skylab» and «Mir» proved the feasibility of a long stay of people performing sophisticated IVA and EVA operations under the space conditions.

Space vehicles of the «Apollo» series, which many times delivered astronauts to the Moon, can be regarded as prototypes of the manned stations. Such stations, fitted for a comprehensive investigation of natural environments, will make it possible to use resources of other planets for the needs of mankind. The presence of scientific expeditions on the planets will require comfortable conditions for work and relaxation. This will require construction of dwelling rooms, laboratory and production compartments, storage premises, shelters, etc. Such structures may be of a substantial volume, and their delivery from Earth will present a problems, as this involves an obvious contradiction between required dimensions of structures and limited capacities of transport containers of carrier rockets.

The engineering community developed a number of approaches to fabrication of transportable structures. The first of them is to make folded parts from flexible materials, such as a high-strength air-tight fabrics or films. Such structures are subdivided into air-supported and air-bearing ones.

Air-supported structures can be employed only under earth conditions, as they require that positive pressure be continuously maintained inside a shell. Examples of such structures are exhibition pavilions, tennis courts, temporary storehouses, etc. Of notice among them is a pneumatic plane developed by the Goodyear Aerospace Corporation for the US Army. In a folded state it has a size of a big suitcase, while in the state ready for flight its wing span is 8.5 m and internal pressure is $0.6 \cdot 10^6$ Pa.

* (1999) *Avtomaticheskaya Svarka*, No. 10, pp. 81–85.

Space engineering did not disregard pneumatic structures either. Thus, G.T. Schjeldahl Company developed satellites «Echo I» and «Echo II» for NASA. They were spherical structures with a diameter of 30.0 and 37.5 m, made from aluminised mylar film 0.0125 mm thick. The satellites were launched into orbit in August 1960 and January 1964. They were transformed into a working state using chemical gas formers. Another example of a soft pneumatic structure is the docking module which allowed Soviet cosmonaut A. Leonov to perform EVA in 1965 [1, 2].

The second method used to construct large-size structures is the modular assembly method. This method was first employed for docking manned and cargo spacecraft. It proved itself to be especially advantageous in construction of long-service orbital stations of the «Salyut» and «Mir» types. Today this method is employed for construction of the ISS. As proved by many years of used, in construction of long-service space objects, preference should be given to metal structures.

At the end of 1966, by the initiative of the PWI, Prof. S.P. Korolyov formulated a task to develop all-welded metal shells capable, when launched to the orbit, of decreasing their volumes to a size of transport containers of carrier rockets. These structures were intended for construction of dwelling and laboratory compartments, docking modules, passages, storage and production premises.

Almost all leading specialists of the PWI participated in the competition opened at the Institute. The most interesting and realistic proposals were submitted by V.M. Balitsky, Candidate of Technical Sciences. They were used as the basis for further efforts on making structures with a transformable volume. The following requirements had to be met in their development: complete factory readiness, air-tightness, minimum possible sizes in the transport state, convenience of assembly after delivery, transformation pressure at a level of working pressure during operation. Theoretical and laboratory analyses allowed selection of the two variants, most feasible from the technology standpoint, out of a large number of those considered. These variants included transformation of a cylindrical shell into a folded surface and transformation of a truncated cone surface.

Substantiation of the possibility of targeted change of shape of closed shells can be based on principles of the theory of isothermal bending of the surface and mirror reflection. The theory suggests that the shell can be bent without tension or compression of its material.

The first transformable structure was based on the principle of transformation of a cylindrical shell into a folded surface with triangular panels connected by plastic hinges. Laboratory studies were conducted on shells with a wall thickness of 0.1 mm and diameter of 400 mm. It was established as a result of experiments on identification of the possible range of variation of the shell length that it could be reduced from nine- to eleven-fold. This development was to be realised in the docking module for the long-service OC «Mir».

Design of the module with a diameter of 2000 mm and height of 2200 mm was developed in collaboration with the S.P. Korolyov RSC «Energiya»[2]. It was 500 mm high in the transport state. Prototype of the docking module was manufactured and tested at the plant of the RSC «Energiya» with participation of the specialists of the PWI (Figure 1). The tests showed complete correspondence of the structure to the design characteristics. Unfortunately, the module was not installed on board the OC «Mir» for a number of reasons and because of weight and size restrictions at the station.

The developed theoretical model of transformation of closed surfaces led to another all-welded air-tight transformable structure being suggested, which was based on the principle of transformation of the truncated cone surface. The height of the cone during the transformation process can be varied over a wide range, i.e. from 10 to 40-fold.

Investigations conducted by the PWI were imbedded into a unique and sufficiently simple technology for transformation of a conical shell into a corrugated disk with annular corrugations. The investigations were carried out on shells with a wall thickness of 0.1–

Figure 1 Docking module in folded (a) and working (b) states.

0.3 mm and diameter of 400 mm. Commercial titanium alloy and stainless steel were used as working materials. Experiments showed good reproducibility of the developed shell transformation method, simplicity of required fixture and comparatively low cost. Successive transformation of the corrugated disk into a conical shell is shown in Figure 1 on p. 455 and Figure 2 on p. 456.

The technology enables joining of a required number of separate corrugated elements into a single structure, which acquires required dimensions and volume, configuration and architecture after transformation. The prototype of the structure of a transformable volume with a diameter of 400 mm has height 65 mm in the transport state and 2000 mm in the working state. We think it interesting to develop a standard module of shell structures of the above two types to be used as base line ones for construction of volumes of planetary stations.

Space exploration cannot manage without reliable communication systems, one of the main elements of which are aerial devices. Quality of an aerial depends in many respects upon its geometrical dimensions. This often generates the need for deployable structures. A team of specialists of the PWI has been working on this problem for a number of years.

Aerial systems with a pointed parabolic reflector were developed. Their reflecting surface is made from a flexible metal canvas of a special weave. This makes it possible to decrease many times the aerial aperture diameter during transportation. Figure 2 shows a sharply focused parabolic aerial in the transport and working states, having an aperture diameter of 2500 mm. Comprehensive investigations showed that electrodynamic characteristics of the aerial were at a level of values of the reflectors with a rigid reflecting surface. Multiple deployments and foldings have almost no effect on performance. It is possible to manufacture such aerial systems with a size of up to 5000 mm.

Two variants of a folded annular aerial with a diameter of 20,000 mm, i.e. made from elastic grooved chords and of the hinge-rod type, were developed in collaboration with the S.P. Korolyov RSC «Energiya»[2] specialists V.G. Osipov, V.G. Parfyonov, P.G. Chernyavsky and Yu.D. Kravchenko.

The first variant, i.e. annular contour composed of two elastic grooved chords connected by elastic braces, is characterised by the fact that its deployment after caulking occurs due to the energy accumulated by elastic chords as a result of folding of the structure. X-ray radiation is provided by a peripheral copper chord with a cross section of 50 mm^2, via which a modulated low-frequency current of 40–50 A flows [3].

Figure 2 Parabolic aerial in transport (a) and working (b) conditions.

Laboratory optimisation of the processes of deployment and formation of a required shape turned out to be very difficult. No methods for reducing weight of parts of such a large length under earth conditions were available. Therefore, the testing process was divided into two stages: I — deployment of elastic chords from spools, II — formation of the contour of a regular annular shape due to elasticity of the chords.

The first stage was performed using a rope suspension with an easily moving trolley. Numerous experiments showed that there was a sufficient amount of energy accumulated by the elastic chords released in deployment of the structure. The second stage required creation of pseudo zero-gravity. This was done using a method of counteracting a mass with spherical thin-walled rubber probes filled with helium. Volume of each of 630 spheres amounted to 1 m^3.

The method of suspension by spheres made it possible to create such conditions for the structure, which were very close to zero-gravity. The slightest effect on the structure caused a change in its shape. In investigations, the annular contour was given both expected and «fancy» shapes. The character, intensity and time of restoration of the preset shape by the contour were registered after removal of disturbing effects. In all the cases the contour steadily acquired the regular annular shape. One of the stages of these tests is shown in Figure 3.

These structures were tested in orbital flight on board the cargo spacecraft «Progress». The transformation process was recorded by photo, film and TV devices. Analysis of the information obtained made it possible to reveal imperfections which did not show up in ground investigations, and make corrections in the contour formation system.

The second variant involved a hinge-rod contour. A specific feature of this structure is a system of forced deployment using wire drives of titanium nickelide with a shape memory effect. Fragments and the entire structure in the full-scale form with a rope suspension and in the pseudo zero-gravity were tested under laboratory conditions [4, 5].

The experiment on deployment of two annular contours with a diameter of 20,000 mm, using drives of titanium nickelide, was conducted on board the cargo spacecraft «Progress». Inaccuracy in calculation of the expected initial temperature did not allow the experiment to

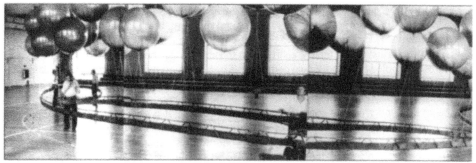

Figure 3 Annular aerial in the suspended state.

be carried out in flight conditions, but owing to prompt correction from Earth both contours acquired the regular annular shape in the heating mode. Further dynamic tests proved in full all the design characteristics [6].

Therefore, many years of research and development in the field of large-size transformable structures demonstrated the promising character of this area, and there is no doubt that their results will be applied in construction of new orbital long-service objects and exploration bases on planets nearest to Earth.

1. Yarimovych, M.I. and Forbes, F.W. (1969) Expandable structures for space applications, In: *Proc. of 8th Int. Symp. on Space Technol. and Sci.*, Tokyo, 1969. Tokyo, pp. 59–63.

2. Schmid, M. (1985) The extendable and retractable mast ERM, In: *Proc. of 2nd Eur. Space Mech. and Tribol. Symp.*, Weersburg, 1985. Paris, pp. 29–33.

3. Podgorny, A.N., Denisov, V.P., Berzhaty, V.I. et al. *Extendable hoist.* USSR author's cert. 1563155, Int. Cl. B 64 G 1/22, publ. 08.01.90.

4. Paton, B.E., Bulatsev, A.R., Zagrebelny, A.A. et al. *Transformable structure.* USSR author's cert. 242664, Int. Cl. B 64 D 1/22, priority 04.10.85, publ. 01.09.86.

5. Paton, B.E., Bulatsev, A.R., Mikhajlovskaya, E.S. et al. (1991) *See pp. 484–495 in this Book.*

6. Paton, B.E., Semyonov, Yu.P., Belousov, P.M. et al. (1992) *See pp. 463–468 in this Book.*

Experimental Studies of Deployment, Forming, Rigid and Dynamic Characteristics of Annular Frame Structures Installed on Board the Cargo Spacecraft «Progress-40»[*]

B.E. Paton[1], Yu.P. Semyonov[2], P.M. Belousov[2], V.Ya. Borisenko[26,] E.Yu. Burmenko[1], S.E. Vakulenko[26], E.P. Vyatkin[2], V.V. Gajdajchuk[27], Yu.I. Grigoriev[2], V.I. Gulyaev[27], B.E. Gutskov[2], M.V. Dzhanikashvili[28], I.S. Efremov[2], V.L. Koshkin[2], Yu.D. Kravchenko[2], E.V. Medzmariashvili[28], I.S. Pilishenko[1], V.N. Samilov[1], A.G. Chernyavsky[2], N.L. Shoshunov[2] and A.T. Yakobashvili[28]

Space exploration for the benefit of the economic development is associated with construction of large-size structures (LSS) in the near-earth orbits. Large-size structures are the structures with dimensions from several tens to thousands of meters, such as solar energy concentrators, different types of aerials, truss structures (for location on them of facilities or panels of solar orbital electric stations, etc.). These structures can serve as the base for

[*] (1992) *Space science and technology*, Issue 6, pp. 69–75.

building of large orbital systems. They will be needed in a period of industrialisation of space, which is the comprehensive and integrated problem, as characterised by the founder of cosmonautics K.E. Tsiolkovsky.

As to the principle of their construction, space LSS can be subdivided into two types: structures assembled from individual components or raw materials and automatically transformable (deployable) structures. Optimisation of such LSS under earth conditions requires building of expensive test rigs and, at the same time, development of analytical models of their deployment and behaviour under natural conditions. In this connection, it is very topical now to optimise under space conditions the process of unfolding and forming of annular LSS, as well as investigate their characteristics under dynamic conditions of a spacecraft.

In a period from 3 to 5 March, 1989, the USSR conducted a space experiment on board the spacecraft «Progress-40» and OC «Mir». The experiment was aimed at verification of the possibility of deployment and forming of two annular LSS under orbital flight conditions. The structures had a diameter of 20 m each. The links were unfolded using drives of titanium nickelide with a shape memory effect. Other aims of the experiment were to study characteristics of the deployed LSS in dynamic modes of a spacecraft and check correspondence of calculated and experimental characteristics of the structures during the experiment [1–3].

The NPO «Energiya»[2], PWI, SDB GPI[28], KICI[27], A.I. Mikoyan KIFPI[26] and other organisations participated in the development, fabrication and optimisation of an annular LSS.

This article is dedicated to description of the main results of the experiment.

Description of the experiment. Deployment of annular LSS installed in special beds on the external surface of the cargo vehicle «Progress-40» was done by the commands sent from Earth after the vehicle departed from the OC «Mir» to a distance of 70–80 m. Orbit of the «Mir» was close to a circle with a height of 350–400 km over the surface of Earth and inclination of $51.6°$. The departure was controlled from the FCC, as well as by the crew of the station using a TV camera installed on the cargo vehicle along axis X.

The first command was to arrange the annular LSS installed in plane II, and in 5 min the second command was sent to arrange the structure in plane IV of the cargo vehicle. Additional heating of the drives was performed at subsequent turns.

As a result, two annular LSS were deployed, having the form of almost regular circles with a diameter of 20 m each. The process of unfolding and forming was watched and fixed by the crew of the «Mir» (commander A. Volkov, flight engineer S. Krikalyov, researcher V. Polyakov). The crew observed the deployed structures many times during further independent flight of the cargo spacecraft «Progress-40», recorded them on photo and video films and transferred to Earth individual fragments of the video records.

During the next two days dynamic operations were performed by the cargo spacecraft «Progress-40», behaviour of the deployed structures was watched at FCC from a TV camera installed in plane II of the spacecraft. Upon completion of the scheduled program of studies a braking pulse was sent and the spacecraft «Progress-40» was landed following the regular program.

Design peculiarities of the annular LSS. Two aggregates with annular LSS were installed on the module of components for refueling of the cargo spacecraft «Progress-40». The aggregates consisted mainly of three functional units: bed, cover and annular LSS proper. The general view of the aggregates is shown in Figure 1.

The bed is intended for transportation of the annular LSS contained in it and is a riveted frame structure of an aluminium alloy (Figures 1 and 2), which is physically divided into two equal parts arranged at an angle of $135°$ to each other. Each half is fitted with a ridged supporting elements to maintain links of the frame in a transport position. The bed has two flanges to install explosive bolts, and four brackets to fix connections of power circuits and telemetry. The bed is attached to the frames of the module using supporting beams.

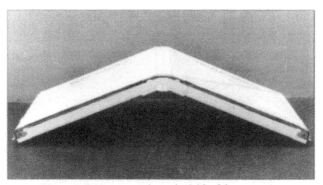

Figure 1 Bed, cover and annular LSS of the aggregate.

The bed cover (Figures 1 and 3) is intended for maintaining the annular LSS in a transport position and releasing it prior to deployment. The cover is a riveted structure based on a titanium alloy sheet 0.4 mm thick. Flanges are installed at the ends of the cover to ensure connection to the bed using the explosive bolts. To maintain the required heat conditions, the external surface of the cover is coated with white enamel, and porous rubber strips are glued onto the internal surface in the zone of interaction with the LSS links. The annular LSS is a hinged-link mechanism (number of mobile links is 78, length is 0.8 m). The link unit is fitted with a lock mechanism for a deployed position. The structure in the deployed form is a regular polygon with an inscribed circle 20 m in diameter.

A wire drive (wire with a diameter of 1.9 mm) made from a material with the memory shape effect (titanium nickelide) is used for deployment of each kinematic pair. In a region of the link unit the wire drive interacts with a groove of the pulley made from a fluoroplastic material, the pulley being rigidly connected to one of the links without the possibility of slipping. Ends of the drives are secured using the fixation units installed at the tips of the neighbouring links to ensure a reliable insulation, as the power circuits of the drive are connected to them. The drives are connected to each other in series using cables pulled inside the link profiles. The power circuits are secured to the external surface of the links.

When the electric current flows via the wire drive, the latter is heated to a temperature of beginning of reverse martensitic transformations of titanium nickelide ($\sim$ 75–85 °C) and starts restoring its initial shape (length). A longitudinal force (restoration force) develops in it. This results in a deployment moment M_{dep} formed in each link unit ($M_{dep} = p_r \, r_p$, where p_r is the restoration force formed in the wire drive and r_p is the pulley radius).

Figure 2 Beginning of deployment of the LSS links.

Figure 3 Deployment of the root and peripheral links.

More than 20 deployments of the LSS were carried out on board the cargo spacecraft «Progress-40» during the process of preparation of the structure for the experiment. They proved correctness of the selected designs and performance of the wire drives of titanium nickelide to be used as part of the deployed multilink assembly.

Main results of the experiment. Unique information, including tens of photos, film and video records and a large scope of telemetry data, was obtained in the course of the experiment. This information requires careful processing. However, even its preliminary analysis shows that the annular LSS are serviceable and can be employed as a load-carrying frame for various facilities to be deployed in the orbit.

Functioning of the annular LSS using drives of a material with the shape memory effect under conditions of the orbital flight can be subdivided into three basic stages: unfolding, i.e. moving apart of links of the structure relative to each other under the effect of forces of wire drives of titanium nickelide in hinges; forming, i.e. acquiring by the structure of the final stable annular shape; and dynamics, i.e. elastic oscillations of the structure caused by dynamic operations performed by spacecraft.

In-process and then repeated analysis of video records made by the crew of the «Mir» gives the following picture of deployment of structures. Slow moving apart of the links began 60–80 s after sending a command to switch on the power for drives of the first structure, in which the root links moved faster than the peripheral ones (Figure 4a). The structure formed an extended dual chain of the links, which resulted in a certain trend to moving apart of both branches of this chain. As a result of 5 min heating of the wire drives the structure acquired the shape of a slightly distorted dual chain about 25 m long (Figure 4b).

Sending a command to switch off the first structure and feeding power to the second structure. Its deployment was similar to that of the first structure, but moving apart of branches of the dual chain of the links was more clearly defined. After 5 min of heating of the titanium nickelide drives the second structure acquired the form of a very extended irregular ellipse.

Therefore, the process of deployment of each structure lasted not more than 5 min. Deployment of annular LSS from the beginning to the end was fixed by the crew of the «Mir» using film, photo and video equipment.

Forming of the structures occurred as a result of a longer heating of the wire drives. The second LSS took the preset annular shape after the 10 min supply of power to the wire drives. The fact of forming of this structure was visually fixed by the «Mir» crew, of which they reported to FCC, made a few photos and transferred to Earth a video record of the cargo spacecraft with a fully deployed annular LSS (Figure 5).

Then the preset annular shape of the structure was recorded from the TV picture transferred to FCC, confirmed by the report of the «Mir» crew and fixed by photo equipment.

Figure 4 Shape of the structure after 5 minutes heating of wire drives in forming of the contour:
a —first seconds; b — after 5 minutes.

Therefore, both structures acquired the required shape of a ring. Then dynamic and rigid characteristics of the structures, as well as controllability of the cargo vehicle with the deployed LSS on board, were studied. The following problems were solved at this stage of the experiment: frequency and decrement of the main tone of oscillations of LSS were experimentally determined; ability of the structures to restore the preset shape after performing turns and correction pulses by the cargo vehicle was checked; and efficiency of the designed program control of a turning manoeuvre, leading to a full stop of relative movement and stabilisation of LSS after spacecraft finishes the turn, was confirmed.

Oscillations of the annular LSS, caused by dynamic operations performed by the cargo spacecraft, were observed using a TV camera installed on board the spacecraft. Preliminary analysis of characteristics of the deployed annular LSS in dynamic operations performed by the spacecraft «Progress-40» showed that in general they corresponded to those calculated and determined by theoretical studies using special mathematical models in preparation of the experiment.

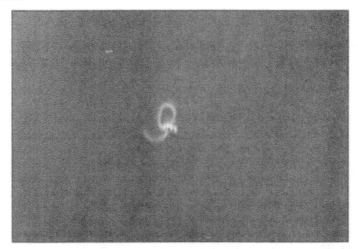

Figure 5 Video picture of the cargo spacecraft with one fully deployed annular large-size structure.

The developed mathematical models were used to calculate 20 lower values of the spectrum of frequencies of natural oscillations of the system and plot the respective movement paths. Mathematical modelling allowed for inertia properties of both annular frame structures and the spacecraft «Progress-40». The calculated value of a period of natural oscillations for the first shape was $T_c = 179$ s. Value of this period measured in the full-scale experiment was $T_e = 190$ s. Some difference between the actual and calculated values of duration of the period was caused by neglecting compliance of a unit of fixation of the annular structure to the vehicle and design peculiarities of the hinge locks.

The program of control of turning manoeuvres relative to each of the three main central axes of inertia of the system, including acceleration conditions, free rotation and deceleration with suppression of relative oscillations of rings by lower forms, was developed on the basis of analysis of dynamic characteristics of the structure. Dynamic tests proved efficiency of the developed program control.

Results of the obtained dynamic characteristics of annular LSS require a more detailed study.

Conclusions

1. A space experiment on deployment of annular LSS was conducted for the first time in the world on board the spacecraft «Progress-40». Unfolding of the links was done using drives of titanium nickelide characterised by the shape memory effect.

2. Characteristics of the deployed LSS in dynamic operations performed by the spacecraft were determined on the base of measurement results. In general, these characteristics corresponded to the calculated ones.

3. Practical feasibility of construction of LSS in orbit of artificial satellites of Earth, using drives of materials with the shape memory effect, is shown.

1. Armand, N.A., Semyonov, Yu.P., Chertok, B.E. et al. (1988) Experimental study in ionosphere of Earth of radiation of the frame aerial in a range of very low frequencies on board the orbital complex «Mir–Progress-28–Soyuz TM-2», *Radiotekhnika i Elektronika*, **33**, No. 11, pp. 25–31.

2. (1987) *Active experiment: space plasma-wave laboratory*, Booklet, Moscow, 12 pp.

3. Gvamichava, A.S. and Koshelev, V.A. (1984) *Construction in space*, Moscow, Znaniye, 16 pp.

Peculiarities of Using Welding Processes for Fabrication and Repair of Large-Sized Structures in Space*

B.E. Paton[1], V.F. Lapchinsky[1], A.R. Bulatsev[1], V.M. Balitsky[1] and V.N. Samilov[1]

Space exploration gave a strong impetus to development of various areas of science and technology, including materials treatment technology. Welding technologies have a high importance. If before recently welding has had a limited application in space engineering, now the situation has changed. We have already developed projects of construction of the next generation of spacecraft, i.e. heavy orbital stations with large crews, large radio telescopes, aerials, reflection and absorption screens, and engineering systems for utilization of solar radiation. Given this, the interest in the problems of their repair and reconditioning

* (1989) *Proceedings of IIW Conf. on Welding under Extreme Conditions*, Helsinki, 1989. Helsinki, Pergamon Press, pp. 181–188.

in flight increases with an increase in time of operation, whereas the problems of deployment, assembly and construction become increasingly topical with an increase in mass and dimensions of structures. In any case, welding processes seem to be indispensable for space applications, where conditions of performing welding operations are fundamentally different from the earth ones.

Space environment is characterised by the following three basic peculiarities.

Zero gravity. In a liquid or gaseous environment it causes absence or suppression of buoyancy, convection and a number of other physical phenomena associated with gravity and differences in density of materials or their phases. At the same time, the surface phenomena in liquids become more pronounced. Peculiarity of zero gravity is that cosmonauts have to work without any support, which is very important for performing manual operations.

High space vacuum. The level of the ambient pressure $(1 \cdot 10^{-2} - 1 \cdot 10^{-4}$ Pa) in a region of low orbits, where large orbital stations are flying now and will fly in the future, is not a surprise for welders. This range of pressures has been well mastered by the earth industry, which uses such joining methods as electron beam welding and diffusion bonding. However, space vacuum is unusual in view of an extremely high rate of evacuation, which is close to infinite.

Presence of contrast light–shadow boundaries. This peculiarity leads to the fact that a workpiece may be in a very wide range of temperatures (about 150–500 K), and the mass transfer process, which in space is less intensive, indicates that zones with a large difference in temperatures may be located very close to each other on a workpiece.

This list of peculiarities, although very short, shows that space in a combination of factors is characterised by extreme conditions for performing welding operations.

No methods for treatment of molten metals under zero gravity conditions were available before the beginning of the investigations, and no mentioning of the possibility of carrying out any process in space could be found in literature. Thus, out of a wide variety of methods available for making permanent joints, it was necessary first of all to chose those which would offer the highest promise for application under such extraordinary conditions. Therefore, researchers were ruled both by specific welding evaluation criteria (practicability, flexibility, simplicity) and by mandatory criteria particularly for space applications (the highest possible reliability, safety, low power consumption, minimal weight and dimensions of hardware, etc.).

Given these requirements, it was established that EBW was the most suitable, reliable, versatile and efficient source for heating materials in space. So, all investigations that followed were conducted using electron beam as a heat source.

In a period from 1969 till 1988 the PWI completed investigations under several programmes to study peculiarities and possibilities of practical application of joining processes in space, i.e. welding, brazing, cutting and coating were studied. The major part of the experiments was first conducted on board a flying laboratory, which enabled a short-time (up to 30 s) zero gravity to be created. For that purpose, special sets of test equipment for automatic and manual welding were installed inside a cabin of the flying lab. Then the results obtained were checked directly in space.

15 Soviet cosmonauts, V.N. Kubasov, G.S. Shonin, V.V. Ryumin, V.A. Lyakhov, L.I. Popov, V.V. Kovalyonok, Yu.V. Malyshev, G.M. Strekalov, V.P. Savinykh, S.E. Savitskaya, V.A. Dzhanibekov, L.D. Kizim, V.A. Soloviov, Yu.V. Romanenko and A.I. Lavejkin, performed the works in space.

Analysis of the results obtained in the course of these research programs and personal observations of experimenters and cosmonauts made it possible to establish a number of interesting principles of a general character, which were used as the basis for development of specific technologies.

It was established that zero gravity affects most intensively the welding processes which are associated with the presence of a liquid phase. The space technology for making permanent joints in some cases may be substantially different from its earth analogue. Analysis of these differences requires at least a brief consideration of some general physical notions.

Molecules of a liquid in a gravitational field are known to have potential energy E_g, associated with gravity, in addition to kinetic (thermal and chemical) energy E_k. In the majority of cases this energy is much lower than the chemical or thermal one. However, there are some physical phenomena which occur at the molecular level (surface tension, adhesion, wetting, capillary pressure), the energy characteristics of which are sufficiently well comparable with E_g. Therefore, in low-gravity fields these phenomena show up much stronger than near the Earth surface, and surface tension and wetting become the dominant effects in the welding processes. The following forces affect a unit length of the interface, dl, between these three phases: $E_{v-l} = \delta_{v-l}dl$; $E_{s-l} = \delta_{s-l}dl$; $E_{s-v} = \delta_{s-v}dl$ (δ_{v-l}, δ_{s-l} and δ_{s-v} are the surface tension coefficients at interfaces between the vapour and liquid, solid and liquid and solid and vapour phases, respectively). The impact of these forces depends upon the distribution of temperatures over the interface between these phases.

It is a well known fact that the surface tension coefficient on the interface between the liquid and vapour phases strongly depends upon the temperature. However, as far as the other two interfaces are concerned, it is characterised by a sufficient thermal stability. Dependence of δ_{v-l} upon the temperature can be approximately described by the following expression:

$$\frac{d\delta_{v-l}}{dT} = -K\left(\frac{\rho}{M}\right)^{\frac{2}{3}}, \qquad (1)$$

where ρ is the density of the liquid; M is its molecular weight; and K is the constant coefficient.

As it can be seen, $\dfrac{d\delta_{v-l}}{dT} < 0$. Naturally, at the presence of temperature gradient dT/dT_1

along the interface between the vapour and liquid phases the surface tension gradient is as follows:

$$\Delta\delta = \int\left(\frac{\partial\delta_{v-l}}{\partial T}\frac{dT}{dl}\right)dl. \qquad (2)$$

In turn, this generates a gradient of the surface tension force, dE_{v-l}/dT, the vector of which is directed toward the lowest temperature and causes a corresponding mass transfer.

Distribution of forces on the interface between the liquid and solid phases is totally different. The adhesion forces, which vary with variations in free energy E_{free}, take place here. Variation in this parameter related to a unit surface can be written as follows:

$$\Delta E_{free} = \delta_{v-l} - \delta_{s-l} - \delta_{s-v}. \qquad (3)$$

It follows from the condition of thermal stability of δ_{s-v} and δ_{v-l}, and decrease in δ_{s-v} with an increase in temperature, that the gradient of the free energy, dE_{free}/dT, is formed along this interface at the presence of the temperature gradient. This generates a gradient of the adhesion force, dF_a/dl. However, in this case the relationship between forces is more complicated, as the value and direction of the adhesion force also depend upon the relationship between δ_{v-l}, δ_{s-l} and δ_{s-v}, which is characterised by a contact angle

$$\cos v = \frac{\delta_{s-v} - \delta_{s-l}}{\delta_{v-l}}. \tag{4}$$

The value of this vector of force varies from maximum (at π) to zero (at $\pi/2$) within a range of $\pi \geq v \geq \pi/2$ (non-wetting), and it is directed, like in the previous case, toward low temperatures. In a range of $0 \leq v \leq \pi/2$ (wetting) a maximum of the vector of force corresponds to $m = 0$ (complete wetting), and it is directed toward a higher temperature.

Of interest is a particular case where the liquid has several interfaces with the solid phase, instead of just one. Here it is necessary to take into account not only the distribution of temperatures, but also the angle between solid surfaces. It can be seen from formulae (3) and (4) that the force formed in a non-wetting liquid is directed to divergence of the angle, and that formed in a wetting liquid has an opposite direction. Mass transfer in a liquid will be directed accordingly. Therefore, this creates conditions for movement of molten metal and its localisation either in the coldest or hottest zone. It becomes clear from the above considerations that two opposite flows can be formed in the liquid pool, one directed toward the cold zone and the other — toward the hot zone. We observed the above phenomena in investigation of behaviour of the liquid pool in zero gravity on optically transparent salt models. Investigation and active control of these effects are imperative conditions for development of a perfect technology for application in space.

The above principles are well known and show up on Earth very insignificantly. There role here is considerably suppressed by gravity forces. In zero gravity, a pronounced manifestation of the surface forces is prevented by stirring of molten metal under the influence of the electron beam and dissipation effects generated by viscosity and thermal diffusivity. Most molten metallic materials are characterised by low kinematic viscosity, moderate thermal diffusivity and high surface tension. Therefore, under zero gravity conditions the process of development of a stable mass transfer caused by the above surface forces lasts normally a few seconds or fractions of a second.

Consider some examples of active manifestation of the surface tension forces under zero gravity conditions.

In zero gravity it is difficult to burn through a hole in a sheet material in welding without backing, because of the low concentration of thermal energy. Diameter of the weld pool can be dozens of times larger than thickness of the material at a large amount of molten metal present. This phenomenon often involves certain difficulties in electron beam cutting. To overcome them, it is necessary that an optimal temperature field be formed. The positive effect in cutting and burning through is that molten metal is not removed from the cut cavity in the form of drops (this could have been very dangerous), but tends to flow to the cut edges under the surface tension effect. At the presence of the appropriate temperature field, it is possible even to provide its preferable localisation at one of the edges. The use of the controlling effect of the temperature field in zero gravity in the case of brazing and coating by the method of thermal evaporation of materials from crucibles is very efficient. This can ensure a direct spread of brazing filler metals and a reliable detachment of vapours from the liquid phase.

Drastic light–shadow boundaries and high temperature differences observed in space create substantial difficulties. This peculiarity of space required development of special criteria for evaluation of quality of welding operations and flexible programming of welding and brazing conditions. However, in general, these difficulties are similar to those which welders face on Earth while welding preheated or heavily cooled workpieces.

Based on the above peculiarities, all welded or brazed joints produced with automatic devices using the technology optimised for space application are not in the least inferior to those produced on Earth. This is equally applicable to coatings produced by the method of thermal evaporation and condensation of materials. Investigations into peculiarities of the technology of welding in space were accompanied by an intensive development of basic principles for fabrication in space of large-size welded structures.

Large-size objects can be fabricated in space by three methods.

Firstly, they can be assembled from separate prefabricated modules. Dimensions and weight of the modules are determined by the load-carrying capacity of the carrier rocket and by general characteristics of the transport spacecraft. At present this method is used for assembly of Soviet orbital space systems, such as «Mir».

Secondly, space objects can be fabricated on Earth and delivered to space in a compact folded state. On the service site they are brought up to the required working state by transformation of their shape and size.

Thirdly, space objects are assembled from individual and, as a rule, adapted elements. At the beginning the second method was of the highest interest for fabrication of large-size structures in space, as it enabled the structures to be manufactured under factory conditions, where their quality was thoroughly controlled.

Consider this method in more detail by an example of two types of structures (metal shells and spatial trusses).

Transformation of shells is most difficult. Changes in shape of thin-walled metal shells are based on geometrical laws of surface bending. Two classes of isometric transformations can be distinguished. One of them is used for closed surfaces attached to a profile or having ties on their areas. Such surfaces can be transformed only by local bending.

Open surfaces, having no ties on their areas or near their boundary which limits displacements, are characterised by wider capabilities in terms of isothermal shape transformation. Transformation of such surfaces can be done using not only local bending, but also global deformations, leading to a change in curvature of the entire surface area.

The PWI has accumulated extensive experience in industrial application of the laws of geometrical bending of surfaces and their manipulation to fabricate welded metal panels for tanks, coils and other similar pieces. However, its direct use for the fabrication of space objects should be quite limited. As a rule, such cases require extraordinary designs, which may combine close and open shells.

The experience of fabrication and testing of such systems of the shells under the open space conditions, including vacuum, shows that they hold high promise for space applications. Besides, one of their main advantages is in a high degree of factory readiness, which ensures almost ideal quality of fabrication.

The same principle of a high degree of factory readiness was used as the basis also for the fabrication of spatial truss structures in space. The main element of the first experimental transformable space truss structures was a lever-hinge link, working to torsion and made so that a hinge journal, if necessary, could be welded to the axle.

The PWI designed, on the basis of this element, a tetragonal transformable truss structure with a maximum length of 15 m and a construction height of 0.42 m.

The truss is made of sections consisting of two sizes of rods (longitudinal and diagonal) and cross pieces with grooves which form triangles of forces in a deployed state. All longitudinal rods have a polygonal hinge joint in the middle, which is fixed in the deployed state and provides compact packing of the truss in the transport position. Therefore, the 15 m long truss is folded into a pack 500 mm long. As proved by static and dynamic laboratory tests, the truss is characterised by a very high rigidity, good frequency-dynamic properties and can be employed for handling a number of application problems in new spacecraft. At the same time, tests of the truss structure under earth conditions failed to give a sufficiently full picture of its behaviour in zero gravity under different kinds of loading. This is attributable to complexity of decreasing weight of such extended systems and an extremely low accuracy of their mathematical modelling. Therefore, it was decided to make and test the truss structure directly in space.

The PWI developed and manufactured a special device for folding and unfolding of the truss structure. In a transport position, it is a rigid cylinder with a diameter of 700 mm and 1500 mm long, which houses the 15 m long truss in a compact packing.

A special platform with research instruments is mounted on the outside edge (end) of the truss. The principle of operation of the device consists in a section-by-section unfolding or folding of the truss. In the process of unfolding or folding of each section the previously installed section of the truss with the instrument platform, located at the end, is reliably fixed with special clamps. Hinge joints of longitudinal rods can be welded after unfolding of each section. This increases rigidity of the truss, but makes it irreversible. Folding or unfolding of the truss can be performed in three modes: automatic (all operations are carried out by a command from the control panel), partially mechanised and manual.

The experiment on construction of the truss structure in space was conducted in 1986 on board the OS «Salyut-7». It should be noted that a simplified version of the device, without welding units, was used in the course of that experiment, as manufacture of the full-scale welded truss structure was considered inexpedient at the first stage of the experiments.

That experiment was performed by cosmonauts L.D. Kizim and V.A. Soloviov. Its program was intended for two runs of EVA of the cosmonauts.

The first EVA on the 28th of May was of a preparatory character, in which the cosmonauts brought out and installed out of board of the station all the required equipment and instruments. The second main EVA was accomplished on the 31st of May. A truss more than 13 m long was constructed. To evaluate its dynamic and strength characteristics, the truss was subjected in several planes to the effect of low-frequency mechanical oscillations, the amplitude and frequency of which were fixed by the instruments. Not only behaviour of the truss proper was studied, but also characteristics of the entire system «orbital station–bracket with the concentrated mass on the outside edge (end)» was investigated. Dynamic characteristics of the truss were fixed by special seismic sensors, the readings of which were fed to a recording device located inside the station. The next step of work during the second EVA was the development of the technologies for welding and brazing of truss assemblies of different designs. Welding and brazing were performed using the VHT.

Upon completion of all these operations, the cosmonauts removed the truss construction device with the truss folded in it and VHT into the pressurised compartments of the station.

Samples of welded and brazed joints of the truss assemblies, produced in space, were delivered to Earth and analysed. As proved by the results of analysis, sound welded and brazed joints in thin-walled structural members can be produced in space, despite the fact that electric power supplied for that experiment was very low. In addition, the experiments demonstrated advantages of the principle of manufacture of transformable welded structures in space.

Assembly and Welding of Large-Size Structures in Space*

A.R. Bulatsev[1], M.I. Morejnis[1], S.A. Skorobogatov[1], V.I. Motrij[1],
V.N. Samilov[1] and D.V. Beletsky[1]

Progress in exploration of space is associated with fabrication of a new generation of space orbital stations. This is attributable to an urge towards utilisation of the unique possibilities offered by orbital flight conditions, which will allow not only costs of accomplishment of the corresponding space program to be recovered, but also a substantial profit to be made.

The USSR has already accumulated extensive experience of prolonged space flights. In the 1980s several orbital stations of the «Salyut» type were working in space. Now the OC «Mir» is in operation (Figure 1) [1]. At the same time, a number of countries are active in development of designs of more powerful complexes. These are, in particular, the projects of space stations «Mir-2» (USSR) and «Freedom» (USA) (Figure 2) [2].

The above projects provide a very sophisticated architecture of complexes, which will include various elements and systems, such as living, research and production modules, power supply, heat removal, orientation and observation systems. Almost all of the above elements will be located on a truss structure which makes up the framework of the station. In turn, the power supply, heat removal and orientation systems will also comprise truss structures. It is planned to use various types of shells for making pressurised and non-pressurised compartments and storage tanks.

All of the above types of truss and shell structures share one characteristic feature. It is the impossibility of delivering them into orbit in a finished, ready for operation form, because of their large dimensions. This involves considerable difficulties. The situation is aggravated by the fact that during operation these structures should meet many extremely stringent requirements:

- high load-carrying capacity;
- high rigidity level;

Figure 1 Orbital complex «Mir» (USSR).

* (1991) *Proceedings of Conf. on Welding in Space and the Construction of Space Vehicles by Welding*, USA. Miami, Sept. 24–26, 1991. Miami, pp. 70–80.

Figure 2 Orbital station «Freedom» (USA).

- minimum weight;
- low susceptibility to the environment effects;
- long period of operation (up to 30 years for advanced space stations);
- good repairability;
- acceptable cost.

The entire package of questions and problems led to a new area of research, i.e. development of methods and devices for construction, maintenance and repair of large-size structures (LSS) in space.

The PWI started intensive developments in this area late in the 1970s. The issues of paramount importance at that time were as follows:

- estimation of loading forces;
- selection of design of a structure as a whole and its individual members and units;
- selection of special structural materials;
- selection of the types of joints in structural members;
- architectural appearance of structures;
- methods for their deployment, assembly and erection.

It was shown that these questions had no unambiguous generalised answer. Technical solutions made in each particular case had to be optimised with allowance for future application of the structure and its assigned service characteristics.

Consider several types of large-size structures, methods for their construction and associated equipment developed by the PWI during the last decade.

Figure 3 Tubular-rope truss structure.

Serious calculations and investigations were conducted to estimate the possibility of construction and utilisation of tubular-rope truss structures in space, which are to be unfolded from a transport position into the working one using truss-assembly units. Prototypes of such trusses and a unit for their unfolding were developed. One such truss had a trihedral section, rigid cross pieces and flexible diagonal braces on its faces (Figure 3). Longitudinal members of the truss were formed from tubular sections of the STEM type. Transverse members were joined into a triangular diaphragm using node sleeves for tubular sections and hinged joints to fix rope braces located in the corners. The diaphragms in a set with rope braces formed the transverse frame of the truss which could be folded into a compact pack. The number of diaphragms in the pack determined the length of a truss being constructed. In the transport position tubular sections in the form of a coiled flat strip were placed in forming devices, and a pack of the transverse frame was placed in a container of the truss-assembly unit (Figure 4). The process of construction of the truss comprised the following stages: transformation of tubular sections from the transport to working position, inserting them into node sleeves of all diaphragms of the pack, step-by-step shaping of the truss in the section formation zone and fixation of the transverse frame diaphragms on the tubular sections. The shape of the truss was controlled and tension of the rope braces was adjusted during the process of section formation. Permanent joints in the connections of the truss between the diaphragm sleeves and tubular sections were made by brazing combined with plug welding. The truss and unit passed the ground tests and showed great promice. The good load-carrying capacity of such a truss with its small weight turned out to be especially attractive. At the same time, the unit for construction of the tubular-rope truss was characterised by a comparatively high weight, which made its utilisation in space under specific production conditions highly problematic.

Figure 4 Truss-assembly unit holding a tubular-rope truss structure.

Trusses of this type have a drawback. They are assembled in space from semi-finished products in the automatic mode. Naturally, this requires a thorough optimisation of the technology and hardware, as even the smallest mistakes may be fatal.

From this standpoint, structures which are fully made on Earth and only deployed and, if necessary, welded or brazed in space are highly advantageous. This approach guarantees high accuracy and quality of manufacture, and makes truss-assembly units much simpler.

In 1984–1985 the PWI designed and manufactured the tetrahedral transformable truss with a construction height of 0.45 m (Figure 5) [3, 4]. This structure is based on a hinge-rod double link designed with the possibility of welding of a hinge, if necessary. The truss is built from individual sections comprising rods of two sizes (longitudinal and diagonal) and cross pieces, forming fixed load-carrying triangles in the deployed state. All the longitudinal rods have a broken hinge in the centre, which is fixed in the deployed state and allows compact packing of the truss for transport. The 15 m long truss is thus folded into a pack 500 mm high. Static and dynamic laboratory tests of the truss showed that it had sufficiently high stiffness and good frequency-dynamic characteristics, and could be used to solve a number of applied problems on modern space vehicles.

A special unit (Figure 6) was designed and manufactured for folding and unfolding of the truss. It is a rigid cylinder 700 mm in diameter and 1500 mm long with a compact pack of the 15 m long truss located inside. A special platform for research equipment is installed on the external end of the truss.

The principle of operation of the unit is a section-by-section unfolding (and folding) of the truss. During the process of unfolding or folding of each new section, the already built part of the truss with the research equipment platform located on its end is reliably fixed using special clamps.

The experiment on construction of the truss in space was carried out in 1986 on the «Salyut-7» by cosmonauts L. Kizim and V. Soloviov [1]. The programme for the experiment provided for two EVAs of the cosmonauts. The first EVA was of a preparatory character, in which the cosmonauts brought out and installed out of board of the station all the required equipment and instruments. The second main EVA was accomplished on the 31st of May. The truss 13 m long was constructed during the second EVA. Low-frequency mechanical

Figure 5 Tetrahedral transformable truss in a deployed position.

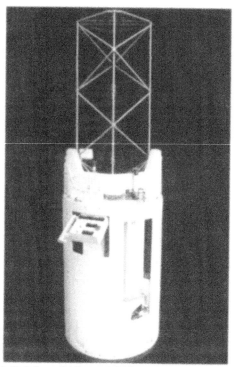

Figure 6 Unit for unfolding and folding of the transformable truss

Figure 7 Tetrahedral transformable truss deployed in open space.

oscillations, whose amplitude and frequency were recorded by instruments, were induced in some planes to estimate dynamic and strength characteristics of the truss. The behaviour was investigated not only of the truss itself, but also of the entire mechanical system «orbital station–bracket with a concentrated mass on the external end». A photo of the truss built in space is shown in Figure 7. Dynamic characteristics of the truss were registered using special seismic sensors and transmitted into the station to a recording device. The next stage of the work during the second EVA consisted in optimisation of the technology for welding and brazing of different designs of fragments of the truss. Welding and brazing were performed using an electron beam VHT.

Samples of welded and brazed joints in the truss connections produced in space were afterwards delivered to Earth and analysed. Results of the analysis showed that, even at the extremely limited electric power allocated for the experiment, very high quality of welded and brazed joints can be achieved in space on thin-walled structural members. The principle of manufacture of transformable welded structures in space was recognised as a highly promising one.

Trusses and unfolding-folding units of a similar type were built by the PWI to equip power supply systems for the space technological module «Kristall». There each truss served as a load-carrying structure for solar batteries. As compared with the above truss, the new one had an increased load-carrying capacity and improved performance. Immediately after launching, both solar batteries were deployed in the automatic mode each to a length of 5 and 10 m, while after 10 days, i.e. after docking of the module with the «Mir», to a full length of 15 m (Figure 8) [5].

Figure 8 Orbital complex «Mir» with solar batteries deployed to a length of 15 m.

Investigations completed by the PWI both on Earth and in space proved the efficiency of the method for transformation of shape and dimensions of structural members for the fabrication of LSS. Structures built using this method can be employed as independent modules and as components of other assemblies of larger size.

The experience gained in deployment and operation in space of such extended truss structures proved their high promising character and performance.

Shells represent the other class of transformable structures. Shell structures of a transformable volume allow pressurised modules to be made in orbit by deploying them from the transport state to the working one with increase in volume of a factor of «a few tens». Shells are very difficult to transform. Analytical studies of shape transformation of thin-walled metal shells, based on geometrical principles of bending of surfaces, made it possible to distinguish two classes of isometric transformations. One of them concerns closed surfaces fixed on a contour or having ties in their areas. Such surfaces can be transformed only by local bends. Non-closed surfaces, having no movement limiting ties in their areas or at their boundaries, are characterised by wider capabilities in terms of isometric transformation. Transformation of such surfaces can be done not only by local bends, but also by global deformations, leading to a change in curvature of the surface over their entire areas. The PWI has accumulated much experience in commercial application of the principles of geometrical bending and deployment of surfaces for building of metal panels for storage tanks, flat-coiled pipes and other similar parts. Partially, this experience can be extended to space shell structures. Examples of welded transformable structures, which can find application for building of both non-pressurised and pressurised space LSS are shown in Figure 9.

As far as truss structures are concerned, it should be noted that each of the types considered above has both advantages and drawbacks, sometimes mutually exclusive. An optimal combination of positive properties was achieved, in our opinion, in the building of a load-carrying truss structure, which can be used as the base line for orbital stations «Mir-2» or «Freedom». This truss is assembled of members which are prefabricated with high accuracy on Earth. Permanent joining of the members is performed in space. Choice of this type of joint is based on the fact that they provide high fatigue strength and rigidity of the truss for the entire period of operation.

The ground prototype of the highly mechanised installation «Stapel» (Figure 10) was built to study methods for construction of such trusses [5]. It provides for construction of a truss in several stages. The first stage of the process flow diagram (preliminary assembly of the rod-type members of the truss) is performed manually by two operators, using auxiliary tools. Moving along the guides in properly equipped work places (Figure 11), the operators

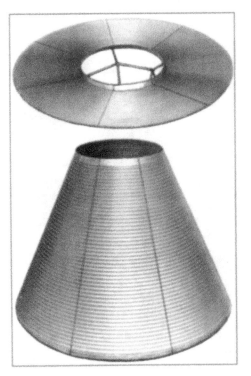

a

b

Figure 9 Transformable shell in a folded (a) and unfolded (b) states.

Figure 10 Highly mechanised installation «Stapel»: *1* — rod-type truss LSS; *2* — containers for rod materials; *3* — diagonal rod of truss LSS; *4* — process equipment for making permanent joints.

Figure 11 Position of an operator in a work place of the installation «Stapel» under neutral buoyancy conditions.

successively service all the working zones where the truss members are to be installed. Then a permanent joint is made between the preliminarily assembled node and rod members using automatic process equipment (second stage of the process flow diagram). And at the third stage the assembled fragment of LSS is moved to the working position, thus leaving the assembly zone of the staple for assembly of the next sections of the truss.

Each of the stages of the above flow diagram was separately investigated under ground conditions in order to simplify and facilitate the tests.

Preliminary assembly of structural members was optimised by the operators under neutral buoyancy conditions (Figure 12). This enabled the possibility of performing all operations of the assembly cycle with actual rod and node members to be checked, the working procedure of the operators to be optimised and the productivity achieved to be estimated. As a result reliable data were obtained proving that more than 21 running metres of a truss could be constructed by cosmonauts during one 5 hour EVA.

Technological experiments on making permanent joints were conducted under vacuum conditions. Different methods for making permanent joints in rod and node members were studied using the specially developed automatic electron beam hardware. The results obtained comfirmed that the available hardware and technology ensure quality joining of the LSS members.

Issues associated with movement of the assembled truss were handled using the «Stapel» installation under laboratory conditions with compensation for gravity forces. Also, a flow diagram was optimised for automatic operation of the final control elements, which provided a stepwise movement of the «Stapel» installation with respect to the truss LSS. Computer modelling was extensively used at all stages of the investigations.

The results obtained prove the possibility of using the «Stapel» installation not only as a means of construction of a truss LSS, but also for maintenance of this structure. Upon

Figure 12 Installation «Stapel» used for optimisation of preliminary assembly under neutral buoyancy conditions.

Figure 13 Design of the space unit for construction and maintenance of a truss LSS.

completion of construction of the truss, it can perform transportation functions to move payloads, lay down service lines, etc.

The ground tests enabled development of a design of a space unit (Figure 13) for construction and maintenance of a truss LSS, which can be taken out to space as part of any space module. It is characterised by wider capabilities, as compared with the above .

Therefore, the investigations completed by the PWI made it possible to build an orbital-production base required for construction of the main types of LSS in space.

1. (1987) *Cosmonautics in the USSR*, Moscow, 360 pp.

2. Gauthier, D.J. (1989) Steps toward a real International Space Station, *Ad Astra*, No. 9, pp. 28–29.

3. Paton, B.E., Lapchinsky, V.F., Bulatsev, A.R. et al. (1989) *See pp. 468–473 in this Book.*

4. Paton, B.E. (1989) Schweiβen im Weltraum, *Verlag des Deutschen Verbandes für Schweisstechnik*, Düsseldorf, DVS, pp. 112–115.

5. Syromyatnikov, V.S. (1991) APAS and a new class, *Pravda*, January 30.

Truss Structures in Orbital Complexes[*]

B.E. Paton[1], A.R. Bulatsev[1], E.S. Mikhajlovskaya[1], A.A. Zagrebelny[1], V.P. Nikitsky[2], A.V. Markov[2] and I.V. Churilo[2]

The concept of construction of various structures in open space is determined by the scope of the experience accumulated and depends upon many factors, such as the attitude of specialists to the problem as a whole, financial and technical support. The principle of the so-called modular structures gained ground during a period when issues associated with the feasibility of a prolonged stay of a man in space were handled. The first docked space vehicles «Soyuz»–«Soyuz» and «Soyuz»–«Apollo», as well as stations of the «Salyut» series, can be regarded as prototypes of the modern space stations. Design peculiarities of stations of this series did not provide for sufficient utilisation of their external surfaces for carrying out systematic research. The need to realise numerous national and international space research programs made it necessary to develop the infrastructure of the orbital stations, increase their energy resource and build various special-purpose and versatile outside platforms. A distinctive feature of the OC «Mir» is the availability of plenty of various-application superstructures and truss structures on the external surfaces of the modules.

This article considers issues associated with utilisation of large-size truss structures (LTS) for construction in space, methods for their building under open space conditions and raising their significance in fabrication of promising long-life large-size space objects.

It should be noted, based on specific features of construction in space, that spatial structures of a complicated developed configuration should have high strength and rigidity, low weight and high efficiency, and should be simple to assemble and repair during the process of their long-time operation. Truss structures meet most completely the main requirements imposed on construction of large objects in orbit. The more so in that on Earth these structures have long been in use, their load-carrying capacity has been thoroughly calculated and methods of their manufacture and assembly have been optimised. In addition, some of them are classical examples of engineering workmanship (open-work silhouettes

---

[*] (1991) *Proceedings of Conf. on Welding in Space and the Construction of Space Vehicles by Welding*, USA, Miami, Sept. 24–26, 1991. Miami, pp. 30–41.

of arch bridges, Eiffel tower, etc.). Speaking about space, LTS can be termed as objects which cannot be delivered to the place of operation in a working state; they have to be deployed or erected in orbit (aerials, ducts, solar cell power systems, load-carrying frames of stations, etc.). Since the very beginning of construction of various-purpose truss structures in space the trend has been to utilise new materials (along with traditional metallic materials), such as composite materials with polymeric, ceramic or metal matrices [1, 2], various carbon-reinforced plastics [3, 4], etc. Composite materials have such advantages as high strength and stiffness to mass ratio, dimensional stability at temperature gradients, isotropic mechanical properties, absence of sensitivity to UV radiation and humidity, etc. They can be employed in different cases depending upon the application. It should be noted at this point that non-metallic components of truss structures are joined to each other using metallic transition pieces [5–7] to ensure the required accuracy during the construction process. Members of truss structures can be joined mechanically (lock, bolted joints, etc.), or by welding or brazing. The applications of LTS can be divided into two types: masts (brackets) and beams, while by type of construction they are divided into transformable, deployable directly in space from the transport position into the working one, and assembled in orbit from individual and (or) transformable members delivered from Earth. Different types of LTS are employed in existing projects, some of which [8, 9] are sophisticated architectural complexes with living, research and production modules, orientation and observation systems. Given that assembly of LTS in orbit is a qualitatively new and technically complicated stage of exploration of space, specialists are of the opinion that both ground-based fundamental research and regular targeted experiments under actual conditions of a space flight [10] are equally important.

For a long time (30 years) the S.P. Korolyov RSC «Energiya»[2] and the PWI were involved in collaborative projects on development of various types of LTS. Some of them passed earth and field tests, while others have been long used as part of flight space complexes. The initial stage of numerous systematic experiments on optimisation of structural members and technology for construction of LTS was completed with an integrated experiment «Mayak» [11], as a result of which a tetrahedral transformable truss-mast structure (Figure 1) with special research equipment located on its end was deployed and extended to 13 m on board the OS «Salyut-7» in May 1986. Physically, the truss structure is based on a hinge-rod link design. The structure consists of separate sections comprising longitudinal and diagonal rods and cross-pieces of Al-6.2Mg alloy. In the deployed state the sections form rigid load-carrying triangles. All longitudinal rods have a foldable hinge in the centre, which allows the 15 m long truss to be compactly folded into a pack 500 mm high (transport position). Unfolding and folding of the truss is performed section-by-section using a special manipulator. The rigidity of the truss can be increased by welding hinges of the longitudinal rods during unfolding of the next section. However, this will exclude the possibility of any further folding.

A tetrahedral truss with longitudinal rods of a special profile, having an increased load-carrying capacity and improved performance, compared with the above structure, was used to build the power system for the «Kristall» module of the «Mir». This truss is a support structure for the solar battery (SB) panels, unfolding and folding of which are performed using a truss deployer. During an independent flight of the «Kristall» module two SB were extended to 5 and 10 m each, and 10 days after its docking with the «Mir» they were deployed to a full length of 15 m. These batteries were successfully employed in space from June 1990 (Figure 2). During that period they were folded, moved to a new place and again unfolded to their full length.

Efforts to increase the capacity of the power system of the «Mir» were made under the international «Mir»–NASA and Shuttle–«Mir» programs. Successful docking of the «Atlantis» spacecraft with the «Mir» was completed in November 1995. Docking was done via

Figure 1 Truss mast «Mayak» deployed at the OS «Salyut-7» (1986).

Figure 2 Reusable solar battery in operation (the photo was made from board the «Soyuz» spacecraft by French austronaut J.-L. Chretien during inspection of the OC «Mir»).

Technical parameters	«Mayak»	ESB	RSB
Overall dimensions of the device for truss unfolding and folding, mm	1100×800	2900×1400×1000	842×840×795
Weight per running metre, kg	0.9	4.4	2.1
Length in an unfolded state, mm	15000	18000	5000, 10000, 15000
Number of intermediate positions of the truss	2	1	2
Type of application	Reusable	Single	Reusable
Method for deployment	Electric drive ($U = 27$ V) Manual	Manual	Electric drive ($U = 27$ V)
Time of deployment to full length, min	15	> 60	20

the docking module (DM), comprising re-equipping SB (ESB) and reusable SB (RSB), which were transported attached to its external surface. Immediately after docking with the station, Russian cosmonauts moved to the service location and deployed ESB to its full length (18 m) using the «Topol-CB» truss deployer. A trihedral truss with longitudinal rods of a tubular section wound on spools in the form of strips, was used as a load-carrying structure. As far as RSB is concerned, it should be noted that this battery was in orbit for two years in a folded and restrained state as part of DM, after which it was moved to a work place and successfully unfolded to its full length. Proper selection of materials of hinges of LTS and use of a special lubricant led to avoidance of their probable jamming under conditions (deep vacuum and pressure) which could have been caused by «cold welding». Some general characteristics of the above trusses of a transformable type are given in the Table above.

Having accumulated sufficient experience in building and long-term operation of transformable trusses for heavy-weight solar panels, the specialists of the S.P. Korolyov RSC «Energiya»[2] and the PWI developed LTS and mechanisms for their unfolding and folding characterised by a higher load-carrying capacity. For example, for the power system of the Russian Module of the ISS, the PWI designed a device with a transformable truss 32 m long to be used as a load-carrying structure of the SB panels with an area 4 times in excess of that of the current ESB and RSB.

Field tests of the overall mock-up of such an unfolding-folding device with the truss used as the load-carrying SB structure were also conducted by American specialists. An emergency situation occurred after successful unfolding (Figure 3), which hampered the process of folding the load-carrying truss. Presumably it was caused by the impact of a temperature gradient resulting from transition of the spacecraft from the shadow to the solar side of the orbit.

An example of a truss-mast structure assembled from individual members in orbit is «Sofora», which is a tetrahedral truss structure with a cross sectional area of 0.5 × 0.5 m, 14.5 m long, with a weight of 5 kg per running metre. It was assembled and mounted by cosmonauts in 1991 on the external surface of the «Kvant» module of the «Mir» during four EVA. One of the specific features of this structure is utilisation of a material with shape memory effect as connecting rods (sleeves) to produce a permanent thermomechanical joint [12]. In September 1992 a propulsion system was installed at the end of the truss to provide roll guidance of the OC «Mir».

Figure 3 Full-scale tests of the overall mock-up of the unfolding (folding) device for the load-carrying
truss SB structure for the «Freedom» station (US «Discovery» space vehicle, 1984).

Truss structures used as the load-carrying frame of future space stations, allowing further
extension of their infrastructure, should meet the increased requirements for strength and
stiffness. This is caused by the fact that, being the basic component of the entire space object,
the frame should be designed for the same service life. Because of the impossibility of
delivering these structures to orbit in the working state, they should be assembled in orbit
from individual and (or) transformable elements. Load-carrying truss beams serving as the
station frame are intended for inhabited and non-inhabited modules and different types of
shell structures to be fixed on them [13–15]. All kinds of sensors, as well as flight and
experimental power systems (SB, solar gas-turbine units, concentrators), heat exchangers,
different types of aerials, booms, transformable structures, meteorite traps and screens,
containers for permanent and temporary storage of the EVA hardware, mobile astronomical
platforms and integrated platforms for preparation and launching of self-contained satellite
systems, etc., are mounted on them to provide continuous monitoring and evaluation of the
working state of the load-carrying trusses. Therefore, the truss itself, with its electric and
information cable systems as well as different-application ducts fixed on it, becomes a
continuous-action centralised R&D complex, a true test bed for systematic research to be
conducted in the fields of space engineering, technology and materials science, natural
resources of Earth, environmental monitoring, natural and technogenic environments around
the station at different distances from it and processes of degradation of structural materials.
Experimental studies [16] and analysis of actual flights [17] showed that various external
effects induced high vibrations in load-carrying extended truss structures, and, when a station
moves from shadow into an illuminated part of the orbit, thermoelastic axial stresses which,
according to calculations [18], amount to 30 % of the tensile strength of the material, have
a serious impact on functioning of the main equipment. Therefore, construction of the
load-carrying extended truss structures in orbit should be preceded by a careful selection of
materials, comprehensive calculations of strength characteristics using efficient solutions on
decreasing the thermal effect [17] and suppressing vibrations [19, 20].

Assembly-erection operations and deployment of LTS in space can be performed both manually and using automatic devices, manipulators and robots [21, 22].

Certain experience in assembly and erection of load-carrying LTS is available [23, 24]. However, it concerns simulation of these processes on earth under conditions that simulate space conditions (short-time zero gravity, neutral buoyancy tank).

Specialists of the PWI and S.P. Korolyov RSC «Energiya»[2] made a substantial contribution into the bank of world experience on optimisation of the technologies for manual assembly of load-carrying LTS and joining their members by welding and brazing [25, 26]. For a long time the above institutions were involved in the collaborative development efforts, which resulted in building of the working mock-up of the assembly-erection device «Stapel» [27]. Its design is determined by a combination of the following technological operations to be performed for construction of LTS:

• assembly of rod and nodal members of the truss by operators, and joining these members (mechanically or using welding technologies);

• movement of a finished fragment of LTS to a position, where the next portion to be assembled comes to the operators' work zone.

The schematic of one of the initial variants of design of the «Stapel» device is shown in Figure 4. The basic principle of construction of LTS in this case is movement of finished sections of the truss, measuring $3 \times 3 \times 3$ m, assembled manually by the cosmonaut-operators, along a base rigidly fixed to the external surface of the station (conveyor assembly of cars is an analogue of this operation). Development of such LTS was timely and topical, as at that time the «Freedom» (Figure 5) and «Mir-2» projects were underway, where the frames were planned to be made of high-capacity load-carrying trusses more than 100 m long and with a cross section of 5×5 and 3×3 m, respectively, equipped with many docking terminals

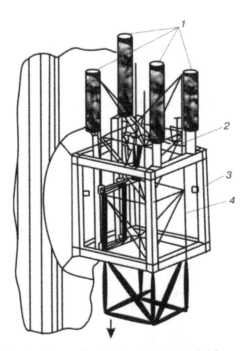

Figure 4 Schematic of the assembly-erection device «Stapel» intended for construction of the frame of the OC «Mir» (variant): *1* — container with rods; *2* — operator's platform; *3* — base; *4* — rod elements of the truss.

Figure 5 Project of the American orbital station «Freedom« (variant of 1986–1992, not realised).

to accommodate pressurised and non-pressurised modules of different applications, and to receive cargo and people delivered to the station by cargo and passenger spacecraft.

The process of manual assembly of rod and nodal members of the truss structure with a cross section of 3×3 m using the «Stapel» device was optimised by operators during neutral buoyancy tests. The rod member comprises two opposite hollow tubular cones made from Al-6.2Mg alloy. The tests showed that, providing that the work stations are properly arranged, two operators during one EVA (5 h) can assemble a tetrahedral truss 45 m long. The weight of 1 running metre of this truss is about 9.6 kg. The tests on producing permanent (brazed or welded) joints in rod and nodal members of LTS, carried out in vacuum ($1 \cdot 10^{-4}$–$5 \cdot 10^{-5}$ mm Hg) using the specially developed automatic electron beam equipment, were also a success. Tests of the system for movement of the finished truss relative to the base using the «Stapel» device were conducted under laboratory conditions, involving compensation for gravity forces and including optimisation of the stepwise movement. These tests also showed that the mechanisms tested could be employed not only as a means for construction of LTS, but also as an elevator for handling of payloads, laying of service lines, etc., after completion of construction of the truss frame.

American specialists conducted tests on a procedure for manual assembly of a truss structure with a cross section of 5×5 m both in a neutral buoyancy tank and on board one of the Shuttle space vehicles (Figure 6).

Figure 7 shows a schematic of the other modification of the assembly-erection device «Stapel», characterised by a different principle of assembly of LTS, namely the process of construction of the truss is carried out by two operators using two independent platforms. Each of the platforms is moved along the edge of the already assembled section of LTS for assembly of the next sections (in analogy with a track-laying machine). Physical connections between the platforms, LTS and the orbital station proper are provided through a frame rigidly attached to the load-carrying girder of the station casing, which ensures a reliable fixation of the truss being constructed. In this case the assembly-erection device «Stapel» can be employed to build structures of different configurations, such as straight-line, cross-shaped, closed, flat, box-section, etc.

The ISS project, the construction of which begins within the next few days, involves a modular design with partial utilisation of load-carrying transformable truss structures as intermediate frames between modules both in the American and Russian Modules of the

Figure 6 Optimisation of manual assembly of the load-carrying truss structure under field conditions (US Shuttle «Atlantis», 1985).

station (Figure 8). Experts involved in the investigation of the ISS modules even now note, and the the many years of experience in operation of the «Mir» prove that a radical physical change in architectural appearance of future generations of the stations is required. This is caused by the fact that ISS, like previous stations, is characterised by mutual shadowing of areas of the neighbouring SB modules at a change of the angle of incidence of solar rays, heating of the neighbouring regions of the station by a reactive jet of the orbit control system engines [28], hard-to-control orientation of ISS, contamination of the outer atmosphere of the modules and difficult access to the station by transport space vehicles. Despite the fact that physically the ISS project successfully solves the problems of development of an infrastructure, we have to state with regret that its design is based on a previous construction concept, and that ISS in fact has retained all the drawbacks characteristic of the currently functioning station «Mir».

At the same time, there are projects [29–33] which share the concept of spatial isolation for construction of orbital complexes. Some of them [30] suggest that a space station should comprise a load-carrying frame in the form of a closed girth truss structure (Figure 9), housing a base unit with modules located at the centre. It can be easily seen from Figure that this construction has a number of advantages, such as absence of shadowing of modules and SB, substantial decrease in the level of contamination of the station and its systems due to engines of the spacecraft servicing the station, increase in viability of the load-carrying truss in the case of failure of some of its fragments, easy-to-control orientation and, therefore, economical operation. All technical capabilities are available now for construction of such stations. This is proved by the results of many years of experience in assembly and erection

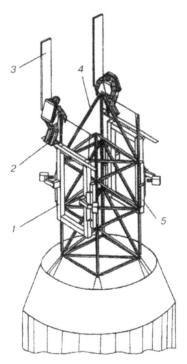

Figure 7 Schematic of assembly-erection device «Stapel» intended for construction of frame of the «Mir-2» station (variant): *1* — platform A; *2* — welding operator's platform; *3* — container with rods; *4* — rod members of the truss; *5* — platform B.

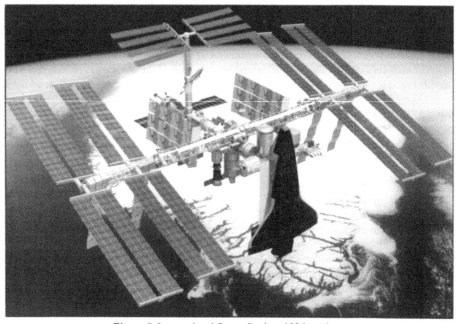

Figure 8 International Space Station, 1996, project.

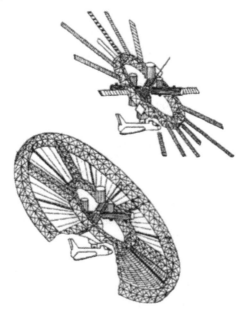

Figure 9 Orbital station of a new generation (XXI century, one of the projects).

of truss structures, as presented in this article. However, we have to admit objectively that their realisation can be done only in the future.

There is no doubt that construction of ISS, which has been started this year, is among the global projects of the XXI century. However, even now it is clear that in future projects we will hardly manage without LTS, which will be used as frames of vast power stations, launching complexes for flight to other planets, etc. We would like to believe that the situation described in this article will be a turning point in the concept of orbital construction, and that the ideas and practical experience of its authors will be embodied in reality.

1. Tenney, D.R., Sykes, G.F. and Bowles, D.E. (1985) Composite materials for space structures, In: *Proc. of 3rd Eur. Symp. on Space Environment*, Noordwijk, 1985. Paris, pp. 9–21.

2. Roy, P. and Mamode, A. (1985) Metals matrix composite materials for structural applications, *Ibid.*, pp. 185–190.

3. Babel, H.W., Shumate, T.R. and Thompson, D.F. (1986) Microcrack resistant structural composite tubes for space applications, In: *Proc. of 18th Int. SAMPE Techn. Conf.*, Seattle, 1986. Covina, Vol. 18, pp. 429– 439.

4. Bowles. D.E. and Tenney, D.R. (1986) Composite tubes for the space station truss structure, *Ibid.*, pp. 414–428.

5. Franz, J. and Laube, H. (1986) Strength of carbon fibre composite: titanium bonded joints as used for SPAS-type structures, In: *Proc. of Conf. on Compos. Des.*, Noordwijk, 1985. Paris, pp. 301–308.

6. Russell, C., Poorman, R., Jones, C. et al. (1991) Consideration of metal joining processes for space fabrication construction and repair, In: *Proc. of 23rd Int. SAMPE Techn. Conf. on Adv. Mater. Afford Processes*, Kianesha Lake, New York, 1991. Covina, pp. 555–567.

7. Bachkov, M.Yu. and Martyushov, V.F. (1990) Clamps and fixing rods to perform technological operations in open space, In: *Problems and prospects of space industrialisation*, Proc. of 24th Tsiolkovsky Readings, Kaluga, 1989. Moscow, pp. 81–87.

8. Riel, F.D. and Morata, L.P. (1992) Space station Freedom preintegrated truss configuration, *AIAA Pap.*, No. 1308, pp. 1–12.

9. Brekke, M. and Duncan, E.F. (1994) International Space Station Alpha payload, *Ibid.,* No. 4668, pp. 1–9.

10. Witt, W.P. (1989) Ground and flight experimentation in future large space systems development, In: *Space — New Commun. Opportun.*, Proc. of 34th Ann. AAS Int. Conf., Houston, 1987. San Diego, pp. 327–331, 333–335.

11. Paton, B.E., Zagrebelny, A.A., Ivanov, V.M. et al. (1989) The first stage of work on construction of large-size structures for the benefit of industrial mastering of space, In: *K.E. Tsiolkovsky and space industrialisation*, Proc. of 23rd Tsiolkovsky Readings, Kaluga, 1988. Moscow, pp. 80–83.

12. Semyonov, Yu.P., Belousov, P.M., Blagov, V.D. et al. (1993) Experiment «Sofora» — integrated investigations on construction of large-size truss structure in orbit, In: *Proc. of Int. Conf. on Large-Size Space Structures*, Nizhnij Novgorod, 1993. Moscow, p. 83.

13. Szyszkowski, W. and Glockner, P.G. (1990) The use of membrane structures in space, *Int. J. Space Structure,* **5**, No. 2, pp. 106–129.

14. Terentiev, Yu.K. (1996) New discovery of air space structures, In: *Proc. of 29th Tsiolkovsky Readings*, Kaluga, 1994. Moscow, pp. 125–126.

15. Samilov, V.N., Burmenko, E.Yu., Pilishenko, I.S. et al. (1990) Large-size structures of transformable volume and technology for their manufacture, In: *Abstr. of pap. of Sci.-Techn. Conf. on Large-Size Space Structures*, Sevastopol, 1990. Moscow, pp. 79–80.

16. Gudramovich, V.S., Baranov, N.G., Galkin, V.F. et al. (1990) Experimental studies of fragments of large-size space structures under static and dynamic loading, *Ibid.*, p. 41.

17. (1991) New European arrays believed adequate for Habble, *Aerospace Daily*, No. 9, pp. 71–72.

18. Luts, J., Allen, D. and Haysler, H. (1988) Finite-element model for analysis of thermal elasticity of large composite space structures, *Aerokosm. Tekhnika*, No. 6, pp. 57–65.

19. Horner, J.B., Rutterman, D.J., Meckle, P.H. et al. (1994.) Effect of actuator coupling on active vibration control of flexible structures, *Contr. and Dyn.*, **17**, No. 1, pp. 214–217.

20. Li Junbao and Zhang Lingmi (1997) Vibration control of a space truss structure with a piezoelectric active member, In: *Transact. of Nahjing Univ. Aeron. and Astron.*, **14**, No. 2, pp. 177–185.

21. (1990) NASA developing telerobotic system to automate assembly in space, *Aviat. Week and Space Technol.*, **133**, No. 10, pp. 197–199.

22. Bogomolov, V.P., Koval, A.D., Senkevick, V.P. et al. (1990) Space robots — problems and prospects of development , In: *Abstr. of pap. of Sci.-Techn. Conf. on Large-Size Space Structures*, Sevastopol, 1990. Moscow, pp. 30–31.

23. (1984) NASA work-platform studies may lead to real space construction, *NASA Activ.*, **15**, No. 9, pp. 12–13.

24. Yuzov, N.I. (1992) Problems of ground optimisation of large-size truss structures in neutral buoyancy, In: *Proc. of Seminar on Large-Sized Truss Structures*, St.-Petersburg, 1991. St.-Petersburg, pp. 17–18.

25. Paton, B.E., Semyonov, Yu.P., Skorobogatov, S.A. et al. *Unit of transformable volume.* USSR author's cert. 300642, Int. Cl. E 04 H 12/18, priority 10.10.88, publ. 01.09.89.

26. Dzhanibekov, V.A., Zagrebelny, A.A., Gavrish, S.S. et al. (1991) *See pp. 184–190 in this Book.*

27. Paton, B.E., Kryukov, V.A., Gavrish, S.S. et al. *Astronaut's work station device.* Pat. 5.779.002, USA, priority 14.02.96, publ. 14.07.98.

28. Pochelle, W.C., Hughes, J.R., Leahy, K.S. et al. (1995) Plume impingement heating of International Space Station (ISS), *AIAA Pap.*, No. 2132, pp. 1–27.

29. Lang, M. *Modular space station.* Pat. 5.156.361, USA, Int. Cl. B 64 G 1/10, publ. 20.10.92.

30. Paton, B.E., Semyonov, Yu.P., Kirilenko, O.V. et al. *Space vehicle.* Pat. 2072951, Russia, Int. Cl. B 64 G 1/1, priority 03.12.92, publ. 10.02.97.

31. Malyshev, G.V. and Kulkov, V.M. (1994) Concept of vertical layout of long-time orbital station, In: *Proc. of 28th Tsiolkovsky Readings*, Kaluga, 1993. Moscow, pp. 54–63.

32. Aldrin, B. *Space station facility.* Pat. 5.134.789, USA, Int. Cl. B 64 G 1/10, publ. 09.02.93.

33. (1997) Power from space for a sustainable world, Coll. pap. of 47th IAF Congr. on Enlarging Scope Appl., Beijing, 1996. *Acta Astronautica*, **40**, No. 208, pp. 345–357.

Space Transformable Reusable Solar Batteries[*]

B.E. Paton[1], A.R. Bulatsev[1], M.I. Morejnis[1], B.I. Perepechenko[1], D.V. Beletsky[1], N.K. Poshivalov[1], A.A. Zagrebelny[1], V.S. Syromyatnikov[2], V.P. Nikitsky[2], E.M. Belikov[2], E.G. Bobrov[2], Yu.I. Grigoriev[2], A.V. Markov[2] and S.A. Gorokhov[2]

In the last months of 1998, mankind launched one of the most expensive projects of the elapsing millennium — construction of a cyclopean space structure, i.e. International Space Station, with a developed infrastructure and tremendous and highly promising research potential. The ISS, which will be 110 m wide, 80 m long and with a total weight of 450 t, will be equipped with two independent high-capacity power systems, American and Russian, allowing performance of power-consuming experiments and research which were impossible on stations of previous generations («Salyut», «Skylab» and «Mir»). The substantial amount of energy resources of the Station will be generated by solar batteries (SB) with a surface area of hundreds of square metres, erection and deployment of which, because of their large dimensions, will have to be done directly in orbit. Building of such unique power stations is a difficult engineering problem. The efforts in this area were initiated by Soviet and US specialists in the middle of the 1980s. They planned to test prototypes of new hardware under service conditions on board the OC «Mir», which was being constructed in those years, and on board the shuttle series spacecraft.

At the end of May 1990 the Soviet Union launched the «heavy» module «Kristall» to the «Mir». The module was equipped with two recently developed SB with wings to be extended to 5 m after separation of the nose fairing to supply life-giving energy (3.4 kW) to the module in order to ensure its independent flight and docking with the station. (Failure to deploy these wings would have led to a loss of one of the most expensive research modules full of the most sophisticated power-extensive research and technology instruments). At the end of the independent flight of the module, prior to its docking with the station, the span of

* (1999) *Avtomaticheskaya Svarka*, No. 10, pp. 86–96.

each wing was increased to 10 m by commands transmitted from Earth, which raised the power potential of the module to 6.8 kW.

A few days after the launch, the «Kristall» module successfully docked with the station and, using a special airborne manipulator, was moved from the axial to the side docking module, where it remains to date. At the same time the batteries were deployed to their full length (15 m each) and connected to the integrated power system of the station. This led to a substantial increase (30.6 %)* in its total power and enabled widening of the scope of research scheduled for performance during the rest of the service life of the station [1]. So, the first task posed for developers of such a high-power SB (usable area of each battery was 36 m²) was successfully solved. The second task, not less complicated, posed for scientists and engineers who created this extended structure, which in fact determined the overall dimensions of the station, was to ensure the possibility of controlling the span of the wings, i.e. fold them and unfold either fully or partially in accordance with the dynamics of flight of the station, as well as in docking «heavy» vehicles to it or in reduction of power consumption. All these operations had to be performed during many years of service in the aggressive space environment. In addition, it was necessary to provide for the possibility of disassembling the wing after expiry of its assured service period and replacing it. This means that the structure of the unfolding-folding device (UFD) had to be light-weight, reliable and controllable both from Earth and directly from on board of a station.

It should be noted that until that time both Soviet and foreign manned and unmanned vehicles used SB which could be deployed only once, and which could not be replaced during service. Meanwhile, often there were cases where equipment of a vehicle functioned normally and the vehicle itself could have been in operation for a long time performing its tasks, if not for rapid degradation of solar cells and impossibility of their replacement, which led to its loss.

Therefore, the two major challenges for developers of a new system were to build UFD and a load-carrying transformable mast, i.e. the SB carrier.

In summer 1986, on board the OS «Salyut», cosmonauts L. Kizim and V. Soloviov conducted experiment «Mayak» on unfolding and folding a tetrahedral hinged truss structure 13 m long with a construction height of 0.45 m [2]. The experiment initiated a new stage in exploration of space as a human habitat, i.e. construction with direct participation by building operators.

Specialists who developed equipment for that experiment had to build a light-weight load-carrying transformable extended structure, capable of being folded and unfolded to a preset length many times and carry research hardware to long distances from the space vehicle to provide monitoring of the environment. For that, a replaceable platform with research instruments, and connected to the space station via electric and telemetry cables, was fixed to the top of the truss folded in a container.

This truss is a hinge-lever structure [3] based on a many times repeated octahedron (Figure 1a). Two racks parallel to the truss axis are added to apices of the octahedron along two edges of the base in parallel to the truss axis. Each section of the truss contains longitudinal and diagonal rods and cross-pieces in the form of a milled I-beam of a complex contour. Taken together, these elements form rigid load-carrying triangles. All the racks parallel to the truss axis are foldable and made in the form of a hinged double-link with an off-centre hinge in the middle of the rod. This allowed the truss to be compactly arranged in the transport position. The rack on the side opposite to the hinge also has an off-centre spring to ensure two fixed positions of the rod, i.e. folded and unfolded. Cross-pieces with

_____
* (1993) Engineering note P31170-106. Ways of increasing power potential of the OC «Mir». The data were prepared by V.V. Teslenko, an associate of S.P. Korolyov RSC «Energiya»².

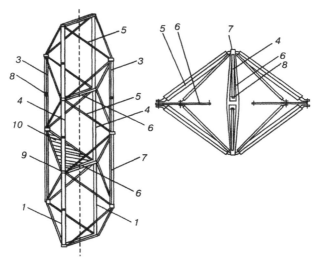

a   b

Figure 1 Transformable truss in the working (a) and packed (b) state: *1* — base edge; *2* — truss axis; *3* — rack; *4* — longitudinal rod; *5* — diagonal rod; *6* — cross-piece; *7* — off-centre hinge; *8* — off-centre spring; *9* — cross-piece axis; *10* — load-carrying triangle.

diagonal rods connected to them make up the truss set. When the truss is arranged in the transport position, four diagonal rods, out of the eight comprising each section, are rotated about axes of the cross-pieces, and four other diagonal rods of the same section connected to them, are rotated about the axis parallel to the axes of the cross-pieces to transform the octahedron into a plane rhombus. Four longitudinal hinged rods are folded along diagonals of the rhombus (Figure 1b). Therefore, the 15 m long truss is transformed into a pack 500 mm high.

In addition to the container with a packed truss, the UFD set also comprised a manipulating device, i.e. truss deployer [4], which formed and extended the truss automatically or manually in a stepwise manner (step 0.45 m) during its unfolding or folding. Figure 2 shows a schematic of the «Mayak» experiment.

In compliance with requirements for the research and experimental equipment to be located on board a space vehicle, prior to its delivery to the station UFD was subjected to a set of standard ground tests for «survivability», safety and ergonomic characteristics, i.e. its compatibility with an operator working in a spacesuit in EVA to assemble it and conduct an experiment.

To meet the above requirements, the developers paid special attention to evaluation of dynamic and strength characteristics of the deployed truss and performance of mechanisms forming it. For that a package of special tests was completed:

• starting, setting up and service life tests (Figure 3) — at the PWI;

• vibration tests and tests to determine natural and resonance frequencies — at the Moscow Aviation Institute;

• ergonomic tests in a neutral buoyancy tank using neutral buoyancy mock-ups of the station and the «Mayak» experiment equipment, and in short-time zero gravity on board the flying lab (FL) IL-76K (Figure 4), as well as tests to determine tolerable static loads (transverse, longitudinal, torsion) for the truss under zero gravity conditions on board the FL at the Yu.A. Gagarin CTC[3], tests to determine tolerable impact loads (random and transverse) for the case of emergency situations (Figures 5 and 6) at one of the test grounds of the S.P. Korolyov RSC «Energiya»[2].

The completed package of ground tests confirmed the validity of the design concepts for performance of the space experiment.

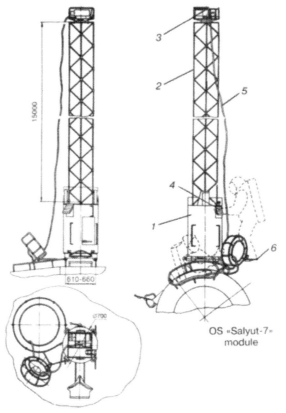

Figure 2 Schematic of the space experiment «Mayak»: *1* — truss deployer; *2* — truss; *3* — payload
platform; *4* — control panel; *5* — cable; *6* — operator's platform.

*During the field experiment, UFD was taken out of the station, located on its external
surface on a platform preliminarily mounted by the cosmonauts and connected to the power
and telemetry systems. Then the operators manually deployed the truss with a platform for
the research equipment installed on its end to a length of 13 m. Probing of the environment
at different distances from the station, measurements of strains in the truss and measurements of loads acting on it were performed in that experiment. After that the truss was folded.
These field tests revealed design advantages and disadvantages of the mechanisms for
formation and extension of the truss and refinement of strength characteristics and load-
carrying capacity of the hinge-rod structure under station flight conditions. Unfortunately,
the short period (4–5 h) of stay of the deployed truss in open space did not allow the
developers to correctly and comprehensively evaluate the effect of dynamics of variations
in temperature conditions on the mechanisms of the truss deployer manipulator, and
especially on the truss proper, which, in the opinion of the specialists in view of the accepted
method of stepwise deployment, could have a substantial impact on quality of functioning
of the equipment remaining under open space conditions for many years.*

*The main conclusion made by the specialists and developers of the equipment was that,
despite the limited scope of the information received, which was attributable to the short
time of the experiment and the small number of telemetry channels allocated for it, the
various-application transformable truss structures could be employed in a new generation
of the future space vehicles.*

The space experiment «Mayak» coincided in time with the beginning of construction
of the multiple-purpose complex «Mir». It was planned that the station would accommodate

Figure 3 Starting and setting up tests of equipment for the space experiment «Mayak» (1985, the PWI).

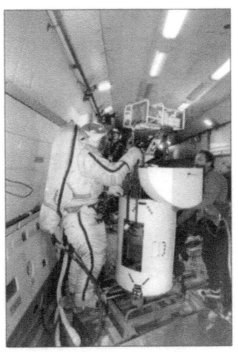

Figure 4 Ergonomic optimisation of UFD under zero gravity conditions on board the flying lab IL-76K (1986, Yu.A. Gagarin CTC[3]).

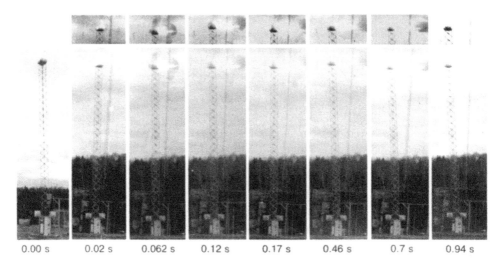

| 0.00 s | 0.02 s | 0.062 s | 0.12 s | 0.17 s | 0.46 s | 0.7 s | 0.94 s |

Figure 5 Cinematic pattern of the tests for determination of a tolerable bending moment for the «Mayak» truss (1986, test ground of the S.P. Korolyov RSC «Energiya»[2]).

Figure 6 Loss in stability of the «Mayak» truss affected by a transverse load of 300 N on a shoulder 15 m long.

Figure 7 Horizontal jig for neutralising weight of the SB panels and the load-carrying truss.

seven modules of different applications with a developed and higher-capacity (compared with stations of the «Salyut» series) power supply system. Nevertheless, even that did not allow performance of power-consuming research and experiments, especially those associated with space technology. Therefore, the chief designer decided to urgently build a new type of flight system for reusable solar batteries (RSB) with a usable area of more than 30 m², which was much in excess of that of the largest span of a single-deployment SB to be used on large space vehicles and modules of the «Mir», both those which had been already introduced into operation and those which still were on staples. The first pair of future advanced batteries of an increased capacity had to be erected on the most «voracious» technological module «Kristall», the assembly of which at that time was in full swing. Undoubtedly, that daring decision imposed a high responsibility on the developers, as the extremely condensed period of design, manufacture and comprehensive cycle of ground tests, binding upon all flight systems, left no room to specialists for any serious irremediable mistakes. A large creative team was formed, the members of which were experts from the PWI and S.P. Korolyov RSC «Energiya»[2]. In addition to the RSB proper, it was necessary to develop also unique test jig equipment, as no experience in working with such extended structures was available. The efforts of the team resulted in designing and manufacturing of horizontal and vertical jigs with systems for differentiated (during unfolding-folding process) neutralisation of weight of the load-carrying truss structure of variable length and the SB panels (Figures 7 and 8). That allowed starting and setting up operations and a vast set of design-optimisation tests to be conducted, including in a thermal vacuum chamber, the strength characteristics of the truss and electrical parameters of SB to be specified, and the required pre-flight preparation of the product to be completed.

Design, manufacture, calculations, ranges and scopes of mock-up, design-optimisation and acceptance tests described in the work statement for development and delivery of the product were specified in the corresponding industry regulatory documents and standards. All this had to be done with no prior experience available in design and operation of such

Figure 8 Vertical jig for neutralising weight of the SB panels and the load-carrying truss.

extended structures, or of any extended and reliable functioning of precision mechanisms of the truss deployer manipulator under conditions of continuous thermal cycling. So it was impossible to theoretically calculate a multiple-hinge (more than 250 hinged joints in a length of 15 m) structure and correctly select materials for friction pairs and their lubrication. Also it was necessary to allow for the fact that the truss deployer with a deployed SB was to be located on a drive to follow the position of the Sun at a velocity of 4°/min, while the station itself would not «sleep» but perform manoeuvres, change orientation, and receive «heavy» modules, passenger and cargo vehicles, i.e. it would «live».

Designs of the transformable truss structure and the RSB truss deployer were based on principles verified in the «Mayak» experiment. The tetrahedral truss folded into a pack and accommodated in a container of the truss deployer (Figure 9) is extended in a stepwise manner to a length of 15 m, pulling out an «accordion» of solar panels from its own container fixed to the truss deployer, using a manipulation platform comprising fasteners and fixing members which form and support the truss being deployed. The upper panel is connected to the retractable truss using a cross-arm fixed on the truss end, while the lower panel is fixed in the container. A rope runs through the peripheral sides of the panels and the ends of the cross-arm to the tension drive of the truss deployer to prevent twisting of the wing of a deployed battery in rotation of RSB during tracking the Sun or swinging at random disturbances, e.g. emergency docking of «heavy» vehicles onto the station. The rope is tensioned with a small (up to 5 N) continuous force. It relaxes in deployment of the battery and is tightened in its folding. The angle between panels of SB in the unfolded position is

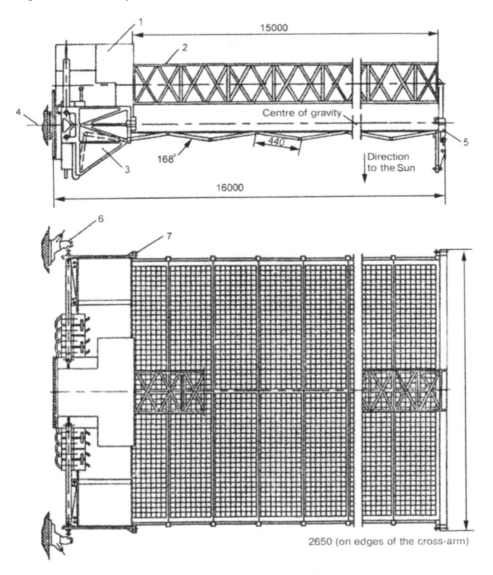

Figure 9 Truss deployer for RSB: *1* — container for arrangement of the truss; *2* — truss being deployed; *3* — container for folding of panels; *4* — RSB rotation axis; *5* — cross-arm; *6* — mechanism for holding RSB (2 pcs); *7* — mechanism for holding the cross-arm.

168°. It is determined by the linear dimensions of the load-carrying truss and the panels. After full deployment each panel should «grip» the rope and thus fix itself in the working position. In folding of the truss, a special device located at the upper end of the container for the panels de-fixes the nearest panel, while the rest of the panels fixed on the rope push it to the receiving device of the container. This is the procedure of ordered packing of the SB panels.

Experimental verification of strength, rigidity and dynamic characteristics of the truss and its load-carrying capacity was carried out on a vertical jig for weight neutralising using dimension-weight mock-ups of the SB panels for all of the assigned service positions of the load-carrying truss, i.e. in a folded state and with the truss extended to 5, 10 and 15 m.

Figure 10 Tests of a fragment of the load-carrying RSB truss to cycling fatigue life.

The truss was also subjected to strength tests by applying repeated-static (at a frequency of 0.1–0.2 Hz) and cyclic (at a frequency of 2.5 Hz) loads.[*]

A fragment of the truss was tested to cyclic fatigue life (Figure 10) of its individual members and the structure as a whole (the number of assigned cycles to fracture at the axis of the root chord of the truss was $1.5 \cdot 10^6$).

It was established as a result of these tests that the rigidity characteristic of the truss [$EJ_{x,y} = (8\text{–}12) \cdot 10^8$ N·cm^2, $GI_p = (1\text{–}2) \cdot 10^8$ N.cm^2], its load-carrying capacity ($M_{bend} > 1000$ N·m and $M_{tors} > 150$ N·m of the bending and torsion moments acting both simultaneously and separately), as well as frequency characteristics ($f = 0.80\text{–}0.75$ Hz for 5 m and $f = 0.47\text{–}0.32$ Hz for 10 m with a logarithmic attenuation decrement not less than 0.6) were close to the calculated ones and in compliance with service requirements. Overloads on the SB panels were continuously monitored during the strength tests and were not in excess of a tolerable level of 0.35 g. Complete unfolding and folding of SB were performed after each test cycle, which eventually allowed determination of the guaranteed service life of SB.

Serious consideration was given also to estimation of the load-carrying capacity of the truss in its emergency state, i.e. loss of stability of one or several supporting chords (incomplete, unfixed deployment of a hinge(s)). These tests were separated into a special cycle and performed on a vertical weight-neutralising jig for different lengths of trusses ($l = 4.5$ m with weight-neutralising force $P = 150$ N and $l = 10$ m with weight-neutralising force $P = 250$ N).

The truss was tested by stepwise loading with a lateral force applied to the its apex, which induced a bending moment of up to 1000 N·m and torsion moment of up to 150 N·m at the truss base. Deformations of the truss at the load application locations were measured during the loading process.

---

[*] The work was performed at TsNIIMASh.

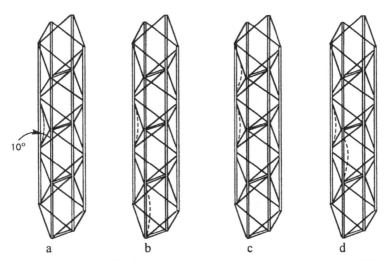

a   b   c   d

Figure 11 Evaluation of the load-carrying capacity of the truss in the emergency position of a longitudinal rod of one of the sections (a), two longitudinal rods in different sections (b), two longitudinal connected rods in neighbouring sections (c) and two longitudinal rods in one sections (d).

Analysis of the test results showed that the truss retained load-carrying capacity in the following cases:

• application of the maximum static bending moment of up to 900–1000 N·m and a transverse moment of up to 100–200 N·m in a direction parallel or normal to the SB panels, as well as at an angle of 45° to them, with one of the sections being in the emergency position (longitudinal rod bent under to an initial angle of 10°) (Figure 11a);

• application of a bending moment of up to 1000 N·m in the case of the emergency position of simultaneously two longitudinal rods located in different planes and sections (Figure 11b);

• application of a bending moment within a range of 360–900 N·m in the case of the emergency position of simultaneously two longitudinal rods connected to each other and located in neighbouring sections (Figure 11c);

• application of a bending moment of no more than 200 N·m in the case of the emergency position of simultaneously two longitudinal rods located in one sections at the same level in the direction of application of the load (Figure 11d).

It was established that the truss did not only retain its load-carrying ability at the unfixed position of one of the chords, but also achieved a failure-free performance with several chords being in the emergency state. However, in such cases folding of SB cannot be done without participation of the operators working in EVA. The most labour- and time-consuming and expensive tests were those for verifying performance of all devices of RSB under conditions which comprehensively simulated the actual ones in all stages of operation of the product [5]. These tests were conducted using a vertical weight-neutralising jig located in the thermal vacuum chamber with a volume of 650 m^3 (Figure 12), which was equipped with a space simulator (pressure of 6.6·10^{-5} Pa), cold black space simulator (temperature of cold screens below −186 °C), main radiation simulator (exposure of 1400 W/m^2), body radiation simulator (directed heat flow of 140 W/m^2), measurement-calculation system and controlling test system.

Simulation of thermal parameters of natural conditions was made by sun (55 min)–shadow (35 min) cycling and by long-time impact of shadow (14 h) and sun (17 h).

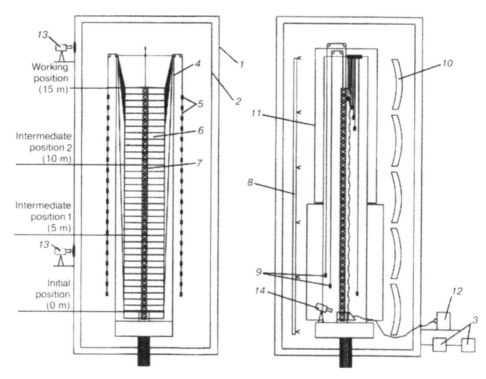

Figure 12 Schematic of tests of RSB in a thermal vacuum chamber: *1* — thermal vacuum chamber
($V = 650$ m^3); *2* — refrigerator; *3* — vacuum system; *4* — weight neutralising jig; *5* — weight
neutralising system for SB panel; *6* — SB; *7* — load-carrying truss; *8* — body radiation simulator;
9 — weight neutralising system for the truss; *10* — main radiation simulator; *11* — screens; *12* — control
panel for RSB UFD; *13, 14* — external and internal TV cameras.

Multiple unfolding-folding cycles of SB at boundary and rated supply voltage of 23, 27
and 34 V were performed after each cycle of cooling (SB panels — to −65 °C) and heating
(SB panels — to +90 °C). The total number of the unfolding-folding cycles was 47. That
made it possible to obtain experimental proof of the required life of RSB (20 full unfolding-
folding cycles).

The set of ground tests enabled developers to reveal advantages and drawbacks of the
product. Under conventional earth conditions the product performed without failure. To be
fair, these tests detected insufficient rigidity of the UFD manipulator platform with the
mechanisms located on it for gripping and fixation of the load-carrying truss in its formation
(unfolding) and packing (folding). In the case of inaccurate weight neutralisation under earth
conditions or random lateral disturbances under the flight conditions, this could have led to
loss of the truss and SB or, to be more exact, to failure of its gripping by the fixation
mechanisms. Also, it was found out that because of the absence of a rigid fixation of the
panels at a set relative angle of 168° in the deployed state, the mechanism for a controlled
ordered arrangement of SB in the receiving container during packing failed to perform its
functions. This resulted in an emergency situation with folding of the battery. The problem
was that a panel located immediately near the receiving container could not be pushed into
it because of non-rigid fixation of the rest of the panels, but instead it hung over the container
(unfortunately that defect had not been detected at the proper time when the functioning of
the product had been tested using a horizontal weight-neutralising jig). The rest of the panels,
which were not rigidly fixed, began to randomly change their inter-panel angle, i.e. behave

in a non-controllable manner. And whereas in the process of sophisticated and extensive design-optimisation tests and design debugging the developers successfully handled the problem of rigid fixation of the platform by adding a number of extra devices, and thus ensured failure-free deployment and retention of SB, they failed to immediately solve the problem of rigid fixation of each of the deployed panels (36 in all) and, therefore, ordered piece-by-piece packing in the receiving container within a short time. The chief designer of the station made a compromise decision, i.e. to deploy batteries in the automatic mode, as had been specified in the work statement, and perform packing manually in folding SB by picking and arranging the nearest panel in the receiving container. Such a decision should be considered as forced, being caused by the terms of launching the «Kristall» module to orbit.

Calculations and investigations of strength and rigidity characteristics of the load-carrying truss, as well as tests of its parts, showed that the engineering intuition and experience gained during the process of building a similar structure in preparation for the «Mayak» experiment did not let the developers down. The truss withstood all the above dynamic and static tests without any problems.

To the delight of the developers the truss did not let them down either during the «Rezonans» experiment conducted several months after commissioning of the two «Kristall» module SB wings of a new generation, fully deployed in the automatic mode. The point of the experiment was as follows: a cosmonaut stepped onto a running track located in one of the compartments of the station and, by selecting the rate of running and swinging the station, set the root part of the RSB truss into oscillation. Meanwhile, the second cosmonaut was watching the behaviour of the truss and battery through a window which had marks deposited on its glass, allowing determination of the amplitude of oscillations at the beginning and end of the truss. Despite the fact that this experimental procedure was very simple, it caused some concern to the specialists responsible for safe operation of the station.

All their doubts were removed after the truss proved itself flexible, strong and reliable, and the results of the flight experiment «Rezonans» and ground tests coincided almost completely.

A new generation of the extended controllable SB was built but, as noted above, developers of that product had the task of ensuring the possibility of its movement to a new preliminarily prepared location or its full replacement after expiry of the assured period.

In 1995, according to the approved international missions «Mir»–NASA and Shuttle–«Mir», it was decided to move one of the active RSB to the «Kvant» module, as well as fold the other one to 3.5 m and lay it up (Figure 13). That was required to clear the docking module to meet the spacecraft shuttle, while the latter had different configuration and dimensions, as compared with the space vehicles «Soyuz» and «Progress».

The tasks associated with folding of the battery were successfully handled by cosmonauts V. Dezhurov and G. Strekalov during EVA, while from the inside the American astronaut N. Thagard provided a stepwise control of RSB using a special panel.

In November 1995 the third and the last RSB of this series was delivered to the «Mir» to replace the battery on the «Kvant» module, which had been in operation for 5.5 years and decreased its energy parameters more than twice.

The product was delivered to the station on board the «Atlantis» spacecraft as part of the docking module, fixed to its external surface in the transport position (Figure 14).

For a number of reasons the RSB delivered remained in the closed state for more than 2 years. And only an accident that led to a partial loss of SB of the «Spektr» module made it necessary to urgently compensate for a decrease in the energy level for the entire station and replace the worked-out RSB. It should be noted that such a long storage of packs with the truss and panels in the rigidly fixed transport position under conditions of deep vacuum and low temperatures was not studied by developers and was of serious concern. Therefore, as was specified in the programme of putting this product into operation, the

Figure 13 RSB battery in a form folded to 3.5 m.

cosmonaut-operators should be definitely present in the deployment zone of the battery during its unfolding. As it became clear later on, that was very appropriate. Folding of the worked-out battery went smoothly, but during unfolding of a new battery in the automatic mode, after the truss was pulled out to 4–5 m, the safety clutch of the deployment traction drive worked to cause an authorised shutdown of the RSB UFD manipulator. The high professionalism of cosmonauts A. Soloviov and P. Vinogradov and American astronaut D. Wolf, who performed manual control, made it possible to get out of that critical situation, i.e. pull out a few next steps of the truss and then coming back to the automatic mode, fully deploy the battery and put it into operation.

    In addition to the third set of RSB, a higher power deployable re-equipping SB (ESB) [6] 18 m long was delivered to the «Mir». That battery was mounted on a trihedral tubular

Figure 14 Transportation of RSB on board the Shuttle «Atlantis».

Table 1 Specifications of RSB and load-carrying trusses

Parameters	Experiment			Service systems			
	«Lockheed»	«Mayak»	RSB	ESB	LRSB*	ISB*	«Lockheed–Marietta»*
System mass, kg:	130.4	–	591	597	–	480	1111
UFD deployment mechanism	–	76	90	150	185	–	415
Load-carrying truss	–	13.5	30	80	165	–	–
SB panels	–	–	210	140	–	160	696
Panel container	–	–	90	80	–	60	–
Other units	–	–	171	147	–	–	–
Truss length, m	30.9	15	15	18	32	19	35
Area of solar cells, m^2	123	–	36	42	136	42	300
Specific mass, kg/m^2	1.06	–	16.4	14.2	–	11.4	3.7
Power, kW	13.7	–	5	6	20	6	30.8
Time of deployment using different methods, min	14 (electric drive)	15 (by 3 methods)**	20 (electric drive)	150 (manual drive)	(electric drive)	30 (by 3 methods)**	15 (electric drive)
Maximum bending moment, N·m	–	320	1000	1500	2300	1500	–
Torsion moment, N·m	–	50	140	200	–	150	–
First natural frequency, Hz	–	0.6	0.32–0.44	0.78	–	0.35–0.44	–
Oscillations decrement	–	–	0.8.–.1.2	0.14–0.16	–	–	–

*Design data. **Manual deployment, using electric drive and combined deployment.

truss [7–9], according to a different unfolding-folding diagram. Immediately after the delivery, Russian cosmonauts Yu. Usachev and Yu. Onufrienko placed and assembled ESB on a specially provided drive of the «Kvant» module and deployed it manually to full length.

Similarly to the RSB case, this model of SB is used now as a baseline to design and build improved transformable solar batteries (ISB)* for the Russian Module of ISS. Design of a new large-size reusable SB (LRSB), based on a similar transformation principle as RSB, has been completed. The length of the wing of this battery is 32 m.

The American module of ISS should be equipped with a transformable SB developed by the «Lockheed–Marietta» Company. Each such battery is 35 m long and 12 m wide. Its prototype is a full-scale active mock-up [10], which passed full-scale tests in 1984 during

---

* Data on ISB and «Lockheed-Marietta» battery for this article were prepared by D.M. Surin and O.S. Sarychev, associates of the S.P. Korolyov RSC «Energiya»[2].

an independent flight of spacecraft «Discovery» [11, 12]. Specifications of these modifications are given in the Table 1.

All the extended SB described in this article share one design peculiarity, which is the presence of a load-carrying transformable truss structure and truss deployer, as well as location of solar cells on separate panels folded like an accordion.

The expected service life of space stations is 15–30 years, while solar cells degrade within 5–7 years. Therefore, it is a big challenge now for developers of such long-time objects to increase the resistance of solar cells, improve their repairability and make them easy to replace. Specialists from many countries all over the world have tried hard to find solution for this problem, planning to build in the near future national long-time manned space stations, which will be equipped, naturally, with high-power SB. In the last years experts from the USA [13, 14], Russia [15], Japan, China, France and Germany [16–20] have made large efforts to develop thin-film SB, which have a smaller weight, can be coiled and are simpler to unfold and fold. In the near future such thin-film SB will be used in power systems of ISS. On the 10th of November, 1998, Russian cosmonauts G. Padalko and S. Avdeev installed a fragment of the experimental thin-film SB on the docking module of the OC «Mir» and started its full-scale tests by measuring electrical parameters. Similar investigations are conducted by the US specialists.

This article suggests one of the practical approaches to solving the problem of power supply to an orbital station, the topicality of which is well illustrated by the fact that the total usable area of SB of the designed and being already constructed ISS is about 10,000 m^2. Now these batteries have to be delivered to the station, deployed, operated, maintained and periodically «regenerated».

1. Syromyatnikov, V.S. (1991) APAS and a new class, *Pravda*, January 30.

2. (1991) In open space. TASS reports. *Ibid.*, May 29.

3. Paton, B.E., Bulatsev, A.R., Zagrebelny, A.A. et al. *Transformable structure*. USSR author's cert. 242664, Int. Cl. B 64 D 1/22, priority 04.10.85, publ. 01.09.86.

4. Paton, B.E., Semyonov, Yu.P., Bulatsev, A.R. et al. *Truss deployer*. USSR author's cert. 242792, Int. Cl. B 64 D 1/22, priority 25.11.85, publ. 01.09.86.

5. Chollvibul, R.W. (1987) Thermal deflection of a deployable and retractable structural mast, *Acta Astronautica*, **15**, No. 11, pp. 905–911.

6. Stesin, V.V., Zagrebelny, A.A., Polinov, Yu.S. et al. *Retractable rod of an unclosed profile*. USSR author's cert. 275177, Int. Cl. H 01 Q 64/02, publ. 22.04.70.

7. Gritsaenko, V.P., Berzhaty, V.I., Bobrov, E.G. et al. *Mechanism for retraction of the rod*. USSR author's cert. 274370, Int. Cl. B 64 G 1/22, publ. 03.05.88.

8. Podgorny, A.N., Denisov, V.P., Berzhaty, V.I. et al. *Extendable hoist*. USSR author's cert. 1563155, Int. Cl. B 64 G 1/22, publ. 08.01.90.

9. Kurkin, V.I. and Terentiev, Yu.K. (1987) Deployable space structures of ballistic and retractable elastically transformable members, In: *Design and fabrication of space devices*, Moscow, Nauka, pp. 72–76.

10. Paton, B.E., Bulatsev, A.R., Mikhajlovskaya, E.S. et al. (1991) *See pp. 484–495 in this Book*.

11. Coval't, C. (1984) Discovery fulfils sail mission objectives, *Aviat. Week and Space Technol.*, **121**, No. 11, pp. 40–42.

12. (1979) Experimental solar array wing Lockheed, *Space World*, No. 33, pp. 172–182.

13. Raushenbach, et al. *Ultra light-weight folding panel structure*. Pat. 4.384.163, USA, publ. 17.05.83.

14. Hamakawa, Yo. and Takakura, H. (1986) Space photovoltaic solar breeder by graphopi-taxial growth of thin film solar cells, *Acta Astronautica*, **14**, No. 3, pp. 439–444.

15. (1998) Flight of orbital complex «Mir», *Novosti Kosmonavtiki*, No. 23/24, p. 6.

16. Lendereux, P. (1986) La Construction de la Space Station debute dans un an, *Air et Cosmos*, No. 1094, pp. 33–37.

17. Schmid, M. (1985) The extendable and retractable mast ERM, In: *Proc. of 2ⁿᵈ Eur. Space Mech. and Tribol. Symp.*, Weersburg, 1985. Paris, pp. 29–33.

18. Foster, C. and Krishnakumal, S. (1986/1987) A class of transportable demountable structures, *Space Structure*, **2**, No. 3, pp. 129–137.

19. Onada, J. (1988) Two-dimensional deployable truss structures for space applications, *J. Spacecraft and Rockets*, **25**, No. 2, pp. 109–116.

20. Hedgepeth, J. and Miller, R. (1988) Structural concepts for large solar concentrators, *Acta Astronautica*, **17**, No. 1, pp. 79–89.

The April issue of Journal «Novosti Kosmonavtiki» for 2000 contains an article by I. Lisov «All wrinkles of Earth», which describes in detail a high-frequency radar mapping of the earth's land (SRTM) from board the manned spacecraft «Endeavour» (mission STS-99), which was performed during the period from 11 till 21 February. The experiments were conducted using an expandable truss that allowed a set of receiving arrays of the radar system to be carried out from the cargo compartment of the spacecraft and extended to a distance of up to 61 m (see the Figure).

The truss extended from the cylindrical container 1.36 m in diameter and 2.92 m long consisted of 87 sections of square cross section with a side of 79.25 cm and diagonal of 112 cm. In the folded state in the container, each section was 1.59 cm thick, and the height of the entire pack of the sections was up to 1.28 m. In the unfolded form the truss was 69.75 m long.

Spars of the truss are made of a carbon-reinforced plastic, and the diagonal members are made from stainless steel, titanium and Invar alloy. Sections are rigidly fixed in an unfolded state using locks on the diagonal members.

The mass of the truss is 75 kg (1 running metre weighs 1.25 kg). Service lines, including coaxial fibre-optic cables and pipelines for the working medium of the propulsion system with a weight of above 200 kg, pass along the truss.

The weight of the container with the truss is 700 kg. The cost of the truss proper is 35 million USD. The truss was developed and manufactured by the Able Engineering Company Inc., California, USA.

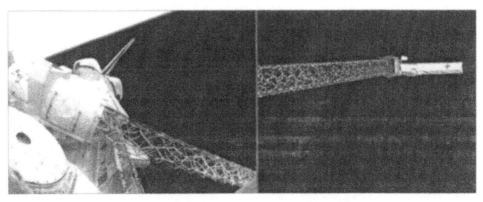

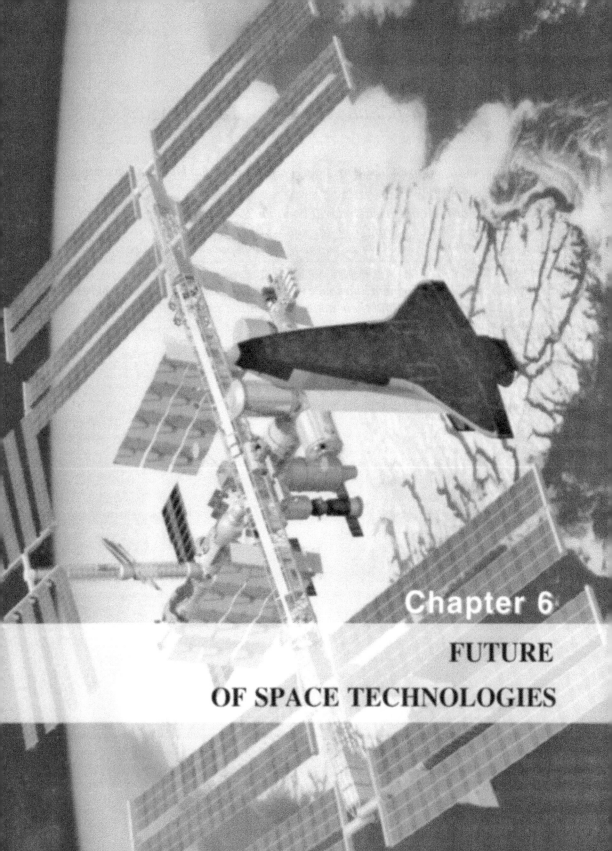

Chapter 6

FUTURE
OF SPACE TECHNOLOGIES

Chapter 6. Future of Space Technologies

Life is endless, so are intelligence and perfection of mankind. Its advancing is eternal: Konstantin E. Tsiolkovsky.

Space exploration has been and will continue to be an important element of scientific progress, starting from the study of Earth and planets of the Solar systems, and then to the investigation of distant space — these are the tasks of future generations.

Further penetration of man into space will be inseparably linked with progress of space technologies and their active application for establishment of new large-scale facilities in orbit.

Welding and Related Technologies in Space and Ocean: Their Exploration in the XXI Century[*]

B.E. Paton[1]

The main task of the welding industry is the designing and manufacture of welded structures. At present structures from steel, non-ferrous metals and alloys are made welded. The welding is intruded more deeply into the manufacture of products from synthetic materials. Technologies of welding dissimilar metals and alloys are developed and welding of composite materials has been mastered. The most heavy-duty civil and military machines and engineering constructions are manufactured in a welded version. Every year the requirements are growing for the quality of welded structures, their reliability and service life.

The scale of welding application can be judged from the data of statistics: about 70 % of all the steel rolled stock, produced in the world, is used in welded structures; the annual volume of the world's production of welding consumables and equipment will reach this year 40 billion USD. Millions of welding, cutting, flaw detection and other operators are involved in the welding manufacturing in different countries of the world.

Of course, all this refers to the production and construction under Earth conditions. However, in all the centuries the people tried to conquer outer space, and at all the stage of the mankind's evolution, the Universe attracted mankind (near and deep space).

The second half of the XX century became the epoch of active intrusion of man into space, into depths of oceans and seas. And this is natural and logical. At the same time, owing to the achievements of scientific and technical progress, the illusive hope appeared in man to win full power over the environment, over nature. This philosophy is very dangerous. It can lead to a global disaster. We should always remember this.

Undoubtedly, in the coming XXI century, space research directed to the solution of Earth problems will be developed intensively. The profound fundamental investigations of the origin of life on the planet and human evolution will be continued. Further human penetration into the space will make it possible to achieve such advances of vital importance as a global information system, production of extraterrestrial resources, space biotechnology, space energy potential, technology of production of semiconductor single crystals, etc.

At present the International Space Station is being constructed in orbit by joint efforts. The problems of exploration of the Moon, its mineral and energy resources are considered real. The conquest of Mars is still to be acheived. To realize these far-reaching plans, it is necessary to construct quite new structures, and to develop radically new materials. Here, the service characteristics of different structures will depend mainly on their space environment. The conditions of human activity in orbit will also be radically changed.

New space engineering will be created both on earth and in orbital conditions. And here, an important role will belong to welding and related quite new processes and technologies, such as welding, cutting and coating deposition.

Taking into account the above, it can be stated that large, extremely complicated works will be expected in space in the coming XXI century. And here the welding technologies used for the creation of complicated equipment and instruments under earth conditions will have great importance. They are existing partially now, but it is necessary to develop new processes of welding, cutting, brazing and deposition of different coatings for the further exploration of space. New exotic materials will appear in the XXI century. And absolutely new technologies will be required for their treatment and joining. Welding science and engineering have been sufficiently prepared for the fulfilment of these works.

---
[*] (2000) From the materials of article, published in *Nauka i Zhizn*, No.6, pp. 2–9.

A different situation arises with the creation of materials and technologies directly in space, where the conditions are complex and should not be oriented to extreme conditions. Not only microgravity and vacuum are typical of space, but also the constant thermal cycling which is inevitable during works in orbit. The living, and, mainly, the work of a man in such extreme conditions are associated with great difficulties, and, sometimes, are impossible. Therefore, spacesuits and highly-reliable life-support systems are required. The spacesuits for the work in the open space were tested many times in operation. They have existed for about 40 years.

Difficulties are encountered in space with the supply of electrical energy, especially, when we deal with its long-time consumption. To generate electrical energy under space conditions, huge solar batteries and powerful accumulators will be required, and in future other sources, nuclear in particular, will be used. They can function in autonomous unmanned orbital station-platforms, and the radiated flux of energy will be received at the manned station.

Some technological operations in space can be performed only by remote control of the working processes using different robots and manipulators. Some models of these devices are already operating in the US shuttles and at the Russian complex «Mir». In particular circumstances, the operator-cosmonaut can participate in their operation, including the operation in the open space.

Welding in space will find wide application only if it will be possible to develop virtually perfect methods of non-destructive testing of welded joints and procedure of diagnostics of welded structures. Comprehensive data banks, which are capable of selecting automatically the welding conditions, and computer simulation, should contribute to this. In general, it is impossible to create new technologies of welding and to use them in space without computerization.

Construction-erection and repair works in space are very complicated and differ radically from those used under earth conditions. Therefore, during continuation of works, connected with the use of electron beam technology in space, it is necessary to improve the equipment and to increase its power efficiency. All this will make it possible to weld and cut metal of almost any thickness.

Laser technology is very interesting for application in open space. It differs from the electron beam technology, at least, by two important advantages: first, there is no accelerating voltage and, secondly, fibre optics can be used. It means, that it is possible to install a stationary laser and to transmit the light beam to any point of the orbital station outside (in open space), as well as inside. The drawbacks of laser technology (low efficiency of the process as compared with electron beam and comparatively larger mass and dimensions of the installation) can be eliminated, however, further efforts will be required. In the near future laser technology, will find, probably, application in space orbits. The diode laser will be, especially, challenging.

There are no obstacles also for the application of resistance welding in space (in particular, spot welding was tested as far back as the sixties). This refers also to seam (roller) welding. It is necessary to design some kind of «Universal» hardware with a changeable tool for spot and roller welding, as well as the devices protecting the operator from possible splashes of metal. Resistance welding will be used in space during construction of different structures and for repair. It can find application in construction of objects on the Moon.

Different welded transformable structures, manufactured on Earth, are of interest. They can be delivered to the orbital station or to the Moon in a compact form. The shell structures are capable «of being blown out» using an excessive inner low pressure and to acquire a preset shape and dimensions. Using welding for joining separate transformable elements it is possible to assemble larger and intricate constructions in space. The transformable unfolding and folding structures, consisting of unified sub-assemblies, will also find application. They can be used both in the orbital space stations and also in the Moon.

Thus welding technologies in space is not science fiction. They are quite real and waiting for their application in practice. I believe that they will be demanded even at the beginning of the XXI century, in the course of progressing of works at the international orbital space station.

The scale and integrated nature of the above-described problems make us to speak about integration of the world's welding science and engineering. This is especially important when national funds for the science are reduced. Using the joint efforts it is necessary to create large international projects which will be realized on the basis of specialization and cooperation of many welding and other research centres of the leading countries of the world.

Space Technologies on the Threshold of the Third Millennium[*]

B.E. Paton[1]

The flight of Yuri Gagarin into space in April, 1961, was certainly one of the major events of the millennium, when the era of man's conquest of near-earth space actually began. Going beyond the limits of gravity pull was due not only to the desire to learn the unknown, inherent to man, but also to the search for new additional energy resources, as well as the ability to use the specific conditions of space for a successful solution of various applied and scientific problems. By the start of the 1970s some practical problems had already been solved, using space, namely pictures of atmospheric phenomena had been taken, diagnostics of mineral deposits had been conducted, satellite communications had been established, etc. Moreover, the results of medico-biological experiments in numerous manned flights confirmed that human activity can be successfully carried out in space. With the start of functioning of «Salyut» and «Skylab» space stations in the near-space orbit, it became possible to conduct systematic studies of near space atmosphere, scientific and process experiments under zero-gravity conditions, as well as monitor the surface of Earth, etc. Solving quite a number of problems related to human activity in space, was required simultaneously:

- erection of large-sized structures in orbit;
- construction of long-term space stations and providing them with power resources;
- performance of EVA by cosmonauts for maintenance of the space objects;
- use of lunar resources in the interests of the national economy;
- industrialization of space, in particular, taking environmentally polluting productions into orbit;
- lighting of northern regions, additional sources of energy, global satellite communications, etc.;
- use of near-earth orbit as the starting orbit in interplanetary flights;
- solving the «space debris» problem.

This will certainly take many decades and depends not only on the level of technical development of the space industry, its creative potential, but also on financial support of the scientific and technical programs being developed. Despite the fact, that the combination of these factors was not always favourable over the past years, we still can speak of considerable successes of cosmonautics. Expansion of the programs, related construction of long-term orbital space stations and of large-sized structures in space, necessitated the implementation of concrete highly specialized technologies in orbit (assembly, mounting, welding, cutting, brazing, etc.). The very first welding experiment, conducted in space in 1969, initiated a new trend in the field of technological developments, namely space technologies. The range of

* (2000) *The Paton Welding Journal*, No. 4, pp. 2–4.

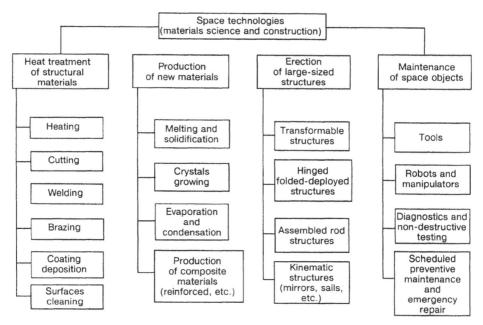

technological processes, conducted in orbit, has become much wider since then, while space technology formed its own areas of biotechnology, agrotechnology, etc. The number of these specific areas will become greater, as the problems, facing the researchers become more profound and broader. However, significant achievements of the production aspect of space technology can already be noted. The main spectrum of technological operations in this area, which have been implemented in near-earth orbit over the last 30 years and have found concrete implementation in the hardware, is shown in the diagram.

A systemic approach to solving the priority problems of the space industry, allowed the OC «Mir» to be created in a short time by the start of the 1980s, through the efforts of the leading experts. The long-term operation of this complex guaranteed successful performance of process experiments and research in the field of materials science, fundamental physics, astronomy, etc. The unique properties of space (microgravity, super-pure dynamic vacuum, low and high temperatures, unusual radiation environment) opened up great prospects for producing materials with a more perfect structure than on Earth and better physico-chemical and mechanical properties. Therefore, it is natural that the attention of space materials science in the initial period of investigations was focused on studying the phenomena that obviously depend on gravity and have been quite well studied on Earth, namely heat and mass transfer, diffusion, surface tension and wettability, capillary phenomena, phases formation, melting and solidification, single crystal growth, etc. The result of theses fundamental studies will be determination of the ability to produce materials with unique properties, difficult (or even impossible) to achieve on Earth. The following materials should be named among those whose fabrication under space conditions can be rational and cost-effective:

• structurally-perfect semiconductor single crystals of a high purity and large dimensions;

• eutectic alloys, consisting of two components, differing in their composition and properties, with unlimited mutual solubility in the molten state and insignificant solubility in the solid state;

• high-strength high-temperature foam materials with a low specific weight;

• superhard metal alloys with greatly different densities of their components;

• optically transparent materials;

- metal alloys, having an area of immiscibility in the liquid state or alloys with non-uniform nucleation;
- composite materials of the type of metal–metal, metal–metal oxide, ceramic and organic substances that do not mix on Earth;
- various coatings for the optical and electronic industry.

Improvement of the properties of materials (electrical resistance, magnetic susceptibility, mechanical strength, permeance, structural perfection, etc.), produced in space, is due to reduction of convection and mass transfer, higher dispersion of the phases, absence of sedimentation, presence of the diffusion mechanism of melt homogenizing in solidification, uniform mixing of gases with metals and of components, immiscible on Earth, and the possibility of conducting crucibleless melting in a cleaner vacuum atmosphere. Intensive development of space materials science in the new millennium is only possible, when perfect equipment and advanced research procedures are used, such as, for instance, displays of nucleation, growth and coalescence in immiscible alloys, using X-ray devices, monitoring the growth rate with fluxmeters, temperature stabilization, using thermal tubes, monitoring temperature and convection in infrared radiation, various methods of levitation of liquid volumes, etc.

The long-term (approximately 25 to 30 years) operation of future orbital stations, involving inevitable damage to the structure and failures of equipment, implies the ability of conducting effective and safe EVA by the cosmonauts. The programs of planning and engineering-technical preparation for EVA include orbital station maintenance and repair, assembly-mounting work and provision of normal functioning of the scientific hardware located on the outer surface of the space station. Robots, automatic and controllable manipulators, and various apparatus for cosmonauts' movement outside the station can be used for performance of the above operations in open space.

The current stage of development of cosmonautics is associated with the era of space optimization. An intensive process of international integration and putting together joint space programs goes on against the background of world crisis. The already started construction of the ISS is an example of international cooperation in the space industry. Creation of large scientific-technological laboratories in orbit will expand the prospects for study of new states of the matter (ultra-microparticles, plasma-crystals, etc.), which is only possible at microgravity. Further progress of those areas of cosmonautics which already now yield real profit, will lead to creation of global networks of communication, TV, information, weather forecasting, navigation, Earth's nature monitoring, etc. Close is the time when large-scale production, vitally important for humanity, will take place in near-earth orbits. Powerful space energy generation systems, namely space solar power stations which will make a significant contribution to power generation on Earth, should become the necessary basis for space industrialization. It is intended to perform power transmission through the radio beam that easily passes through the atmosphere and is guided to the terrestrial receiving aerial by special means. Perfection of the means of payload delivery to orbit, creation of an ergonomic system of transportation in orbit proper, will allow fitting permanent Lunar outposts and using its resources for the needs of terrestrial production. Mastering of principally new flying vehicles (for instance, solar sail, etc.) for reaching the velocity required for leaving the Solar system, will be the prerequisite for a successful exploration of the planets in other galaxies. The progress made by mankind in space exploration is certainly great, and the pace of development of various areas of space technology will only be increasing in the future, thus confirming its revolutionary impact. Commercialization of space programs which has become very obvious recently, certainly raisers concerns. Nonetheless, we still believe, that through the efforts of international cooperation in the space area, it will be possible to implement more than one major project for studying fundamental space phenomena, construction of industrial facilities in the terrestrial orbit and lunar outposts, making flights to other planets and that the world experience of space technology, gained over the past years, will be used in all these developments.

Challenging Projects

Challenging Projects for Undertaking Works in the Russian Module of International Space Station

«RESTAVRATSIYA»

Customer S.P. Korolyov RSC «Energiya»[2]

Main performer PWI[1]

Co-performers Regional R&D Centre of Voronezh State University of Technology;
Yu.A. Gagarin CTC[3]

Space object RM ISS

Technology and design of hand tool are developed for performance of repair-restoration operations of thermal barrier coatings of external surfaces of RM ISS modules. Comprehensive tests of operating tool mock-up are made

«MORFOS»

Customer National Space Agency of Ukraine

Main performer Institute of Metal Physics of the NAS of Ukraine

Co-performers PWI[1]; «Fonon»[9]; Central RI of machine-building

Space object RM ISS

Investigations of processes of directed crystallization using method of direct observation of crystal–melt interface in microgravity conditions. In-process video recording. In-process information about parameters using PC. Tests of operating electrical mock-up are performed

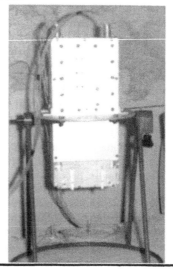

«LUCH–ZONA»
Customer National Space Agency of Ukraine
Main performer PWI[1]
Co-performers IPSC[18];
Institute of Electrophysics of the NAS of Ukraine;
Institute of Super-Hard Materials of the NAS of Ukraine
Space object RM ISS

Furnace for study of producing perfect semiconductor single crystals in microgravity conditions using the method of crucible-free zonal melting with a disc-type electron beam.
Technological parameters of the process are optimized in the mock-up

«EKRAN»
Customer S.P. Korolyov RSC «Energiya»[2]
Main performer PWI[1]
Co-performers «Fonon»[9];
Institute of Technical Mechanics of the NAS of Ukraine (Dnepropetrovsk); IPSC SD RAS[8];
Regional R&D Centre of Voronezh State University of Technology; Yu.A. Gagarin CTC[3]
Space object RM ISS

Creation of zone of stable superhigh vacuum (level of rarefaction $1 \cdot 10^{-14}$–$1 \cdot 10^{-12}$ mm Hg) in orbital flight conditions for conducting technological investigations. This is reached at the expense of undisturbed running-on flow along a specialized polymeric transformable screen, removed from station at 20–30 m distance and mounted normal to the vector of flight speed. It is envisaged that the research equipment will be mounted in the zone of superhigh vacuum. Mock-ups of separate sub-assemblies are made

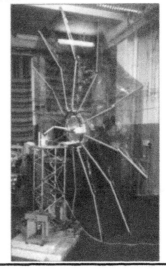

«GIBRID»
Customer S.P. Korolyov RSC «Energiya»[2]
Main performer PWI[1]
Space object RM ISS

Creation of versatile tool, an integrated laser-arc plasmatron, for realization of hybrid processes of welding, cutting and brazing of stainless steel, copper, titanium and aluminium alloys in atmosphere and vacuum. Power source and hybrid laser-plasma torch are now tested

ANNEX

Annex

Our book is a collection of articles covering a wide spectrum of research and development activities in the field of space — from technologies and materials science to fabrication of structures.

The book is meant not only for specialists involved in development and management in the field of space technology. Because of the technical character of the book, we include in this Section a dictionary of special terms and expressions, as well as extra references for readers who may express interest in the problems posed and wish to get a deeper insight into the issues of space exploration.

References

Chapter 1

1. *At the transition of two centuries. 1996–2001* (2001) Moscow, S.P. Korolyov RSC «Energiya», 1320 pp.

2. *Avtomaticheskaya Svarka* (1999) No. 10 (special issue devouted space technology).

3. Baker, D. (1973) Skylab — a spaceborne materials laboratory, *Engineering*, **11**, No. 213, p. 810.

4. Brodsky, Z.F., Klimuk, P.I., Loktev, A.L. et al. (2000) *Aerospace epoch. Memorial events*, Moscow, 224 pp.

5. Kubasov, V.N., Nikitsky, V.P. and Lapchinsky, V.F. (1994) Twenty-five years since the first welding experiment in space, *Avtomatich. Svarka*, No. 11, pp. 3–6.

6. Malinovsky, B.M. (2002) *Professor Boris Paton — the life-work*, Kiev, Naukova Dumka, pp. 156–169.

7. Moderne Werkstoffe im Luft- und Raumfahrtzeugbau (2000) *Blech Rohre Profile*, **47**, No. 5, p. 13.

8. Paton, B.E. and Lapchinsky, V.F. (1997) *Welding and related technologies in space*, English ed., Cambridge Int. Sci. Publ., 121 pp.

9. Paton, B.E. and Lapchinsky, V.F. (2000) *Welding and related technologies in space*, Japanese ed., Sanpo Publ., 189 pp.

10. Paton, B.E. and Semyonov, Yu.P. (1983) To orbits in future, *Pravda*, No. 227, November 29, p. 5.

11. Paton, B.E., Vavilova, I.B., Negoda, O.O. et al. (2001) *Ukraine in constellation of space countries of the world*, Kiev, NAN Ukr., 94 pp.

12. *Professor S.P. Korolyov. Scientist. Engineer. Man. Creative portrait as recollected by his contemporaries* (1986) Coll. articles, Moscow, Nauka, 519 pp.

13. *Soviet and Russian cosmonauts. XX century* (2001) A handbook. Ed. by Yu.M. Baturin, Moscow, Novosti Kosmonavtiki, 405 pp.

14. *Space: technologies, materials, structures* (2000) Coll. sci. pap. Ed. by B.E. Paton, Kiev, PWI, 528 pp.

15. *S.P. Korolyov and his work. Light and shades in history of cosmonautics* (1998) Moscow, Nauka, 716 pp.

16. *Yu.A. Gagarin Cosmonauts Training Centre of the Russian State Research Institute* (2000) Ed. by P.I. Klimuk, Moscow, Kladez–Books, 271 pp.

Chapter 2

1. Aladiev, S.I. (1986) Towards the problem of rate of growth of crysyals on Earth and in weightlessness, *Kosmich. Issledovan.*, **24**, No. 6, pp. 947–950.

2. Anzawa, S., Abe, Y., Takahashi, K. et al. (1995) Development of materials processing facilities and experiments for space flyer unit: development of GHF and MHF, *Ishikawajima-Harima Eng. Rev.*, **35**, No. 2, pp. 133–138.

3. Avduevsky, V.S., Agafonov, M.S., Grishin, S.D. et al. (1987) Experimental and theoretical investigations of convection in conditions of action of small accelarations, In: *K.E. Tsiolkovsky and problems of space industrialisation*, Proc. of 20th Tsiolkovsky Readings, Kaluga, 1985. Moscow, pp. 3–12.

4. Avduevsky, V.S., Leskov, L.V. and Feonychev, A.I. (1988) State-of-the-art and prospects of theoretical investigations in the field of space engineering, In: *22nd Gagarin Sci. Readings on Cosmonautics and Aviation*, Moscow, 1987. Moscow, pp. 36–44.

5. Barmin, I.V. and Egorov, A.V. (1986) Prospects of development of equipment for space engineering, In: *K.E. Tsiolkovsky and problems of development of science and technology*, Moscow, pp. 63–66.

6. Barmin, I.V., Egorov, A.V., Maslennikov, V.G. et al. (1984) Problems of producing items under microgravity, In: *Ideas of K.E. Tsiolkovsky and problems of space manufacturing*, Proc. of 18th Tsiolkovsky Readings, Kaluga, 1983. Moscow, pp. 136–143.

7. Bayuzick, R.J. (1988) Space processing of metals and alloys, In: *Coll. techn. pap. of Symp. on Commer. Opportunities Space*, Taipei, 1987. Washington, pp. 137–160.

8. Berrenberg, T., Rex, S., Kaureauf, B. et al. (1999) Concept for a dedicated facility for directional solidification of transparent model substances fluid science laboratory, In: *Proc. of 2nd Eur. Symp. on Util. of ISS*, Noordwijk, 1998. Noordwijk, pp. 347–350.

9. Berzhaty, V.I., Zvorykin, L.L., Ivanov, A.I. et al. (1999) Prospects of realisation of vacuum technologies in conditions of orbital flight, *Avtomatich. Svarka*, No. 10, pp. 108–116.

10. Bewersdorff, A. and Feuerbacher, B. (1985) Materials science experiment under microgravity — Spacelab 1, *Europhys. News*, **16**, No. 7/8, pp. 14–16.

11. Bulloch, Ch. (1984) Materials processing in space. Plenty of prophets, but what about propits? *Intervaia*, **39**, No. 7, pp. 679–683.

12. Chassay, R. and Carswell, B. (1987) Processing materials in space: the history and the future, *AIAA Pap.*, No. 392, pp. 1–11.

13. Cheshlya, Yu.V. (1984) Prospects and problems of producing alloys with a directed quasi-eutectic structure in weightlessness, In: *Ideas of K.E. Tsiolkovsky and problems of space manufacturing*, Proc. of 18th Tsiolkovsky Readings, Kaluga, 1983. Moscow, pp. 72–77.

14. Cohen, A. (1986) Space technology today, *J. Vac. Sci. and Technol.*, **A4**, No. 3, pp. 263–267.

15. Cohen, H. (1984) Space reliability technology: a historical perspective, *IEEE Trans. Reliab.*, **33**, No. 1, pp. 36–40.

16. Denisov, A.G., Kuznetsov, I.A. and Makarenko, V.A. (1981) Equipment for molecular-beam epitaxy, In: *Reviews on electron engineering*, Series 7, Issue 17, No. 829, pp. 15–17.

17. Engelhard, W. (1987) Made in space, *Maschinenmarkt*, **93**, No. 28, pp. 53–58.

18. Eppler, D.B. and Callawey, R.K. (1994) Managing EVA technology: the present and the future, *AIAA Pap.*, No. 4618, pp. 1–11.

19. Fordyce, J.S., Grisaffe, S.J. and Stephens, J.R. (1989) Space 2010, *Adv. Mater. and Proc. Inc. Metal. Progr.*, **135**, No. 1, pp. 69–71.

20. Fujimori, Y. and Shimaoka, T.K. (1995) NASDA microgravity programs, *AIAA Pap.*, No. 391, pp. 1–11.

21. Graham, S.J. and Rhome, R.S. (1994) Achievements in microgravity: ten years of microgravity research, *Ibid.*, No. 344, pp. 1–19.

22. Grebnyuk, V.G. and Lapchinsky, V.F. (1986) Peculiarities of design of secondary electric supply of electron beam units, In: *Problems of electromagnetic compatibility of power semiconductor converters*, Tallinn, AN ESSR, pp. 71–72.

23. Grishin, S.D., Leskov, L.V. and Savichev, V.V. (1982) Problems of physics of weight-lessness and prospects of space manufacturing, In: *K.E. Tsiolkovsky and problems of space manufacturing*, Proc. of 16th Tsiolkovsky Readings, Kaluga, 1981. Moscow, pp. 18–22.

24. Heyman, J. (1993) Space technology: materials science, *News Bull. Astronaut. Soc. West. Austral.*, **18**, No. 9, pp. 89–90.

25. Hodinazova, J. (1978) Perspektivy vyvoje kosmiznych technologii do roku 2000, *Ustr. Ved. Techn. a Ekon. Inform.*, No. 1608, 82 pp.

26. Holmes, P. and Stalker, R.J. (1993) Trends in space technology, Coll. pap. of 8th Nat. Space Eng. Symp., Queensland, 1993. *Nat. Publ. Inst. Eng. Austral.*, **7**, No. 93, pp. 38-45.

27. Khryapov, V.T., Markov, E.V., Kulchitsky, N.A. et al. (1984) State-of-the-art, prospects and problems of development of space semiconductor materials science, In: *Ideas of K.E. Tsiolkovsky and problems of space manufacturing*, Proc. of 18th Tsiolkovsky Readings, Kaluga, 1983. Moscow, pp. 64–71.

28. Kicza, M. and Feeley, T.J. (1994) Space science: a new direction, *AIAA Pap.*, No. 4593, pp. 1–10.

29. Kline, R.L. (1995) Technology as a driver for improved space products, Coll. techn. pap. of 44th IAF Congr., Graz, 1993. *Acta Astronautica*, **35**, No. 9/11, pp. 601–605.

30. Kohler, H.W. (1980) Metallurgie im Weltraum- zur Situation, *VDI-Z.*, pp. 12-21.

31. Kowal, S. (1979) Naroddziny technologii kosmiczney, *Astronautyka*, **21**, No. 1, pp. 16–17.

32. Kuribayashi, M. (1989) Space materials science, *Chem. Eng. Jap.*, **53**, No. 10, pp. 712–713.

33. Later, P., Nodle, W. and Chay, Ya. (1990) Use of automated molecular-beam epitaxy for growing thin semiconductor layers, *Elektronika*, **53**, No. 19, pp. 9–13

34. Leger, L., Visentine, J. and Santos-Mason, B. (1987) Selected materials issues associated with space station, *SAMPE Quart.*, **18**, No. 2, pp. 48–54.

35. Maloney, L.D. (1991) Aerospace: America's technology driver, *DES News*, **47**, No. 16, pp. 60–62, 64.

36. Marsha, T.R. (1994) Scientific achievements of 10 years of spacelab: an overview of the missions, *AIAA Pap.*, No. 340, pp. 1–21.

37. Martin, G.L. and Phomt, R.C. (1995) Microgravitational investigations under space station conditions, *Ibid.*, No. 388, pp. 1–11.

38. *Materials science in space. A contribution to the scientific basis of space processing* (1989), Moscow, Mir, pp. 39–59.

39. McKannan, E. (1978) Materials processing experiment capabilities in space, *US Dep. Commer. Nat. Bur. Stand. Spec. Publ.*, No. 520, pp. 10–13.

40. Microgravity (1995) *ESA Bull.*, No. 82, pp. 129-131.

41. Mishina, L.V., Krylov, A.N. and Zvorykin, L.L. (1991) *Kinetic modeling of flows near complex form bodies. Rarefied gas dynamics*, New York, pp. 1391–1397.

42. Monti, R. (1985) Space processing, *Earth-Orient. Appl. Space Technol.*, **5**, No. 1/2, pp. 129–138.

43. Motegi, T. (1994) Use of microgravity conditions for production of materials, *Rep. China Inst. Technol.*, No. 41, pp. 191–203.

44. Nusinov, M.D. (1982) *Effect and modeling of space vacuum*, Moscow, 205 pp.

45. Osipian, Yu.A. and Regel, L.L. (1985) «Salyut-6–Soyuz», materials science and technology, In: *Proc. of Int. Meet. on Some Results of Investigations in the Field of Space Materials Science*, Riga, 1985. Moscow, pp. 5–11.

46. Pizzano, F. (1986) Reliability and maintenance simulation of the Hubble space telescope, In: *Proc. of Ann. Reliab. and Maint. Symp.*, Las Vegas, 1985. New York, pp. 83–87.

47. Poss, L.J. (1989) Technology directions-introduction, In: *Proc. of 26th Gooddard Met. Symp. on Technol. and Civ. Future in Space*, Greenbelt, 1988. San Diego, pp. 121–123.

48. Produktion neuartiger Werkstoffe im Weltraum (1986) *Die Materialforschungen beim Battelle-Institut Astronautik*, **23**, No. 1, pp. 22–23.

49. Recherche en micrograviti (1994) *ESA Bull.*, No. 80, pp. 80–81.

50. Regel, L.L. (1988) Investigations of gravity effect on crystal growth: achievements and prospects, *Acta Astronautica*, **17**, No. 11/12, pp. 1241–1244.

51. Reginsky, K., Lamin, M.A., Mashanov, V.I. et al. (1995) Shield intensity oscillations in resonance condition during MBE growth of Si on Si (111), *Surf. Sci.*, No. 327, pp. 93–99.

52. Rhome, R.C. (1994) The continuing evolution of NASA's microgravity-science program, *AIAA Pap.*, No. 105, pp. 1–12.

53. Shulym, V.F. and Sorokin, I.V. (1999) Space engineering in Japan (Review), *Avtomatich. Svarka*, No. 10, pp. 117–119.

54. *Space materials science* (1989) Ed. by V.S. Avduevsky, Moscow, Mir, 221 pp.

55. Stesin, V.V., Dubenko, G.P., Zagrebelny, A.A. et al. *Method of vacuum-proof joints of flexible pipes with a sealed chamber*. USSR author's cert. 261051, Int. Cl. F 02 C B 29 B, priority 28.06.68, publ. 06.01.70.

56. Taylor, K., Watkins, J. and Gallowey, P. (1994) Overview, history and current status of United States' materials processing in space flight activity, *AIAA Pap.*, No. 111, pp. 1–10.

57. Testardi, L.R. (1986) Material processing in space: an overview, In: *Space Sci. and Appl.*, New York, pp. 153–158.

58. Tewksbaury, S.K. and Hornak, A. (1994) Can optoelectronic and silicon chips be monolitically integrated? *Laser Focus World*, May, pp. 151–156.

59. US, European firms offer microgravity processing service (1992) *Aviat. Week and Space Technol.*, **132**, No. 14, p. 29.

60. Vinit, N. and Josuhiho, K. (1989) Space processing, *Aerospace Amer.*, **27**, No. 12, p. 76.

61. Werner, D. (1996) Raumfahrtforschung als Quelle fur neue Produkte, *Ind.-Anz.*, **118**, No. 20, p. 26.

62. Whitten, R.P. and Gabris, E.A. (1994) Commercial space processing, *AIAA Pap.*, No. 389, pp. 1–9.

63. Yamanaka, T. (1986) Technology for the coming space manufacturing age, *J. Jap. Soc. Aeronaut. and Space Sci.*, **34**, No. 384, pp. 18–25.

64. Yurchenko, N.N. and Yurchenko, O.N. (2001) *Electric supply systems of board technological equipment operating in space*, Kiev, IED, 143 pp.

Chapter 3

On-Land Testing

1. Abe, G., Yasuda, C., Fujiwara, M. et al. (1990) Aircraft experiment to research on physical phenomena under microgravity, *Mitsubishi Heavy Ind. Tech. Rev.*, **27**, No. 6, pp. 1–7.

2. Agapakis, J.E. and Masubuchi, K. (1985) Remotely manipulated and autonomous robotic welding fabrication in space, *J. Soc. Photo-Opt. Instrum. Eng.*, No. 580, pp. 68–77.

3. Bachurko, V.I. (1989) Selection of characteristics of adequacy in testing of assembly-erection operations with functional blocks in buoyancy, In: *K.E. Tsiolkovsky and space industrialisation*, Proc. of 23rd Tsiolkovsky Readings, Kaluga, 1988. Moscow, pp. 124–133.

4. Bachurko, V.I., Dodukh, V.N. and Kovalenko, L.M. (1991) Study of the process of deployment of multilink transformable structure in simulated conditions of weightlessness, In: *Space industry: from experiments to industrial scales*, Proc. of 25th Tsiolkovsky Readings, Kaluga, 1990. Moscow, pp. 73–83.

5. Bachurko, V.I., Lokshin, M.L., Pais, E.V. et al. (1987) Testing of a fragment of large-sized folded structures in buoyancy, In: *16th Gagarin Sci. Readings on Cosmonautics and Aviation*, Moscow, 1986. Moscow, pp. 237–238.

6. Burriesci, L. and Steakley, B. (1984) Cryogenical optical systems in the Lockheed sensor test facility, *J. Soc. Photo-Opt. Instrum. Eng.*, No. 509, pp. 140–151.

7. Diplocr, B.R. (1991) Environmental test facilities at RAL, *SLRC Bull.*, **4**, No. 7, p. 18.

8. Goncharov, V.A., Biryukov, V.M., Dubashevskaya, I.V. et al. (1990) Digital simulation of growth of semiconductor single crystals in weightlessness, In: *Problems and prospects of space industrialisation*, Proc. of 24th Tsiolkovsky Readings, Kaluga, 1989. Moscow, pp. 46–50.

9. Hayashi, T., Oshima, K., Hashimoto, M. et al. (1986) Space simulation chamber, *Uchu Kagaku Kenkyu Hokoku*, No. 36, pp. 1–46.

10. Holl, S., Roos, D. and Wein, J. (1991) Numerical simulation of controlled directional solidification under microgravity conditions, Coll. pap. of 41st IAF Congr., Dresden, 1990. *Acta Astronautica*, **25**, No. 8/9, pp. 553–559.

11. Jost, R.J. (1982) The NASA space environment simulation laboratory, In: *Artif. part. beams space plasma stud.*, Proc. of Conf. of NATO Adv. Res. Inst., Geilo, 1981. New York, London, pp. 331–337.

12. Kimura, S., Oruyama, T. and Tsuchiya, S. (1996) Experiment on antenna assembling mechanism on STS-VII. Ground test on assembling mechanism and teleoperation system, *J. Commun. Res. Lab.*, **43**, No. 2, pp. 127–136.

13. Langbun, D., Grosbach, R. and Heide, W. (1990) Parabolic flight experiments on fluid surfaces and wetting, *Appl. Microgravity Technol.*, **2**, No. 4, pp. 198–211.

14. Masubuchi, K. and Najama, M. (1991) Fusion welding experiments under low-gravity conditions using aircraft, In: *Proc. of Conf. on Welding in Space and the Construction of Space Vehicles by Welding*, Miami, 1991. Miami, pp. 186–197.

15. Microgravity environment to be constructed at Tsukuba science city (1992) *Techo Jap.*, **25**, No. 6, p. 68.

16. Modelling of processes proceeding in space (1994) *Chem. Eng. Jap.*, **58**, No. 3, pp. 204–206.

17. Nishida, M. and Watanabe, J. (1988) Development of equipment for producing atomic oxygen for simulation of real flight conditions of space vehicles, *J. Jap. Soc. Aeronaut. and Space Sci.*, **36**, No. 417, pp. 38–44.

18. Paton, B., Bulatsev, A., Perepechenko, B. et al. *Attachment of work station for conductance of experiments in space.* Pat. 20353, Ukraine, Int. Cl. B 64 G 9/00, priority 14.02.95, publ. 15.09.00.

19. Patrushev, V.I., Bachurko, V.I., Golotyuk, S.V. et al. (1986) Assembly of large structures in orbit and in hydromedium, In: *K.E. Tsiolkovsky and space manufacturing*, Proc. of 19th Tsiolkovsky Readings, Kaluga, 1984. Moscow, pp. 75–82.

20. Patrushev, V.I., Bachurko, V.I., Lokshin, M.L. et al. (1987) Problems of development and testing of elements of folded structures in hydromedium, In: *16th Gagarin Sci. Readings on Cosmonautics and Aviation*, Moscow, 1986. Moscow, p. 231.

21. Plester, V. (1995) Microgravity research during aircraft parabolic flights: the 20 ESA campaigns, *ESA Bull.*, No. 807, pp. 57–68.

22. Regel, L.L., Parfeniev, R.V., Vidensky I.V. et al. (1987) Properties and structure of materials produced in experiment «Alkutest» using unit «Kristallizator» (on-land tests), In: *16th Gagarin Sci. Readings on Cosmonautics and Aviation*, Moscow, 1986. Moscow, pp. 211–212.

23. Regel, L.L., Shalimov, V.P. and Dryuchenko, D.D. (1986) *System of on-land testing of experiments on space materials science using unit «Kristallizator»*, Moscow, IKI, 11 pp.

24. Sawaoka, L. (1991) Software and hardware for conductance of experiments in the field of space materials science in Japan, *J. Jap. Soc. Mech. Eng.*, **94**, No. 871, pp. 495–499.

25. Shalimov, V.P. (1994) Is centrifuge the means for space technology on Earth, In: *Proc. of 28th Tsiolkovsky Readings*, Kaluga, 1993. Moscow, pp. 94–103.

26. Shields, N. (1984) Assembly and maintenance of space-based systems: human operator simulations for future missions, In: *Proc. of Conf. on Appl. Rob. Aerosp. Ind.*, St. Louis, 1984. Deaborn, 13 p.

27. Siolanger, G.V. and Bareiss, L.E. (1986) Martin Marietta atomic oxvgen-beam facility, In: *Proc. of 18th Int. SAMPE Techn. Conf.*, Seattle, 1986. Covina, Vol. 15, pp. 722–731.

28. Sonnensimulator fur Satellitentests (1986) *Luft- und Raumfahrt*, 7, No. 4, pp. 122–123.

29. Stesin, V.V., Masalov, Yu.N., Lapchinsky, V.F. et al. *Device for education and training of welder.* USSR author's cert. 48398, Int. Cl. B 23 K, priority 04.07.68, publ. 06.10.69.

30. Stesin, V.V., Paton, B.E., Paton, V.E. et al. *Device for investigation of materials, mechanisms and processes with simulation of space conditions.* USSR author's cert. 48700, Int. Cl. B 23 K, 21 H 30/17, priority 30.09.68, publ. 17.10.69.

31. Vedernikov, A.A., Komarov, V.F., Melikhov, I.V. et al. (1987) Experiments «Nucleatsiya», «Drejfy» and «Khalkantit» on crystallization from water solutions in conditions of weightlessness (on-land tests), In: *16th Gagarin Sci. Readings on Cosmonautics and Aviation*, Moscow, 1986. Moscow, pp. 210–211.

32. Veisfeld, L.O., Porshnev, G.P. and Semyonov, A.G. (1995) New technologies of on-land testing of items of space engineering in vacuum chambers with simulation of heat processes in the conditions of Mars atmosphere, In: *Abstr. of pap. of Russian Sci.-Techn. Conf. on Innovation of High Technologies for Russia*, St.-Petersburg, 1995. St.-Petersburg, Part 3, p. 43.

33. Wada, B.K., Kui, K.P. and Gleiser, R.J. (1986) On-land tests of large space structures, *Aerokosm. Tekhnika*, No. 9, pp. 189–195.

34. Wildeman, K.J. and Plaeger, G.R. (1989) Cryogenic vibration test setup for space qualification, *Cryogenics*, **29**, No. 5, pp. 559–562.

35. Zavarykin, M.P. and Zorin, S.V. (1988) On-land simulation of vibrational convection in weightlessness, In: *Digital and experimental simulation of hydrodynamic phenomena in weightlessness*, Sverdlovsk, pp. 85–92.

Welding, Cutting and Brazing of Metals

36. Anderson, D.A., Tannehill, J.C. and Pletcher, R.H. (1984) *Computational fluid mechanics and heat transfer*, Vol. 1, New York, Hemisphere P.C., 392 pp.

37. Bebermeir, H. (1994) Substrate of solar battery HST: after-flight check-out of technology — recent results of reseach and experimental-design works for assessment of welded joints, In: *Photovolt. energy convers.*, Proc. of 1st IEEE World Conf., 24th IEEE Spec. Conf., Waikoloa, 1994. Piscataway, Vol. 2, pp. 1986–1989.

38. Bulatsev, A.R., Lapchinsky, V.F., Zagrebelny, A.A. et al. *Method of position brazing of tubular structure and pipeline in hard-to-reach place*. Pat. 27794, Ukraine, Int. Cl. B 23 K 1/00, priority 06.08.93, publ. 16.10.00.

39. Drabovich, Yu.I., Maslobojshchikov, N.N., Yurchenko, N.N. et al. *Transistor*. Pat. 66-1602959, Japan, Int. Cl. H 01 L 25/02, publ. 12.06.91.

40. Drabovich, Yu.I., Maslobojshchikov, N.N., Yurchenko, N.N. et al. *Transistor*. USSR author's cert. 730213, Int. Cl. H 01 L 25/02, publ. 30.09.86.

41. Drabovich, Yu.I., Pazeev, G.F. and Shelyagin, V.D. *Device for protection of transistor inverter from overloading*. USSR author's cert. 1534619, Int. Cl. H 02 H 7/12, priority 28.03.86, publ. 07.01.90.

42. Drabovich, Yu.I., Yurchenko, N.N., Shevchenko, P.N. et al. (1985) Electric supply system of versatile hand electron beam tool, In: *Problems of space technology of metals*, Kiev, Naukova Dumka, pp. 65–67.

43. Duley, W.W. (1983) *Laser processing and analysis of materials*, New York, Plenum Press, 504 pp.

44. Dunkerton, S., Dunn, B.D. and Fernie, W.B. (2001) Welding and joining in a space environment, *Rivista Ital. della Saldatura*, No. 5, pp. 615–626.

45. Dyachenko, V.V. and Olshansky, A.N. (1968) Effect of controlled atmosphere pressure on welding process variables, *Svarochn. Proizvodstvo*, No. 11, pp. 5–9.

46. Erdmann-Jesnitzer, F. (1981) Welding in space, In: *Proc. of JOM-1 Conf. on Joining of Metals*, Helsingor, 1981. Helsingor, pp. 319-330.

47. Friedler, K., Stickler, R, and Siegfried, E. (1981) Brazing under microgravity in a resistance heated furnace, In: *Materials processing in the reduced gravity environment of space*, Proc. of Conf., Boston, 1981. Boston, pp. 639–649.

48. Fujii, H., Aoki, Y. and Nogi, K. (2001) Electron beam and gas tungsten arc welding under microgravity, *Transact. JWRI*, **30**, No. 1. pp. 105–109.

49. Hart, R.E.Jr., Piszcror, M.E. and Patton, J.V. (1985) Welded and soldered interconnections survive 60,000 (LEO) thermal cycles, In: *Proc. of 18th IEEE Photovolt. Spec. Conf.*, Las Vegas, 1985. New York, p. 1759.

50. Irving, B. (1991) Electron beam welding, soviet style: a front runner for space, *Welding J.*, **70**, No. 7, pp. 55–59.

51. Kazakov, N.F. and Braun, A.G. (1986) Diffusion welding is the space technology of the future, In: *K.E. Tsiolkovsky and space manufacturing*, Proc. of 19th Tsiolkovsky Readings, Kaluga, 1984. Moscow, pp. 56–60.

52. Kazakov, N.F., Rusin, S.P. and Kazakov, S.P. (1986) Prospects of diffusion welding application in space, *Ibid.*, pp. 83–88.

53. Keanini, R.C. and Rubinsky, B. (1990) Plasma arc welding under normal and zero-gravity conditions, *Welding J.*, No. 6, pp. 41–49.

54. Khomich, N.S. and Neznamova, L.O. (1999) Magnetic-abrasive cleaning of surface, *Avtomatich. Svarka*, No. 10, pp. 120–121.

55. Kurrin, B.F. (1990) Welding in space, *Weld. Des. and Fabr.*, No. 5, pp. 22–24.

56. Kuvin, B.F. (1990) Welding in space: questions remain, *Weld. Des. and Fabr.*, 63, No. 5, pp. 22–24.

57. Lancaster, J.F. (1984) *The physics of welding*, Oxford, Pergamon Press, 297 pp.

58. Lapchinsky, V.F. (1974) *Welding in space. Technology of electric fusion welding of metals and alloys*, Moscow, Mashinostroyeniye, pp. 686–690.

59. Lapchinsky, V.F. (1981) Welding in space, In: *Welding in the USSR*, Moscow, Mashinostroyeniye, pp. 487–494.

60. Masubuchi, K. (1987) Welding in space, *J. Light Metal Weld. and Constr.*, 25, No. 1, pp. 2-5.

61. Matsuzawa, Y. (1989) Space welding in space stations, *Join Assem. Eng.*, 5, No. 9, pp. 66-75.

62. Mokhnach, V.K., Shelyagin, V.D., Stesin, V.V. et al. *Device for electron beam treatment.* USSR author's cert. 1345491, Int. Cl. B 23 K 15/00, priority 19.11.85, publ. 15.06.87.

63. Nishikawa, H. and Ohji, T. (2001) Welding in space, *Welding Technique*, No. 1, pp. 69–75.

64. Nishikawa, H., Yoshida, K., Ohji, T. et al. (2000) Fundamental characteristics of GHTA under low pressure, *Quart. J. Jap. Weld. Soc.*, 18, No. 2, pp. 272–279.

65. Nogi, K. and Aoki, Y. (1997) Behaviour of bubbles in welding for repair in space, *Materials and Design*, 18, No. 4/6, pp. 275–278.

66. Nogi, K., Aoki, Y. and Fujii, H. (1999) Special features of EBW and non-consumable electrode arc welding under microgravity conditions, *Avtomatich. Svarka*, No. 10, pp. 39-43.

67. Nogi, K., Aoki, Y., Fujii, H. et al. (1998) Weld formation in microgravity, *ISIJ Int.*, 38, No. 2, pp. 163–170.

68. Paton, B.E. (1990) Technologische Gesichtspunkte des Schweissens im Weltraum, *Schweißen und Schneiden*, No. 3, pp. 117–120.

69. Paton, B.E., Pokhodnya, I.K., Marchenko, A.E. et al. *Method of consumable electrode arc welding.* USSR author's cert. 43647, Int. Cl. B 23 K, priority 12.10.67, publ. 26.09.68.

70. Paton, B.E., Pokhodnya, I.K., Marchenko, A.E. et al. *Torch for arc welding in vacuum.* USSR author's cert. 43792, Int. Cl. B 23 K, priority 12.10.67, publ. 14.10.68.

71. Paton, B.E., Zagrebelny, A.A.,Gavrish, S.S. et al. (2002) Extremal technology: risk and its minimizing, In the press, *The Paton Welding J.*, No. 11, 12.

72. Pirjola, R., Viljanen, A., Pulkkinen, A. et al. (1999) Space weather risk in power systems and pipelines, Abstr. of pap. of 24th Gen. Assem. on Space and Planet Sci., *Geophys. Res. Abstr.*, 1, No. 3, p. 652.

73. Russell, C. and Zagrebelny, A. (2001) Evaluation of the universal hand tool for welding applications in space, In: *Proc. of 12th Ann. Aeromat Conf.*, Long Beach, 2001. ASM, 18 pp.

74. Sasabe, K. and Siegfried, E. (1985) Flow processes in capillary gaps during brazing (in vacuum in microgravity), *Schweißen und Schneiden*, **37**, No. 11, pp. 585–590.

75. Seyffarth, P. and Krivtsun, I.V. (2002) *Laser-arc processes and their applications in welding and material processing*, London, Taylor & Fransis, 184 pp.

76. Shelyagin, V.D., Sakharnov, V.A., Tserkus, I.N. et al. *Multibeam electron gun*. USSR author's cert. 1408652, Int. Cl. B 23 K, priority 22.09.86, publ. 08.03.88.

77. Shelyagin, V.D., Sakharnov, V.A., Tserkus, I.N. et al. *Multibeam electron gun*. USSR author's cert. 1408653, Int. Cl. B 23 K, priority 22.09.86, publ. 08.03.88.

78. Siewert, T.A., Heine, R.W., Adams, C.M. et al. (1977) The Skylab brazing experiments, *Welding J.*, **56**, No. 10, p. 2915.

79. Stesin, V.V., Bakaeva, V.I., Dudko, D.A. et al. *Torch for constriction arc welding*. USSR author's cert. 304086, Int. Cl. B 23 K 17/00, B 23 K 19/16, priority 04.01.70, publ. 03.03.71.

80. Stesin, V.V., Belfor, M.G., Vojtsekhovich, E.E. et al. *Device for wire feeding*. USSR author's cert. 330925, Int. Cl. B 23 K 9/12, D 21 F 23/00, priority 15.12.70, publ. 07.03.72.

81. Stesin, V.V., Dubenko, G.P., Dudko, D.A. et al. *Torch with filament cathode for constriction arc welding*. USSR author's cert. 261136, Int. Cl. B 23 K, priority 28.06.68, publ. 06.01.70.

82. Stesin, V.V., Dubenko, G.P., Zagrebelny, A.A. et al. *Handle for hand tool used by a man in a spacesuit with excessive pressure*. USSR author's cert. 48141, Int. Cl. B 23 K, priority 22.08.68, publ. 08.09.69.

83. Stesin, V.V., Dudko, D.A., Zagrebelny, A.A. et al. *Torch with filament cathode for constriction arc welding*. USSR author's cert. 271279, Int. Cl. B 23 K, priority 28.06.68, publ. 04.03.70.

84. Stesin, V.V., Paton, B.E., Malkin, Yu.I. et al. *Device for resistance spot welding*. USSR author's cert. 47369, Int. Cl. B 23 K, 21 H 30/17, priority 21.10.68, publ. 08.07.69.

85. Stesin, V.V., Paton, B.E., Paton, V.E. et al. *Device for research welding works in open space conditions*. USSR author's cert. 48689, Int. Cl. B 23 K, 21 H 30/17, priority 08.07.68, publ. 17.10.69.

86. Stesin, V.V., Tishura, V.I., Sakharnov, V.A. et al. *Hand tongs for resistance spot welding*. USSR author's cert. 336934, Int. Cl. B 23 K 11/10, B 23 K 11/28, priority 25.01.71, publ. 28.01.72.

87. Suita, Y., Tsukuda, Y., Terajima, N. et al. (2000) GHTA welding experiments under sumilated space environment in flying laboratory, *Quart. J. Jap. Weld. Soc.*, **18**, No. 2, pp. 228–235.

88. USA–Ukraine. Tests of experiment on space welding (1996) *Novosti Kosmonavtiki*, **6**, No. 16, p. 10.

89. Vanschen, W. (1990) Schweissversuche im Kosmos an den Raumschiffen «Sojus-6» und «Skylab», *Blech Rohre Profile*, **37**, No. 10, pp. 716–717.

90. Watson, J.K. (1986) Engineering considerations for on-orbit welding operations, *J. Astronaut. Sci.*, **34**, No. 2, pp. 121–132.

91. Zhbanov, O.L., Shelyagin, V.D., Tserkus, I.N. et al. *High-voltage converter transformer*. USSR author's cert. 1333110, Int. Cl. B 23 K 15/00, priority 13.08.84, publ. 22.04.87.

92. Zhbanov, O.L., Shelyagin, V.D., Tserkus, I.N. et al. *High-voltage converter transformer*. USSR author's cert. 1333111, Int. Cl. B 23 K 15/00, priority 13.06.84, publ. 22.04.87.

Coating Deposition

93. Beauchamp, W.T., McLean, B. and Larro, M. (1994) Qualification test results for bluered reflecting solar cell covers and other new products for the solar power market, In: *Photovolt. energy convers.*, Proc. of 1ˢᵗ IEEE World Conf., 24ᵗʰ IEEE Spec. Conf., Waikoloa, 1994. Piscataway, Vol. 2, pp. 2062–2065.

94. Bogorad, A., Bowman, C., Seehra, S. et al. (1992) The effects of conducting breaks on electrostatic discharges on optical solar reflector panels, Coll. pap. of IEEE Ann. Conf. on Nuclear and Space Radiation, New Orleans, 1992, Part 1, **39**, No. 6, pp. 1790–1796.

95. Demidenko, E.B., Yurchenko, N.N. and Grebenyuk, V.G. *Device for protection of transistor inverter from overloads and short-circuits in the loading circuit*. USSR author's cert. 955346, Int. Cl. H 02 H 7/12, publ. 30.08.82.

96. Drabovich, Yu.I., Yurchenko, N.N., Shevchenko, P.N. et al. *Transistor inverter*. USSR author's cert. 788313, Int. Cl. H 01 L, publ. 15.12.80.

97. Grande, M. and Burton, W.M. (1986) Aluminium mirror coatings in space: a study of the decrease in ultraviolet normal incidence reflectance produced by controlled oxidation of evaporated aluminium mirror surfaces, *Surface and Interface Anal.*, **9**, No. 1/6, p. 518.

98. de Groh, K.K. and Smith, D.C. (1997) Investigation of teflon FEB embrittlement on spacecraft in low earth orbit, In: *Proc. of 7ᵗʰ Int. Symp. on Space Environment*, Toulouse, 1997. Noordwijk, pp. 255–256.

99. Guerard, F. and Guillaumon, J.-C. (1997) Thermal control paints and various materials for space use, *Ibid.*, pp. 437–458.

100. Guillaumon, J.-C. (1997) Thermal control coating under development at CNES, *Ibid.*, pp. 427–434.

101. Hribar, V.F., Bauer, J.L. and O'Donnel, T.P. (1986) Electrically conductive, black thermal control coatings for spacecraft application. Silicon martix formulation, In: *Proc. of 18ᵗʰ Int. SAMPE Techn. Conf.*, Seattle, 1986. Covina, Vol. 18, pp. 272–286.

102. Kuminecz, J.F. *Spray applicator for spraying coatings and other fluids in space*. Pat. 4.519.545, USA, Int. Cl. B 05 B 9/047, publ. 28.05.85.

103. Osantowski, J.F., Keski-Kuha, R.A.M. and Herzig, H. (1991) Technology of optical coatings for extreme UV-range, Coll. pap. of 28ᵗʰ COSPAR Plen. Meet. on Astrophys. FUV and EUV Wavelengths, Hague, 1990. *Adv. Space Res.*, No. 11, pp. 202–208.

104. Staleri, B.J., Avery, J.E., Burgess, R.M. et al. (1987) Thin film tandem GaAs/CuInSe₂ solar cells for space power applications, In: *Proc. of 19ᵗʰ IEEE Photovolt. Spec. Conf.*, New Orleans, 1987. New York, pp. 280–284.

105. Toupikov, V.I., Khatipov, S.A., Chernyavsky, A.I. et al. (1997) Degradation of mechanical and electrophysical properties from films under combined action of far ultraviolet and thermal cycling, In: *Proc. of 7ᵗʰ Int. Symp. on Space Environment*, Toulouse, 1997. Noordwijk, pp. 77–85.

106. Weldable primer coating (1990) *Weld. Rev.*, **9**, No. 4, p. 226.

107. Wetheimer, M.R., Czeremuszkin, G., Gerny, J. et al. (1997) Plasma-deposited coatings and surface fluorination protection of spacecraft materials against atomic oxygen erosion, In: *Proc. of 7ᵗʰ Int. Symp. on Space Environment*, Toulouse, 1997. Noordwijk, pp. 343–402.

108. Whitaker, A.F. (1991) Coating could protect composites from hostile space environment, *Mater. Perform.*, **30**, No. 9, pp. 48–50.

109. Wickersham, C.E., Foster, E.L. and Stickford, G.H. (1981) Reactively sputter-deposited high-emissivity tungsten carbide/carbon coatings, *J. Vac. Sci. and Technol.*, **18**, No. 2, pp. 223–225.

110. Zorchenko, V.V., Palatnik, L.S., Nikitsky, V.P. et al. (1986) Special features of the approach to predicting the service life of materials of structures under space conditions, In: *Problems of space technology of metals*, Kiev, PWI, pp. 60–66.

Melting and Crystallisation

111. Agafonov, M.S., Levtov, V.L., Leskov, L.V. et al. (1984) Technological experiments under conditions of short-time zero-gravity, In: *Ideas of K.E. Tsiolkovsky and problems of space manufacuring*, Proc. of 18th Tsiolkovsky Readings, Kaluga, 1983. Moscow, pp. 106–115.

112. Aladiev, S.I. and Okhotin, A.S. (1982) Effect of variable density on the rate of growth of crystals from vapour phase, In: *Space technology and materials science*, Moscow, pp. 34–38.

113. Andrews, J.B., Briggs, C.J. and Robinson, M.B. (1990) Containerless low gravity processing of immiscible gold-rhodium alloys, In: *Proc. of 7th Eur. Symp. on Mater. and Fluid Sci. Microgravity*, Oxford, 1989. Paris, Noordwijk, pp. 121–126.

114. Arnold, W., Jacqmin, D., Gaud, R. et al. (1990) Convection phenomena in low-gravity processing: the GTE GaAs space experiment, *AIAA Pap.*, No. 0409, pp. 1–7.

115. Barmin, I.V., Egorov, A.V. and Senchenkov, A.S. (1989) Experiments on crucible-free zonal melting in weightlessness, In: *18th Gagarin Sci. Readings on Cosmonautics and Aviation*, Moscow, 1988. Moscow, pp. 261–262.

116. Barmin, I.V., Volkov, Yu.L., Gromakov, Yu.G. et al. (1983) Peculiarities of realization of process of crucible-free zonal melting of semiconductors in the conditions of microaccelerations, In: *K.E. Tsiolkovsky and problems of space manufacturing*, Proc. of 17th Tsiolkovsky Readings, Kaluga, 1982. Moscow, pp. 8–17.

117. Baumgartl, J., Gewald, M., Rupp, R. et al. (1990) The use of magnetic fields and microgravity in melt growth of semiconductors: a comparative study alloys, In: *Proc. of 7th Eur. Symp. on Mater. and Fluid Sci. Microgravity*, Oxford, 1989. Paris, Noordwijk, pp. 47–58.

118. Benz, K.-W. (1985) Einkristallzuchtungen in der Schwerelosigkeit, 1984–1985, *Lab. Prax.*, Sonderpubl. Labor 2000, pp. 16–19.

119. Benz, K.-W. (1988) German crystal growth experiments in space, *Oyo Butsuri*, **57**, No. 10, pp. 1505–1515.

120. Benz, K.-W. (1990) Factors controlling crystal perfection during growth under microgravity, In: *Proc. of 7th Eur. Symp. on Mater. and Fluid Sci. Microgravity*, Oxford, 1989. Paris, Noordwijk, pp. 59–62.

121. Bewersdorff, A. and Egry, I. (1990) Containerless processing in space. Requirements analysis, *Ibid.*, pp. 62–68.

122. Brant, N.B., Belov, A.G., Devyatkova, K.V. et al. (1983) Investigation of peculiarities of formation of crystalline structures, grown under microgravity conditions, *Vesti MGU*, Series Physics, Astron., **24**, No. 1, pp. 3–8.

123. Burfeindt, J. and Kemmerle, K. (1987) Schwerelosigkeint als technologisches Werkzeug fur Produktion und Verarbeitung, *Konstrukteur*, **18**, No. 10, pp. 18–20.

124. Burkhanov, G.S., Burov, I.V., Barmin, I.V. et al. (1987) Structure of pseudoalloy of Pb–Zn system, solidified under conditions of decreased gravitation, In: *16th Gagarin Sci. Readings on Cosmonautics and Aviation*, Moscow, 1986. Moscow, p. 214.

125. Castellani, L., Gondi, P. and Barbieri, F. (1981) Dispersion characteristics of material, slagged in a metallic vessel after melting and solidification at different levels of gravitation, In: *Proc. of 21st Int. Sci. Congr.*, Rome, 1981. Rome, pp. 143–152.

126. Day, D.E. and Ray, C.S. (1986) Research on containerless melts in space, In: *Oppor. acad. res. low-cravity environment*, New York, pp. 165–192.

127. Deryeterre, A. and Froyen, L. (1984) Melting and solidification of metallic composites, In: *Proc. of RIT/ESA/SSC Workshop Conf. on Effect. Gravity Solidificat. Immiscible Alloys*, Jarva Krog, 1984. Paris, pp. 65–67.

128. Drabovich, Yu.I., Yurchenko, N.N., Shevchenko, P.N. et al. (1986) Stabilized power unit for a zonal melting, In: *Problems of space technology of metals*, Kiev, PWI, pp. 96–99.

129. Duffar, T., Dussere, P. and Abadie, J. (1995) Crucible-semiconductor interactions during crystal growth from the melt in space, Coll. pap. of Symp. of COSPAR Sci. Com. during 30th COSPAR Sci. Assem. on Micrograv. Sci., Hamburg, 1994. *Adv. Space Res.*, **16**, No. 7, pp. 199–203.

130. Duguette, D.J. (1978) Metallurgical studies in Skylab and Apollo-Soyuz flights, *US Dep. Commer. Nat. Bur. Stand. Spec. Publ.*, No. 520, pp. 15–16.

131. Fredriksson, H. (1983) The effect of the natural convection on the solidification of metallic alloys, In: *Abstr. of pap. of 4th Eur. Symp. on Mater. Sci. Microgravity*, Madrid, 1983. Madrid, p. 93.

132. Gelfgat, Yu.M. and Gorbunov, L.A. (1985) About homogenizing of melts of multi-component systems under zero-gravity conditions, *Izv. AN SSSR*, Series Physics, **49**, No. 4, pp. 667–672.

133. German, R., Shneider, G., Kruger, H. et al. (1982) Experiments on growing crystals of solid solutions Bi-Sb under microgravity conditions, In: *Space technology and materials science*, Moscow, pp. 17–26.

134. Gorbatko, V.V. (1982) Technological experiment «Halong» on growing crystals of semiconductor compounds and experiment «Imitator» on measurement of temperature profiles of furnace «Kristall» in orbital station «Salyut», *Kosmich. Issledovan.*, **20**, No. 2, pp. 310–312.

135. Grishin, S.D. and Leskov, L.V. (1982) About crystallisation of melts in zero-gravity, In: *Problems of mech. and heat exchange in space engineering*, Moscow, pp. 207–224.

136. Herl, D.T.J. (1989) Growth of crystals, In: *Space materials science: introduction into fundamentals of space technology*, Moscow, pp. 371–391.

137. Hofmann, P., Klett, R., Sickinger, P. et al. (1992) Crystal growth experiments on Russian unmanned spacecraft, In: *Abstr. of pap. of 43rd IAF Congr. and 29th COSPAR Plen. Meet.*, Washington, 1992. Washington, p. 642.

138. Ivanov, L.I. and Pimenov, V.N. (1981) Experiments with metallic materials in orbital complex «Salyut-6–Soyuz», *Fizika i Khimiya Obrab. Materialov*, No. 1, pp. 59–64.

139. Ivanov, L.I., Kubasov, V.N., Pimenov, V.N. et al. (1977) About peculiar features of melting composite material in zero-gravity conditions, *Ibid.*, No. 5, p. 117.

140. Jonsson, R., Wallin, S., Holm, P. et al. (1983) UP, GAS, TKK Space Shuttle. Materials science experiments in Sweden. Sounding rockets. GAS, Space Shuttle, In: *Proc. of 6th ESA Symp. on Eur. Rocket and Balloon Programmes and Relat. Res.*, Interlaken, 1983. Paris, pp. 387–393.

141. Kasich-Pilipenko, I.E., Lapchinsky, V.F., Ovchinnikov, Yu.V. et al. *Solar furnace.* USSR author's cert. 1040879, Int. Cl. B 23 K, priority 26.08.81, publ. 10.05.83.

142. Khryapov, V.T., Markov, E.V., Prokopiev, E.V. et al. (1983) Investigations in the field of technology of semiconductor materials science, made using unit «Kristall», In: *K.E. Tsiolkovsky and problems of space manufacturing*, Proc. of 17th Tsiolkovsky Readings, Kaluga, 1982. Moscow, pp. 24–31.

143. Kinoshita, K. and Yamada, T. (1995) $Pb_{1-x}Sn_xTe$ crystal growth in space, *J. Cryst. Growth*, **147**, No. 1/2, pp. 91–98.

144. Kubasov, V.N., Ivanov, L.I., Pimenov, V.N. et al. (1977) Investigation of phase formation in diffusion interaction of tungsten-rhenium alloy with molten aluminium under zero-gravity conditions, *Fizika i Khimiya Obrab. Materialov*, No. 5, p. 129.

145. Kuribayashi, K. (1991) Growing crystals under microgravity conditions, *J. Jap. Soc. Precis. Eng.*, **57**, No. 5, pp. 787–790.

146. Langbein, D. and Roth, U. (1986) Das Verhalten von Teilchen und Blasen bei konfektionsfreier, gerichteter Erstarrung, *Z. Flugwissenschaft und Weltraumforschung*, **10**, No. 5, pp. 357–367.

147. Lenski, H. (1990) Advanced facilities for crystal growth, *Micrograv. Quart.*, **1**, No. 1, pp. 47–52.

148. Les resultats des premieres experiences scientifiques francaises d'elaboration des materiaux dan l'espace (1982) *COSPAR Inf. Bull.*, No. 93, pp. 83–85.

149. Maslyaev, S.A., Pimenov, V.N. and Sasinovskaya, I.P. (1987) About the mechanism of solidification of alloys of Al–Cu system in zero-gravity, In: *16th Gagarin Sci. Readings on Cosmonautics and Aviation*, Moscow, 1986. Moscow, pp. 208–209.

150. McKannan, E. (1997) Results from metallurgical flight experiments, In: *Mater. Sci. Space Appl. Space Process*, New York, pp. 383–398.

151. Nikitsky, V.P., Ivanov, A.I., Markov, A.V. et al. (1999) Technology of producing materials in space and its hardware in experiments in station «Mir» and Russian module of ISS, *Avtomatich. Svarka*, No. 10, pp. 100–104.

152. Osipian, Yu.A., Tatarchenko, V.V. and Shcherbina-Samojlova, M.B. (1982) Experiments on study of metals and alloys in conditions of decreased gravitation in orbital complex «Salyut–Soyuz», In: *Space materials science*, Proc. of 6th Sci. Readings on Cosmonautics, devoted to the memory of outstanding Soviet scientists-pioneers of space exploration, Moscow, 1982. Moscow, pp. 6–19.

153. Ossipyan, I.A. and Tatarchenko, V.A. (1988) Crystal growth from the melt by capillary shaping technique, *Adv. Space Res.*, **8**, No. 12, pp. 17–34.

154. Pant, P., Wijngaard, J., Haas, K. et al. (1991) Physical properties of manganese-bismuth alloys, produced under conditions of a low gravity force, *Aerokosm. Tekhnika*, No. 4, pp. 24–29.

155. Pant, P., Wijngaard, J.H. and Haas, C. (1990) Physical properties of manganese-bismuth specimens produced in microgravity, *J. Spacecraft and Rockets*, **27**, No. 4, pp. 369–372.

156. Pimenov, V.N., Kubasov, V.N. and Ivanov, L.I. (1984) Influence of zero-gravity state on the crystallization of the metallic materials, *Acta Astronautica*, **11**, No. 10/11, pp. 687-690.

157. Polezhaev, V.I., Bello, M.S., Verezub, R.A. et al. (1991) *Convective processes in zero-gravity*, Moscow, Nauka, 240 pp.

158. Praizey, J.P., Jamin-Changeart, F. and Sainct, H. (1986) Metallurgy laboratory for Columbus, In: *Abstr. of pap. of 37th IAF Congr. on Space: New Opport. People*, Innsbruck, 1986. IAF-261.

159. Presnyakov, A.A., Aubakirova, R.K. and Degtyareva, A.S. (1996) Behaviour of metallic melts under microgravity conditions in orbital station «Mir», *Doklady AN Resp*. Kazakhstan, No. 1, pp. 15–21.

160. Ratke, L. (1988) Immiscible alloys under microgravity conditions, *Adv. Space Res.*, **8**, No. 12, pp. 7–16.

161. Ratke, L. (1995) Coarsening of liquid Al–Pb dispersions under microgravity — a EURECA experiment, Coll. pap. of Symp. of COSPAR Sci. Com. during 30th COSPAR Sci. Assem. on Micrograv. Sci., Hamburg, 1994. *Adv. Space Res.*, **16**, No. 8, pp. 95–99.

162. Regel, L.L. (1986) The analysis of new results in the field of materials science in space, In: *Abstr. of pap. of 37th IAF Congr. on Space: New Opport. People*, Innsbruck, 1986. IAF-281.

163. Regel, L.L., Parfeniev, R.V., Vidensky I.V. et al. (1986) Growth of semiconductor Te crystals, Te–Se, Te–Si alloys and directed crystallisation of eutectic Al–Cu alloys under zero-gravity, *Ibid.*, IAF-283.

164. Regel, L.L., Vidensky, I.V., Morozov, V.E. et al. (1986) Structure and properties of silver-germanium alloys, produced under microgravity conditions in Soviet-Indian experiment «Overcooling» in unit «Isparitel-M», In: *15th Gagarin Sci. Readings on Cosmonautics and Aviation*, Moscow, 1985. Moscow, pp. 210–211.

165. Savitsky, E.M., Durov, I.V. and Burkhanov, G.S. (1981) Investigation of effect of zero-gravity factor on structure and properties of alloys with special physical properties, In: *10th Gagarin Sci. Readings on Cosmonautics and Aviation*, Moscow, 1980. Moscow, p. 288.

166. Second EVA (1986) *Pravda*, June 1.

167. Shalimov, V.P., Zemskov, V.S. and Raukhman, M.R. (1993) Formation and dynamics of bubbles in experiment on crucible-free zonal melting of indium antimonide under microgravity conditions, In: *Proc. of 27th Tsiolkovsky Readings*, Kaluga, 1992. Moscow, pp. 31–38.

168. Slobozhanin, L.A. (1983) Problems of stability of molten material in zonal melting, In: *K.E. Tsiolkovsky and problems of space manufacturing*, Proc. of 17th Tsiolkovsky Readings, Kaluga, 1982. Moscow, pp. 18–23.

169. Suzuki, Y., Kodama, S., Ueda, O. et al. (1995) Gallium arsenide crystal growth from metallic solution under microgravity, Coll. pap. of Symp. of COSPAR Sci. Com. during 30th COSPAR Sci. Assem. on Micrograv. Sci., Hamburg, 1994. *Adv. Space Res.*, **16**, No. 7, pp. 95–98.

170. Tatarinov, V.A., Meshcheryakova, I.O., Markov, S.V. et al. (1990) Directed crystallization of profiled single crystal of germanium onboard the orbital station «Mir», In: *Problems and prospects of space industrialisation*, Proc. of 24th Tsiolkovsky Readings, Kaluga, 1989. Moscow, pp. 37–45.

171. Verezub, N.A., Zubritskaya, I.N and Egorov, A.V. (1985) Investigation of peculiarities of producing some semiconductor systems in unit «Splav», *Izv. AN SSSR*, Series Physics, **49**, No. 4, pp. 687–690.

172. Vits, P. (1984) Experiment technology for microgravity research on Spacelab and other systems like TEXUS and MAUS, In: *Proc. of 14th Int. Symp. on Space Technol. and Sci.*, Tokyo, 1984. Tokyo, pp. 1643–1646.

173. Wiedemeier, H., Trivedi, S.B., Zhong, X.-R. et al. (1986) Crystal growth and transport rates of the GeSe-xenon system under microgravity conditions, *J. Electrochem. Soc.*, **133**, No. 5, pp. 1015–1021.

174. Witt, A.F. (1988) Electronic materials processing and the microgravitv environment, In: *Coll. techn. pap. of Symp. on Commer. Opportunities Space*, Taipei, 1987. Washington, pp. 201–211.

175. Yamanaka, T., Azuma, H., Kawasaki, K. et al. (1986) Floating furnace for supercooling processing, In: *Abstr. of pap. of 37th IAF Congr. on Space: New Opport. People*, Innsbruck, 1986. IAF-267.

176. Zemskov, V.S., Ivanov, L.I., Savitsky, E.M. et al. (1980) Main results of experiments under zero-gravity and some problems of space materials science, *Izv. AN SSSR*, Series Physics, **49**, No. 4, pp. 673–680.

177. Zemskov, V.S., Raukhman, M.R., Kozitsina, E.A. et al. (1991) Experiments on directed crystallisation of indium antimonide in ampoules in artificial satelites «Cosmos-1744» and «Foton», *Fizika i Khimiya Obrab. Materialov*, No. 5, pp. 46–52.

178. Zemskov, V.S., Raukhman, M.R., Kozitsina, E.A. et al. (1994) Experiments on growing crystals of indium antimonide under zero-gravity conditions, In: *Proc. of 28th Tsiolkovsky Readings*, Kaluga, 1993. Moscow, pp. 72–79.

179. Zubritsky, I.A., Maksimovsky, S.N. and Rodo, Ch. (1982) Growing of crystals $Ga_xIn_{1-x}P$ from the solution under zero-gravity conditions, In: *Space materials science*, Proc. of 6th Sci. Readings on Cosmonautics, devoted to the memory of outstanding Soviet scientists-pioneers of space exploration, Moscow, 1982. Moscow, pp. 30–43.

Chapter 4

1. Drabovich, Yu.I. and Yurchenko, N.N. *Device for bridge-type generator start*. USSR author's cert. 570985, Int. Cl. H 01 L, publ. 30.08.77.

2. Drabovich, Yu.I. and Yurchenko, N.N. *Method for control of switching-over transistors*. USSR author's cert. 955414, Int. Cl. H 01 L, publ. 30.08.82.

3. Drabovich, Yu.I. and Yurchenko, N.N. *Transistor inverter. USSR author's cert. 577627*, Int. Cl. H 01 L, publ. 25.10.77.

4. Drabovich, Yu.I., Maslobojshchikov, N.N., Yurchenko, N.N. et al. *Transistor*. Pat. 2525393, France, Int.Cl. H 01 L 25/02, 23/34, publ. 20.07.84.

5. Drabovich, Yu.I., Maslobojshchikov, N.N., Yurchenko N.N. et al. *Transistor*. Pat. 3201296, FRG, Int.Cl. H 01 L 25/02, 23/52, H 03 R 17/56, H 02 H 7 /00, H 05 K 10/00, publ. 12.06.86.

6. Drabovich, Yu.I., Pazeev, G.V., Yurchenko, N.N. et al. *Transistor invertor*. USSR author's cert. 470048, Int. Cl. H 01 L, publ. 05.05.75.

7. Drabovich, Yu.I., Yurchenko, N.N. and Shevchenko, P.N. *Transistor inverter*. USSR author's cert. 668053, Int. Cl. H 01 L, publ. 15.06.79.

8. Drabovich, Yu.I., Yurchenko, N.N., Shevchenko, P.N. et al. *Transistor inverter*. USSR author's cert. 584418, Int. Cl. H 01 L, publ. 15.12.77.

9. Oraevsky, V.N., Mishin, E.V. and Ruzhin, Yu.Ya. (1989) Artificial injection of energy particles in near-earth space, In: *Electromagnet. and plasma processes from the Sun to Earth core*, Moscow, pp. 77–86.

10. Paton, B.E., Lankin, Yu.N., Masalov, Yu.A. et al. *Device for electron beam welding and cutting*. USSR author's cert. 57784, Int. Cl. B 23 K, priority 26.06.69, publ. 11.08.71.

11. Sagdeev, R.Z. and Zhulin, I.A. (1995) Active experiments in the ionosphere and magnetosphere, *Vestnik AN SSSR*, No. 12, pp. 84–91.

12. Shelyagin, V.D., Lebedev, V.K., Mokhnach, V.K. et al. *Power high-voltage transformer.* USSR author's cert. 368656, Int. Cl. H 01 F 27/36, B 23 K 15/00, priority 12.03.71, publ. 26.01.73.

13. Winckler, J.R., Malcolm, P.R., Arnoldy, R.L. et al. (1989) An electron beam experiment in the magnetosphere, *EOS*, **70**, No. 25, pp. 657, 666–668.

Chapter 5

1. Aristarkhov, V.D. and Klein, A.A. (1992) Optimisation of mechanism of deployment of folded parabolic antenna, In: *Space systems with flexible links and transformable structures*, Moscow, pp. 92-98.

2. Beauchmamp, P.M. and Rodgers, D.H. (1995) New concepts for inflatable structures applied to spacebome radars, *AIAA Pap.*, No. 3795, pp. 1–8.

3. Bekey, I. (1988) Space construction results: the EASE/ACCESS flight experiment, *Acta Astronautica*, **17**, No. 9, pp. 987–996.

4. Belyakov, I.T., Sychev, Yu.K. and Martyushov, V.F. (1989) Assembly of truss structures in open space, In: *18th Gagarin Sci. Readings on Cosmonautics and Aviation*, Moscow, 1988. Moscow, p. 245.

5. Bernasconi, M.C. and Rits, W.J. (1989) Inflatable space rigidized support structures for large spacebome optical interferometer systems, Coll. pap. of 40th IAF Congr., Torremolinos, 1989. *IAF Prepr.*, No. 338, pp. 1–6.

6. Bowden, M.L. and Woolery, B.K. (1992) Space station solar array deployment mast, In: *Abstr. of pap. of 43rd IAF Congr. and 29th COSPAR Plen. Meet.*, Washington, 1992. Washington, p. 210.

7. Bulaev, O.I. and Isachenkov, V.E. (1984) Investigation of variants of cylindrical transformable shells, In: *Ideas of K.E. Tsiolkovsky and problems of space manufacturing*, Proc. of 18th Tsiolkovsky Readings, Kaluga, 1983. Moscow, pp. 48–55.

8. Bulatsev, A., Gutsal, P., Linevich, Yu. et al. *Device for joining connections of bar three-dimensional truss.* USSR author's cert. 1531540, Int. Cl. B 64 G, priority 22.04.88, publ. 22.08.89.

9. Bulatsev, A., Vasiliev, V., Zagrebelny, A.A. et al. *Container of connection elements.* USSR author's cert. 316255, Int. Cl. T 04 B 7/00, priority 10.08.89, publ. 01.07.90.

10. Bulatsev, A.R., Zagrebelny, A.A., Nikitsky, V.P. et al. *Truss.* USSR author's cert. 292136, Int. Cl. E 04 B, priority 30.05.88, publ. 01.04.89.

11. Bunakov, V.V. (1985) Deployment of space structures of large surfaces using ice drive, In: *17th Gagarin Sci. Readings on Cosmonautics and Aviation*, Moscow, 1983–1984. Moscow, p. 250–254.

12. Burmenko, E.Yu., Goliusov, T.A., Gonchar, O.Yu. et al. *Method of manufacture of items having a shape of a single-cavity hyperboloid with corrugated walls.* USSR author's cert. 1438885, Int. Cl. B 21 D 15/00, priority 20.04.87, publ. 23.11.88.

13. Burmenko, E.Yu., Pilishenko, I.S., Samilov, V.N. et al. *Method of manufacture of flexible variable-section branch pipes.* USSR author's cert. 1357109, Int. Cl. B 21 D 51/24, priority 02.07.86, publ. 07.12.87.

14. Bush, H.G., Herstrom, C.L., Heard, W.L. et al. (1991) Design and fabrication of an erectable truss for precision segmented reflector application, *J. Spacecraft and Rockets*, **28**, No. 2, pp. 251–257.

15. Chalykh, A., Matveev, V., Nikiforov, A. et al. (1997) About mechanism of surface roughness development on polyimide films during anisotropic etching by fast atomic oxygen, In: *Proc. of 7th Int. Symp. on Space Environment*, Toulouse, 1997. Noordwijk, pp. 243–246.

16. Dayton, L. (1990) Blowup space probes shrink the cost of missions, *New Sci.*, **125**, No. 1699, p. 39.

17. Denisova, A.N. *Device for assembly of three-dimensional structures*. USSR author's cert. 1601302, Int. Cl. E 04 G 21/26, publ. 23.10.90.

18. Dornheim, M.A. (1999) Inflatable structures taking to flight, *Aviat. Week and Space Technol.*, **150**, No. 4, pp. 60–62.

19. Drozdov, Yu.N. and Seok-Sam Kim (1999) Friction units in open space, *Problemy Mashinostr. i Nadyozhn. Mashin*, No. 3, pp. 54–55.

20. Fedorchuk, S.D. (1988) Determination of limiting flexibility of elements of space bar structures by a preset level of reducing their rigidity, In: *Experim.-theoret. investigations of antenna structure and deep water basements*, Moscow, pp. 40–44.

21. Finckenor, J. and Thomas, F. (1990) Mechanical joints and large components for pathfinder in-space assembly and construction, In: *Coll. techn. pap. of 31th AIAA/ASME/ASCE/AMS/ASC Conf. on Struct., Struct. Dyn. and Mater.*, Long Beach, 1990. Washington, Part 1, pp. 476–490.

22. Fiore, J., Kramer, R., Larkin, P. et al. (1994) Mechanical design and verification of the TOPEX/Poseidon deployable solar array, In: *Coll. techn. pap. of 35th AIAA/ASME/ASCE/AMS/ASC Conf. on Struct., Struct. Dyn. and Mater.*, Hilton Head, 1994. Washington, Part 1, pp. 125–135.

23. Freedom (1990) *New Bull. Astronaut. Soc. West. Austral.*, **15**, No. 5, pp. 51–52.

24. Freeland, R.E., Bilyeu, G.D. and Veal, G.R. (1996) Development of flight hardware for a large, inflatable-deployable antenna experiment, *Acta Astronautica*, **38**, No. 4/8, pp. 251-260.

25. van Casteren, J. and Ferri, P. (1989) EURECA flight operations, *ESA Bull.*, No. 60, pp. 25–32.

26. Genta, G. and Brusa, E.J. (1966) Project AURORA: preliminary structural definition of the spacecraft, *J. Brit. Interplanet Soc.*, **49**, No. 8, p. 296.

27. Geyer, F. (1990) Mechanisms to assemble space structures in extravehicular environment, In: *Proc. of 4th Eur. Space Mech. and Tribol. Symp.*, Cannes, 1989. Noordwijk, pp. 125–128.

28. Glazkov, Yu.N., Golotyuk, S.V., Ivanov, V.M. et al. (1985) Some technological problems of designing large space structures, In: *17th Gagarin Sci. Readings on Cosmonautics and Aviation*, Moscow, 1983–1984. Moscow, pp. 259–260.

29. Goldsworthy, W.B. (1983) The development of a composite beam building machine for on-site construction of large space structures, In: *Proc. of Conf. on Space Manuf.*, Princeton, 1983. San Diego, pp. 177–182.

30. Golovkov, S. (1998) Antenna for cosmonauts is made in Georgia, *Novosti Kosmonavtiki*, **8**, No. 13, p. 33.

31. Gralewski, M.R., Adams, L. and Hedgepeth, J.M. (1992) Deployable extendable support structure for the Radarsat synthetic aperture radar antenna, In: *Abstr. of pap. of 43rd IAF Congr. and 29th COSPAR Plen. Meet.*, Washington, 1992. Washington, pp. 98–99.

32. Guest, S.D. and Pellegrino, S. (1996) A new concept for solid surface deployable antennas, *Acta Astronautica*, **38**, No. 2, pp. 103–113.

33. Gvamichava, A.S. and Fedorchuk, S.D. (1995) *Rational designing of large-sized space structures*, Moscow, 25 pp.

34. Heard, W.L., Bush, H.G., Watson, J.J. et al. (1988) Astronaut/EVA construction of space station, In: *Coll. techn. pap. of AIAA SDM Int. Space Conf.*, Williamsburg, 1998. Washington, pp. 39–46.

35. Hedgepeth, J.M. (1989) Application of high-fidelity structural deployment analysis to the development of large deployable trusses, In: *Coll. pap. of 40th IAF Congr.*, Malaga, 1989. IAF Prepr., No. 339, pp. 1–16.

36. Henry, J.-P. *Mechanism for the automatic extention performing a rotary movement of itself.* Pat. 927083, USA, Int. Cl. B 64 G 1/44, publ. 03.11.87.

37. Inagaki, S. (1994) Space large-scale structure, *J. Math. Sci.*, **32**, No. 12, pp. 49–55.

38. Inflatable-structure space station concept revived (1989) *Flight Int.*, **136**, No. 4194, p. 19.

39. Kam, D.J. (1988) Retractable advanced rigid array, In: *Proc. of 20th IEEE Photovolt. Spec. Conf.*, Las Vegas, 1988. New York, Vol. 2, pp. 860–867.

40. Kaszubowski, M., Martinovic, Z. and Cooper, P. (1990) Structural dynamic characteristics of a Space Station Freedom first assembly flight concept, *AIAA Pap.*, No. 0748, pp. 1–11.

41. Kato, S., Sakai, I., Muragishi, O. et al. (1989) Research and development of a reflector structure employing inflatable elements, In: *Coll. pap. of 40th IAF Congr.*, Malaga, 1989. IAF Prepr., No. 334, pp. 1–10.

42. Kinteraya, G.G., Medzmariashvili, E.V. and Datashvili, L.Sh. (1999) The 5–30 m deployable high-precision light-weight space antenna reflectors and the ground-based stand-test complex for asssembly and testing large deployable space structures, In: *Proc. of Eur. Conf. on Spacecraft Structures, Materials and Mech. Testing*, Braunschweig, 1999. Noordwijk, pp. 11–18.

43. Kislitsky, M.I. (1992) About evaluation of effectiveness of space platforms, In: *Proc. of Seminar on Large-sized Truss Structures*, St.-Petersburg, 1991. St.-Petersburg, p. 42.

44. Kitamura, K. (1988) On present and future space structures. Their materials and construction, *Transact. JWRI*, **17**, No. 1, pp. 209–222.

45. Kitamura, T., Okazaki, K., Natori, M. et al. (1988) Development of a hingeless mast and its applications, *Acta Astronautica*, **17**, No. 3, pp. 341–346.

46. Koizumi, T., Yamamoto, K., Kurafuji, Y. et al. (1987) Large deloyable structures for space applications, *Mitsubishi Denki Giho*, **61**, No. 3, pp. 41–44.

47. Kulik, A.D., Shichanin, V.N., Samilov, V.N. et al. *Deployable contour antenna.* USSR author's cert. 1184409, Int. Cl. E 04 B, priority 24.10.83, publ. 08.06.85.

48. Kurkin, V.I., Terentiev, K.I. and Kurmanaliev, T.I. (1987) Deployable space structures from flexible ballastic and extendable elastically-transformable elements, In: *Design and technology of manufacture of space instruments*, Moscow, pp. 72–76.

49. Melnikov, V.M. and Krivolapova, O.Yu. (1996) Control of unfolding and folding of large-sized structures being formed by centrifugal forces, *Kosmich. Issledovan.*, **34**, No. 6, pp. 663–665.

50. Meurant, R.C. (1990) Structure, form and meaning in microgravity — the integral space Habitation, *Int. J. Space Structure*, **5**, No. 2, pp. 90–105.

51. Misawa, M., Yasaka, T. and Miyake, S. (1989) Analytical and experimental investigations for satellite antenna deployment mechanisms, *J. Spacecraft and Rockets*, **26**, No. 3, pp. 181–187.

52. Miura, K., Furuya, H. and Suzuki, K. (1985) Variable geometry truss and space crane arm, *Acta Astronautica*, **12**, No. 7/8, pp. 599–607.

53. Miura, K., Natori, M., Sakamaki, M. et al. (1984) Simplex mast: an extensible mast for space application, In: *Proc. of 14th Int. Symp. on Space Technol. and Sci.*, Tokyo, 1984. Tokyo, pp. 387–362.

54. Nichimura, T. (1987) Cutline of flexible structures in space, *J. Soc. Instrum. and Contr. Eng.*, **26**, No. 10, pp. 829–834.

55. Nikolsky, V.V. (1992) Effectiveness of large-sized structures, In: *Proc. of Seminar on Large-sized Truss Structures*, St.-Petersburg, 1991, p. 43.

56. Nozaki, K., Matsumoto, S., Saito, T. et al. (1992) A concept study of inexpensive solar power satellites using rotational force for self-development spacecraft, In: *Abstr. of pap. of 43rd IAF Congr. and 29th COSPAR Plen. Meet.*, Washington, 1992. Washington, p. 180.

57. Onoda, J., Watanabe, H., Ichida, K. et al. (1988) Two-dimensionally deployable SHDF truss, *Rep. Inst. Space and Astronaut. Sci.*, No. 633, pp. 1–15.

58. Pai, S. and Chamis, C.C. (1994) Probabilistic progressive buckling of trusses, *J. Space-craft and Rockets*, **31**, No. 3, pp. 466–474.

59. Paton, B.E., Bernadsky, V.N., Samilov, V.N. et al. *Three-dimensional truss for space constructions*. USSR author's cert. 1058217, Int. Cl. E 04 B, priority 19.04.82, publ. 01.08.83.

60. Paton, B.E., Bulatsev, A.R., Zagrebelny, A.A. et al. *Method of line assembly of bar three-dimensional truss and device for its realization*. USSR author's cert. 280862, Int. Cl. B 64 G, priority 05.10.87, publ. 01.08.88.

61. Paton, B.E., Dudko, D.A., Samilov, V.N. et al. *Transformable module*. USSR author's cert. 183142, Int. Cl. E 04 B, priority 19.04.82, publ. 07.01.83.

62. Pilishenko, I.S., Samilov, V.N., Burmenko, E.Yu. et al. *Collapsible frame antenna*. USSR author's cert. 1410803, Int. Cl. E 04 B, priority 02.07.86, publ. 15.03.88.

63. Pilishenko, I.S., Samilov, V.N., Burmenko, E.Yu. et al. *Drum for winding of elastic closed strip*. USSR author's cert. 1410804, Int. Cl. E 04 B, priority 02.07.86, publ. 15.03.88.

64. Romanov, O.Ya., Nikolsky, V.V. and Likhachev, A.N. (1992) Large-sized bar structures of space objects, problems and application, In: *Proc. of Seminar on Large-sized Truss Structures*, St.-Petersburg, 1991. St.-Petersburg, pp. 3–4.

65. Runavot, J. (1991) French «platform-related» technological experiments on board Soviet space station, *Acta Astronautica*, **25**, No. 7, pp. 375–378.

66. Samilov, V.N., Balitsky, V.M. and Sukhenko, I.V. *Transformable structure*. USSR author's cert. 69093, Int. Cl. E 04 B, priority 18.12.76, publ. 12.02.73.

67. Samilov, V.N., Balitsky, V.M., Sukhenko, I.V. et al. *Transformable structure*. USSR author's cert. 46212, Int. Cl. E 04 B, priority 16.10.67, publ. 17.04.69.

68. Samilov, V.N., Burmenko, E.Yu., Gryanik M.V. et al. *Collapsible reflector.* USSR author's cert. 1660089, Int. Cl. H 01 Q 15/20, priority 04.04.89, publ. 30.06.91.

69. Samilov, V.N., Gonchar, O.Yu. and Burmenko, E.Yu. (1986) Development of transformable structure of lock module and technology of its construction, In: *Problems of space technology of metals*, Kiev, PWI, pp. 67–71.

70. Samilov, V.N., Pilishenko, I.S. and Burmenko, E.Yu. *Method of manufacture of corrugated products.* USSR author's cert. 1461561, Int. Cl. B 21 D 13/10, priority 28.01.87, publ. 28.02.89.

71. Samilov, V.N., Sukhenko, I.V. and Balitsky, V.M. *Method of creation of transformable structures.* USSR author's cert. 67993, Int. Cl. E 04 B, priority 24.08.70, publ. 08.01.73.

72. Saprykin, Yu.I., Lapchinsky, V.F., Stesin, V.V. et al. *Three-dimensional truss.* USSR author's cert. 331728, Int. Cl. E 04 B, priority 04.06.91, publ. 01.11.91.

73. Semyonov, Yu.P., Ryumin, V.V. and Nikitsky, V.P. (1995) State-of-the-art and prospects of construction of structures in space, In: *Proc. of Int. Conf. on Welded Structures*, Kiev, 1995. Kiev, pp. 213–220.

74. Shalin, R.E., Minakov, V.T., Deev, I.S. et al. (1997) Study of polymer composite specimens surface changes after the long-term exposure in space, In: *Proc. of 7th Int. Symp. on Space Environment*, Toulouse, 1997. Noordwijk, pp. 375–383.

75. Shubin, V.S., Bashkirova, T.A., Aronsky, V.M. et al. *Method of manufacture of discs with circular corrugations.* USSR author's cert. 995983, Int. Cl. B 21 D 13/10, priority 05.12.82, publ. 17.02.83.

76. Soosaar, K. (1989) Structural design and validation challenges of large space systems, In: *Space — New Commun. Opport.*, Proc. of 34th Ann. AAS Int. Conf., Houston, 1987. San Diego, pp. 313–317.

77. *Spacecraft structures and mechanisms* (1995) Ed. by T.P. Sarafin, Kluwer A.P., 850 pp.

78. Stesin, V.V., Zagrebelny, A.A., Polinov, Yu.S. et al. *Extendable cylindrical rod.* USSR author's cert. 275177, Int. Cl. H 01 Q, priority 17.10.68, publ. 03.07.70.

79. Suezawa, E. (1989) Space station and colonies, In: *Welding in space. Technology of assembling and joining*, Tokyo, pp. 66–75.

80. Sunahara, J. (1987) Parameter estimation and optimal shape control for LSS, *J. Soc. Instrum. and Contr. Eng.*, **26**, No. 10, pp. 863–870.

81. Sved, J. (1985) Assembly and maintenance of space platforms, *J. Brit. Interplanet Soc.*, **38**, No. 7, pp. 319–327.

82. Sychev, Yu.K. and Martyushov, V.F. (1989) Application of truss structures in creation of habitable objects in space, In: *K.E. Tsiolkovsky and space industrialisation*, Proc. of 23rd Tsiolkovsky Readings, Kaluga, 1988. Moscow, pp. 114–118.

83. Takahashi, K., Nagata, H. and Kudo, I. (2000) Simulation for deployment of an inflatable disc in orbit, *J. Spacecraft and Rockets*, **37**, No. 5, pp. 707–708.

84. Takamatsu, K. and Onoda, J. (1991) New deployable truss concepts for large antenna structures or solar concentrators, *Ibid.*, **28**, No. 3, pp. 330–338.

85. Taran, V.M. *Deployable structure of space object.* USSR author's cert. 1086680, Int. Cl. B 64 G 1/22, 1/44, publ. 15.09.94.

86. Tauber, W. (1986) Development of an antenna structure for a deployable offset antenna, In: *Proc. of Conf. on Compos. Des. Space Appl.*, Noordwijk, 1985. Paris, pp. 329–336.

87. Terentiev, Yu.K. (1994) Prospects of use of inflated large-sized structures in space, In: *Proc. of 28th Tsiolkovsky Readings*, Kaluga, 1993. Moscow, pp. 68–70.

88. Toss, R.A. and Bradley, O.H. (1985) Thermal analysis of the ACCESS space truss, In: *Proc. of Conf. and Expo on Space Techn.*, Anaheim, 1985. Dearborn, pp. 7/1–7/10.

89. Turkin, I.K., Safronov, V.S. and Starovojtov, V.V. (1996) Development of software-methodical complex of optimum designing of thin-walled shell structures of pespective flight vehicle with allowance for extreme load, In: *Proc. of 29ᵗʰ Tsiolkovsky Readings*, Kaluga, 1994. Moscow, pp. 94–100.

90. Unda, J., Weisz, J., Rivacoba, J. et al. (1994) Family of deployable/retractable structures for space application, *Acta Astronautica*, **32**, No. 12, pp. 767–784.

91. Usyukin, V.I., Zimin, V.N., Krylov, V.V. et al. (1991) Mechanics of self-folding truss antenna, In: *20ᵗʰ Gagarin Sci. Readings on Cosmonautics and Aviation*, Moscow, 1990–1991. Moscow, p. 240.

92. Vyazhlinsky, A.D. and Ostroumov, B.V. (1992) Design and development of structures of large-span panels in the form of complex branched rope systems, In: *Space systems with flexible links and transformable structures*, Moscow, pp. 68–76.

93. Wada, B. (1990) Adaptive structures: an overview, *J. Spacecraft and Rockets*, **27**, No. 3, pp. 330–337.

94. Waters, T.L. and Waters, M.T. (1990) Hyperboloidal deployable space antenna, In: *Proc. of 3ʳᵈ Pacif. Int. Symp. on Adv. Space Sci. Technol. and Appl. (PISSTA)*, Los Angeles, 1989. San Diego, pp. 127–134.

95. Wehrli, Ch., Frohlich, C. and Romero, J. (1995) The European Retrievable Carrier (EURECA) as an observing platform from a user's point of view, Coll. pap. of Symp. of COSPAR Sci. Com. during 30th COSPAR Sci. Assem. on Micrograv. Sci., Hamburg, 1994. *Adv. Space Res.*, **16**, No. 8, pp. 51–54.

96. Wilson, M.L., McConachi, I.O. and Johnson, G.S. (1988) Space structures built in space, *Mod. Plast. Int.*, **18**, No. 9, pp. 60, 62, 64, 66, 68.

97. Yakushchenko, V.F. and Kuzmin, A.B. (1992) Modelling of sequence of assembly of large-sized structures, In: *Proc. of Seminar on Large-sized Truss Structures*, St.-Petersburg, 1991. St.-Petersburg, pp. 38–39.

98. Yakushchenko, V.F., Tsygankov, O.S. and Kuzmin, A.B. (1992) Evaluation of technological adaptability of manual assembly of elements of large-sized structures relative to conditions of EVA in orbit, In: *Proc. of Seminar on Large-sized Truss Structures*, St.-Petersburg, 1991. St.-Petersburg, pp. 40–41.

99. Zajtsev, G.P., Astapov, V.Yu., Volkov, I.T. et al. (1985) Peculiar features of designing and manufacture of junctions of self-constructing three-dimensional structures, In: *17ᵗʰ Gagarin Sci. Readings on Cosmonautics and Aviation*, Moscow, 1983-1984. Moscow, p. 263.

100. Zhuravin, Yu. (1999) Ukrainian module for ISS, *Novosti Kosmonavtiki*, **9**, No. 1, pp. 26–27.

101. Zimin, V.N., Koloskov, I.M. and Meshkovsky, V.E. (2000) About calculation of dynamic characteristics of deployable truss large-sized space structure, In: *Appl. problems of mechanics of aerospace systems*, Abstr. of pap. of All-Russia Conf., N.E. Bauman MSTU, Moscow, 2000. Moscow, p. 82.

102. Zwanenburg, R. (1985) A deployable and retractable strongbank structure, In: *Proc. of 2ⁿᵈ Eur. Space Mech. and Tribol. Symp.*, Weersburg, 1985. Paris, pp. 239–245.

103. Zwanenburg, R. (1993) The Columbus solar array mechanisms, In: *Proc. of 5ᵗʰ Eur. Space Mech. and Tribol. Symp.*, Noordwijk, 1992. Paris, pp. 49–54.

Chapter 6

1. Anselmo, J.C. (1998) In orbit, *Aviat. Week and Space Technol.*, **149**, No. 2, p. 17.

2. Arkov, P.F., Berzhaty, V.I., Demina, E.A. et al. (1998) Problems and prospects of technical investigations in manned space complexes, *Kosmonavtika i Raketostr.*, No. 12, pp. 108–115.

3. Arkov, P.F., Kuznetsov, A.A., Markov, A.V. et al. (1999) Semiconductor materials science and production in space conditions, *Avtomatich. Svarka*, No. 10, pp. 56–58.

4. Dunn, B.D. (2000) New materials in space, *Materials World*, **8**, No. 1, p. 25.

5. Goree, J. (1994) Plasma crystals, *AIAA Pap.*, No. 699, pp. 1–7.

6. Motegi, T. (1994) Use of microgravity conditions for production of materials, *Rep. China Inst. Technol.*, No. 41, pp. 191–203.

7. New discoveries unobtainable in gravity — one of FMPTs results (1993) *STA Today*, **5**, No. 10, p. 5.

8. Osipian, Yu.A. (1998) Problems and prospects of space technology and materials science, *Kosmonavtika i Raketostr.*, No. 12, pp. 116–123.

9. Santoli, J. (1996) Project AURORA: a preliminary study of a light, all-metal solar sail, *J. Brit. Interplanet Soc.*, **49**, No. 8, p. 296.

10. Starkov, V.V. (1996) Solar power systems of large and small forms, *Mashinostroitel*, No. 10, pp. 53–54.

11. The space industry (1998) *News Bull. Astronaut. Soc. West. Austral.*, **23**, No. 11, pp. 118–120.

12. Weltraumerprobte giutager im auromobilan (1994) *Techn. Rdsch.*, **86**, No. 48, p. 8.

13. Werner, D. (1996) Raumfahrfrorschung als Quelle fur neue Produkte, *Ind.-Anz.*, **118**, No. 20, p. 26.

Terminology

AMBIENT OUTER ATMOSPHERE (AOA) — around space vehicles is a complex dynamic formation, including gaseous, aerosol and finely-dispersed phases. The main sources of AOA formation are desorption and diffusion of gases and vapours, adsorbed and absorbed by the materials of the external coatings, destruction and evaporation of the external coating materials, gas jets of operating engines, evolutions of gases and vapours during functioning of the life support system, gas escape from pressurized compartments, evolution of aerosol and dispersed particles during engine starting and cut-off, as well as through drainage systems, evolution of dispersed particles during vibration of structural elements and mechanical impacts. During performance of astrophysical, geophysical, materials science and technical experiments on board OS «Salyut» and OC «Mir» it was found that the processes in AOA significantly contaminate the external surface of the space vehicles, having a negative influence on the condition of the measuring instruments of the control and orientation systems, as well as on the readings of the instrumentation and scientific equipment. In AOA stationary («background») condition, when no dynamic operations are conducted for a long time, the pressure of the AOA gas phase at orbital complex surface is $\sim 10^{-6}$–10^{-5} mm Hg by the results of flight measurements. During dynamic operations, when the propulsion systems are functioning, the pressure in AOA abruptly rises by 2 to 4 orders of magnitude compared to the background conditions, and then is relaxed to the initial condition. The time for relaxation of pressure disturbance is determined by the cyclogram of the engines operation, as well as the dimensions of the surface sections of structural elements exposed to the impact of these engine jets.

AERODYNAMIC LOADS — loads applied to the structural elements of a flying vehicle (aircraft, rocket) or descent vehicle that are induced by the aerodynamic forces during movement through the atmosphere. Aerodynamic forces are manifested in the form of distributed forces and frictional forces. During the rocket flight through the active flight leg the aerodynamic loads first rise during acceleration and, having reached a certain maximum value (usually at the altitude of 10–12 km), drop practically to zero after the rocket has left the dense layers of the atmosphere.

AIRLOCK MODULE, DOCKING COMPARTMENT — special pressurized compartment of a space vehicle or orbital station used for the cosmonauts going out for space walks without de-pressurizing the cabin (working compartments or living quarters) or for transition from one compartment into another one (if the compartments have different pressure or composition of the atmosphere). An airlock module has two or more pressurized manholes, one of which opens into the space vehicle cabin; other manholes are used for the cosmonauts' EVA. Airlock module was used for the first time in 1965 in «Voskhod-2» space vehicle during cosmonaut A. Leonov's space walk.

ALLOYING COMPONENTS — metallic additives (ferroalloys) or chemical compounds based on alloying elements, added either to the base alloy or the weld to provide the required physico-mechanical properties of the alloy or the weld metal, respectively.

ALTITUDE CHAMBER — hermetically sealed air-tight chamber for artificially changing the atmospheric pressure. Distinction is made between altitude chambers of a low (vacuum) and high (compressive) pressure; stationary and mobile. An altitude chamber which allows the temperature to be varied is called a thermal pressure chamber. Low pressure altitude chambers are used for altitude testing and training, testing high-altitude protective clothing and equipment, in space medicine, etc. Altitude chambers simulating the conditions of space, so-called space simulation test chambers, are used for ground-based testing of space vehicles or their compartments and elements. Altitude chambers volume can be from several hundred dm^3 up to tens of m^3. Pressure can vary from atmospheric pressure to 0.1 μPa and lower. Level (or depth) of vacuum to 0.1 Pa pressure is formed by mechanical vacuum pumps, namely up to 1 mPa by diffusion pumps and up to 10 μPa by applying nitrogen traps, which are flasks immersed in liquid nitrogen, and below 10 μPa by getter pumps in small-volume altitude chambers or by applying screens, cooled by liquid nitrogen or helium for large-volume altitude chambers [$1 \cdot 10^{-2}$ mm Hg/bar = 1.3 Pa].

ANISOTROPY — inhomogeneity of physico-mechanical properties of the material in different directions.

AUGER-SPECTROSCOPY — a section of electron spectroscopy, where the methods are based on measurement of the energy and intensity of currents of Auger-electrons, emitted by atoms, molecules and solids at Auger effect. Auger-electron energy is determined by the nature of the atoms, emitting them and their chemical environment, thus allowing determination of atoms in the compounds and obtaining information about their chemical condition. Auger spectroscopy is used both for fundamental studies and for elementary analysis. Auger-spectroscopy became the most widely applied for elementary analysis of the subsurface layer of a solid of the thickness of several atomic layers. Layer-by-layer atomization of the studied sample with inert gas ions is used to obtain data on deeper-lying layers. The sensitivity of Auger-spectroscopy methods is ~10^{12} at/cm^2. In the majority of cases modern Auger-spectrometers can operate in the scanning mode and provide information about the distribution of individual elements over the sample surface. Auger-spectrscopy with slow electron diffraction also provides information on the composition of the subsurface layers of single crystal samples and on their structure and changes.

AUSTENITE — non-magnetic solid solution of carbon in γ-iron. Austenite is a stable modification of iron alloys with carbon at temperatures above the critical point A_3. Austenite, present in steel at normal temperature, is called residual austenite.

BINARY ALLOY (TWO-COMPONENT) — alloy including, one alloying component, added to the alloy base to give it the required functional properties.

CLUSTER-TYPE VACUUM UNIT — a unit, consisting of several super high-vacuum chambers, connected to each other by vacuum mechanisms for substrates transportation without taking them out into the atmosphere. The substrate route is determined by the sequence of processing operations (applying various films, performed in different chambers; ion etching of the surface; doping; heat treatment, etc.). The cost of such units is very high — up to several tens mln US dollars.

COAGULATON — reduction of the degree of dispersity, i.e. increase of the dimensions of particles of any disperse system under the impact of intermolecular attractive forces, due to temperature rise, adding some salt, as well as the impact of the electric field (electrocoagulation). It is rational to use coagulation as one of the methods for cleaning sanitary waters in the life-support system of the orbital station, as it causes precipitation of a sediment — coagulate — from the colloidal solution or formation of a continuous solid-phase structure (gelatinization).

CONDENSATION — material transition, as a result of its cooling or compression, from a gaseous into the condensed state (liquid or solid). Vapour condensation is only possible at a temperature below the critical temperature for the given material. The same amount of heat evolves during condensation as was spent for evaporation of the condensed material. Condensation is widely applied in power engineering, chemical technology and cryogenic engineering, etc. In gas turbine construction the processes of evaporation and vacuum condensation are used to produce high-temperature and thermal barrier coatings on parts of the hot section of turbines of various-purposes.

CONVECTION — heat transfer by material flows in liquids, gases or loose media. Free convection arises in the gravity field at non-uniform heating (heating from below) of fluid or loose materials. Under the impact of the buoyancy force, the heated material moves relative to the less heated material in the direction, opposite to that of the gravity force. Convection leads to equalizing of the material temperature. Convection intensity depends on the temperature difference between the layers, heat conductivity and viscosity of the medium. At forced convection material displacement is mostly provided, using a pump, mixer or some other device. Convection is very common in nature, in the lower layer of the Earth's atmosphere, in the ocean, in the Earth interior, in the stars.

CORROSION — a phenomenon and process of chemical transition (e.g. oxidation) or spontaneous undesirable fracture of items of metals (and other materials) under environmental impact. Corrosion resistance is determined by the material composition and structure, presence of mechanical stresses, condition of the surface, corrosiveness (presence of sulphur, sulphates, chlorides, moisture, etc.) of the atmosphere.

COSMOS, SPACE — synonym of the astronomical definition of the Universe. Earth with its atmosphere is often excluded from the notion of «cosmos». A distinction is made between near space, including circumterrestrial space and deep space — the world of stars and galaxies.

CROSSOVER — minimal diameter of the electron beam.

CRYOGENIC — cooled to low temperatures, for instance, at immersion into liquid nitrogen or helium.

DAMPING DECREMENT — qualitative characteristic of oscillation damping rate.

DENDRITES — arborescent crystals, formed in solidification of solutions or metal melts under conditions of a directed heat removal. Dendritic columnar structure can be observed in the weld metal. This structure is particularly pronounced in single-pass welds of large cross-section.

DIFFRACTION — X-ray diffraction is a phenomenon arising at elastic scattering of X-ray radiation in crystals, amorphous bodies, liquids or gases, and consists of the appearance of deflected (diffracted) beams propagating at specific angles to the initial beam. X-ray diffraction is due to spatial coherence of secondary waves, arising in scattering of primary radiation by electrons of various atoms.

DIFFUSION — spontaneous process of displacement of the atoms or molecules or microscopic particles (Brownian motion) in gases, liquids or solids, resulting from equalization of the activities of the atoms, molecules or colloidal particles in an initially inhomogeneous system arising from their chaotic thermal motion.

DOPING OF SEMICONDUCTORS — dosed addition of impurities or inducing structural defects in a semiconductor to change its electrical properties. Doping of semiconductors is usually performed in the processes of growing single crystals and epitaxial structures.

DOPING PROFILE — doping impurity distribution over the thickness of a film or crystal.

DOUBLETS — splitting of X-ray diffraction lines into two components, related to the nature of the characteristic X-ray radiation, or to the change of the object structure (in this case to inhomogeneity of the solid solution).

ECHO-METHOD (METHOD OF REFLECTION) — method of ultrasonic flaw detection, where the size and location of the defect are determined by applying ultrasound to the controlled item and recording the ultrasonic beam reflected from the defect.

ELECTRON BEAM — flow of electrons (the transverse dimensions of which are usually much smaller than their length), emitted by one source and moving along close trajectories concentrated in a certain region of space.

ELECTRON BEAM WELDING (EBW) — fusion welding process in which the metal is heated by an electron beam. Equipment for electron beam welding is called the electron beam unit. It basically consists of the working chamber, vacuum system, the electron beam gun with the high-voltage direct current source, system of control and monitoring of the welding process, system of assembling the parts to be welded and their displacement in the chamber.

ELECTRON MICROSCOPE — an instrument for observation and photographing of many times (up to 10^6 times) enlarged image of objects, which uses, instead of light rays, the flows of electrons, accelerated up to high energies (30–100 keV and more) under deep vacuum conditions. Transmission, reflection, emission and scanning electron microscopes are used. The first scanning microscopes were built in 1938–1942, operating on the principle of scanning, i.e. successive displacement of a thin electron beam (probe) from point to point over the object. By mid-1960s the scanning electron microscopes reached a high level of technical perfection, and since that time they began to be widely applied in research.

EMMISSION SYSTEM — in the electron beam gun of the welding unit — part of the gun, including the cathode assembly and accelerating electrode.

EPITAXY — ordered oriented growth of one crystal on the surface of another one (single crystal substrate). A distinction is made between heteroepitaxy, when the materials of the substrate and the growing crystal are different, and homoepitaxy (autoepitaxy) when they are the same. Oriented growth of a crystal inside the bulk of another one is called endoepitaxy. Epitaxy mainly proceeds by precipitation from liquid solutions or melts, solidification from the gaseous phase by chemical reactions or vapour phase condensation in vacuum. It depends on the purity and defectiveness of the substrate surface.

EROSION — fracture of the surface of a material (part), resulting from long-term mechanical impact (shock or contact) by abrasive particle flows, hard carbon, sand, cement dust or by electrical discharges. Erosion is usually observed on the surfaces of parts in a turbine hot section as a result of the eroding impact of the fuel combustion products (velocity of 300 m/s; pressure of 20 atm, temperature of 1500 K), containing fine particles in the used air. Erosion resistance is evaluated by a reciprocal of the loss of mass, amount of removed material or depth of the erosion cavity in the damaged area.

EXTRA-VEHICULAR ACTIVITY (EVA) — activities (operations), performed by the crew outside the space vehicle pressurized compartments.

FACTORS OF SPACE ENVIRONMENT — zero gravity, space vacuum, sharp light–shadow boundaries, ionizing radiation, micrometeorite particles.

FOCAL SPOT — in electron beam welding (melting, evaporation) is the heated spot, corresponding to the electron beam diameter at the point of its falling on the surface of a workpiece or remelted billet.

FREE-MOLECULAR FLOW — mode of flowing of highly rarefied gas, when the mean free path length of gas particles l is essentially greater than the characteristic size L of the body around which the gas flows. This flow mode is in place under the condition of l/L $\mu \geq 3$ (l/L ratio is called Knudsen number, Kn). For existing space vehicles, the latter condition is usually fulfilled at orbital flight altitudes in the Earth's atmosphere above 250 km.

FREE PATH (MEAN FREE PATH LENGTH, l) — average distance covered by a neutral particle of gas between two successive collisions with other particles. In the case of elastic collisions, when just the direction and velocity of movement of a gas particle are changed, $l = (\sqrt{2}n\sigma)^{-1}$. Here n is the number of gas particles in a unit of volume, σ is the effective cross-section of collision. In the case of rigid elastic spheres the diameter of effective cross-section of collision is equal to the sum of the diameters of these spheres.

HEATED SPOT — active spot in fusion welding. A section of the welded item surface, through which energy is applied to the latter from the heat source, namely electric arc, welding flame, electron beam, etc.

HETEROJUCTION — contact of two semiconductors of different composition. Range of energy values (energy gap) that the electrons cannot have in a semiconductor crystal, mobility of the charge carriers, their effective masses, etc. are changed on the interface. In a heterojunction, change of these properties occurs over a distance of the order of the width of the space-charge region.

HETEROSTRUCTURE — semiconductor structure with several heterojunctions. As the width of the energy gap, mobility of the charge carriers, their effective masses, etc. are changed on the heterojunction interfaces, it becomes possible to effectively control the movement of the charge carriers and their recombination, as well as the light flows.

HETEROSYSTEM — inhomogeneous thermodynamic system consisting of parts (phases) differing in their physical properties or composition. Adjacent phases of the heterosystem are separated from each other by physical interfaces, on which one or several properties of the system change abruptly (composition, density, crystallographic structure, electric or magnetic moment, etc.).

HOMOGENIZED SYSTEM — a homogenized system in which the properties (composition, density, pressure, etc.) continuously change in space. Gas mixtures, liquid or solid solutions and melts can be homogeneous.

HYDRAULIC TESTING LABORATORY — engineering facility designed for zero-gravity simulation in a hydraulic medium, incorporating a water tank, process equipment and registering instruments to support cosmonauts' training, testing and research, mock-ups of manned space vehicle and stations, systems of water preparation, spacesuit cooling, etc.

INDICATRIX — auxiliary surface characterizing the dependence of any property of the medium on direction. In order to construct the indicatrix, radius-vectors are drawn from one point, their length being proportional to a quantity characterizing the given property in a given direction, for instance, conductivity, refraction index, and modulus of elasticity.

INHIBITORS — preparations added to a system to delay undesirable chemical reactions, or to practically stop them. Corrosion inhibitors: addition of relatively small amounts of inhibitors to a medium contacting the metal surface, which markedly lowers the corrosion rate. Applied in rocket engineering to lower the rate of corrosion damage of metals by rocket fuel elements. Often used are inhibitors, which lower the corrosiveness of the medium by forming protective films on the metal surface, consisting of the products of inhibitor interaction with the fuel, metal or its corrosion products. Hydrogen fluoride, sulphuric and phosphoric acids are examples of inhibitors.

INJECTION — injection of carriers is the penetration of non-equilibrium (excess) charge carriers into a semiconductor or dielectric under the impact of the electric field.

INVERTOR — a unit for performing inversion.

LASER HOLOGRAPHIC CONTROL — designed for comparison of the pre-loaded welded joints with the reference joints (not loaded); reveals anomalies in the pre-loaded joints.

LIGHT GUIDE — a closed device for directed transmission (channeling) of light. In open space its transmission is only possible within line-of-sight distance and involves losses due to initial divergence of radiation, absorption and atmospheric dispersion. Transition to the light guide significantly reduces the energy losses in the transmission of light over large distances, and allows light energy transmission along curvilinear propagation paths.

LOAD-CARRYING STRUCTURE — structure, to the elements of which external loads are applied.

MASS-SPECTROMETRY, MASS-SPECTROMETRIC ANALYSIS — method of material investigation by determination of the mass of atoms and molecules present in its composition and of their quantity. The totality of mass values and their relative contents is called the mass-spectrum. Mass-spectrometry users vacuum separation of ions with different ratios of mass m to charge e under the impact of electric and magnetic fields. Therefore the examined material is first of all ionized. In the case of liquid and solid materials they are first evaporated and then ionized. The high accuracy and sensitivity of mass-spectrometry as a method of analysis is widely used in chemistry for elementary and structural molecular analysis. In physico-chemical investigations mass-spectrometry is applied to determine the energy of ionization, evaporation heat, energy of the atoms bond in the molecules, etc.

MASS TRANSFER — process of material transfer from one phase into another in non-equilibrium binary or multicomponent systems.

METAL AGEING — change of the metal properties resulting from internal processes, usually proceeding more slowly at room temperature (natural ageing) and more intensively at higher temperature (artificial ageing). The term «ageing of metals» is mostly understood to mean change of the alloy properties as a result of decomposition of oversaturated solid solutions — so-called precipitation (age) hardening. Subsequent decomposition of the solid solution at room temperature, higher temperature or after cold deformation is metal ageing.

The capability of a number of alloys to undergo ageing allows the production of materials with very high physical properties, namely hardness, strength, magnetic and other properties.

METALLOGRAPHIC INVESTIGATIONS, METALLOGRAPHIC ANALYSIS — investigation of grain structure (texture) of metal on sections or fractures, performed as a rule by photographing the studied surface with the magnification of ×5 up to ×20000 and more.

METEORITE HAZARD — risk of collision of the space vehicle with the hard particles of interplanetary environment (space dust, meteorite bodies) that may lead to disturbance of the normal operation of the space vehicle. Its collisions with relatively large meteorite particles (of more than 10^{-4}–10^{-5} g mass) may result in penetration of the space vehicle shell, and its de-pressurizing if no protective measures are taken. Penetration of particles into the pressurized compartments in a number of cases may lead to explosive processes and fires (in the case of an oxygen atmosphere), their penetration into tubing filled with liquid — to generation of shock waves and structure failure, etc. The study of meteorite hazards is conducted in two directions: studying meteorite material and studying the interaction of the space vehicle shell with meteorite particles.

MICROSTRUCTURE — material structure, visible in a toolmaker's microscope on a sample surface, ground and polished to mirror-like lustre, and etched with special reagents. The process of microstructural studies is called microstructural analysis.

MODELLING (physical) — replacement of investigation of some object or phenomenon by experimental studies of a model having the same physical nature. Modeling is resorted to not only for economic reasons, but also because full-scale testing is very difficult or absolutely impossible to implement when the dimensions of the actual object or the magnitudes of its other characteristics (pressure, temperature, process rate, etc.) are too large (or small).

OPERATOR'S WORK PLACE — platform with the hardware, means of observation, fixation and control, allowing the cosmonaut to perform various jobs (operations).

ORBITAL STATION — space vehicle, functioning for a long-time in circumterrestrial, circumlunar or circumplanetary orbit. An orbital station can be manned (with a crew of cosmonauts) or can function in the automatic mode. The purpose of an orbital station is to address the scientific and applied tasks, namely studying circumterrestrial space and Earth (planet) from the orbit of the Earth artificial satellite, conducting meteorological, astronomical, radioastronomical and other observations, medico-biological experiments, investigation of the behaviour of materials and equipment in a space flight, etc. Orbital stations can also be used as bases for assembly of heavy space structures in orbit, designed for flight to other planets of the Solar system. Repair-restoration work on board orbital stations, the replacement of apparatus and systems at the end of their service life, by new ones delivered by cargo ships, servicing the orbital stations, ensure long-term functioning of the orbital stations. The flight altitude of a near-earth orbital station is 200–500 km.

PERVEANCE — characteristic conductance P of the electron-optical system. This parameter is entirely determined by the gun design and unambiguously relates the anode current and voltage of the gun: $P = \dfrac{I_a}{U_a^{3/2}}$. In terms of physics, it means that with the increase of the radiator current above the value given by this expression the beam diameter when it passes through the anode diaphragm becomes greater than the admissible diameter and the peripheral part of the beam hits the edge of the anode diaphragm.

PLASTIC RELAXATION — usually used with the addition of «mechanical stresses» or as «plastic relaxation of heteroepitaxial stresses». This process is understood to be the introduction of defects into an elastically-stressed crystal to lower the mechanical stresses down to zero (complete relaxation). In our case imperfect dislocations are introduced into a highly compressed (4 %) thin film of germanium on silicon (plane stressed state), forming an ordered network on the film–substrate interface. The crystalline cell of a film becomes cubic and the stresses are relieved.

PRESSURIZED COMPARTMENT — space vehicle compartment with a pressurized shell where a certain pressure of gas (air) is maintained. Distinction is made between pressurized compartments accommodating the instrumentation and hardware (for instance, an instrument module) providing favourable conditions for instrument operation (pressure, temperature, humidity, etc.) and crew pressurized compartments (for instance, descent vehicle, service compartment), where the atmosphere required for supporting the vital functions of the crew is maintained. Tightness of the casing and detachable joints are provided by applying materials with a higher-density structure, high-performance welding and brazing processes, as well as application of tight joints.

PRIMARY (PAY) LOAD — mechanical load, normally applied to a structure in service (structure dead weight, weight of cargo in the jack, etc.). Part of the primary load for which a structure is designed, is called payload (it does not include the weight of the structure proper or weight of the jack).

PROTECTORS — specially-shaped items of an alloy of an active metal (magnesium, zinc, aluminium) fastened to the object (storage tank, vessel bottom, etc.) and protecting it from corrosion, having the role of the anode in a corrosion galvanic cell.

REPAIR-RESTORATION OPERATIONS — a package of work to support or restore operable condition or serviceability of the hardware. Major repairs may require complete dismantling of the items and flaw detection in them, replacement or repair of all the faulty parts, assembly, integrated checking, adjustment and testing. Medium repair consists in restoration of service properties by replacement or restoration of individual parts of the item and conducting appropriate starting-up and adjustment testing.

SCANNING ELECTRON MICROSCOPES (SEM) with heater cathode are designed for studying massive objects with a resolution from 50 up to 200 A. SEM accelerating voltage can be adjusted in the range from 1 up to 30–50 kV. Two or three electron beams are used to focus a narrow electron probe on the sample surface. Magnetic deflecting coils scan the probe over a preset area on the object. Interaction of the probe electrons with the object gives rise to several kinds of radiation, namely secondary and reflected electrons; electrons that have passed through the object (if it is thin); X-ray radiation (braking or characteristic); light radiation, etc. Any of these radiations may be registered by an appropriate detector, converting radiation into electric signals that are fed to the electron beam tube (EBT) after amplification and modulate its beam. Scanning of EBT beam is synchronised with the electronic probe scanning in SEM, and a magnified image of the object is observed in EBT screen (magnification is equal to the ratio of the frame height in EBT screen to the width of the scanned surface of the object). The image is photographed directly from EBT screen. The main advantage of SEM is the highly informative nature of the instrument due to the ability to observe an image using signals from different detectors. SEM enables the study of microrelief, chemical composition distribution over the object, *p–n* barriers, and X-ray structural analysis, and many other operations to be conducted.

SEDIMENTATION — directed movement of particles of a disperse phase in the field of action of gravitational or centrifugal forces in a liquid or gaseous media.

SEGREGATION — non-uniformity of an alloy chemical (phase) composition. It develops during material solidification (in this case, it is called liquation), as well as during heat treatment. Limited solubility of components in the alloys in the solid state results in their ageing in cooling, or precipitation hardening, related to precipitation of dissolved atoms (excess phases) along the boundaries of grains, subgrains, on dislocations or along certain crystallographic planes of the grains of a polycrystalline material.

SEMICONDUCTORS — substances intermediate between metals and dielectrics in terms of conductivity. Unlike metals, semiconductor conductivity grows exponentially with the temperature rise, this being attributable to the increase of the number of free electrons (charge carriers) due to breaking up of their bonds with the semiconductor atoms. The bond of the electrons with the atoms can be broken up in a semiconductor also at different external impacts, namely light, fast particle flows, strong electric field, etc.

SOLAR ENERGY CONCENTRATOR — a device for concentration (focusing) of solar energy applied in solar power units, rocket engines, units for welding, brazing and other technological processes in space for which temperature increase is required. A solar energy concentrator has a mirror reflecting surface, usually in the form of a paraboloid of revolution. The light flux, falling parallel to the optical axis, is concentrated in the paraboloid focus, forming a focal spot with a high energy density

SOLIDUS LINE — curve on the constitutional diagram of multicomponent systems, expressing the dependence of the temperature of equilibrium co-existence of the liquid and solid phases on the solid phase composition.

SPACE MONITORING — observation of the condition of the environment, using TV images, photos, multispectral photos, etc., taken from the space vehicle, with the aim of environmental control and protection. In addition, space monitoring includes using the space vehicle for collection of some data from the ground and sea stations and their transmission to the data processing centres. Space monitoring enables the areas and nature of environmental change to be revealed with minimal time inertia; tracing and mapping the spread of man's influence; assessing the intensity of the processes and amplitudes of ecological shifts; studying the interaction of technogenic systems (cities, channels, etc.) with the surrounding natural systems.

SPACE VACUUM — the state of gas at pressures below atmospheric pressure. The notion of «vacuum» also covers gas in a free space, for instance, cosmic space. Space vacuum is a rarefaction existing in cosmic space. Space vacuum is high compared to that achievable in industrial laboratories. Absolute vacuum, i.e. space containing absolutely no gas or dust particles, is non-existent in cosmic space just as it is unattainable in engineering. It is conditionally assumed that space vacuum begins beyond the upper part of the atmosphere at the altitude of several thousands kilimeters. For practical purposes, however, it can be assumed, that space vacuum starts significantly lower. Space vacuum is one of the most important elements of the combined specific environment of orbital station flight. The influence of space vacuum is manifested in the change of the space vehicle heat exchange conditions, in the appearance of the corona discharge, in material sublimation and change of their mechanical properties, and in loss of lubrication. Because of vacuum man can only stay in cosmic space in pressurized compartments of the space vehicle or in spacesuits fitted with the life-support system.

SPECTRAL ANALYSIS — determination of the qualitative and quantitative composition of a material based on studying its spectra. It enables determination of the elementary, isotopic or molecular composition of the material. Emission spectral analysis is most often used.

STREAKS — narrow bands in the electron diffraction patterns at electron diffraction by reflection from the crystal surface layers, often characterizing its atomic smoothness (the number of streaks and their spacing provides information on the structure of an elementary surface cell).

TEST BENCH, JIG — a special facility (device) for subassembly and complete assembly of objects, conducting research, retrofitting, control, special and acceptance testing of the object as a whole or its separate elements and assemblies.

TERMINATOR — a boundary line between the sunlit and unlit parts of the surface of a cosmic body (e.g. a planet). In the absence of an atmosphere (Moon, Mercury), the terminator coincides with the line of contact of the direct solar rays with the surface of the body. Therefore, in the terminator area the surface relief details, casting much longer shadows, are particularly clearly defined. In a sufficiently dense atmosphere (as on Earth and Venus) the solar rays are significantly distorted as a result of refraction, and the true terminator shifts towards the dark hemisphere. The sunlit region of the atmosphere is separated from the unlit part by a band of penumbra smoothly deepening towards the nocturnal hemisphere. This gives rise to a wide range of twilight phenomena in the terminator vicinity. The planet is encircled by a zone of twilight penumbra, forming a gradual light and heat transition from the diurnal to the nocturnal hemisphere. When observed from space a brightly coloured halo of dawn rises above the terminator, where the structure is determined by the condition of the atmosphere. The halo is used to study the altitude structure of the atmosphere .

THREE-DIMENSIONAL REFLEXES — light spots in an electron diffraction pattern at electron diffraction by reflection, that originate from electron diffraction by a volume of crystalline sections of the surface relief (electrons may pass through a considerable volume of the substrate or film).

TRAINING FACILITY, SIMULATOR — engineering facility for training operators, allowing them to acquire the skills and experience of using the hardware, coping with emergencies and contingencies, and interacting with the «ground» flight control personnel. An integrated simulator enables acquiring skills to fulfill all the flight tasks, beginning with pre-start operations, actions during orbital injection, orbiting, docking approach and docking with space vehicles, performance of scientific experiments, preparation for descent, and descent and landing. A specialized simulator is designed for addressing one or several tasks, as a rule with higher accuracy and degree of detail than would apply in an integrated simulator.

TEXTURE — prevailing orientation of crystal grains in polycrystals or molecules in amorphous bodies, liquid crystals or polymers, along an axis or in the plane of symmetry, resulting in anisotropy of material properties. A texture may develop during forming under the action of elastic stresses, thermal impacts, energy and magnetic fields, etc., or with a combination of these factors (for instance, thermomechanical and thermomagnetic processing of materials). Texture forms at directional solidification, epitaxial growth, adsorption, phase transformations, vacuum and electrolytical deposition of coatings, at crystallization and deformation of polymer materials, metal pressing and other treatment of materials.

PHASE — state of a material in thermodynamic equilibrium, differing by physical properties from other possible equilibrium states (other phases) of the same material. Transition of a material from one phase into another — phase transition — is related to qualitative changes of material structure and properties. For instance, gaseous, liquid or solid (crystalline) states (phases) of a material differ by the nature of movement of particles (atoms, molecules) and presence or absence of an ordered structure. Various crystalline phases differ from each other by the type of the crystalline structure, electric conductivity, electric and magnetic properties, presence or absence of superconductivity, etc. Liquid phases differ from each other by component concentration, presence or absence of superfluidity, anisotropy of elastic and electric properties. In hard alloys the crystalline structure phases can differ by density, modulus of elasticity, melting temperature and other properties.

UPPER ATMOSPHERIC EMISSION — Earth's self-radiation at more than 40 km altitudes.

WORKING CONDITIONS — conditions of running various technological processes, structure functioning under actual production conditions (unlike laboratory conditions).

X-RAY INSPECTION, RADIOGRAPHY — revealing inner defects in welds by exposing the object of control to X-rays (transmission) and producing a visible image on a fluorescent screen (X-ray examination) or on a sensitive X-ray film after its development. Ability of X-ray radiation to penetrate in a different manner through different materials is responsible for the light and shadow contrast of the image. It is based on the law of weakening of the intensity of X-ray radiation, penetrating into a material of different thickness $I = I_0 e^{-\mu x}$, where I_0 is the intensity of X-ray radiation that penetrated through layer x on the material surface; x is the thickness of the layer, cm; μ is the linear coefficient of intensity variation in the studied material.

X-RAY SPECTROMETRY — analysis, using X-ray spectra of the studied material.

ZERO GRAVITY — condition of a material body, in which the acting field of gravity does not induce mutual pressure of one part of the body on the other, or their deformation. Only short-term zero gravity is achievable on Earth. It is observed during the first 1–2 s of the body free fall, when air resistance still practically does not affect the body movement. Longer-term zero gravity is achieved in an aircraft flight by a special trajectory, namely a parabola, i.e. going from a horizontal flight first to the ascending section of the parabola with subsequent movement through its apex and then along its descending part. This method (so-called flight along a Kepler parabola) provides zero gravity for 30–50 s, but it is preceded by overloads. A number of technological processes can proceed only under conditions of zero gravity, namely growing some high-quality crystals, producing foam materials, etc.

Reference literature, was used when compiling this section:

1. Cosmonautics (1985) Encyclopedia. Ed. by V.P. Glushko, Moscow, Sov. Entsiklopediya.
2. Physical encyclopedia (1984) Ed. by A.M. Prokhorov, Moscow, Sov. Entsiklopediya.
3. Reference book on welding (1974) Ed. by K.K. Khrenov, Kiev, Naukova Dumka.

4. Explanatory dictionary on chemistry and chemical technology (1987) Ed. Yu.A. Lebedev, Moscow, Russky Yazyk.

5. Yuzov, N.I., Kryuchkov, B.I. and Shuvalov, V.A. (1998) Extra-vehicular activity of the
   cosmonauts, Zvezdny Gorodok, Moscow district.

6. Encyclopedia of inorganic materials (1977) 2 volumes, Kiev, Ukr. Sov. Entsiklopediya.

Index

Adhesion 3–6, 10, 37, 40, 45, 50, 53, 73, 216, 259, 297, 324, 470

Ambient outer atmosphere (AOA) 77–80, 82, 387, 388, 547

Burn-through of metal 125, 183, 261, 278–282

Compatibility 85, 87, 88, 185, 205, 207, 497

Control system 48, 92–94, 96, 98, 100, 103, 106, 109, 123, 125, 175, 307, 324, 327, 330, 378, 391, 392, 491

Crucible 10, 14–16, 42, 46, 55, 57, 58, 60–64, 82, 85, 90, 91, 93, 98, 106, 120, 181, 182, 186, 202, 210, 299, 320, 322–328, 331, 336–342, 345, 349, 350, 352, 353, 355–357, 366, 367, 372–374, 381–385, 400, 402, 404, 407–410, 471, 519, 521

Deep vacuum 4, 52, 81, 154, 179, 213, 221, 229, 385, 388, 487, 507, 555

Degradation of material 39, 74, 77, 378

Diffusion 53, 57, 59, 61–64, 72, 78, 154, 159, 221, 225, 226, 261, 263, 267, 273, 275, 291, 324, 333, 340–343, 353, 355, 357, 360–363, 365, 374, 380, 398, 404, 518, 519, 549

Dispersed particles 80, 214, 216

Doublet 333, 353, 354, 412, 550

Electron beam 6, 14, 17, 19, 28, 29, 31, 35, 37, 42, 44, 45, 47–52, 55, 80, 86, 89, 92, 93, 98, 100, 102, 103, 106–108, 120, 141–146, 152–156, 158, 160–163, 166, 167, 177, 179, 180–183, 185, 186, 190, 191, 193, 194, 204, 206, 210, 212, 227, 229, 236, 239, 240, 243, 244, 259–267, 269, 270, 272, 273, 275, 277, 278, 283, 288, 298–303, 317, 318, 322, 324–326, 328, 329, 336, 345, 349, 372, 376, 382, 398–402, 407, 409, 413, 417, 419, 421, 423–426, 429, 436–439, 443, 444, 469, 471, 516, 521, 550

equipment (device, hardware, installation, unit) 7, 50, 55, 106, 109, 159, 164, 165, 168, 175, 176, 186, 189–191, 201, 211, 212, 245, 298, 398, 417–420, 422, 425, 434, 438–440, 445, 482, 490 (see also Equipment)

gun 9, 14, 16, 48, 52, 55, 93, 98, 101, 104, 106, 142, 164, 168, 174–176, 180, 181, 186, 187, 190–194, 210, 230, 241, 259, 267, 268, 311, 317, 322, 323, 329, 381, 399, 400, 417–419, 421–423, 425, 434–436, 439–441, 443

tool 14, 29, 31, 91, 104, 105, 109, 141, 142, 146–148, 151–153, 175, 178, 180–182, 184, 187, 190–194, 201, 211, 277, 299, 311, 312, 314, 317, 318, 327, 349, 382 (see also VHT, «Universal»)

methods (see also Technological processes)

brazing 23, 27, 188, 195, 277, 311

coating deposition 188, 195, 259, 311, 326, 372, 376, 379, 380

cutting 8, 13, 23, 27, 29, 38, 43, 45, 46, 93, 154, 155, 159–161, 164, 167, 188, 195, 212, 311, 471

evaporation 20, 321, 322, 324, 349, 372, 378, 381, 383, 390

melting 22, 159, 187, 407–413

welding (EBW) 3, 8, 9, 12, 19, 23, 26, 27, 41, 44–46, 50, 52, 54, 55, 71, 76, 77, 86, 88, 89, 93, 104, 105, 146, 151, 154, 155, 159–162, 164, 166, 168, 174–183,

List of Authors

Printed and bound by CPI Group (UK) Ltd, Croydon, CR0 4YY

23/10/2024

01778249-0005